LEGIONELLA

Proceedings of the 2nd International Symposium

SPONSORS

Center for Infectious Diseases
Centers for Disease Control

Health Effects Research Laboratory
U.S. Environmental Protection Agency

Department of Medicine
Emory University School of Medicine

National Institute of Allergy and Infectious Diseases
National Institutes of Health

American Society of Heating, Refrigerating
and Air-Conditioning Engineers

EXECUTIVE COMMITTEE
Clyde Thornsberry, Chairman

Don J. Brenner
James C. Feeley
LeRoy J. Pickles

Claire V. Broome
W. Lee Hand
Hazel W. Wilkinson

PROGRAM ADVISORY COMMITTEE
James C. Feeley, Chairman

James M. Barbaree
William B. Baine
Harry N. Beaty
Ron J. Fallon
David W. Fraser
Walter Jakubowski
A. William Pasculle
Jonas A. Shulman
John O'H. Tobin

Christopher L. R. Bartlett
Paul H. Edelstein
Carl B. Fliermans
Herman Friedman
Pieter Meenhorst
Milton Puziss
John K. Spitznagel
Washington C. Winn, Jr.

Peter Skaliy and William B. Cherry, Consultants

SUBCOMMITTEES OF THE PROGRAM ADVISORY COMMITTEE

Clinical Features and Laboratory Diagnosis
Paul H. Edelstein, Chairman

Microbiology
Don J. Brenner, Chairman

Pathophysiology
Washington C. Winn, Jr.,
    Chairman

Immunology
Herman Friedman, Chairman

Epidemiology
Claire V. Broome, Chairman

Ecology and Environmental Control
Walter Jakubowski, Chairman

LEGIONELLA

Proceedings of the 2nd International Symposium

Editors:

CLYDE THORNSBERRY
ALBERT BALOWS
JAMES C. FEELEY
WALTER JAKUBOWSKI

American Society for Microbiology • Washington, D.C.
1984

Library of Congress Cataloging in Publication Data

International Symposium on Legionella (2nd : 1983 : Washington, D.C.)
  Legionella : proceedings of the 2nd International Symposium on Legionella.

  Includes indexes.
  1. Legionella pneumophila—Congresses. 2. Legionnaires' disease—Congresses. I.
Thornsberry, Clyde. II. American Society for Microbiology. III. Title.
QR201.L44I58 1983   616.9′2   83-21499

ISBN 0-914826-58-1

LEGIONELLA:
PROCEEDINGS OF THE 2nd
INTERNATIONAL SYMPOSIUM

Contents

MICROBIOLOGY

PATHOLOGY AND PATHOPHYSIOLOGY

IMMUNOLOGY

EPIDEMIOLOGY

ECOLOGY AND ENVIRONMENTAL CONTROL

Methods

Habitat

Ecological Interactions with Other Organisms

Disinfection

PREFACE

The mystique surrounding Legionnaires disease reached its peak in 1977–1978. The etiological agent had been isolated and named, the peculiar spectrum of disease in humans was documented, the natural habitat of legionellae was at least partially determined, the descriptive and analytical epidemiology had been defined in some detail, and some interesting attributes of *Legionella pneumophila* and other species (antigenic patterns, growth, nutritional, pH, and oxygen requirements, physiology, and cellular fatty acid composition) were described. In November 1978 an International Symposium on Legionnaires Disease was held. This was a much-heralded event that received broad coverage by television and the lay press. A mix of over 500 scientists, epidemiologists, physicians, engineers, and health workers from several different countries attended this Symposium, and it proved to be an excellent forum for the exchange of scientific and technical information. The publication of the proceedings of that Symposium was well received as a useful document for disseminating the data and information presented and for "priming the research pump" in terms of providing direction and ideas of what questions still needed partial or complete answers.

Jay Sanford in his summary remarks at the First Symposium indicated that the meeting was the true opening of a new era because of the more in-depth involvement of the scientific community. That has not changed. Scientists from throughout the world met at the 2nd International Symposium in June 1983 to define the problems, to present and discuss new data, and to determine the course of future research on *Legionella*. From the medical side we assembled microbiologists, immunologists, epidemiologists, pathologists, infectious disease specialists, and other clinicians; from the environmental side we assembled environmental microbiologists, ecologists, and heating, air-conditioning, refrigeration, sanitary, and other environmental engineers.

Our original notion, which has been borne out by the material presented, was that the 2nd Symposium could be divided into six sections: (i) Clinical Features and Laboratory Diagnosis, (ii) Microbiology, (iii) Pathology and Pathophysiology, (iv) Immunology, (v) Epidemiology, and (vi) Ecology and Environmental Control. Within each of these sections we felt it appropriate to begin with one or more "State of the Art" lectures and to end with a summary lecture. For some sections we also felt the need for round table discussions of the problems and how they might be resolved by further study. Although we realized the difficulties associated with publishing what transpired at round tables, we have included a resumé of each round table discussion and hope that they will serve as a springboard for productive research in the future.

The listing of the cosponsors of the Symposium reflects the breadth of the studies that are ongoing with legionellae and legionellosis: the American Society for Heating, Refrigerating, and Air-Conditioning Engineers; the Department of Medicine of the Emory University School of Medicine, Atlanta; the National Institute of Allergy and Infectious Diseases of the National Institutes of Health; the Health Effects Research Laboratory of the U.S. Environmental Protection Agency; and the Center for Infectious Disease of the Centers for Disease Control. We thank the members of these institutions for their encouragement and for their administrative and financial support.

We also wish to thank the members of industry who supported the meeting by either contributions or exhibits, in particular Miles Pharmaceuticals, BioDx, and Remel.

Finally we wish to give special thanks to the members of the various committees who made arrangements for the meeting, put together the program, and reviewed manuscripts.

This meeting can be well described as timely—much has been accomplished in the last 5 years. As a prime example, in 1978 at the First Symposium, Don Brenner announced the naming of *Legionella pneumophila*; at this meeting he announced that there are now more than 20 known species. We believe this Symposium and these Proceedings will aid in the continuation of the research activities on *Legionella* and legionellosis.

Clyde Thornsberry
Centers for Disease Control
Atlanta, Georgia

Albert Balows
Centers for Disease Control
Atlanta, Georgia

James C. Feeley
Centers for Disease Control
Atlanta, Georgia

Walter Jakubowski
Environmental Protection Agency
Cincinnati, Ohio

CLINICAL FEATURES AND LABORATORY DIAGNOSIS

STATE OF THE ART LECTURE

Laboratory Diagnosis of Legionnaires Disease

PAUL H. EDELSTEIN

Infectious Disease Section, Medical and Research Services, Veterans Administration Medical Center, Wadsworth Division, Los Angeles, California 90073, and The University of California-Los Angeles School of Medicine, Los Angeles, California 90024

In the 4 years since the First International Symposium on Legionnaires Disease, considerable progress has been made in the laboratory diagnosis of these infections (1, 6, 8, 26). One of the major advances has been in the improvement of media and consequently in the ability to identify cases by actual isolation of the organism. Recognition of new species and serogroups has made culture isolation even more important because of its independence from bacterial immunological characteristics. Serological diagnostic methods, because of antigenic diversity, have been recognized as being more complex and demanding than was originally thought. Despite a 4-year span, there is still trans-Atlantic divergence regarding the best way to prepare antigens for antibody testing. Direct and indirect immunofluorescent (FA) methods for detection of *Legionella pneumophila* in sputum have proven to be exceptionally helpful in early diagnosis but still pose problems of both sensitivity and specificity. Detection of soluble antigens in urine, described in the original symposium, has now been refined so that it is a valuable method of diagnosis and may supercede some of the other methodologies.

Studies of the utility of diagnostic tests for *Legionella* infections have been stymied by the lack of easily applied definitions of disease. This makes absolute determinations of the sensitivity and specificity of laboratory tests particularly difficult and has placed more reliance on tests of relative sensitivity and specificity, for example, the sensitivity of direct FA detection of antigen in those patients with culture-proven infection. These limitations have placed restrictions on progress in laboratory diagnosis and make it mandatory that extreme caution be used in study interpretation.

The development of charcoal-yeast extract (CYE) medium, and then buffered CYE (BCYE) medium, by James Feeley and his colleagues has made possible both the routine isolation of *L. pneumophila* from clinical specimens and basic studies on *Legionella* spp. (18). It is now recognized that earlier generations of media, such as FG and MHIH, were suboptimal for primary isolation. The addition of α-ketoglutarate to BCYE medium (BCYEα) seems to improve growth even more, and the addition of antimicrobial agents to BCYEα medium makes possible the isolation of *L. pneumophila* and some other species from sputum (3, 5).

Isolation of *L. pneumophila* from sputum strikes some as unusual. However, there are more than 60 positive sputum cultures reported in the literature (3, 5). In addition, selective media have been useful for isolation from contaminated respiratory secretions of *L. longbeachae, L. wadsworthii, L. dumoffii*, and possibly *L. micdadei* (7, 17; P. Edelstein, unpublished data). Almost all of these positive cultures grew only with the use of BCYEα formulations containing antimicrobial agents. The most commonly used and effective selective agents are polymyxin B, anisomycin, and either cefamandole (BMPAα medium) or vancomycin (PVA medium) (3, 5). Since some of the *Legionella* spp., especially *L. micdadei*, are inhibited by cefamandole but not by vancomycin, the vancomycin-containing medium is better for the isolation of these more sensitive species (17). The disadvantage of PVA medium is that it is not as inhibitory to other oral flora as is BMPAα medium. Low-pH pretreatment of contaminated clinical specimens also is of benefit, as reported by Buesching and colleagues (2, 3). In my laboratory, we are currently plating all specimens on BCYEα, BMPAα, and PVA media, with and without low-pH pretreatment. Rarely, heat pretreatment may be of additional selective benefit (9).

Extrapulmonary *Legionella* infections have been documented, by means of culture or direct FA, with increased frequency (26). Culture of "sterile" abscesses, liver biopsies, bone marrow, and blood from patients with perplexing febrile illnesses should probably now be routinely performed on media appropriate for *Legionella* isolation.

The sensitivity of culture diagnosis is uncertain, in part because of the unavailability of good selective media and techniques in studies of previous outbreaks and in part because of the

3

lack of an easily applied definition of disease. Culture is more sensitive for *L. pneumophila* detection than is direct FA testing, by a factor of 1.2 to 4 (3, 6, 8, 26). This is not surprising, since the lower limit of detectability of *L. pneumophila* in seeded sputum by direct FA techniques is about 10^4 bacteria per ml (P. Edelstein, unpublished data). It appears as if the sensitivity of culture methods for other species is lower, perhaps due to use of nonoptimal techniques (6, 17). The specificity of culture diagnosis appears to be exceptionally high, with no evidence for positive cultures in asymptomatic persons (6).

A plethora of methods have been introduced for serological diagnosis of *Legionella* infections. The indirect FA method is still the reference method, as it is the best standardized (11, 12, 14, 24). Other methods, which either are easier to use in a small laboratory not equipped with a specially illuminated microscope or can be automated, are gaining some favor. Unfortunately, few of these other methodologies have been rigorously standardized by the use of epidemic sera or sera from well-defined cases; some studies also suffer from "incorporation bias," in which the result of the test being studied is used to determine whether the patient has disease (19). Regardless, both enzyme immunoassay and microagglutination techniques appear promising (10, 13, 15, 27).

The concern expressed in the first *Legionella* symposium about cross-reactivity in mycoplasma infections now appears to have been discounted (20). Still needed is a comprehensive study of serological reactions in hospitalized patients with a variety of acute bacterial pneumonias.

Still somewhat controversial after 4 years of debate is the best method of antigen preparation for the indirect FA serological test. Results of three studies have shown little difference between Formalin and heat fixation, but many still feel, without much objective evidence, that the Formalin-fixed antigen is more specific (11, 14, 23; E. D. Renner, J. C. Zimmerman, and G. L. Lattimer, Abstr. Annu. Meet. Am. Soc. Microbiol. 1983, C102, p. 328). With other methods of antigen preparation, primarily for enzyme immunoassay, soluble cell fraction preparations have been found to be better than sonicated cell antigens (10, 27).

The major problem in serological diagnosis is the existence of multiple antigenic types with little cross-reactivity. Optimal yield can be achieved only with multiple antigen preparations; almost one-half of positive serological reactions were missed with the use of only a single antigen in a study by Wilkinson and colleagues (25). The use of polyvalent antigen preparations for screening purposes may be necessary until more automated methods are standardized or until serogroup-common, genus-specific antigens can be developed (12, 25). Little is known of the absolute sensitivity or specificity of serological testing except with *L. pneumophila* serogroup 1, making interpretation of test results difficult; Wilkinson and colleagues estimate a sensitivity of 80% and a specificity of 96% when using seroconversion criteria for multiple antigens (25). The value of serological tests in the diagnosis of sporadic cases of *Legionella* infection has not been studied. But because of low prevalence, the positive predictive values will likely be low (25).

The question of rapid serological diagnosis hinges in part on the question of test specificity. Taylor and Harrison report that up to 40% of cases of Legionnaires disease can be diagnosed on the first hospital day (after approximately 5 days of illness) on the basis of serological testing of a single specimen (22), because of the high specificity of their antigen preparation. This has not been found in studies using heat-killed antigen (8, 16, 26). Whether this can be confirmed when diagnostic methods other than serological ones are used awaits further study. Also, whether early serological diagnosis can be achieved by using a very specific and sensitive immunoglobulin M enzyme immunoassay awaits further study (10).

Direct or indirect FA methods for detection of *Legionella* spp. in respiratory tract secretions or tissues have proven to be an especially useful diagnostic technique (4). The major advantage is the rapidity of testing, making same-day diagnosis possible. A major change from the first *Legionella* symposium is the realization that sputum is an excellent specimen to test by direct FA methods (8, 26). The direct FA conjugates for *L. pneumophila* serogroups 1 through 4, made by the Centers for Disease Control, are highly specific. The specificity of the polyvalent reagent for *L. pneumophila* serogroups 1 through 4 is about 99.9% for sputum examination in my laboratory (6). Practically, this means that a positive test is diagnostic of disease even when the prevalence of disease is very low. Unfortunately the sensitivity of direct FA testing, as assessed by culture diagnosis, ranges from 25 to 70%, with 60% probably being average (3, 6, 8, 26). This, combined with little or no cross-reactivity between serotypes, means that a negative test has no diagnostic significance. The sensitivity and specificity of the direct FA conjugates for *Legionella* spp. other than *L. pneumophila* are undefined, making results of testing with these reagents still experimental. Although multivalent conjugate pools are available for these other species, extreme caution must be used in interpretation of results until they are clinically evaluated.

The use of monoclonal antibodies in clinical antigen detection has not been studied. Their probable very high specificity would decrease even further the sensitivity of immunological tests. A high-avidity, genus-specific, species-nonspecific monoclonal antibody would of course be the diagnostician's dream.

Diagnosis of Legionnaires disease by urine testing for soluble bacterial antigen is a very promising method (21). This can be accomplished by radio- or enzyme immunoassay or by latex agglutination. It appears to be very specific and at least as sensitive as direct FA testing for disease detection. This method has the same problem as other immunological tests, that of tremendous antigenic diversity and limited cross-reactivity. However, once commercially available, it may replace other immunological means of diagnosis.

This research was supported by the Medical Research Service of the Veterans Administration.

LITERATURE CITED

1. Balows, A., and D. W. Fraser (ed.). 1979. International Symposium on Legionnaires' Disease. Ann. Intern. Med. 90:489–703.
2. Bopp, C. A., J. W. Sumner, G. K. Morris, and J. G. Wells. 1981. Isolation of Legionella spp. from environmental water samples by low-pH treatment and use of a selective medium. J. Clin. Microbiol. 13:714–719.
3. Buesching, W. J., R. A. Brust, and L. W. Ayers. 1983. Enhanced primary isolation of Legionella pneumophila from clinical specimens by low-pH treatment. J. Clin. Microbiol. 17:1153–1155.
4. Cherry, W. B., B. Pittman, P. P. Harris, G. A. Hebert, B. M. Thomason, L. Thacker, and R. E. Weaver. 1978. Detection of Legionnaires disease bacteria by direct immunofluorescent staining. J. Clin. Microbiol. 8:329–338.
5. Edelstein, P. H. 1981. Improved semiselective medium for isolation of Legionella pneumophila from contaminated clinical and environmental specimens. J. Clin. Microbiol. 14:298–303.
6. Edelstein, P. H. 1983. Culture diagnosis of Legionella infections. Zentralbl. Bakt. Parasitenkd. Infektionskr. Hyg. Abt. 1 Orig. Reihe A 255:96–101.
7. Edelstein, P. H., D. J. Brenner, C. W. Moss, A. G. Steigerwalt, E. M. Francis, and W. L. George. 1982. Legionella wadsworthii species nova: a cause of human pneumonia. Ann. Intern. Med. 97:809–813.
8. Edelstein, P. H., R. D. Meyer, and S. M. Finegold. 1980. Laboratory diagnosis of Legionnaires' disease. Am. Rev. Respir. Dis. 121:317–327.
9. Edelstein, P. H., J. B. Snitzer, and J. A. Bridge. 1982. Enhancement of recovery of Legionella pneumophila from contaminated respiratory tract specimens by heat. J. Clin. Microbiol. 16:1061–1065.
10. Elder, E. M., A. Brown, J. S. Remington, J. Shonnard, and Y. Naot. 1983. Microenzyme-linked immunosorbent assay for detection of immunoglobulin G and immuno-

globulin M antibodies to Legionella pneumophila. J. Clin. Microbiol. 17:112–121.
11. Fallon, R. J., and W. H. Abraham. 1979. Scottish experience with the serologic diagnosis of Legionnaires' disease. Ann. Intern. Med. 90:684–686.
12. Fallon, R. J., and W. H. Abraham. 1982. Polyvalent heat-killed antigen for the diagnosis of infection with Legionella pneumophila. J. Clin. Pathol. 35:434–438.
13. Farshy, C. E., G. C. Klein, and J. C. Feeley. 1978. Detection of antibodies to Legionnaires' disease organism by microagglutination and micro-enzyme-linked immunosorbent assay tests. J. Clin. Microbiol. 7:327–331.
14. Harrison, T. G., and A. G. Taylor. 1982. Diagnosis of Legionella pneumophila infections by means of formolised yolk sac antigens. J. Clin. Pathol. 35:211–214.
15. Harrison, T. G., and A. G. Taylor. 1982. A rapid microagglutination test for the diagnosis of Legionella pneumophila (serogroup 1) infection. J. Clin. Pathol. 35:1028–1031.
16. Kirby, B. D., K. M. Snyder, R. D. Meyer, and S. M. Finegold. 1980. Legionnaires' disease: report of sixty-five nosocomially acquired cases and review of the literature. Medicine (Baltimore) 59:188–205.
17. Muder, R. R., V. L. Yu, R. M. Vickers, J. Rihs, and J. Shonnard. 1983. Simultaneous infection with Legionella pneumophila and Pittsburgh pneumonia agent. Clinical features and epidemiologic implications. Am. J. Med. 74:609–614.
18. Pasculle, A. W., J. C. Feeley, R. J. Gibson, L. G. Cordes, R. L. Myerowitz, C. M. Patton, G. W. Gorman, C. L. Carmack, J. W. Ezzell, and J. N. Dowling. 1980. Pittsburgh pneumonia agent: direct isolation from human lung tissue. J. Infect. Dis. 141:727–732.
19. Ransohoff, D. F., and A. R. Feinstein. 1978. Problems of spectrum and bias in evaluating the efficacy of diagnostic tests. N. Engl. J. Med. 299:926–930.
20. Renner, E. D., C. M. Helms, N. H. Hall, W. Johnson, Y. W. Wong, and G. L. Lattimer. 1981. Seroreactivity to Mycoplasma pneumoniae and Legionella pneumophila: lack of a statistically significant relationship. J. Clin. Microbiol. 13:1096–1098.
21. Sathapatayavongs, B., R. B. Kohler, L. J. Wheat, A. White, and W. C. Winn, Jr. 1983. Rapid diagnosis of Legionnaires' disease by latex agglutination. Am. Rev. Respir. Dis. 127:559–562.
22. Taylor, A. G., and T. G. Harrison. 1981. Formolised yolk sac antigen in early diagnosis of Legionnaires' disease caused by Legionella pneumophila serogroup 1. Lancet ii:591–592.
23. Wilkinson, H. W., and B. J. Brake. 1982. Formalin-killed versus heat-killed Legionella pneumophila serogroup 1 antigen in the indirect immunofluorescence assay for legionellosis. J. Clin. Microbiol. 16:979–981.
24. Wilkinson, H. W., D. D. Cruce, and C. V. Broome. 1981. Validation of Legionella pneumophila indirect immunofluorescence assay with epidemic sera. J. Clin. Microbiol. 13:139–146.
25. Wilkinson, H. W., A. L. Reingold, B. J. Brake, D. L. McGiboney, G. W. Gorman, and C. V. Broome. 1983. Reactivity of serum from patients with suspected legionellosis against 29 antigens of Legionellaceae and Legionella-like organisms by indirect immunofluorescence assay. J. Infect. Dis. 147:23–31.
26. Winn, W. C., Jr., and A. W. Pasculle. Laboratory diagnosis of infections caused by Legionella species. Clin. Lab. Med. 2:343–369.
27. Wreghitt, T. G., J. Nagington, and J. Gray. 1982. An ELISA test for the detection of antibodies to Legionella pneumophila. J. Clin. Pathol. 35:657–660.

Clinical Features of Legionellosis

HARRY N. BEATY

Department of Medicine, University of Vermont, Burlington, Vermont 05401

At the time of the First International Symposium on Legionnaires Disease in 1978, a great deal of new information was communicated about all aspects of this recently recognized disease and its etiological agent. Five years later, we have learned little that is new about the clinical spectrum of this infection or its treatment. On the other hand, major advances have been made in understanding the microbiology of *Legionella pneumophila*, its source in nature, and the pathophysiology of infections it produces.

Recognition of a number of species within the genus *Legionella* that have been linked directly or indirectly to human disease make the adoption of a series of clinical definitions necessary. For the purposes of this discussion, the term legionellosis refers to all clinical syndromes produced by organisms classified within the genus *Legionella*. Legionnaires disease is the pneumonic form of legionellosis caused by *L. pneumophila*, and Pontiac fever is the self-limited, nonpneumonic, respiratory illness caused by the same organism. Other known forms of legionellosis include pneumonia that is caused by *L. micdadei* and other species of *Legionella* and rare manifestations of disease caused by infection with *L. pneumophila*. Thus far, Legionnaires disease, which has been the cause of a number of outbreaks and many sporadic cases of pneumonia, is far and away the most common form of legionellosis.

The diagnosis of this group of infections depends upon isolation of the organism, demonstration of bacteria by direct immunofluorescent techniques, documentation of an appropriate rise in serum antibody, or identification of bacterial antigens in body fluids. In laboratories with extensive experience, *L. pneumophila* can be isolated from the sputum of 40% of patients with Legionnaires disease. The organism also can be recovered from lung tissue, pleural fluid, and, on rare occasions, blood. Direct immunofluorescent stains are highly specific for the individual species of *Legionella* and can be used to demonstrate organisms in sputum or tissue. By far the most widely available and commonly used method of establishing the diagnosis of legionellosis involves measuring a fourfold or greater rise in antibody titer by using indirect fluorescent-antibody assays or their equivalent. With the use of proper techniques and controls, the sensitivity of these methods is about 80%, with a specificity of 95% or greater. Their major disadvantage lies in the delay that is inherent in establishing a diagnosis on the basis of rising antibody titers. Along with culture of the organism or its demonstration in sputum or tissues by specific stains, demonstration of antigen in urine holds promise for establishing an early diagnosis. As yet, however, this requires application of techniques that are not routinely available.

CLINICAL FEATURES

Legionnaires disease. The clinical features of the two most common forms of legionellosis, Legionnaires disease and Pontiac fever, are shown in Table 1. In the case of Legionnaires disease, the data are derived from a review of approximately 400 cases reported from various centers; the data on Pontiac fever are from the initial outbreak that occurred in 1968 (6, 9). Obviously, there is a range of frequencies for each manifestation of Legionnaires disease that is reported. For example, gastrointestinal symptoms occur in 13 to 54% of cases. There also is a wide variance in the frequency of neurological manifestations reported. It is not clear whether these disparities are the result of real differences in the disease produced by individual strains of *L. pneumophila* or the result of variations in data accumulation.

The typical patient with Legionnaires disease has the abrupt onset of high fever, nonproductive cough, chills, and headache or myalgias. Peak symptomatic involvement often occurs within 48 h. A point that is not apparent in a simple listing of signs and symptoms of Legionnaires disease is that these patients often appear quite toxic. Their faces are flushed, and they may appear near prostration. They often are too breathless to talk comfortably, and cough occurs in uncomfortable paroxysms. Within 3 or 4 days, the cough may become productive of small amounts of nonpurulent sputum, which occasionally is streaked with blood.

Although suspicion of Legionnaires disease can be heightened by the presentation of a

TABLE 1. Clinical features of Legionnaires disease
and Pontiac fever

Feature	Frequency (%) in:	
	Legionnaires disease[a]	Pontiac fever[b]
Fever	98	91
Cough	87	57
Chills	73	91
Headache	43	91
Myalgia/arthralgia	43	95
Diarrhea	36	15
Neurological findings	26	23

[a] From Kirby et al. (9).
[b] From Glick et al. (6).

patient with typical features, many individuals have qualitatively and quantitatively different manifestations. For example, some patients have a more prolonged, indolent course, and others may be afebrile or have a paucity of respiratory symptoms. Several studies have documented that the clinical manifestations of Legionnaires disease are not sufficiently different from those of other pneumonias to be of diagnostic value.

Pontiac fever. Pontiac fever is characterized by the abrupt onset of fever, chills, headache, and myalgia. Cough is much less common, as are a variety of other nonspecific symptoms. The illness manifested by patients with this form of legionellosis is similar to influenza and is indistinguishable from a variety of virus-induced respiratory syndromes.

Major differences between Legionnaires disease and Pontiac fever are shown in Table 2. The incubation period of Legionnaires disease is 2 to 10 days, the attack rate apparently is quite low, and, by definition, pneumonia is present. On the other hand, the incubation period of Pontiac fever is quite short, the attack rate in the Pontiac, Mich., outbreak was about 95%, and it is a benign, self-limited illness that lasts 3 to 5 days and is not associated with pneumonia.

Another form of legionellosis that has been reasonably well characterized is the pneumonia caused by *L. micdadei*. Initially, this infection was called Pittsburgh pneumonia because it was first recognized in hospitals in Pittsburgh, Pa., and to date, the accumulated experience in those institutions provides the largest number of cases that have been carefully evaluated (10). Table 3

is adapted from the work of Muder and colleagues and compares various clinical features of this infection and Legionnaires disease. This comparative study did not reveal significant differences in the clinical presentations of patients with these infections. Patients with *L. micdadei* infection were significantly more likely to be immunocompromised, were more likely to have had surgery before developing infection, and had been hospitalized significantly longer before the onset of symptoms than patients with Legionnaires disease. These differences are of limited value in assessing an individual patient, and may not hold true in other institutions.

Smaller numbers of cases of infection caused by other species of *Legionella* have been reported. In all of these instances, pneumonia has been the principal clinical feature, and there is no evidence that any of these species produce unique clinical manifestations of infection.

CLINICAL EPIDEMIOLOGY

Males contract Legionnaires disease two to three times more frequently than do females. The average age of patients with this infection is around 60 years, with a wide range around the average. The disease has been reported in young children, but it is distinctly less common in the first and second decades of life than in the fifth, sixth, and seventh. Serological data from some studies, however, show that individuals in the pediatric age groups are as likely to have antibody titers suggestive of previous infection as adults (1, 5). Coupled with the low incidence of disease in children, these results suggest that infection with *L. pneumophila* is a more benign event among children than among adults.

A majority of patients with Legionnaires disease have one or more underlying conditions that appear to have predisposed them to infection. The most important of these include immunosuppression, malignancy, heart disease, chronic pulmonary disease, and chronic renal failure. Immunosuppression often is a result of treatment of associated conditions with corticosteroids, immunosuppressants, or cytotoxic drugs. Conditions that appear to be less important risk factors but frequently are associated with Legionnaires disease include diabetes, alcoholism, and antecedent surgery. Cigarette smoking has been implicated as a risk factor in most, but not all, of the reported series.

TABLE 2. Differences between Legionnaires disease and Pontiac fever

Disease	Incubation period	Attack rate	Risk factors	Pneumonia	Course
Legionnaires	2 to 10 days	<5%	Multiple	100%[a]	Progressive, sometimes fatal
Pontiac fever	5 to 66 h (mean, 36 h)	>95%	None	Absent	Self-limited, 3 to 5 days

[a] Included in definition.

TABLE 3. Clinical features of Legionnaires disease and *L. micdadei* pneumonia

Feature	Frequency (%) in:	
	L. micdadei pneumonia[a]	Legionnaires disease[b]
Fever	87	98
Cough	83	73
Neurological abnormalities	60	26
Diarrhea	26	36

[a] From Muder et al. (10).
[b] From Kirby et al. (9).

From the limited data available about other pneumonic forms of legionellosis, the overwhelming majority of cases occur in individuals with significant underlying illnesses.

LABORATORY ABNORMALITIES

Radiology. There is no indication that the radiographic abnormality associated with one of the pneumonic forms of legionellosis is significantly different from that seen with any of the others. The radiographic features of Legionnaires disease have been reported extensively (4, 8). Initially, unilateral lobar/segmental or diffuse patchy infiltrates are seen with about equal frequency. The next most common abnormality is a poorly marginated/rounded density not unlike that seen with septic pulmonary embolization. At the time of peak involvement of the radiograph, which usually occurs after several days, spread to contiguous portions of the lung or to the opposite side usually has occurred in the majority of patients. Pleural effusion is seen in 30 to 50% of the cases, but it usually is small and unilateral. Because a significant number of patients have coincidental congestive heart failure, it often is difficult to attribute pleural effusion to infection alone. Cavitation occurs in fewer than 10% of cases. Resolution of the X-ray often lags behind clinical improvement. Only about one-third of patients have complete clearing within 1 month, and some patients are left with permanent residua.

Laboratory tests. A wide variety of laboratory abnormalities have been reported in patients with Legionnaires disease, but the frequency and severity of associated conditions make it impossible to determine with certainty those that are attributable directly to the infection. Leukocytosis, consisting predominantly of increased numbers of granulocytes, is present in the majority of patients. In about 15% of patients, the leukocyte count exceeds 20,000/mm³. A smaller proportion of patients have counts below 5,000/mm³, but this often is attributable to prior treatment of an underlying malignancy. The hematocrit, concentration of clotting fac-

tors, and platelet count generally are unaffected or only moderately abnormal. Modest elevations in blood urea nitrogen and serum creatinine are indicators of mild renal insufficiency that usually is reversible. Significant proteinuria and microscopic hematuria are seen in 25 to 50% of patients. The most common abnormality in serum electrolytes is hyponatremia, with the serum sodium in the range of 115 to 130 meq/liter in 50 to 60% of cases. Mild to moderate hypophosphatemia also is seen in about one-third of patients. Minimal elevations in serum bilirubin, alkaline phosphatase, and transaminases have been reported, but clinically evident jaundice is uncommon. Significant elevations in creatine kinase and lactic acid dehydrogenase have been encountered in a high proportion of patients. This is thought to represent muscle injury and has been associated with histological evidence of rhabdomyolysis in a few instances. When a lumbar puncture has been performed because of impressive neurological findings, analyses of spinal fluid are almost always normal. Hypoxia and hypocarbia are common in these patients, but almost any blood gas abnormality can be encountered, depending on the extent of pneumonia and the presence or absence of underlying pulmonary disease.

COMPLICATIONS

Twenty-five to thirty percent of patients with pulmonary forms of legionellosis require the facilities and services of an intensive care unit; the majority of these are intubated and ventilated mechanically. About 15% of patients have modest hypotension, but a small number have true septic shock that may be associated with the manifestations of disseminated intravascular coagulation. Renal insufficiency usually is transient, but renal failure requiring temporary dialysis has been reported. Pulmonary complications that have been recognized include empyema, lung abscess, and prolonged impairment of function. About one-third of patients have neurological abnormalities secondary to a reversible, diffuse encephalopathy. The most common manifestations of central nervous system involvement are headache, confusion, and delirium. However, ataxia, which occasionally is associated with permanent cerebellar dysfunction, dysarthria, peripheral neuropathy, cranial nerve palsies, and seizures, has been reported. A number of individual cases or small series of Legionnaires disease associated with pericarditis, endocarditis, peritonitis, etc., have been reported. For the most part, bacteria have not been demonstrated in extrathoracic tissues in these instances. Direct fluorescent-antibody studies have identified bacteria in the liver, spleen, and kidneys of patients who died of

Legionnaires disease, but these organisms usually are within reticuloendothelial cells (3). Very rare instances of extrathoracic sites of inflammation from which *L. pneumophila* was recovered have been reported (2, 7). It is assumed that the multiple organ system involvement seen in patients with Legionnaires disease is due to one or more toxins.

TREATMENT

Treatment of Pontiac fever is aimed at providing symptomatic relief. Patients do not require hospitalization, and all patients recover from the infection in 3 to 5 days whether they receive antibiotics or not. Because most cases of this disease have been diagnosed retrospectively in the course of studying outbreaks, it has not been possible to determine whether administration of specific therapy mitigates the severity of the illness.

Treatment of pneumonic forms of legionellosis requires both general and specific measures. Patients who acquire the infection in the community may be managed at home with oral antibiotics and therapy aimed at providing symptomatic relief. Generally, the severity of the illness and its association with significant underlying diseases make hospitalization necessary. Patients frequently require fluid and electrolyte replacement, supplemental oxygen, antipyretics, etc. For many patients, early nutritional support should be instituted. Shock and failure of major organ systems should be treated early and vigorously.

The antibiotic of choice for treatment of Legionnaires disease and other forms of legionellosis is erythromycin. It should be administered intravenously in doses of 2 to 4 g/day for a minimum of 1 week, followed by oral therapy with doses of 2 g/day for 2 weeks. Rifampin, which has striking activity in vitro against legionellae, has been administered in conjunction with erythromycin, but there are no studies that have proven greater effectiveness for this combination. Rifampin should not be administered as a single agent, because of the probability of rapid emergence of resistant strains. Sulfatrimethoprim also has a high degree of activity in vitro and may have a role for patients who cannot take erythromycin or are immunocompromised and present with an undiagnosed pneumonia. Other antimicrobial agents are more effective in vitro than clinically. This discrepancy may be due to the intracellular localization of these pathogens, which has the effect of isolating them from many antibiotics that do not penetrate into cells well.

The response to erythromycin may be dramatic, with about one-half of the patients becoming afebrile within 2 days. Ten to twenty percent of patients remain febrile for longer than 5 days, however. Even though patients are afebrile, they may be inordinately weak and easily fatigued for days to weeks. As indicated above, clearing of the chest radiograph lags well behind clinical improvement. Relapse of infection occurs infrequently, particularly if the high-dose, prolonged course of erythromycin has been used. In several instances, relapses have been managed successfully with the administration of a second cycle of erythromycin therapy. Often, rifampin is added as a precaution.

The overall mortality rate in outbreaks of legionellosis that include a high proportion of patients with serious underlying diseases is around 25%. Factors such as age greater than 65 years, presence of septic shock, and coma are associated with an increased mortality rate. Nevertheless, if patients receive at least 48 h of appropriate therapy they usually survive that episode of infection. They may, however, develop superinfections or other complications and die within a few days or weeks. In those instances, legionellosis has to be considered a contributing factor but not the primary cause of death.

CONCLUSION

While we do not know the full spectrum of the clinical manifestations of legionellosis, individual case reports or retrospective analyses of the pneumonic forms of the disease that we currently recognize are not likely to yield many new insights. Prospective studies with properly matched controls, including patients with other forms of pneumonia, could provide data about unique clinical features or alternate forms of therapy. The likelihood of studies of this type being performed is very low. Careful epidemiological and clinical studies may provide useful information about why Legionnaires disease is uncommon in children or why some patients get Legionnaires disease after a given exposure and others develop Pontiac fever. Little is known about the protective effect of prior infection with organisms of the same species and serogroup, organisms of the same species but different serogroup, and organisms of different species. Relatively little progress has been made in identifying the causes of the major extrathoracic manifestations of legionellosis, which is critical if they are to be mitigated by intervention. The answers to most of these questions will depend on clues derived from the various areas of research rather than from observations at the bedside.

LITERATURE CITED

1. **Anderson, R. D., B. A. Lauer, D. W. Fraser, P. S. Hayes, and K. McIntosh.** 1981. Infections with *Legionella pneumophila* in children. J. Infect. Dis. **143:**386–390.

2. **Dorman, S. A., N. J. Hardin, and W. C. Winn, Jr.** 1980. Pyelonephritis associated with *Legionella pneumophila*, serogroup 4. Ann. Intern. Med. **93:**835–837.

3. **Evans, C. P., and W. C. Winn, Jr.** 1981. Extrathoracic localization of *Legionella pneumophila* in Legionnaires' pneumonia. Am. J. Clin. Pathol. **76:**813–815.

4. **Fairbank, J. T., A. C. Mamourian, P. A. Dietrich, and J. C. Girod.** 1983. The chest radiograph in Legionnaires' disease. Radiology **147:**33–34.

5. **Foy, H. F., P. S. Hayes, M. K. Cooney, C. V. Broome, I. Allan, and R. Tobe.** 1979. Legionnaires' disease in a prepaid medical-care group in Seattle, 1963–75. Lancet **ii:**767–770.

6. **Glick, T. H., M. B. Gregg, B. Berman, G. Mallison, W. W. Rhodes, Jr., and I. Kassanoff.** 1978. Pontiac fever, an epidemic of unknown etiology in a health department. I. Clinical and epidemiologic aspects. Am. J. Epidemiol. **107:**149–160.

7. **Kalweit, W. H., W. C. Winn, Jr., T. A. Rocco, and J. C. Girod.** 1982. Hemodialysis fistula infections caused by *Legionella pneumophila*. Ann. Intern. Med. **96:**173–175.

8. **Kirby, B. D., H. Peck, and R. D. Meyer.** 1979. Radiographic features of Legionnaires' disease. Chest **76:**562–565.

9. **Kirby, B. D., K. M. Snyder, R. D. Meyer, and S. M. Finegold.** 1980. Legionnaires' disease: report of sixty-five nosocomially acquired cases and review of the literature. Medicine (Baltimore) **59:**188–205.

10. **Muder, R. R., V. L. Yu, and J. J. Zuravleff.** 1983. Pneumonia due to the Pittsburgh pneumonia agent: new clinical perspective with a review of the literature. Medicine (Baltimore) **62:**120–128.

Clinical Features and Prognosis of Severe Pneumonia Caused by *Legionella pneumophila*

C. MAYAUD, M.-F. CARETTE, E. DOURNON, A. BURE, T. FRANCOIS, AND G. AKOUN

Respiratory Intensive Care Unit, Tenon Hospital, and Laboratory of Microbiology, Claude Bernard Hospital, Paris, France

This clinical study was done to evaluate the clinical patterns and prognosis of Legionnaires disease (LD) in immunocompromised and previously healthy patients.

Patients and methods. This study was part of a global prospective trial including 280 patients (48% previously healthy, 52% immunocompromised) admitted to our respiratory intensive care unit from August 1980 to January 1983 because of either acute respiratory failure or impending rapid extension of pneumonia. All patients were submitted to complete clinical examination and specific investigations for LD.

A total of 21 patients (18 male, 3 female), 18 to 70 years old, were finally included in this study and were divided into two groups. Group 1 consisted of 10 previously healthy patients (mean age, 53 years) admitted for acute respiratory failure. Seven of these patients were smokers. In three, examination for LD revealed an undiagnosed underlying illness: diabetes mellitus ($n = 1$), lung cancer ($n = 1$), and acute myelomonoblastic leukemia ($n = 1$). In these cases, LD was usually sporadic. Group 2 consisted of 11 immunocompromised patients (mean age, 37 years) admitted because of acute respiratory failure or impending rapid extension of pneumonia. Four of these patients were smokers. Their underlying illnesses were renal transplant ($n = 7$), lymphoma ($n = 1$), multiple sclerosis ($n = 1$), acute lymphoblastic leukemia ($n = 1$), and systemic lupus erythematosus ($n = 1$). All were receiving immunosuppressive and cytotoxic chemotherapy. LD was usually nosocomial.

Diagnosis of LD was established by demonstration of *L. pneumophila* by direct immunofluorescence assay, culture, or both in 11 patients and by serological criteria alone (at least a fourfold increase in the indirect immunofluorescent test to a titer of $\geq 1:128$, or a stable titer of $>1:256$ with a compatible clinical history) in 10 patients. Twenty isolates were serogroup 1.

Relevant clinical features of LD. LD had an acute onset ($n = 16$) and often was associated with constant hyperthermia ($\geq 39°C$) ($n = 17$); pneumonia with a radiological alveolar infiltrate(s), consolidation, or both ($n = 20$); vomiting, diarrhea, or abdominal pain ($n = 15$); and confusion, delirium, or stupor ($n = 14$) (Table 1). Penicillin or cephalosporin treatment was ineffective. This association of symptoms, very suggestive of LD or infection due to chlamydiae in our experience, was similar for both groups of patients, except that neuropsychic disorders were more frequent in group 1.

Prognosis and evolution of LD. Some findings associated with fatal outcome of the disease were severe shock, severe hypoxemia, elevated creatine phosphokinase levels, and necessity for dialysis or controlled ventilation (Table 2). These findings were more frequent in previously healthy patients than in immunocompromised patients, although the same number of deaths occurred in both groups.

In our patients, prognosis was essentially re-

TABLE 1. Relevant clinical features of LD with respect to immunological status of patients

Group (n)	Fever ≥ 39°C	Pneumonia	Neuropsychic disorders	Gastrointestinal symptoms	β-Lactam failure
1 (10)	9	10	9	8	9/9
2 (11)	8	10	5	7	5/5

TABLE 2. Indices of severity and fatal evolution of LD with respect to immunological status of patients

Group (n)	No. with shock	Mean PaO^2 (range) (kPa)	Mean creatine phosphokinase level (range) (mU/ml)	No. requiring: Dialysis	Controlled ventilation	No. of deaths
1 (10)	4	5.8 (4.2–7.0)	1,160 (31–2,700)	2	6	2
2 (11)	2	9.4 (4.2–11.9)	430 (9–2,100)	2	2	2

TABLE 3. Indices of severity and fatal evolution of LD with respect to duration of symptoms before diagnosis and treatment

Delay in diagnosis (days)	No. with shock	Mean PaO^2 (range) (kPa)	Mean creatine phosphokinase level (range) (mU/ml)	No. requiring:		No. of deaths
				Dialysis	Controlled ventilation	
<7 ($n = 14$)	0	8.2 (4.9–11.9)	410 (9–2,700)	0	2	0
≥7 ($n = 7$)	6	5.8 (4.2–8.3)	1,560 (300–2,100)	4	6	4

lated to the delay in diagnosis and proper treatment of LD (Table 3).

Conclusions. (i) The characteristics of severe forms of LD appear to be quite standard, whether patients are immunocompromised or not. (ii) The prognosis for LD appears to be essentially linked to the delay in diagnosis and treatment. (iii) In our study, prognosis was not linked at all to the immunological status of the patients.

Immunocompromised patients are carefully managed, and the diagnosis of LD is generally made faster; the mean delay was 4.8 days (range, 1 to 13 days). In contrast, previously healthy adults receive various ineffective antibiotics outside the hospital; diagnosis was delayed by a mean of 7.2 days (range, 4 to 13 days). Ideally, one should compare patients whose LD is diagnosed and treated after an identical delay.

Comparative Radiographic Features of Legionnaires Disease and Other Sporadic Community-Acquired Pneumonias

J. T. MACFARLANE, A. C. MILLER, A. H. MORRIS, D. H. ROSE, and W. H. R. SMITH

Departments of Thoracic Medicine and Radiology, City Hospital, Nottingham, United Kingdom

Legionnaires disease (LD) ranks second to pneumococcal infection as the most common cause of sporadic community-acquired pneumonia in adults admitted to the City Hospital, Nottingham, England (4). We have not found the initial clinical and laboratory features of LD to be sufficiently distinct to allow differentiation from other causes of community-acquired pneumonia. We have investigated whether there are specific radiographic features of LD. In a largely prospective study, we compared the chest radiographs (CXRs) of 49 consecutive adults admitted with LD with those of 147 adults with other forms of community-acquired pneumonia. The latter included 91 patients with pneumococcal pneumonia (PNP) (31 of whom had pneumococcal bacteremia, pneumococcal antigenemia, or both [bact/Ag]), 46 with mycoplasma pneumonia (MP), and 10 with psittacosis pneumonia. LD was diagnosed serologically (44 cases) or by the detection of the organism in lung or respiratory secretion by culture or direct fluorescent-antibody staining (5 cases). MP and psittacosis pneumonia were diagnosed serologically. Pneumococcal infection was diagnosed by the culture of *Streptococcus pneumoniae* from blood, pleural fluid, or lung, by the detection of pneumococcal polysaccharide capsular antigen in serum, urine, sputum, pleural fluid, or lung, or by both methods.

Radiographs were reviewed systematically by two observers. In addition to posteroanterior views, lateral views were available in most instances. Table 1 summarizes the details of the cases studied.

On presentation, CXR shadowing was mainly homogeneous in 82% of LD cases, 81% of bact/Ag PNP cases, and 70% of other PNP cases. Half of the patients with MP had homogeneous shadowing, as did 60% of the patients with psittacosis. A predilection was seen for the lower lobes in all groups. Sixty-one percent of LD cases had unilobar involvement on presentation, a figure similar to that for psittacosis pneumonia and non-bact/Ag PNP. More patients with bact/Ag PNP (65%) and MP (52%) had two or more lobes involved initially.

Pleural fluid was noted on CXRs of 24% of LD

TABLE 1. Details of cases studied

Pneumonia	No. of cases	Mean age (yr)	% Male	Mortality (%)
LD	49	51	75	12
Pneumococcal				
Bact/Ag	31	61	81	32
Other	60	47	89	12
Mycoplasma	46	30	57	0
Psittacosis	10	46	70	0

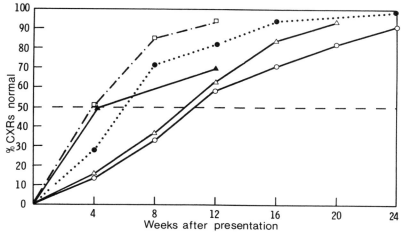

FIG. 1. Radiographic clearance rate of pulmonary shadows for different pneumonias. Symbols: ○, LD (n = 42); △, bact/Ag PNP (n = 19); ●, other PNP (n = 53); □, MP (n = 37); ▲, psittacosis pneumonia (n = 10).

cases, a figure similar to that for the other pneumonias except bact/Ag PNP (52%). Pleural fluid occupied less than one-third of the hemithorax in all LD cases. Some degree of segmental or lobar collapse was seen more frequently with LD (37%) than with psittacosis pneumonia (20%), MP (26%), and PNP (26%). Hilar lymphadenopathy was only recognized with MP (22%). Pulmonary cavitation was not seen with MP but occurred in four patients with PNP, one patient with LD, and one patient with psittacosis pneumonia.

Radiographic deterioration after presentation was more common with LD (65%) and bact/Ag PNP (52%) than with non-bact/Ag PNP (26%), MP (25%), and psittacosis pneumonia (0%). Reasons for radiographic deterioration for the LD patients included extension within the same lobe (43%), spread within the ipsilateral lung (26%), and spread to the opposite lung (28%). More than one reason was present in some cases. Spread within the same lung (22%) and to the opposite lung (22%) was also a feature of bact/Ag PNP but was uncommon with the other pneumonias.

To assess the rate of radiographic resolution, serial CXRs taken at approximately monthly intervals were examined for 161 survivors, including 42 with LD, 19 with bact/Ag PNP, 53 with non-bact/Ag PNP, 37 with MP, and 10 with psittacosis pneumonia. The rate of clearance of pulmonary shadows is shown in Fig. 1. Pulmonary shadows cleared particularly slowly after LD and bact/Ag PNP. Only 60% of the cases had cleared by 12 weeks, and 12% of the LD radiographs were still not clear at 20 weeks. CXR clearance after both MP and psittacosis pneumonia was much faster, with 50% being clear by 4 weeks. Non-bact/Ag PNP improved at an inter-

mediate rate, with 82% clear by 12 weeks. After resolution of pulmonary shadowing, intrapulmonary linear streaky opacities were still present in 29% of LD and 26% of bact/Ag radiographs, suggesting that healing may be associated with fibrosis in such cases. Residual pleural shadows were noted in 37% of the bact/Ag cases but were uncommon with the other groups.

We conclude that there are no unique radiographic features of sporadic community-acquired LD. Common features include unilobar lower lobe involvement on presentation, a small amount of pleural fluid collection in one-fourth of cases, and some degree of pulmonary collapse in more than one-third of cases. Lung cavitation was uncommon, and hilar lymphadenopathy was not seen. Radiographic deterioration occurred in two-thirds of the cases, with spread of shadows both within the same lung and to the opposite lung. Clearance of pulmonary shadows in survivors was slow. The radiographic features we found with sporadic community-acquired LD are similar to those reported for nosocomial LD (3) and epidemic LD (1, 2). Lung cavitation has been reported occasionally (2). Many of the radiographic features were also seen with other types of sporadic pneumonias, particularly with PNP.

LITERATURE CITED

1. Dietrich, P. A., R. D. Johnson, J. T. Fairbank, and J. S. Walke. 1978. The chest radiograph in Legionnaires' disease. Radiology 127:577–582.
2. Fairbank, J. T., A. C. Mamourian, P. A. Dietrich, and J. C. Girod. 1983. The chest radiograph in Legionnaires' disease. Radiology 147:33–34.
3. Kirby, B. D., H. Peck, and R. D. Meyer. 1979. Radiographic features of Legionnaires' disease. Chest 76:562–565.
4. Macfarlane, J. T., M. J. Ward, R. G. Finch, and A. D. Macrae. 1982. Hospital study of adult community-acquired pneumonia. Lancet ii:255–258.

Legionella pneumophila Infection Associated with Arrhythmia or Myocarditis in Children Without Pneumonia

MADDALENA CASTELLANI PASTORIS, GIOVANNI NIGRO, MIRELLA FANTASIA MAZZOTTI, AND MARIO MIDULLA

Batteriologia e Micologia Medica, Istituto Superiore di Sanità, and IV Clinica Pediatrica, Università La Sapienza, Centro CNR Virus Respiratori, Rome, Italy

Legionella pneumophila in adults has been associated with severe pneumonia, with or without concomitant neurological, gastrointestinal, and cardiac disorders, but little is known about the clinical features of this infection in children. To extend the knowledge of *L. pneumophila* infection in childhood, paired sera of 159 pediatric patients with different illnesses of unknown etiology were tested retrospectively for *L. pneumophila* antigens by the indirect fluorescent-antibody test. Three of twenty children aged 2 months to 11 years, admitted to the hospital from June 1963 to March 1979 for cardiac troubles but without respiratory diseases, showed evidence of *L. pneumophila* infection. In all of these patients, illness due to coxsackievirus types B1 to B6, cytomegalovirus, influenza virus types A and B, parainfluenza virus types 1, 2, and 3, adenovirus, respiratory syncytial virus, herpes simplex virus, or *Mycoplasma pneumoniae* was ruled out by complement fixation tests and, in most cases, by isolation attempts.

The first case was a 2.5-year-old girl admitted to the hospital in May 1971 suffering from tachycardia and arrhythmia after fever (38.0 to 38.5°C), vomiting, and diarrhea for a few days. Laboratory findings were within normal limits. The electrocardiogram showed sinus tachycardia (130 beats per min) with monofocal ventricular extrasystoles (Fig. 1). Sera obtained 2 and 4 weeks after the onset of the symptoms showed a fall in titer to *L. pneumophila*, both group 1 and 2, from 128 to 32.

The second case was a 10-year-old boy admitted to the hospital in September 1972 for congestive heart failure after malaise, asthenia, and headache for 3 days and vomiting with abdominal pain for 1 day. At admission temperature was 37.8°C, pulse was 90 beats per min, and blood pressure was 80/50 mmHg. No pathological signs were found on chest examination. Results of laboratory studies included the following: leukocyte count, 16,000/mm^3, with 82% neutrophils; erythrocyte sedimentation rate, 23 mm/h; sodium, 165 meq/liter; potassium, 1.9 meq/liter; alanine aminotransferase, 80 U/liter; and aspartate aminotransferase, 50 U/liter. Urinalysis showed albuminuria. "Configuratio mitralica" appeared on X ray. The electrocardiogram (Fig. 2) revealed prominent U waves, complex ventricular extrasystolic arrhythmia

with multifocal isolated beats or beats grouped in pairs or in tachycardic paroxysms (150 to 170 beats per min), and signs of myocardic damage and hypopotassemia. The child was treated with 5% glucose infusion, quinidine, potassium, antibiotics (ampicillin and dicloxacillin), betamethasone, and adrenocorticotropin. Although slight electrocardiographic abnormalities and cardiac enlargement persisted, the child was discharged 2 months later. In May 1973 the electrocardiogram tracing returned to normal. Sera drawn 17 and 42 days after the onset of the illness showed a fourfold increase in titer, from 16 to 64, to *L. pneumophila* serogroup 2.

The third case was a 7-month-old girl found to have cardiac troubles in March 1979. Three weeks before, she had a 1-day fever (38.0°C) followed later by coryza, pharyngitis, and dry cough. The infant was hospitalized for 3 days. Laboratory findings were normal. Radiological examination showed cardiac enlargement. The electrocardiogram revealed tachycardia (145 beats per min) and arrhythmia with monofocal ventricular extrasystoles. Treatment with oral quinidine and betamethasone was carried out for a few days. The antibody titer of sera obtained during the hospitalization and 24 days later rose from 16 to 64 to *L. pneumophila* serogroup 1.

Acute myocardial diseases are frequently considered to be a possible viral infection, although they often remain of unconfirmed etiology. Cardiac involvement in association with Legionnaires disease has been recognized in only a few instances in adults (2, 3, 4) and never in children, but to our knowledge it has never been seen as a self-limited illness. The three cases described above showed seroconversion or a fourfold fall in titer to *L. pneumophila*. Two of them did not show the currently accepted antibody level for an acute *Legionella* infection, and the third had a fall in titer. However, in the reported cases there was no serological evidence of concurrent

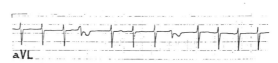

FIG. 1. Electrocardiogram (aVL) showing arrhythmia with monofocal ventricular extrasystoles.

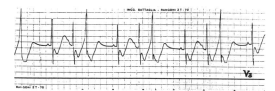

FIG. 2. Electrocardiogram (V₃) showing prominent U waves and complex ventricular extrasystolic arrhythmia with multifocal isolated beats or with beats grouped in pairs. There are signs of myocardic damage and hypopotassemia.

infection by respiratory viruses, coxsackieviruses, or *M. pneumoniae*.

It has been postulated (1) that infection with *L. pneumophila* in children may cause an illness not characterized by symptoms in the lower respiratory tract. One of our patients had severe myocarditis with clinical features of toxic disorders of the myocardium; the other two had self-limited arrhythmia. Evidence of *L. pneumophila* infection in children with acute cerebellar ataxia

infection in children with acute cerebellar ataxia but no pneumonia was recently found (G. Nigro, M. Castellani Pastoris, M. Fantasia Mazzotti, and M. Midulla, Pediatrics, in press). These findings suggest that clinical patterns in children might be undiagnosed because of their nonpulmonary involvement or their mildness. It is therefore suggested that *Legionella* infection should be considered in arrhythmia and myocarditis in children.

These studies were supported in part by a grant from the Consiglio Nazionale delle Ricerche, Progetto finalizzato "Controllo malattie da infezione."

LITERATURE CITED

1. **Andersen, R. D., B. A. Lauer, D. W. Fraser, P. S. Hayes, and K. McIntosh.** 1981. Infections with *Legionella pneumophila* in children. J. Infect. Dis. 143:386–390.
2. **Gross, D., H. Willens, and S. M. Zeldis.** 1981. Myocarditis in Legionnaires' disease. Chest 79:232–234.
3. **Harris, L. F.** 1981. Legionnaires' disease associated with massive pericardial effusion. Arch. Intern. Med. 141:1385.
4. **Landes, B. W., G. W. Pogson, G. D. Beauchamp, R. K. Skillman, and J. H. Brewer.** 1982. Pericarditis in a patient with Legionnaires' disease. Arch. Intern. Med. 142:1234–1235.

Legionnaires Disease in Spain

EMILIO BOUZA AND MARTA RODRIGUEZ-CREIXEMS

Infectious Disease Unit, Centro Especial "Ramón y Cajal," Madrid, Spain

The incidence, morbidity, and mortality of legionellosis are significant, but they are variably estimated in different hospitals and countries. Spain has been frequently referred to in studies of patients who acquired legionellosis while visiting there (2, 5, 6, 9).

This paper reports a survey of serum samples of 239 healthy individuals living in the Madrid area. Of these, 109 were blood donors, and the remaining 130 were sanitation personnel working in two Madrid hospitals. All subjects had at least one indirect fluorescent-antibody (IFA) titer determined, and 123 of them had a second determination between 2 and 5 weeks after the first. We also report the first 45 indigenous cases of legionellosis diagnosed at our hospital between October 1978 and May 1983. Our unit has investigated the possibility of legionellosis in all patients with pneumonia in our care and in a small group of patients with non-pneumonic febrile episodes. We also investigated such possibilities with clinical specimens of patients with pneumonia, hospitalized elsewhere in Spain, which were referred to us. Part of the present information has already been reported in the Spanish medical literature (1, 10, 11).

We accepted the diagnosis of legionellosis when one or more of the following criteria were fulfilled: (i) an increase in IFA titer of at least

fourfold (to at least 1:128) in serum specimens obtained during the acute and convalescent phases of the illness, or single specimens with high or maintained anti-*Legionella* titers (≥1:256 by IFA), with a compatible clinical setting; (ii) visualization of legionellae in clinical specimens with specific fluorescent antibodies (DFA); and (iii) isolation of legionellae from clinical specimens.

IFA and DFA techniques were performed with reagents and instructions kindly provided by the Centers for Disease Control (Atlanta, Ga.). For the isolation of legionellae we routinely used charcoal-yeast extract agar. From June 1982 charcoal-yeast extract agar was replaced by the buffered medium and supplemented with α-ketoglutarate.

Titers of anti-*L. pneumophila* serogroup 1 antibodies in our healthy population were 1:128 in fewer than 7% of the cases. There was no difference between the two groups (blood donors and sanitation personnel) and a titer of 1:128 was never exceeded.

In the 52-month period of our study, 45 cases were diagnosed as legionellosis, and *Legionella* spp. caused approximately 6% of our pneumonias. Patients came from different areas of Spain (Madrid, 27; Barcelona, 5; Bilbao, 4; Benidorm, 1; Cáceres, 1; Cuenca, 1; Jaén, 1; Mallorca, 1;

Oviedo, 1; Santander, 1; Segovia, 1; and Vitoria, 1), and only two of them had been abroad (to the United States and Canada) in the preceding months. The disease was community acquired in 39 patients and nosocomial in the remaining 6. Cases were presented throughout the year, but a clear increase in incidence was observed in the summer months. Of our 45 cases, only 8 were previously healthy. The remaining 37 had underlying diseases which were classified, according to the criteria of McCabe and Jackson (7), as ultimately fatal in 8 cases and nonfatal in 29 cases.

The age of our patients was from 18 to 87 years (mean, 52 years; standard deviation, 17 years), and a clear male predominance was present (39 males and 6 females).

Patients were admitted to the hospital after one period of clinical evolution, which varied between 1 and 22 days (mean, 7.2 days; standard deviation, 4.4 days). Fever was the most common clinical sign (89%), and in more than 85% of the cases temperature was greater than 39°C. Other clinical manifestations included chills (59%), myalgia (50%), and headache (41%).

Clinical manifestations pointing to respiratory tract involvement included cough (32 cases), which in some became productive (27 cases) or accompanied by blood-streaked sputum or hemoptysis (9 cases), chest pain (53%), and dyspnea (42%). Physical findings included moist rales or evidence of a varying degree of consolidation. Physical chest examination was unremarkable in 11% of our cases.

Excluding headaches, 19 of the 45 patients (42%) presented data on admission suggesting central nervous system involvement, including lethargy, confusion, delirium, and disorientation. Three of these patients also had dysarthria, and one had marked ataxia. In one patient, who had extensive pneumonia and nuchal rigidity, a translaryngeal aspiration and two consecutive cerebrospinal fluid samples showed *L. pneumophila* (serogroup 2) in DFA, but unfortunately it was not isolated on culture. In another patient, reported elsewhere (4), who died of legionellosis, *L. pneumophila* serogroup 1 was seen in the brain and other extrathoracic organs.

A total of 16 of our 45 patients (35.5%) had clinical manifestations suggesting digestive tract involvement. Watery diarrhea was present in 13 cases, and there was liver and spleen involvement in 42 and 9% of the cases, respectively. Jaundice was detected in only two cases.

In our series only two patients had renal failure severe enough to require dialysis.

Routine laboratory data were nonspecific and included leukocytosis (64%), elevation of serum glutamic oxalacetic and glutamic pyruvic transaminases (71%), increased alkaline phosphatase

(49%), elevated bilirubin (24%), low inorganic phosphorus (48%), and hyponatremia (31%). Blood urea nitrogen was elevated in 39% of our cases, and proteinuria (81%), hematuria (55%), and casts (39%) were frequently present.

Chest X-ray abnormalities were absent in only one case and this patient had symptoms of Pontiac fever. The remaining 44 patients had radiological findings consistent with pneumonia. Infiltrates were alveolar and remained unilateral

TABLE 1. Diagnosis and evolution of legionellosis[a]

Case	Serology	Visualization	Isolation	Treatment	Evolution
1	Cs	N	—	E	C
2	Cs	P	N	E	C
3	Cs	N	N	Cx	C
4	Cs	P	N	E	D
5	Cs	—	—	E	C
6	Cs	—	—	E	C
7	P	—	—	E	C
8	Cs	—	—	E	C
9	Cs	P(?)	—	R	C
10	—	P	N	I	D
11	—	P	—	I	D
12	P	P	—	I	D
13	Cs	P	—	I	D
14	N	P	—	I	D
15	P	P	—	E	C
16	Cs	N	N	Cx	C
17	Cs	P	—	E	C
18	N	P	N	E	D
19	Cs	—	—	E	C
20	Cs	P	N	E	C
21	Cs	—	—	E, Cx, R	C
22	Cs	—	—	E	C
23	Cs	—	—	E	C
24	Cs	—	—	I	C
25	Cs	—	—	E	C
26	Cs	—	—	E	C
27	P	P	N	R	D
28	N	P	—	E	C
29	Cs	P	N	E	C
30	N	P	N	E	C
31	Cs	N	—	E	C
32	P	P	—	E	D
33	—	N	N	E	C
34	N	P	P	E	C
35	Cs	N	P	E	C
36	Cs	—	—	E	C
37	Cs	—	—	Cx	C
38	Cs	N	N	E	C
39	—	P	—	I	D
40	P	—	—	E	C
41	—	P	P	E	D
42	Cs	—	—	E	C
43	Cs	N	P	E	C
44	P	—	—	I	C
45	Cs	—	—	E	C

[a] Abbreviations: Cs, conversion; P, positive; N, negative; E, erythromycin; Cx, cefoxitin; R, rifampin; I, inadequate; C, cure; D, death; —, not done.

(25 cases) or were bilateral (19 cases). A total of 19 of our patients had infiltrates that remained confined to the originally involved lobe, and the remaining 25 patients had two or more lobes involved. All lobes have been found to be affected, but involvement of the lower lobes was more common. Only 4 of our 44 patients had pleural effusions, and 6 patients had evidence of abscess or cavity formation.

The diagnostic criteria used in our 45 cases are summarized in Table 1. The most widely used test was IFA (41 of 45 patients), and 36 of our cases had diagnostic values in this test (29 seroconversion and 7 seropositivity). Five cases, confirmed by other procedures, were negative (three reached a titer of 1:128 without conversion, one had a titer of 1:64, and the remaining one had only one serum sample available).

DFA examinations of respiratory tract secretions, lung tissues, pleural fluid, and other clinical specimens were performed in 28 of our cases, and they were positive in 19, negative in 8, and doubtful in 1. The doubtful case was a patient who had a chest infiltrate and evidence of endocarditis. Blood cultures were negative, the patient seroconverted, and after several days of therapy with rifampin, aortic valve replacement was performed. DFA on the involved valves had fluorescence, which is considered doubtful by us and by the Centers for Disease Control. In the eight negative cases the specimens were usually obtained after several days of antimicrobial therapy.

We tried to isolate legionellae in 16 of our cases, and we succeeded 4 times (four isolates of *L. pneumophila* serogroup 1). All of our isolates were obtained since June 1982, when we started to use buffered charcoal-yeast extract agar containing α-ketoglutarate.

With serology, visualization, and isolation as the three diagnostic criteria, 30 of our cases had only one positive criterion and the remaining 15 cases had two.

Thirty-one patients received erythromycin (mean duration, 20 days), four received cefoxitin, and three received rifampin. In this group of 37 patients, 5 died (13.5%); 4 of them had received erythromycin for less than 5 days, and 1 had received rifampin. The remaining eight patients received inadequate antimicrobial therapy; six died (75%), and two survived. Our overall mortality was 24%.

In summary, our data suggest that legionellosis is not hyperendemic in the Madrid area of Spain but is a significant cause of morbidity and mortality. The clinical presentation of our cases is similar to that reported previously (3, 8), but the high incidence of pulmonary cavitation in our patients is remarkable. Cefoxitin proved to be a potentially useful drug for our patients, although it has failed in other reported cases.

LITERATURE CITED

1. **Bouza, E., and M. Rodriguez-Creixems.** 1979. Enfermedad de los legionarios. Med. Clin. 73:396.
2. **Boyd, J. F., W. M. Buchanan, T. I. F. McLeod, R. I. Shaw-Donn, and W. P. Weir.** 1978. Pathology of five Scottish deaths from pneumonic illnesses acquired in Spain due to Legionnaires' disease agent. J. Clin. Pathol. 31:809–816.
3. **Fraser, D. W., T. R. Tsai, W. Orenstein, W. E. Parkin, H. J. Beecham, R. G. Sharrar, J. Harris, G. F. Mallison, S. M. Martin, J. E. McDade, C. C. Shepard, P. S. Brachman, and the Field Investigation Team.** 1977. Legionnaires' disease. Description of an epidemic of pneumonia. N. Engl. J. Med. 297:1189–1197.
4. **Gatell, J. M., J. M. Miro, M. Casal, O. Ferrer, M. Rodriguez-Creixems, and J. García-San Miguel.** 1981. *Legionella pneumophila* antigen in brain. Lancet ii:202–203.
5. **Jenkins, P., A. C. Miller, J. Osman, S. B. Pearson, and J. M. Rowley.** 1979. Legionnaires' disease: a clinical description of thirteen cases. Br. J. Dis. Chest 73:31–38.
6. **Lawson, J. H., N. R. Grist, D. Reid, and T. S. Wilson.** 1977. Legionnaires' disease. Lancet i:1083.
7. **McCabe, W. R., and G. G. Jackson.** 1962. Gram-negative bacteremia. Arch. Intern. Med. 110:83–91.
8. **Meyer, R. D.** 1983. *Legionella* infections: a review of five years of research. Rev. Infect. Dis. 5:258–278.
9. **Reid, D., N. R. Grist, and R. Nájera.** 1978. Illness associated with "package tours"; a combined Spanish-Scottish study. Bull. W.H.O. 56:117–122.
10. **Rodriguez-Creixems, M., E. Bouza, F. Soriano, L. Casimir, P. Martí-Belda, L. Buzón, and F. Baquero.** 1981. Enfermedad de los legionarios. Experiencia con 27 casos. Med. Clin. 77:349–355.
11. **Zabala, P., E. Frieyro, E. Bouza, M. Rodriguez-Creixems, L. Barbolla, I. Sanjuán, L. Buzón, C. Masa, R. Pérez-Maestu, and M. N. Fernández.** 1979. Enfermedad de los legionarios en un varón de la provincia de Madrid afecto de leucemia linfoide crónica. Med. Clin. 73:1–4.

Isolation of *Legionella pneumophila* with a Protected Specimen Brush

J. PIERRE, P. BRUN, G. BERTHELOT, J. CHASTRE, C. GIBERT, AND E. BERGOGNE-BEREZIN

Microbiology Department and Intensive Care Unit, Bichat Hospital, 75877 Paris Cedex 18, France

During an outbreak of Legionnaires disease (LD), a protected specimen brush (PSB) (2) was used as a method of sampling for bacteriological study and isolation of *Legionella pneumophila*. The PSB consists of a telescoping double catheter with distal occlusion, protecting a nylon bristle brush (Fig. 1). Sampling was performed through a flexible bronchoscope (C. Gibert, J. Pierre, J. Chastre, R. Bouchama, A. Akesbi, F. Viau, and P. Brun, Program Abstr. Intersci. Conf. Antimicrob. Agents Chemother. 22nd, Miami Beach, Fla., abstr. no. 698, 1982). Distal secretions were also collected for direct examination (3). The brush was removed and vortexed in 1 ml of distilled water, which was considered the undiluted sample. Diluted (10^{-1}) and undiluted samples were plated in duplicate on buffered charcoal-yeast extract medium with and without antibiotics (1); cultures were incubated for 10 days in a 2.5% CO_2 atmosphere at 37°C and observed daily. Quantitative cultures were also performed on standard media.

A total of 45 patients with acute pneumonia, including 8 women and 37 men (mean age, 60.9 ± 17.6 years) were investigated. Pneumonia was defined by clinical data and X-ray consolidation. Underlying disease was generally important and included malignancy, respiratory disease, aftermath of open heart surgery, renal failure, collagen disease, diabetes mellitus, pancreatitis, head injury, and arthritis. LD was confirmed in 20 patients by isolation of the organism or serological data. PSB material culture was positive in eight cases (40%) for *L. pneumophila* serogroup 1, and direct fluorescent-antibody staining was positive in two cases; none of these patients had received erythromycin before sampling, and two were treated with ampicillin. Cultures remained negative in 12 cases, with the direct fluorescent-antibody test positive in 1 case; six patients had received erythromycin before collection by PSB, three others received ampicillin, ampicillin plus gentamicin, and vancomycin plus metronidazole, and the remaining two had not received any antibiotic.

In patients with LD, associated flora was absent on standard media in 14 cases; coagulase-positive staphylococci were isolated in 4 cases, and *Pseudomonas aeruginosa* and acinetobacters were isolated in 2 cases. Quantitation of the organisms present indicated that these species were not responsible for the infection. A total of 25 patients had no evidence of LD as cultures by PSB remained negative and antibodies were not detected in serum specimens obtained 3 weeks or more (12 cases) or 10 days or less (9 cases) after onset of illness. In the 11 non-

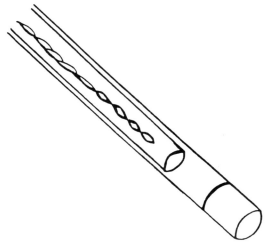

FIG. 1. Protected specimen brush consisting of a telescoping double catheter with distal occlusion protecting a nylon bristle brush. Sampling is done through a flexible bronchoscope.

LD cases for which a definitive diagnosis was made, disease was due to bacteria (5 cases), viruses (3 cases), atelectasis (2 cases), and cardigenic edema (1 case).

To conclude, 45 patients with acute pneumonia were investigated, and LD was diagnosed in 20 cases. Direct fluorescent-antibody staining performed on bronchial aspirates was positive in three cases (15%). Excluding the six patients who had received erythromycin, culture of PSB material was positive in 57.1% of the cases (8 of 14) for *L. pneumophila* serogroup 1. Since PSB provides uncontaminated samples, even in ventilated patients, it may be an alternate method for isolation of *L. pneumophila*.

LITERATURE CITED

1. **Feeley, J. C., R. J. Gibson, G. W. Gorman, N. C. Langford, J. K. Rasheed, D. C. Mackel, and W. B. Baine.** 1979. Charcoal-yeast extract agar: primary isolation medium for *Legionella pneumophila*. J. Clin. Microbiol. **10:**437–441.
2. **Wimberley, N., L. J. Faling, and J. G. Bartlett.** 1979. A fiberoptic bronchoscopy technique to obtain uncontaminated lower airway secretions for bacterial culture. Am. Rev. Respir. Dis. **119:**337–343.
3. **Winn, W. C., Jr., W. B. Cherry, R. O. Franck, C. A. Casey, and C. V. Broome.** 1980. Direct immunofluorescent detection of *Legionella pneumophila* in respiratory specimens. J. Clin. Microbiol. **11:**59–64.

Comparison of Media for Recovery of *Legionella pneumophila* Clinical Isolates

J. D. KEATHLEY AND W. C. WINN, JR.

Medical Center Hospital of Vermont and University of Vermont College of Medicine, Burlington, Vermont 05401

The most widely used medium for isolation of *Legionella* spp. is buffered charcoal-yeast extract agar. The addition of α-ketoglutarate to this medium reportedly enhances the growth and recovery of *Legionella pneumophila* (1). In addition to buffered charcoal-yeast extract agar, media preparations without charcoal have been reported to be useful in the detection and study of legionellae. Two such media are transparent yeast extract-phosphate-hemin agar (4) and enriched horse blood agar (3). The purpose of this study was to evaluate the ability of these media to recover *L. pneumophila*.

Media. Buffered charcoal-yeast extract agar was prepared in-house according to the method of Feeley et al. (2). α-Ketoglutarate (0.1%) was added to half of the medium before autoclaving. Greaves horse blood agar (3) and transparent yeast extract-phosphate-hemin agar were also prepared in-house. Two lots of each medium were prepared.

Commercially prepared buffered charcoal-yeast extract agar was purchased from Regional Medical Laboratories, Inc., Scott Laboratories, Inc., and GIBCO Laboratories. Media from Regional Medical Laboratories included α-ketoglutarate (0.1%); media from Scott Laboratories and GIBCO Laboratories did not. Plates from two different lots were purchased from each company except GIBCO Laboratories, from which only one lot was available. All media were tested for ability to support growth of *L. pneumophila*, using a standardized stock culture inoculum (10^8 to 10^9 CFU/ml) and a standardized spleen cell suspension (10^3 to 10^4 CFU/ml).

Specimens. Specimens included one sputum and eight lung tissue samples, all of which had previously yielded *L. pneumophila*. The sputum was homogenized by vortexing with glass beads. Approximately 0.5 g of each tissue sample was ground in 2 ml of saline. Each plate was inoculated with 0.05 ml of the prepared specimen and then streaked for isolation in a spoked-wheel fashion. Specimens were inoculated onto duplicate plates from each lot of each medium. All plates were incubated at 35°C in air and were examined at 2, 3, 4, 7, 10, and 14 days. Plates were examined for morphologically typical *Legionella* colonies by using a stereoscope. Suspect colonies were reisolated to blood and buffered charcoal-yeast extract agar. *L. pneumophila* isolates grew only on buffered charcoal-yeast extract agar. Direct immunofluorescence was used to confirm the identification of such isolates. The number of *Legionella* colonies on each plate was recorded.

All media supported the growth of *L. pneumophila* when inoculated with the standardized stock culture suspension. Buffered charcoal-yeast extract agar with α-ketoglutarate supported heavy growth. Buffered charcoal-yeast extract agar without α-ketoglutarate supported moderate to heavy growth. Greaves horse blood agar supported growth of a few *L. pneumophila* colonies. Transparent yeast extract-phosphate-hemin agar supported moderate growth of tiny colonies.

L. pneumophila was not recovered on Greaves horse blood agar or transparent yeast extract-phosphate-hemin agar with the standardized spleen cell suspension. Buffered charcoal-yeast extract agar yielded *L. pneumophila* in all cases when inoculated with standardized spleen cell suspension. The numbers of CFU recovered are shown in Table 1.

Results with media inoculated with clinical

TABLE 1. Recovery of *L. pneumophila* from standardized suspensions

Medium	Growth with the following suspension:	
	Stock culture[a]	Spleen cell[b]
Buffered charcoal-yeast extract agar		
Regional Medical Laboratories	Heavy	96 (80–100)
In-house with α-ketoglutarate	Heavy	47 (28–65)
In-house without α-ketoglutarate	Heavy	22 (12–32)
Scott Laboratories	Moderate	16 (8–25)
GIBCO Laboratories	ND	8 (7–8)[c]
Greaves horse blood agar	Few	No growth
Transparent yeast extract agar	Moderate (tiny colonies)	No growth

[a] Heavy, Growth in three zones; moderate, growth in two zones; few, growth in one zone; ND, not done.
[b] Expressed as mean number of CFU from three replicate inocula to two different lots; range is in parentheses.
[c] One lot.

TABLE 2. Recovery of *L. pneumophila* from clinical specimens

Medium[a]	No. positive/total no.				Median no. of colonies at 14 days
	Specimens	Plates at:			
		3 days	4 days	14 days	
1	9/9	27/36	33/36	33/36	74
2	9/9	21/36	31/36	31/36	41
3	9/9	27/36	33/36	33/36	100
4	8/9	1/36	22/36	29/36	26
5	7/9	1/18	10/18	14/18	40
6	3/9	0/34	0/34	4/34	0
7	0/9	—	—	—	—

[a] 1, Buffered charcoal-yeast extract agar with α-ketoglutarate, prepared in-house; 2, buffered charcoal-yeast extract agar without α-ketoglutarate, prepared in-house; 3, Regional Medical Laboratories buffered charcoal-yeast extract agar; 4, Scott Laboratories buffered charcoal-yeast extract agar; 5, GIBCO Laboratories buffered charcoal-yeast extract agar; 6, transparent yeast extract-phosphate-hemin agar; 7, Greaves horse blood agar.

specimens are summarized in Table 2. Buffered charcoal-yeast extract agar prepared in-house, both with (1) and without (2) α-ketoglutarate, and buffered charcoal-yeast extract agar purchased from Regional Medical Laboratories, Inc. (3) recovered *L. pneumophila* from all nine specimens. These media recovered the highest numbers of colonies, and the time until recovery was faster than with the other media. The majority of colonies were recovered after 3 days of incubation, compared with 4 days and often longer for initial detection on the other media. Greaves horse blood agar failed to recover any *Legionella* isolates; transparent yeast extract-phosphate-hemin agar recovered three isolates, all at 14 days.

The results of this study show that the inclu-sion of α-ketoglutarate in buffered charcoal-yeast extract agar enhances the growth and recovery of *L. pneumophila* from clinical specimens. Media without α-ketoglutarate recovered fewer colonies and required longer periods of time for initial detection.

Media without charcoal performed poorly. Greaves horse blood agar was prepared by adding L-cysteine and ferric pyrophosphate before autoclaving. This medium supported growth of *L. pneumophila* only with a very high inoculum. We also prepared a modification of the agar by filter-sterilizing L-cysteine and ferric pyrophosphate solutions and adding them to the agar after autoclaving. This medium also failed to support the growth of *L. pneumophila*.

On transparent agar, only 4 of 34 plates inoculated grew *L. pneumophila*, and growth did not appear until 14 days of incubation. Colonies were very tiny and did not appear similar to the characteristic *Legionella* colonies seen on buffered charcoal-yeast extract agar.

Our results show that buffered charcoal-yeast extract agar remains the best medium for isolation of *L. pneumophila* from clinical specimens and that its performance is enhanced by α-ketoglutarate. Media without charcoal do not perform well in recovery of clinical isolates.

LITERATURE CITED

1. **Edelstein, P. H.** 1981. Improved semiselective medium for isolation of *Legionella pneumophila* from contaminated clinical and environmental specimens. J. Clin. Microbiol. **14**:298–303.
2. **Feeley, J. C., R. J. Gibson, G. W. Gorman, N. C. Langford, J. K. Rasheed, D. C. Mackel, and W. B. Baine.** 1979. Charcoal-yeast extract agar: primary isolation medium for *Legionella pneumophila*. J. Clin. Microbiol. **10**:437–441.
3. **Greaves, P. W.** 1980. New methods for the isolation of *Legionella pneumophila*. J. Clin. Pathol. **33**:581–584.
4. **Johnson, S. R., W. O. Schalla, K. H. Wong, and G. H. Perkins.** 1982. Simple, transparent medium for study of legionellae. J. Clin. Microbiol. **15**:342–344.

Growth of *Francisella tularensis* on Media for the Cultivation of *Legionella pneumophila*

KATHY B. MacLEOD, CHARLOTTE M. PATTON, GEORGE C. KLEIN, AND JAMES C. FEELEY

Division of Bacterial Diseases, Center for Infectious Diseases, Centers for Disease Control, Atlanta, Georgia 30333

Francisella tularensis and *Legionella pneumophila* are fastidious gram-negative bacilli which require the presence of cysteine for growth on artificial media. The culture medium on which *L. pneumophila* was first isolated was Mueller-Hinton agar supplemented with 1% hemoglobin and 1% IsoVitaleX. Since this medium also supports the growth of *F. tularensis*, (R. E. Weaver, personal communication), it appeared that a single medium could be used to grow both *L. pneumophila* and *F. tularensis*. With this in mind, we tested the ability of some of the media developed for the cultivation of *L. pneumophila* to support the growth of *F. tularensis* from guinea pig tissue. We also tested the ability of an *F. tularensis* medium to support the growth of *L. pneumophila*.

The *L. pneumophila* media tested were char-

TABLE 1. Growth of *F. tularensis* on *Francisella* and *Legionella* culture media

Culture medium	Growth[a] of:	
	F. tularensis	*L. pneumophila*
GCBA	80%	<0.01%
BCYE with glucose	100%	93%
BCYE	43%	100%
CYE with glucose	19%	93%
CYE	<0.01%	84%

[a] Expressed as the percentage of CFU per milliliter of inoculum that formed on BCYE with glucose for *F. tularensis* and on CYE for *L. pneumophila*.

coal-yeast extract (CYE) and buffered charcoal-yeast extract (BCYE), with and without 2.5% glucose. The *F. tularensis* medium was glucose-cysteine-blood agar (GCBA), with and without sodium chloride, prepared from scratch as described by Downs et al. (1). The CYE and BCYE media were prepared as described previously (2, 3). We used inocula prepared from infected guinea pig spleens because *L. pneumophila* cultures grown in vivo appear to be more fastidious than those grown on artificial culture media (2). The *L. pneumophila* inoculum was a 10-fold dilution in sterile distilled water of a suspension of guinea pig spleen infected with the Philadelphia 1 strain. The *F. tularensis* inoculum was a 10-fold dilution in sterile saline (0.85%) of guinea pig spleen infected with a glycerol-positive strain of *F. tularensis* isolated from human

blood. Standardized suspensions of guinea pig spleen were prepared as described previously (2). The plates of culture media were inoculated with 0.1 ml of suspension and incubated at 35 to 37°C in an atmosphere containing 2.5% CO_2.

The results are shown in Table 1. The number of CFU per milliliter of inoculum was used as a measure of the growth-supporting ability of the media. The *Legionella* medium BCYE with glucose actually supported growth of *F. tularensis* better than the GCBA medium. The results indicate that *F. tularensis* can be isolated on *Legionella* media and that extreme caution should be used when culturing specimens on *Legionella* media in an area in which tularemia is endemic. In fact, the Centers for Disease Control received a culture isolated on BCYE medium which was submitted for confirmation of *Legionella* spp. but proved to be *F. tularensis* (R. E. Weaver, personal communication).

LITERATURE CITED

1. Downs, C. M., L. L. Coriell, S. S. Chapman, and A. Klaubell. 1947. The cultivation of *Bacterium tularenses* in embryonated eggs. J. Bacteriol. 53:89–100.
2. Feeley, J. C., R. J. Gibson, G. W. Gorman, N. C. Langford, J. K. Rasheed, D. C. Mackel, and W. B. Baine. 1979. Charcoal-yeast extract agar: primary isolation medium for *Legionella pneumophila*. J. Clin. Microbiol. 10:437–441.
3. Pasculle, A. W., J. C. Feeley, R. J. Gibson, L. G. Cordes, R. L. Myerowitz, C. M. Patton, G. W. Gorman, C. L. Carmack, J. W. Ezzell, and J. N. Dowling. 1980. Pittsburgh pneumonia agent: direct isolation from human lung. J. Infect. Dis. 141:727–732.

Results of Legionnaires Disease Direct Fluorescent-Antibody Testing at Centers for Disease Control, 1980–1982

A. L. REINGOLD, B. M. THOMASON, AND J. KURITSKY

Centers for Disease Control, Atlanta, Georgia 30333

Since the recognition of Legionnaires disease (LD) and the isolation of *Legionella pneumophila* serogroup 1, many additional *Legionella* species and serogroups have been described, and their role as human pathogens has been demonstrated. The relative frequency of pneumonia caused by the various serogroups and species has not, however, been ascertained. To examine this question we reviewed the results of LD direct fluorescent-antibody (DFA) testing performed at the Centers for Disease Control (CDC) (1).

We reviewed laboratory reports from the LD DFA laboratory for specimens received between 1 January 1980 and 31 May 1982. Information on the type of specimen, the date received, the age, sex, and state of residence of the patient, and the

result of DFA testing was reviewed. Each patient was counted once, regardless of the number of specimens submitted. In addition, state health department laboratories were surveyed to determine whether they forwarded all specimens received for LD DFA testing to CDC or only a selected subgroup of specimens. All specimens were tested with the following conjugates: *L. pneumophila* serogroups 1 through 6, *L. dumoffii*, *L. bozemanii*, *L. micdadei*, and *L. gormanii* (1). All specimens received after 30 July 1980 also were tested with a conjugate against *L. longbeachae* serogroup 1, and those received after 1 October 1980 were tested with a conjugate against *L. longbeachae* serogroup 2.

Specimens of pulmonary origin from 1,395 patients were received between 1 January 1980 and 31 May 1982. Of these, 141 (10%) were positive, including 44 (6%) of 794 submitted by the 17 states which forwarded all specimens to CDC and 97 (16%) of 601 from the 33 states which forwarded selected specimens to CDC. Among states forwarding all specimens to CDC, Virginia (11.3%) and Alabama (10.1%) had the highest positivity rates, but no regional differences were observed.

Although 61% of the patients tested were male, the positivity rate did not vary with sex (male, 9.5%; female, 9.0%). The positivity rate increased significantly with increasing patient age (χ^2 for linear trend = 5.01, P = 0.025), with the highest rate seen among patients 60 to 69 years old. Eighty percent of the positive samples for which patient age was specified came from patients 50 years of age or older.

For states submitting all specimens to CDC, the positivity rate was highest during the third quarter (July through September), but this difference was not significant. The positivity rate by year declined consistently over time for states forwarding all specimens to CDC, but not for states forwarding selected specimens.

The positivity rate was highest for lung specimens (14%), but was almost as high (11%) for bronchial and tracheal secretions. Expectorated sputum yielded a lower positivity rate (5%), whereas pleural fluid and tissue specimens yielded the lowest rate (3%).

L. pneumophila serogroup 1 was found in 50% of the positive patients, and *L. pneumophila* serogroup 6 was found in another 14%; *Legionella* spp. other than *L. pneumophila* were present in 13% of the patients. The relative frequencies of *L. pneumophila* serogroup 1, other *L. pneumophila* serogroups, and species other than *L. pneumophila* did not vary with the age or sex of the patient, the date the specimen was submitted, the type of specimen, or whether the state forwarded all or selected specimens to CDC.

Our results demonstrate that *L. pneumophila* serogroup 1 was the most common cause of pulmonary infections in patients suspected of having LD whose specimens were submitted to CDC in 1980–1982. *L. pneumophila* serogroups 1 and 6 together accounted for almost two-thirds of such cases, while all non-*L. pneumophila* species combined were present in only 13% of patients with legionellosis.

LITERATURE CITED

1. **Cherry, W. B., and R. M. McKinney.** 1979. Detection of Legionnaires' disease bacteria in clinical specimens by direct immunofluorescence, p. 93–103. *In* G. L. Jones and G. A. Hébert (ed.), "Legionnaires' ": the disease, the bacterium and methodology. Centers for Disease Control, Atlanta, Ga.

Panvalent Antiserum for Detection of Legionellae in Lung Specimens by Indirect Fluorescent-Antibody Testing

SUSAN L. BROWN, WILLIAM F. BIBB, AND ROGER M. McKINNEY

Division of Bacterial Diseases, Center for Infectious Diseases, Centers for Disease Control, Atlanta, Georgia 30333

Although direct fluorescent-antibody (DFA) testing has been used successfully for a number of years to detect legionellae in clinical specimens, the number of known species and serogroups of *Legionella* has now increased to the extent that routine performance of DFA for all serological variants is impractical (1–3). Nine species of *Legionella* have been documented: *L. pneumophila, L. micdadei, L. bozemanii, L. dumoffii, L. gormanii, L. longbeachae, L. jordanis, L. oakridgensis,* and *L. wadsworthii. L. pneumophila* has been divided into seven serogroups and *L. longbeachae* into two. There are at least three additional undocumented serological variants of *L. pneumophila,* represented by strains OR2, Seattle 1, and San Francisco 9. Also, there are other isolates, including IN23, WO44, Mt. St. Helens 7, Mt. St. Helens 9, Lansing 2, and Lansing 3, which may represent new species or new serogroups within established species. These organisms are all potential pathogens and ideally should be included in any screening test of clinical or environmental specimens. DFA testing has proven to be a sensitive and specific method for detecting legionellae in these specimens. However, because of background fluorescence and dilution effects, there is a limit to the number of individual conjugates that may be included in a screening pool. The increasing number of known *Legionella* species and serogroups requires an efficient and sensitive alternative to DFA testing. This study indicates that indirect fluorescent-antibody (IFA) testing of clinical specimens provides such an alternative.

Lung homogenates that were submitted to the Centers for Disease Control reference laboratory from patients with suspected legionellosis from November 1977 through May 1982 were originally screened by DFA. In this study, 498 of these lung homogenates that were preserved by the addition of 10% buffered Formalin to a final concentration of 1% and stored at 4°C were screened by an IFA test using a panvalent antiserum pool containing antibodies to 25 serological variants of *Legionella*. Antibodies to the 25 serological variants were produced in rabbits. Antibody to the rabbit immunoglobulin was produced in goats, and the immunoglobulin was precipitated by ammonium sulfate. Fluorescein-conjugated goat anti-rabbit immunoglobulin was prepared by methods described previously (1). The preparation of rhodamine-labeled normal goat serum to be used as a counterstain was also as described previously (1).

To perform the IFA test, we made smears of lung homogenates on glass microscope slides by applying a drop of well-mixed homogenate within a 1.5-cm circle and removing the excess fluid. The smears were air dried and heat fixed. The smears were stained as follows. First, they were covered with the panvalent serum pool which was diluted in rhodamine-labeled goat serum so that all individual sera would yield a 4+ result at least two twofold dilutions higher than the working dilution with homologous antigens. After incubation for 30 min at room temperature in a humid chamber, the slides were rinsed with phosphate-buffered saline, immersed in phosphate-buffered saline for 15 min, rinsed with water, and air dried. The smears were then covered with the working dilution of goat anti-rabbit immunoglobulin-fluorescein conjugate and incubated, washed, and rinsed as described above. For routine use the conjugate was first diluted in borate-saline so that a 1:4 dilution in rhodamine-labeled normal goat serum would yield the working dilution. After staining, we mounted the dried smears with a cover slip, using buffered glycerol (pH 9), and read for fluorescence as previously reported (1). All specimens that were positive with the panvalent pool were rerun with the appropriate polyvalent pool(s) and finally with the appropriate individual sera. Since earlier studies with other systems have shown that the immunoglobulin M fraction of antiserum sometimes causes cross-reactions that are not observed with immunoglobulin G (4), we treated the individual antisera with 2-mercaptoethanol to inactivate the immunoglobulin M and reran all positive homogenates with individual antisera with which they had been positive. The previously positive homogenates were considered negative if no fluorescing organisms were seen with 2-mercaptoethanol-treated antisera.

Of the total of 498 lung homogenates, 39 (7.8%) were positive and 459 (92.2%) were negative by IFA (Table 1). Four of the positive samples had previously been negative by DFA.

TABLE 1. Results of IFA tests for legionellae on 498 lung homogenates[a]

Test results	No. of lung homogenates	% of total	% of positives
IFA negative			
Total	459	92.2	—
Previously DFA positive[b]	4	0.8	—
IFA positive			
Total	39	7.8	100.0
Previously DFA positive	35	7.0	89.7
Previously DFA negative[c]	4	0.8	10.3

[a] All specimens had been previously screened by DFA for *L pneumophila* serogroups 1 through 6, *L. micdadei*, *L. bozemanii*, *L. gormanii*, and *L. dumoffii*.

[b] These homogenates were repeatedly retested in the present study and were shown now to be DFA and IFA negative.

[c] These homogenates contained organisms that were not previously screened for by DFA.

Of these, one was positive for strain BL540 (*L. jordanis*), one for Lansing 2, one for Lansing 3, and one for Seattle 1 (Table 2). These four organisms represent antigenic types for which DFA reagents were not available at the initial screening. Upon testing with the appropriate DFA reagents, the four homogenates were also positive for these organisms. These represented 10.3% of the positive specimens.

One homogenate positive by DFA with strain LA 1 (*L. pneumophila* serogroup 4) antiserum was positive by IFA with both LA 1 and Lansing 3 antisera. Another homogenate positive by DFA with strain WIGA (*L. bozemanii*) antiserum was positive by IFA with both WIGA and Tucker 1 (*L. longbeachae* serogroup 2) antisera. These were shown to be cross-reactions with the Lansing 3 and Tucker 1 antisera, respectively. The cross-reactions were eliminated by treatment of the Lansing 3 and Tucker 1 antisera with 2-mercaptoethanol.

Table 2 shows several lung homogenates containing organisms that stained with both Knoxville 1 and OLDA antisera and organisms that stained with Knoxville 1, OLDA, and San Francisco 9 antisera. These three strains have all been shown to share various common antigens. Analysis of the antigenic composition of these strains with monoclonal antibodies shows that strains OLDA and San Francisco 9 share antigens which are not present in strain Knoxville 1.

Three homogenates previously positive by DFA for *L. pneumophila* serogroup 1 and one homogenate previously positive by DFA for *L. dumoffii* were negative by IFA (Table 1). These specimens were repeatedly retested by both methods in the present study and were found to

TABLE 2. Specimens positive for legionellae by antigenic type

Antiserum to strain:	No. positive	% of total	% of positives
Knoxville[a] (*L. pneumophila* serogroup 1)	4	0.8	10.3
Knoxville 1-OLDA[b] (*L. pneumophila* serogroup 1)	16	3.2	41.0
Knoxville 1-OLDA-San Francisco 9[c] (*L. pneumophila* serogroup 1)	4	0.8	10.3
Togus 1 (*L. pneumophila* serogroup 2)	4	0.8	10.3
LA 1 (*L. pneumophila* serogroup 4)	1	0.2	2.6
Chicago 2 (*L. pneumophila* serogroup 6)	1	0.2	2.6
Seattle 1 (*L. pneumophila*, undesignated serogroup)	1	0.2	2.6
WIGA (*L. bozemanii*)	1	0.2	2.6
TATLOCK (*L. micdadei*)	1	0.2	2.6
TEX-KL (*L. dumoffii*)	1	0.2	2.6
Long Beach 4 (*L. longbeachae* serogroup 1)	1	0.2	2.6
Tucker 1 (*L. longbeachae* serogroup 2)	1	0.2	2.6
BL 540 (*L. jordanis*)	1	0.2	2.6
Lansing 2 (*Legionella* sp.)	1	0.2	2.6
Lansing 3 (*Legionella* sp.)	1	0.2	2.6

[a] Homogenates positive with Knoxville 1 antiserum only.
[b] Homogenates positive with Knoxville 1 and OLDA antisera.
[c] Homogenates positive with Knoxville 1, OLDA, and San Francisco 9 antisera.

be negative. This finding may be due to the instability of the antigen upon long-term storage.

The results show that IFA is a sensitive and specific method for screening clinical specimens for the presence of legionellae. In a few cases there were unexpected cross-reactions due to immunoglobulin M in the rabbit serum. These cross-reactions can be avoided by treating the whole serum with a reducing agent such as 2-mercaptoethanol. The criteria for interpreting the results of IFA testing are the same as those established for the DFA test (1). Morphology, the intensity of fluorescence, and the number of organisms must be considered in the final judgment of whether a specimen is positive or negative. Rhodamine-labeled normal goat serum used as a diluent for antisera and conjugate

significantly reduced background fluorescence, and the contrast in both lung homogenate smears and tissue sections, which were also run, was as good as that obtained with DFA reagents.

The number of positive specimens was too small to draw any firm conclusions about the relative distribution of the various species and serogroups of *Legionella*. However, some general observations may be made (Table 2). *L. pneumophila* serogroup 1 represented 61.5% (24 homogenates) of the positive specimens. This prevalence of serogroup 1 infections has been a consistent finding in other studies. Non-*L. pneumophila* infections constituted eight (22.8%) of the positive specimens. Four (10.3%) of the positive specimens were detected with antisera against the newly discovered organisms *L. jordanis*, Seattle 1, Lansing 2, and Lansing 3. It should be noted that three of the four new positive specimens are non-*L. pneumophila*; these constituted 37.5% of the non-*L. pneumophila* positive specimens.

The finding of four lung specimens that were positive with sera to newly discovered *Legionella* variants further emphasizes the need to test for as many *Legionella* antigenic variants as practical in any diagnostic test. IFA appears to provide a simple, sensitive, and specific means to this end. This method allows a large number of antigenically distinct organisms to be screened for in one test, significantly reducing the time needed for reading slides. Only one anti-species fluorescein conjugate is needed rather than a separate conjugate for each antigenically distinct organism as would be needed for DFA.

LITERATURE CITED

1. Cherry, W. B., and R. M. McKinney. 1979. Detection in clinical specimens by direct immunofluorescence, p. 92–103. *In* G. L. Jones and G. A. Hébert (ed.), "Legionnaires' ": the disease, the bacterium, and methodology. Centers for Disease Control, Atlanta, Ga.
2. Cherry, W. B., B. Pittman, P. P. Harris, G. A. Hebert, B. M. Thomason, L. Thacker, and R. E. Weaver. 1978. Detection of Legionnaires disease bacteria by direct immunofluorescent staining. J. Clin. Microbiol. 8:329–338.
3. McKinney, R. M. 1980. Serological classification of *Legionella pneumophila* and detection of other *Legionella* by direct fluorescent antibody staining. Clin. Immunol. Newsl. 1:1–3.
4. Thomason, B. M., and J. G. Wells. 1971. Preparation and testing of polyvalent conjugates for fluorescent-antibody detection of salmonellae. Appl. Microbiol. 22:876–884.

Acute-Phase Immunological Testing for *Legionella* Pneumonitis

JOHN B. CARTER AND SARALEE CARTER

Henrotin Hospital and Northwestern University Medical School, Chicago, Illinois 60610

In an attempt to provide acute-phase (same-day) diagnosis of *Legionella* pneumonitis, 937 patients admitted to an acute-care medical service with clinical variations of atypical pneumonia (fever, pulmonary infiltrates, and nondiagnostic sputum Gram stain or culture) were tested during a 48-month period for immunological evidence of active *Legionella* infection.

The testing procedure consisted of modifications of traditional indirect fluorescent-antibody (IFA) techniques (3) that included routine detection of both immunoglobulin G (IgG) and IgM antibodies for *Legionella pneumophila* serogroup 1 by using monospecific IgG- and IgM-specific fluorescein isothiocyanate conjugates (Wellcome Research Laboratories). Developmental work indicated that polyspecific fluorescein isothiocyanate conjugates would not reliably detect solitary IgM antibody and would not separate an IgM antibody titer of active infection from IgG antibody titers reflecting previous exposure. Trial use of polyvalent antigen substrate (*L. pneumophila* serogroups 1 through 4) (Centers for Disease Control) resulted in an unacceptably high incidence of false-positive IgM-specific antibody reactions. IgM antibody false-positivity did not occur with the use of monovalent serogroup 1 antigen (Centers for Disease Control). Testing of commercially available positive control sera (BioDx) for *L. pneumophila* serogroups 1 through 6 showed that sera containing antibody specific for all strains cross-reacted with the serogroup 1 antigen, albeit occasionally at a lower titer. Further procedure modifications included a 1-hour incubation of patient serum and antigen substrate at 37°C and reading slides at 1,000× oil immersion, features necessary for the optimal detection of IgM-specific antibodies. Serum titers routinely covered low- to high-range dilutions (8, 16, 32, and 64 for IgM conjugate; 8, 32, 128, and 512 for IgG conjugate) so that initial lower antibody titers (8 to 32) could be followed into midrange "conversions" (64 to 128), a feature not possible with a low-range (<64) cutoff for negative results.

Figure 1 records the results of testing of initial serum samples from 937 patients with clinical variations of atypical pneumonia. All samples were submitted during the acute phase of illness, most within the first week after admission. Approximately 84% of the cases initially tested resulted in titers reported as "no serological evidence of active *Legionella* infection." Approximately 8.2% of the cases (77 cases) had initial test results reported as "serological evidence of active *Legionella* infection." All positive case reports were correlated with clinical presentation and routine microbiological findings to ensure the validity of the report. The 7.4% of cases with IFA results in the "borderline" zone were reported as "nondiagnostic results; management depends on clinical situation or further study; suggest follow-up titers in 7 to 10 days if diagnosis remains uncertain." At least six borderline cases later converted to positive titers, but an uncertain number of these cases likely represented active *L. pneumophila* infection as, in the absence of an alternate diagnosis, the clinical tendency was to treat the patient for *Legionella* infection without waiting for confirmatory results. Patients who responded favorably to treatment (either causally or coincidentally) tended to be discharged and lost to follow-up study. Successfully treated patients available for follow-up study showed evidence that early intensive antibiotic therapy blocked an antibody titer rise in many cases. IgM *Legionella* titers occasionally persisted for several months, emphasizing the importance of clinical correlation of diagnostic reports. At least three patients reported as having 256 to 512 IgG and negative IgM *Legionella* antibody titer had proven non-*Legionella* pulmonary disease, suggesting that high IgG titers in the absence of IgM antibody may reflect past infection.

Concurrent studies included IFA testing of all samples after the first year (702 patients) for serological evidence of *Mycoplasma pneumoniae* infection (2) as part of an "atypical pneumonia profile," which identified 14 positive cases (2.0%) with no evidence of *Legionella*-*Mycoplasma* cross-reactivity. *Legionella* IFA testing of 108 healthy control subjects showed all subjects to be in the negative range, although 25% had low-titer (≤64) IgG *Legionella* antibody. Testing of 43 patients with diagnosed bacterial pneumonia showed all cases to have negative *Legionella* IFA results, with 25% showing low-titer IgG *Legionella* antibody. The mortality rate of cases of known outcome was 3%, with all fatal cases occurring in patients with severe underlying illness. The low mortality rate is attributed in part to the awareness of the medical staff of the *Legionella* diagnostic program and to rapid treatment of all cases of

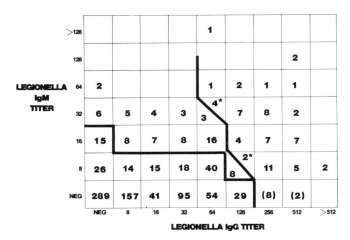

* BIOPSY CONFIRMED CASES
( ) OCCASIONAL NEGATIVE CLINICAL CORRELATION

FIG. 1. Results of *Legionella* IFA testing of the acute serum from 937 cases of suspected atypical pneumonia. Lower left section shows 793 (84.4%) negative results, upper left section shows 70 (7.4%) borderline results, and right section shows 77 (8.2%) positive (diagnostic) results.

presumed *Legionella* infection.

The rapid availability of an acute-phase *Legionella* IFA procedure resulted in modifications of clinical management of patients with presumed atypical pneumonia. Increased medical-staff awareness of sporadic *Legionella* infection led to increased use of erythromycin as the primary treatment for pneumonitis before laboratory results were available. Excellent correlation of test results with clinical findings and treatment response soon led to discontinuance of invasive diagnostic procedures. (Pulmonary biopsy or tracheal aspiration was performed on 6 of the first 10 reported positive cases for confirmation.) Increased frequency of testing identified a subset of cases of *Legionella* infection with subacute presentation, longer course, and slower response to treatment (1). Appreciation of a broader clinical spectrum of *Legionella* infection led to increased ordering of the acute-phase *Legionella* IFA test, sometimes with a marginal regard to recommended criteria for testing and a consequently decreased incidence of positive results, but with no decrease in the absolute case frequency (12% positive in the first 30 months of testing and 8.5% positive after 48 months).

The study suggests that *Legionella* antibody responses occur sufficiently early to be detected by a sensitive IFA procedure in the acute phase

of illness. The acute-phase *Legionella* IFA procedure rapidly excluded nearly all negative cases, detected clearly positive cases for early, specific therapy, and identified borderline cases for further study. The incidence of sporadic *Legionella* infection was sufficient to warrant routine testing in a full-service infectious disease laboratory. Consideration of cases with lower IgG titers and significant IgM titers in clinically compatible illness increased the diagnostic sensitivity of the procedure and permitted clinical judgment in roughly 90% of cases based on testing of the acute serum sample. Well-controlled modifications of the traditional *Legionella* IFA procedure are valid in an acute-care setting; however, each laboratory must establish its own decision criteria based on clinical correlation of its own test results.

LITERATURE CITED

1. **Carter, J. B.** 1979. Legionnaires' disease at a community hospital. Ann. Intern. Med. **91**:794.
2. **Carter, J. B., and S. L. Carter.** 1983. Acute-phase, indirect fluorescent antibody procedure for diagnosis of *Mycoplasma pneumoniae* infection. Ann. Clin. Lab. Sci. **13**:150–155.
3. **Wilkinson, H. W., D. D. Cruce, B. J. Fikes, L. P. Vealy, and C. E. Farsky.** 1979. Indirect immunofluorescence test for Legionnaires' disease, p. 111–116. *In* G. L. Jones and G. A. Hébert (ed.), "Legionnaires' ": the disease, the bacterium and methodology. Centers for Disease Control, Atlanta, Ga.

Commercial *Legionella* Direct and Indirect Immunofluorescence Reagents

THENA M. DURHAM, CHARLOTTE T. HALE, WILLIAM K. HARRELL, AND HENRY M. COLVIN

Center for Infectious Diseases, Centers for Disease Control, Atlanta, Georgia 30333

The Centers for Disease Control (CDC) currently recommends the direct and indirect immunofluorescence (FA) tests as the reference serological methods for diagnosing legionellosis. Because *Legionella* FA reagents were not in commerical distribution before the enactment date (28 May 1976) of the Food and Drug Administration (FDA) Medical Device Amendments, the reagents were automatically categorized as class III products. This is the most stringent level of regulatory control, requiring the product manufacturer to substantiate the efficacy and safety of the product via statistically validated clinical trial data as well as to follow regulations for good manufacturing practices (class I) and product standards (class II).

Because appropriate clinical data were difficult to obtain, commercial manufacturers did not enter the market, and CDC became the major provider of *Legionella* FA reagents. Not wishing to continue in this role, CDC petitioned the FDA for reclassification of the *Legionella* FA products, and the reagents were subsequently placed in class I. Thereafter, manufacturers were expected to file their own petitions (510K applications) with the FDA, wherein they could claim "substantial equivalence" to the CDC-produced reagents and thereby enter the market.

To date, the following manufacturers have submitted 510K applications to the FDA: Zeus Technologies, Inc., Raritan, N.J.; Litton Bionetics, Inc., Kensington, Md.; and BioDx, Inc., Denville, N.J. The status of these applications as of 3 June 1983 was as follows: the FDA had approved the submissions from Zeus Technologies for both direct and indirect FA reagents; the submissions from Litton Bionetics had been placed on hold awaiting additional information from the manufacturer; and the submission from BioDx for direct FA reagents had been approved, whereas the submission for indirect FA reagents was still being reviewed.

To assist manufacturers in developing reagents that are reliable and of high quality, the CDC has developed interim specifications for *Legionella* direct and indirect FA reagents (1). These parameters are used to assess the performance of each product submitted for evalua-

tion as well as the adequacy of the labeling which accompanies that product (1). Only those products for which homlogous CDC reference reagents are available are accepted for evaluation. The following manufacturers have submitted products to CDC for courtesy evaluation: BioDx; Difco Laboratories, Detroit, Mich.; Litton Bionetics; and Zeus Technologies (Table 1, see next page).

All of the BioDx reagents (kits excluded) met the applicable CDC specifications. The Difco conjugates for *L. pneumophila* serogroups 2 and 3 were satisfactory; however, the serogroup 1 conjugate was underreactive. All of the Litton Bionetics direct FA reagents and most of their IFA reagents were satisfactory. The Litton serogroup 6 IFA antigen was extremely filamentous and morphologically atypical; therefore, the Litton individual IFA antigen for serogroup 6, as well as the polyvalent suspension and slide containing this antigen, was unsatisfactory. The Litton IFA normal serum control was also unsatisfactory because it was packaged as a serum dilution rather than as whole serum. All of the Zeus direct FA antigens were satisfactory; conversely, all of the direct FA conjugates were underreactive. All of the Zeus IFA sera were satisfactory; however, the Zeus IFA antigens were unsatisfactory because problems were encountered with adherence of the bacterial cells to the surface of microscope slides. The Zeus IFA antigens contained no yolk sac when submitted for evaluation. The anti-human conjugates submitted by Litton Bionetics and Zeus Technologies were both satisfactory.

To reiterate, these were courtesy evaluations performed on commercial reagents still in various stages of the production process. Therefore, the evaluation results summarized above are not applicable to any similar final product. To date, no commercial *Legionella* FA reagents in final packaging have been submitted to CDC and found to be satisfactory.

LITERATURE CITED

1. **Diagnostic Products Evaluation Branch.** 1975. Specifications and evaluation methods for immunological and microbiological reagents, vol. 1, Bacterial, fungal, and parasitic, 4th ed. Centers for Disease Control, Atlanta, Ga.

TABLE 1. Commercial *Legionella* FA reagents submitted to CDC for courtesy evaluation

Manufacturer	Direct FA		Indirect FA		
	Conjugates	Antigens	Conjugates	Antigens	Sera
BioDx, Inc.[a]	*L. pneumophila* serogroups 1 through 6; polyvalent (serogroups 1 through 6)	*L. pneumophila* polyvalent slide (serogroups 1 through 6)	—	—	*L. pneumophila* monkey sera, serogroups 1 through 6
Difco Laboratories	*L. pneumophila* serogroups 1, 2, and 3	—	—	—	—
Litton Bionetics, Inc.	*L. pneumophila* serogroups 1 through 6; polyvalent (serogroups 1 through 6); FITC[b]-labeled normal globulin	*L. pneumophila* serogroups 1 through 6	Anti-human FITC globulin	*L. pneumophila* serogroups 1 through 6; polyvalent antigen suspension (serogroups 1 through 6); polyvalent slide (serogroups 1 through 6)	*L. pneumophila* monkey sera, serogroups 1 through 6; normal monkey control
Zeus Technologies, Inc.[c]	*L. pneumophila* serogroups 1 through 6; *L. bozemanii, L. dumoffii, L. gormanii, L. micdadei, L. longbeachae* serogroups 1 and 2; polyvalent A (*L. pneumophila* serogroups 1 through 4); polyvalent B (*L. pneumophila* serogroups 5 and 6, *L. dumoffii, L. longbeachae* serogroup 1; polyvalent C (*L. bozemanii, L. micdadei, L. gormanii, L. longbeachae* serogroup 2)	*L. pneumophila* serogroups 1 through 6; *L. bozemanii, L. dumoffii, L. gormanii, L. micdadei, L. longbeachae* serogroups 1 and 2	Anti-human FITC globulin	*L. pneumophila* serogroups 1 through 6; polyvalent A (serogroups 1 through 4)	*L. pneumophila* monkey sera, serogroups 1 through 4; normal human control

[a] The BioDx direct FA kit cannot be accepted for evaluation until CDC develops a multiple-component product specification for the *Legionella* direct FA reagents.

[b] FITC, Fluorescein isothiocyanate.

[c] Reagents originated at Springwood Microbiologicals, Evans, Ga., and were transferred to Carr-Scarborough Microbiologicals, Atlanta, Ga., and then to Zeus Technologies, Raritan, N.J. Springwood Microbiologicals is no longer in business; Carr-Scarborough Microbiologicals is no longer producing *Legionella* FA reagents.

Correlation Between the Microagglutination and Indirect Fluorescent-Antibody Tests for Antibodies to 10 *Legionellaceae* Antigens

MICHAEL T. COLLINS,† KLAUS LIND, OLE AALUND, JEANNETTE McDONALD,‡ AND WILHELM FREDRIKSEN

Department of Microbiology, Colorado State University, Fort Collins, Colorado 80523, and State Serum Institute and Royal Veterinary and Agriculture College, Copenhagen, Denmark

The microagglutination (MA) and indirect fluorescent-antibody (IFA) tests were performed as previously described (5, 11), using *Legionella pneumophila* serogroups (SG) 1 through 6, *Fluoribacter (Legionella) bozemanae, Fluoribacter (Legionella) dumoffii, Fluoribacter (Legionella) gormanii,* and *Tatlockia (Legionella) micdadei* antigens. MA titers were determined on sera from 273 patients with pneumonia to all 10 legionella antigens. IFA titers were measured on sera from all 273 patients to *L. pneumophila* SG 1 through 4, from 85 patients to *L. pneumophila* SG 5 and 6, and from 75 patients to *F. bozemanae, F. dumoffii, F. gormanii,* and *T. micdadei.*

MA test positivity was defined on the basis of the prevalence of MA titers measured on a normal control population (n = 200) to all 10 legionella antigens. Positive MA titers were those equal to or greater than the upper limit of normal (ULN) for each antigen. Four cutoffs for ULN were evaluated, 1, 5, 10, and 15% (i.e., a ULN of 15% was defined as the titer exceeded by not more than 15% of the normal sera). IFA titers were considered positive if they were ≥1:256, according to the established criteria for evaluation of single sera by IFA (11).

The ULN MA titers for each legionella antigen at each ULN cutoff level are listed in Table 1. The percentages of MA and IFA results which were in agreement are presented in Table 2. Of the 1,562 pairs of MA and IFA results compared, 86, 90, 92, and 95% agreed when the 15, 10, 5, and 1% ULN criteria for MA test positivity, respectively, were used.

Disagreements were obviously of two types: (i) MA positive with IFA negative or (ii) MA negative with IFA positive. Disagreements of the first type occurred in 180 of the 1,562 (11.5%) MA-IFA comparisons when the 15% ULN MA positivity criterion was used and in 43 (2.8%) instances when the 1% ULN criterion was used. Thus, the rates of disagreement were approximately those predicted by the 15 and 1% sensitivity limits set for the MA test.

Disagreements of the second type (MA negative with IFA positive) were more significant. Of 86 positive IFA titers, the corresponding MA titers were negative in 19 and 37 instances when the 15 and 1% ULN MA cutoffs, respectively, were used. Most of these discrepancies, however, resulted when the MA test was positive to a different antigen than was the IFA or when the MA titer was positive as evidenced by seroconversion but did not reach the more restrictive cutoff titer for MA positivity when the 1% ULN criterion was used.

Determination of the prevalence of MA antibodies in a normal population is crucial to correct interpretation of MA serological results. Thus far, the background prevalence of MA titers in humans has been reported only for *L. pneumophila* SG 1 through 4 (2, 6, 8, 10). MA titer distributions determined in the present study were in general agreement with those reported previously and support the criterion that MA titers of ≥1:32 are suggestive of *L. pneumophila* infection for SG 1 through 4 (6, 8, 10). Based on our findings with other antigenic types of legionellae, this same criterion for MA positivity could be applied to antibody titers against *L. pneumophila* SG 5 and 6, *F. gormanii,* and *T. micdadei,* depending on the desired levels of MA test sensitivity and specificity. Definition of MA titers of ≥1:32 as positive for *F. bozemanae* and *F. dumoffii,* however, would result in 22 and 38% of the normal human population tested being considered infected with these two agents, respectively, a doubtful situation. The higher prevalence of normal background MA titers of ≥1:32 for *F. bozemanae* and *F. dumoffii* could be due to more frequent exposure of people to these two organisms, or, more likely, to expression of antigens on the surface of these legionellae which cross-react with human antibodies to other organisms more common in the environment. The occurrence of such cross-reactions with legionellae is well documented (3, 4, 7, 9, 12–14).

In setting a test level at the point desired to identify those with a specific disease and to omit those without it, one must judge the relative costs of classifying results as false-negatives and false-positives. The prevalence of the disease in the community, the cost of additional examina-

† Present address: Department of Pathobiological Sciences, University of Wisconsin, Madison, WI 53706.

‡ Present address: 112 N. Washington St., Mt. Horeb, WI 53572.

TABLE 1. Upper limit of MA titers in a normal population (n = 200) for 10 legionella antigens at four cutoff levels

Antigen	Titer at the following ULN cutoff level[a]:			
	15%	10%	5%	1%
L. pneumophila SG 1	4	8	8	32
L. pneumophila SG 2	32	64	64	128
L. pneumophila SG 3	8	8	16	32
L. pneumophila SG 4	8	8	8	16
L. pneumophila SG 5	8	8	16	32
L. pneumophila SG 6	16	16	32	128
F. bozemanae	64	64	128	128
F. dumoffii	128	128	256	256
F. gormanii	8	8	8	16
T. micdadei	16	64	64	128

[a] Levels are the maximal percentage of the normal population having a titer greater than or equal to the designated titer.

tions that may be necessary, and the purpose of applying the test must also be considered. Klein and co-workers (8) chose to define the ULN for the MA test with *L. pneumophila* SG 1 as the level of antibody titer exceeded by no more than 15% of the normal sera tested.

Our report evaluates the use of more restrictive ULN criteria on MA and IFA test agreement. As the cutoff titers defining MA positivity to each legionella antigen increased, the number of MA positive/IFA negative disagreements decreased, and the rate of agreement between the two tests increased, reaching 95% when the positive MA level was set at the 1% ULN criterion. The compromise for improved overall test agreement was that the number of MA negative/IFA positive disagreements almost doubled, as described above.

Cross-reactivity of antibodies in human sera with multiple legionella MA antigens was a common finding. However, there was almost always a single antigen to which the agglutina-

tion titer was strongest. Similar findings with the MA test were reported previously (2, 6), and antigenic analysis of *L. pneumophila* by crossed immunoelectrophoresis has demonstrated extensive sharing of antigens between SG (1, 3).

As the complexity of testing for antibodies to legionellae increases with each newly reported antigenic type, the MA test becomes more attractive as a simple, cost-effective method for serological diagnosis of legionellosis. Our studies with normal human sera, however, emphasize the need for definition of MA antibody prevalence to each legionella antigen for correct interpretation of MA serological results. At the present time, the MA test is probably best applied as a screening test, using the 15% ULN criterion for MA positivity, to eliminate the vast majority of pneumonia patient sera from need of IFA testing. Based on the results presented here, very few IFA-positive sera would be missed by MA screening. The diagnostic reliability of the MA test would best be proven by measurement of its sensitivity and specificity, using isolation-proven cases of legionellosis and suitable controls.

LITERATURE CITED

1. **Collins, M. T., S.-N. Cho, N. Høiby, F. Esperson, L. Baek, and J. S. Reif.** 1983. Crossed immunoelectrophoretic analysis of *Legionella pneumophila* serogroup 1 antigens. Infect. Immun. **39:**1428–1440.
2. **Collins, M. T., S.-N. Cho, and J. S. Reif.** 1982. Prevalence of antibodies to *Legionella pneumophila* in animal populations. J. Clin. Microbiol. **15:**130–136.
3. **Collins, M. T., F. Espersen, N. Høiby, S.-N. Cho, A. Friis-Møller, and J. S. Reif.** 1983. Cross-reactions between *Legionella pneumophila* (serogroup 1) and twenty-eight other bacterial species, including other members of the family *Legionellaceae.* Infect. Immun. **39:**1441–1456.
4. **Edelstein, P. H., R. M. McKinney, R. D. Meyer, M. A. C. Edelstein, C. J. Krause, and S. M. Finegold.** 1980. Immunologic diagnosis of Legionnaires' disease: cross-reactions with anaerobic and microaerophilic organisms and infections caused by them. J. Infect. Dis. **141:**652–655.
5. **Farshy, C. E., G. C. Klein, and J. C. Feeley.** 1978. Detection of antibodies to Legionnaires disease organism by

TABLE 2. Rate of agreement of MA and IFA results at four cutoff levels for definition of positive MA titers

Antigen	No. of MA-IFA comparisons	No. of positive IFA titers	% of MA and IFA results which agree at the following ULN MA cutoff:			
			15%	10%	5%	1%
L. pneumophila SG 1	273	39	91	93	93	94
L. pneumophila SG 2	273	1	81	93	93	96
L. pneumophila SG 3	273	10	83	83	94	96
L. pneumophila SG 4	273	19	91	91	91	93
L. pneumophila SG 5	85	1	92	92	95	99
L. pneumophila SG 6	85	4	92	92	92	92
F. bozemanae	75	2	85	85	92	92
F. dumoffii	75	2	93	93	97	97
F. gormanii	75	0	85	85	95	97
T. micdadei	75	6	88	95	95	93
Mean			86	90	92	95

microagglutination and micro-enzyme-linked-immunosorbent assay tests. J. Clin. Microbiol. **7:**327–331.

6. **Helms, C. M., W. Johnson, E. D. Renner, W. J. Hierholzer, Jr., L. A. Wintermeyer, and J. P. Viner.** 1980. Background prevalence of microagglutination antibodies to *Legionella pneumophila* serogroups 1, 2, 3, and 4. Infect. Immun. **30:**612–614.

7. **Klein, G. C.** 1980. Cross-reaction to *Legionella pneumophila* antigen in sera with elevated titers to *Pseudomonas pseudomallei*. J. Clin. Microbiol. **11:**27–29.

8. **Klein, G. C., W. J. Jones, and J. C. Feeley.** 1979. Upper limit of normal titer for detection of antibodies to *Legionella pneumophila* by the microagglutination test. J. Clin. Microbiol. **10:**754–755.

9. **Ormsbee, R. A., M. G. Peacock, G. L. Lattimer, L. A. Page, and P. Fiset.** 1978. Legionnaires' disease: antigenic peculiarities, strain differences, and antibiotic sensitivities of the agent. J. Infect. Dis. **138:**260–264.

10. **Smalley, D. L., and D. D. Ourth.** 1981. Seroepidemiology of *Legionella pneumophila*. A study of adults from Memphis, Tennessee, U.S.A. Am. J. Clin. Pathol. **75:**201–203.

11. **Wilkinson, H. W., D. D. Cruce, and C. V. Broome.** 1981. Validation of *Legionella pneumophila* indirect immunofluorescence assay with epidemic sera. J. Clin. Microbiol. **13:**139–146.

12. **Wilkinson, H. W., B. J. Fikes, and D. D. Cruce.** 1979. Indirect immunofluorescence test for serodiagnosis of Legionnaires disease: evidence for serogroup diversity of Legionnaires disease bacterial antigens and for multiple specificity of human antibodies. J. Clin. Microbiol. **9:**379–383.

13. **Winn, W. C., Jr., W. B. Cherry, R. O. Frank, C. A. Casey, and C. W. Broome.** 1980. Direct immunofluorescent detection of *Legionella pneumophila* in respiratory specimens. J. Clin. Microbiol. **11:**59–64.

14. **Wong, K. H., P. R. B. McMaster, J. C. Feeley, R. J. Arko, W. O. Schalla, and F. W. Chandler.** 1980. Detection of hypersensitivity to *Legionella pneumophila* in guinea pigs by skin test. Curr. Microbiol. **4:**105–110.

Validation of Serological Tests for Legionnaires Disease by Using Sera from Bacteriologically Proven and Epidemiologically Associated Cases

A. G. TAYLOR, T. G. HARRISON, A. BURÉ, AND E. DOURNON

Central Public Health Laboratory, Colindale, London NW9 5HT, England, and Hôpital Claude Bernard, Paris 75019, France

We have previously described an indirect fluorescent-antibody test (IFAT) using formolized yolk sac antigen prepared with the Pontiac strain of *Legionella pneumophila* for the serodiagnosis of Legionnaires disease (LD) (2, 5). This test is currently used in 100 diagnostic laboratories in Great Britain as well as in some laboratories in other European countries. More recently we have described a semiautomatable rapid microagglutination test (RMAT) (3). Results have been published which show that both tests have good specificity (4). As in the evaluation of other serodiagnostic tests for LD (7), the small numbers of sera available from cases of the disease proven by non-serological methods have hindered collection of data on their sensitivity. We have now evaluated the sensitivity of these IFAT and RMAT methods by using sera from 35 bacteriologically proven cases. Diagnosis of LD was made by both isolation of *L. pneumophila* serogroup 1 and immunofluorescent demonstration of the organism in respiratory tract specimens (14 cases), by bacterial isolation only (9 cases), or by immunofluorescent demonstration only (12 cases).

The results obtained with these sera from proven LD patients are shown in Table 1. In the IFAT 29 of 35 patients showed diagnostic titers, i.e., a fourfold rise to a titer of ≥64 (22 cases) or a single or standing titer of ≥128 (7 cases). Four patients not showing diagnostic titers did, however, show titers of ≥16, and as we have previously demonstrated (2) that 97% of control subjects show titers of <16 in this test, a diagnosis of LD would have been strongly suspected.

Serum specimens from 34 of the patients were also available for testing by the RMAT, and 30 of these showed diagnostic titers, i.e., a fourfold rise to a titer of ≥16 (22 cases) or a single or standing titer of ≥32 (8 cases) (Table 1). Control studies indicate that a titer of ≥8 in the RMAT is suggestive of LD (3), and hence a diagnosis of LD in two of the four patients not meeting the diagnostic criteria would have been suspected.

In this series of proven LD cases, the sensitivity (number of serodiagnostic patients/total number of patients) of the IFAT, using formolized yolk sac antigen and the diagnostic criteria specified above, is 83%, and that of the RMAT is 88%. However, in both tests 94% of patients showed titers either diagnostic for or suggestive of LD. One of the two patients who were seronegative in both tests died within 2 days of hospital admission, at which stage of illness antibodies may not have reached detectable levels.

The two tests have been further evaluated for sensitivity by using sera from cases associated with outbreaks of LD where *L. pneumophila* serogroup 1 was isolated from or demonstrated in at least one of the patients (e.g., Kingston Hospital) (1). The results are shown in Table 2.

TABLE 1. Results of IFAT and RMAT in 35 patients with bacteriologically proven LD

No. of patients	Diagnosis by:	IFAT[a]				RMAT[a]			
		Fourfold rise in titer	Single or standing titer	Suggestive titer	Negative titer	Fourfold rise in titer	Single or standing titer	Suggestive titer	Negative titer
14[b]	Isolation and IF[c]	9	3	1	1[d]	8	3	1	1[d]
9	Isolation only	7	0	1[e]	1	7	0	1[e]	1
12	IF only	6	4	2	0	7	5	0	0

[a] The number of patients with the indicated result is shown.
[b] Insufficient serum for RMAT in one patient.
[c] IF, Demonstration of the organism by immunofluorescence.
[d] Patient died within 2 days of hospital admission.
[e] Patient died on the day of hospital admission; the single specimen was taken postmortem.

TABLE 2. Results of IFAT and RMAT in 27 epidemiologically associated cases where at least one case was proven by IF[a], bacterial isolation, or both

No. of patients	Outbreak	IFAT[b]				RMAT[b]			
		Fourfold rise in titer	Single or standing titer	Suggestive titer	Negative titer	Fourfold rise in titer	Single or standing titer	Suggestive titer	Negative titer
8	Kingston Hospital[c]	4	3	1	0	3	3	1	1[d]
4	Hotel (Midlands)[e]	1	2	0	1	1	1	1	1
6[f]	Hospital (London)[g]	5	1	0	0	2	2	1	0
3	Hotel (Portugal)[g]	2	0	1	0	2	0	1	0
6	Hospital (Paris)[g]	4	2	0	0	4	2	0	0

[a] IF, Demonstration of the organism by immunofluorescence.
[b] The number of patients with the indicated result is shown.
[c] Reference 1.
[d] Single specimen taken about 2 months after onset.
[e] Reference 6.
[f] In one patient there was insufficient serum for RMAT.
[g] Unpublished data.

In this series of 27 patients, 24 were diagnostic by the IFAT, in 2 cases the results were suggestive of LD, and only 1 patient was seronegative. Serum specimens from 26 of these patients were also available for testing by the RMAT; in 20 of these a diagnostic result was obtained, in 4 cases titers were suggestive of LD, and 2 patients were seronegative. A possible explanation for the smaller proportion of patients diagnostic in the RMAT than in the IFAT is that the RMAT predominantly detects immunoglobulin M, whereas the IFAT detects all major immunoglobulin classes. Hence, a lower titer with the RMAT may be obtained with late convalescent specimens, where immunoglobulin M levels may have fallen. The sensitivity of the IFAT in this group of epidemiologically associated patients was 89%, and that of the RMAT was 77%; in both tests 92% of the patients showed titers diagnostic for or suggestive of LD.

LITERATURE CITED

1. Fisher-Hoch, S. P., C. L. R. Bartlett, J. O. Tobin, M. B. Gillett, A. M. Nelson, J. E. Pritchard, M. G. Smith, R. A. Swann, J. M. Talbot, and J. A. Thomas. 1981. Investigations and control of an outbreak of Legionnaires' disease in a district general hospital. Lancet i:932–936.

2. Harrison T. G., and A. G. Taylor. 1982. Diagnosis of Legionella pneumophila infection by means of formolised yolk-sac antigens. J. Clin. Pathol. 35:211–214.

3. Harrison, T. G., and A. G. Taylor. 1982. A rapid microagglutination test for the diagnosis of Legionella pneumophila (serogroup 1) infection. J. Clin. Pathol. 35:1028–1031.

4. Taylor, A. G., and T. G. Harrison. 1983. Serological tests for Legionella pneumophila (serogroup 1) infections. Zentralbl. Bakteriol. Parasitenkd. Infektionskr. Abt. 1 Orig. A 255:20–26.

5. Taylor, A. G., T. G. Harrison, M. W. Dighero, and C. M. P. Bradstreet. 1979. False positive reactions in the indirect fluorescent antibody test for Legionnaires' disease eliminated by use of formolised yolk-sac antigen. Ann. Intern. Med. 90:686–689.

6. Tobin, J. O., C. L. R. Bartlett, S. A. Waitkins, G. I. Barrow, A. D. Macrae, A. G. Taylor, R. J. Fallon, and F. R. N. Lynch. 1981. Legionnaires' disease: further evidence to implicate water storage and distribution systems as sources. Br. Med. J. 282:573.

7. Wilkinson, H. W., D. D. Cruce, and C. V. Broome. 1981. Validation of Legionella pneumophila indirect immunofluorescence assay with epidemic sera. J. Clin. Microbiol. 13:139–146.

Seroreactivity Towards *Legionella pneumophila* in Patients with Pneumonia: Possible Cross-Reactions

INGEGERD KALLINGS, SVEN BLOMQVIST, AND GÖRAN STERNER

National Bacteriological Laboratory, Stockholm, and Danderyds Hospital, Danderyd, Sweden

The incidence of *Legionella pneumophila* infections has been estimated to be 1 to 6.4% in the United States (2) and 0.65% of patients with atypical pneumonia in Austria (7). In Sweden, there are 15 to 30 sporadic cases of *Legionella* infections diagnosed annually, and three outbreaks are known to have occurred (3).

In a retrospective survey in southern Sweden, the incidence was 2% in patients with pneumonia (6). The prevalence of antibodies at titers of ≥64 among healthy blood donors is generally low, 0 to 4% in areas not associated with outbreaks. However, in association with outbreaks the prevalence of antibodies is 20% (A. Forsgren and I. Kallings, presented at the 13th International Congress of Chemotherapy, Vienna, Austria, 1983). A prospective study was outlined to establish the prevalence of *Legionella* infections among patients with community-acquired pneumonia seeking attention at a clinic for infectious diseases at Danderyds Hospital, Danderyd, Sweden. The hospital is situated just north of Stockholm. It has 1,200 beds and a catchment area with 700,000 inhabitants.

When a greater-than-fourfold increase in antibody titer to *L. pneumophila* is seen, the case is considered to be serologically confirmed. The diagnostic value, however, of single high titers has been much disputed. A thorough search for other etiological agents might help to solve some problems involved in the evaluation of *Legionella* antibodies.

During a period of 14 months, all patients over 7 years of age seeking attention at Danderyds Hospital for respiratory tract infections were entered into the study if they fulfilled the following criteria: chest X-ray consistent with pneumonia, a history of acute respiratory disease for not more than 2 weeks, and no treatment in a hospital during the preceding month. A total of 437 patients from whom paired serum samples were possible to obtain fulfilled these criteria. Three-quarters of the patients were treated in the hospital; one-quarter were treated as outpatients.

Serum was drawn on admission and 1 to 3 weeks after onset of disease. Sera were examined for antibodies to *L. pneumophila* serogroups 1, 2, and 4 by indirect immunofluorescence, using heat-killed antigen prepared at the National Bacteriological Laboratory, Stockholm, Sweden. All sera were also examined for antibodies to *Mycoplasma pneumoniae* (by complement fixation), *Chlamydia psittaci* (by complement fixation), influenza virus types A and B, cytomegalovirus, and respiratory syncytial virus. Sera from patients with antibody titers of ≥64 to *L. pneumophila* were also examined for antibodies to *Haemophilus influenzae* and *Streptococcus pneumoniae* serogroups 1 through 8 by enzyme-linked immunosorbent assay. These sera were also examined for antibodies to *Chlamydia trachomatis* by micro-indirect immunofluorescence. The criterion for serological evidence of *Legionella* infection was a greater-than-fourfold titer rise to ≥128, and that for a presumptive diagnosis was a rise to ≥256.

A total of 133 (30%) of the 437 patients had titers of ≥64 in one or more of the *Legionella* serogroups tested. Significant titer rises against *L. pneumophila* were seen in 10 patients (Table 1). Two of these patients had concomitant titer rises to *H. influenzae* and to *S. pneumoniae*. Fifteen patients had a presumptive diagnosis of infection with *L. pneumophila* with no evidence of other etiological agents. Seven of these 15 patients had titers of ≥1,024. When patients with antibodies to other agents were included, the number with titers of ≥256 to *L. pneumophila* was 26 (Table 2). When the two patients with other agents cultured from the blood are not counted, the incidence of confirmed *Legionella* cases is 8 of 437 (1.8%). When patients with presumptive evidence of legionellosis and no other explanation of their disease are included, the incidence is 5.7%. The clinical illness did not differ from what has been described previously for legionellosis. Most of the patients had been traveling abroad at the possible time for contracting their disease. A total of 64 of the patients with stationary titers of ≥64 to *L. pneumophila* had serological or cultural evidence of another cause of their disease, while in 36 (29%), no etiology of the infection could be established.

The incidence of serologically confirmed *Legionella* cases in this material was the same as has earlier been reported from Sweden, i.e., 1.8% (6). If patients with presumptive serological evidence of legionellosis and no other etiology found are included, the incidence is higher (5.7%). The percentage of patients with stationary antibody titers of 64 to 128 exceeded by far the prevalence of antibodies among healthy Swedish blood donors without known exposure to *L. pneumophila* (24% versus 0 to 4%) (Forsgren and Kallings, presented at the 13th ICC).

TABLE 1. Patients with significant titer rises of antibodies to *L. pneumophila*

Pa-tient	Titer[a] to *L. pneumophila* serogroup:			Concomitant titer rise[b]
	1	2	4	
1		64–256		
2	128–8,192		16–8,192	*H. influenzae* immunoglobulin G (0.9), immunoglobulin M (0.6); *S. pneumoniae*
3	128–8,192			
4			128–4,096	
5	64–1,024		64–512	
6	64–2,048		64–4,096	
7	128–512			
8	32–128			*H. influenzae* immunoglobulin G (0.9), immunoglobulin M (0.5); *S. pneumoniae*
9[c]		32–128		
10[d]		64–256		

[a] The first number is the titer on admission to the hospital; the second number is the titer 1 to 3 weeks after onset of disease.
[b] Measured by enzyme-linked immunosorbent assay.
[c] This patient had growth of *H. influenzae* in blood culture.
[d] This patient had growth of *S. pyogenes* in blood culture.

TABLE 2. Patients with elevated antibody titers (to ≥256) against *L. pneumophila* but without titer rises (presumptive cases of legionellosis)

Pa-tient	Titer[a] to *L. pneumophila* serogroup:			Serological evidence of other agents as cause of disease[b]	Diagnosis
	1	2	4		
11	512–512	1,024–1,024	128–128	None	Legionellosis?
12	2,048–2,048	64–64	64–64	*H. influenzae* immunoglobulin G (0.8), immunoglobulin M (0.7); *C. trachomatis* (32–64)	*H. influenzae* infection
13	128–128	512–1,024		*C. trachomatis* (64–128)	Legionellosis?
14			512–1,024	None	Legionellosis?
15	256–256	128–128	256–256	None	Legionellosis?
16	512–512	32–64	512–512	None	Legionellosis?
17	256–512			Adenovirus (<5–20)	Adenovirus infection
18	128–256	128–128		*M. pneumoniae* (−320)	Mycoplasma
19	512–512			*C. psittaci* (160–80–40)	Ornithosis
20	256–256			*M. pneumoniae* (20–320), parainfluenza I (>5–40)	Mycoplasma parainfluenza virus infection
21[c]	1,024–1,024	1,024–2,048	1,024–1,024	*M. pneumoniae* (5–60)	Mycoplasma
22	512–512	1,024–512	256–256	*M. pneumoniae* (40–320–40)	Mycoplasma
23	128–256	128–256	256–256	Cytomegalovirus (20–40)	Legionellosis?
24	256–256	128–128	128–128	None	Legionellosis?
25[d]	128–128	128–128	2,048–1,024	None	Legionellosis?
26			256–256	None	Legionellosis?
27	256–256	256–256	128–128	*M. pneumoniae* (40–320)	Mycoplasma
28	256–256		512–256	*M. pneumoniae* (80–320)	Mycoplasma
29	256–256		128–128	None	Legionellosis?
30	256–256	256–256	128–128	None	Legionellosis?
31	256–128	256–256	256–256	None	Legionellosis?
32	64–128		256–256	None	Legionellosis?
33	256–256		128–128	*S. pneumoniae*	Pneumococcal infection
34	128–128	128–256	64–128	None	Legionellosis?
35	256–256	128–64		None	Legionellosis?
36	256–256			*M. pneumoniae* (160–160)	Legionellosis? Mycoplasma?

[a] The first number is the titer on admission to the hospital; the second number is the titer 1 to 3 weeks after onset of disease.
[b] Titers are given within parentheses.
[c] Immunoglobulin M titers were 128–128 to serogroup 1, <64–<64 to serogroup 2, and 128–128 to serogroup 4.
[d] Immunoglobulin M titer was 128–128 to serogroup 4.

Several explanations of this and of the concomitant titer rises to other agents in confirmed and presumptive *Legionella* cases may be considered. Antibodies to *L. pneumophila* persist for at least for 2 years in 33% of confirmed cases (4). The discrepancy between the antibody levels in this study and the control material argues against earlier *Legionella* infections as the only explanation of moderately elevated antibody titers among the patients. Anamnestic stimulation was unlikely because most of the patients had not suffered from earlier pneumonias.

There is also the possibility of cross-reactions. In an earlier study, cross-reactions with other common pneumonia agents in sera from patients with confirmed legionellosis in a common source outbreak were not found (5). When sera from culture-verified *Mycoplasma* cases were examined earlier, antibodies to *L. pneumophila* were found in less than 10%. Further, the percentage of patients with seroconversion to *M. pneumoniae* and antibodies to *L. pneumophila* at titers of ≥64 was the same as in the total material (~25%). Of the confirmed *Legionella* cases, none had antibodies to *M. pneumoniae*.

Infections caused by more than one microorganism at the same time could be another explanation of concomitant seroconversions. If that were the case, 2 of 10 patients would at the same time be infected by three bacterial species, e.g., *L. pneumophila*, *H. influenzae*, and *S. pneumoniae*, and 2 of 10 patients would have a *Legionella* infection together with septicemia due to other agents.

A rather high antibody titer to *L. pneumophila*

in four patients (no. 12, 17, 21, and 22) with a diagnosis based on seroconversion to other agents and no recent history of pneumonia pointed towards the possibility of nonspecific B-cell activation. Such a nonspecific stimulation was shown to occur during varicella virus and measles infections (1). It was suggested then that a viral infection induces T-cell activation with the release of B-cell-activating factors. Most likely there is more than one explanation for all of the observations mentioned.

LITERATURE CITED

1. **Arneborn, P., G. Biberfeld, M. Forsgren, and L. V. von Stedingk.** 1983. Specific and non-specific B cell activation in measles and varicella. Clin. Exp. Immunol. **51**:165–172.
2. **Atkinson, A. R., R. J. White, S. K. R. Clarke, and A. D. Blainey.** 1980. Incidence of Legionnaires' disease in a district general hospital. Postgrad. Med. J. **56**:622–623.
3. **Kallings, I., and L. O. Kallings.** 1983. Epidemiological patterns of legionellosis in Sweden. Zentralbl. Bakteriol. Parasitenkd. Infektionskr. Hyg. Abt. 1 Orig. Reihe A **254**.
4. **Kallings, I., and K. Nordström.** 1983. The pattern of immunoglobulins with special reference to IgM in Legionnaires' disease patients during a 2 year followup period. Zentralbl. Bakteriol. Parasitenkd. Infektionskr. Hyg. Abt. 1 Orig. Reihe A **254**.
5. **Nordström, K., I. Kallings, H. Dahnsjö, and F. Clemens.** 1983. An outbreak of Legionnaires' disease in Sweden: report of sixty-eight cases. Scand. J. Infect. Dis. **15**:43–55.
6. **Walder, M., B. Svanteson, J. Ursing, S. Cronberg, T. Johnsson, and A. Forsgren.** 1981. Incidence of *Legionella pneumophila* in acute lower respiratory tract infections. Scand. J. Infect. Dis. **13**:159–160.
7. **Wewalka, G.** 1981. Epidemiologic and Diagnostik von Legionella-Infektionen in Österreich. Zentralbl. Bakteriol. Parasitenkd. Infektionskr. Hyg. Abt. 1 Orig. Reihe A **249**:261–281.

Serogrouping of Legionellae by Slide Agglutination Test with Monospecific Sera

DICK G. GROOTHUIS, HARM R. VEENENDAAL, AND PIETER L. MEENHORST

National Institute of Public Health, Bilthoven, and Department of Infectious Diseases, University Hospital, Leiden, The Netherlands

Serogrouping of isolates of *Legionella pneumophila* from patients and the environment is essential to correlate an infection in a patient with a suspected source. Since conjugated rabbit sera were made available by the Centers for Disease Control (CDC), Atlanta, Ga., the direct immunofluorescence test (DFT) has become widespread and is now generally accepted as the method for serogrouping of isolates.

The DFT has its limitations as it is rather laborious and some isolates react with more than one conjugated antiserum (2). The grouping of the strains is based on historical grounds rather than on antigenic relationships.

The present work was undertaken to establish the antigenic relationships between the serogroups of *L. pneumophila* and the non-*L. pneumophila* legionellae. These relationships can be used to develop a typing scheme similar to that used for salmonellae and to prepare antisera which can be used in a slide agglutination test (SAT) for serogrouping of unknown strains. Cells of legionellae grown on buffered charcoal-yeast extract agar with α-ketoglutarate were harvested by gently scraping the agar plates, suspended in phosphate-buffered saline, and heated in boiling water. Agglutinating antisera were raised in rabbits by injecting intravenously

TABLE 1. Agglutination of unabsorbed rabbit sera with heat-killed cells

Serum	L. pneumophila group:						Strain L1	Strain L2	L. boze-manii	L. mic-dadei	L. du-moffii	L. gor-manii	L. long-beachae	L. jor-danis
	1	2	3	4	5	6								
L. pneumophila 1	11	—	—	—	—	—	—	6	—	—	5	—	—	—
L. pneumophila 2	—	11	7	—	—	—	—	—	—	—	—	—	—	—
L. pneumophila 3	—	7	11	—	—	11	3	—	—	—	—	—	—	—
L. pneumophila 4	—	—	7	11	6	—	11	5	—	—	—	—	—	—
L. pneumophila 5	—	—	—	4	9	—	8	—	—	—	—	—	—	—
L. pneumophila 6	—	—	7	—	—	11	3	—	—	—	—	—	—	—
L. pneumophila L1	—	—	5	8	10	5	11	8	—	—	—	—	—	—
L. pneumophila L2	6	—	—	—	—	—	4	11	—	—	6	—	—	—
L. bozemanii	—	—	—	—	—	—	—	—	11	—	—	—	—	8
L. micdadei	—	—	—	—	—	—	—	—	—	10	—	—	—	—
L. dumoffii	—	—	—	—	—	—	—	6	—	—	11	—	—	—
L. gormanii	—	—	—	—	—	—	—	—	—	—	—	8	—	—
L. longbeachae	—	—	—	—	—	—	—	—	—	—	—	—	8	—
L. jordanis	—	—	—	—	—	—	—	—	4	—	—	—	—	11

[a] Dilution steps 1, 2, 3, 4, 5, 6, 7, 8, 9, 10, and 11 correspond to reciprocal dilutions of 20, 40, 80, 160, 320, 640, 1,280, 2,560, 5,120, 10,240, and ≥20,000, respectively.

the cell suspensions of the type strains of *L. pneumophila* serogroups 1 through 6, *L. micdadei*, *L. bozemanii*, *L. dumoffii*, *L. gormanii*, *L. longbeachae* serogroup 1, and *L. jordanis*. The agglutination was performed in microdilution plates, using the diluted antiserum and gentian violet cell suspensions as the agglutinating antigen. Plates were read for agglutination after incubation overnight at 37°C.

The titer of the sera to homologous and heterologous cell suspensions was tested by checkerboard titrations (Table 1). Reciprocal homologous titers ranged from 2,500 to over 20,000. Antigenic relationships were detected and fur-

TABLE 2. Antigenic relationships within *L. pneumophila* and between other *Legionella* species

Species	Strain	Common antigenic determinants[a]
L. pneumophila 1	Philadelphia 1	a, b
L. pneumophila 2	Togus 1	c
L. pneumophila 3	Bloomington 2	c, d, (e)
L. pneumophila 4	Los Angeles 1	e, g
L. pneumophila 5	Dallas 1E	g
L. pneumophila 6	Chicago 2	d
L. pneumophila L1	Leiden 1	e, f, g
L. pneumophila L2	Leeuwarden 2	a, b, f, (h)
L. bozemanii	WIGA	(i), j
L. micdadei	TATLOCK	
L. dumoffii	ATCC 33279	(a), h
L. gormanii	ATCC 33297	
L. longbeachae	ATCC 33462	i
L. jordanis	ATCC 33623	(j)

[a] Determinants within parentheses are haptens. By subsequent absorptions monospecific sera could be obtained.

ther unraveled by absorptions with cross-reacting cell suspensions one by one, followed by retesting in the agglutination test after every absorption. All sera and antigens were tested in a crossover manner (Table 2). Monospecific sera could be obtained. Based on these results it can be concluded that the serogroup-specific antigens are heat stable. This is supported by the results of Collins et al. (1), who came to the same conclusion based on crossed-immunoelectrophoresis studies.

The SAT was chosen for serogrouping isolates on a routine basis. Live cells grown on buffered charcoal-yeast extract agar with α-ketoglutarate could be serogrouped by using the monospecific sera in the SAT. Sera to be used routinely in the SAT were made monospecific by absorption and dilution for practical reasons.

To evaluate the SAT, 30 coded strains (type strains and other strains previously serogrouped by DFT) were tested by SAT with the monospecific agglutinating antisera. Both methods yielded identical results for 28 strains. Two strains, previously described as Leiden 1 (L1) (2), gave a moderate positive reaction with the conjugated sera for *L. pneumophila* serogroups 4 and 5, but did not agglutinate with one of the monospecific sera in the SAT. The possibility of a new serogroup was considered. Therefore, antiserum was raised against one of the L1 strains. The serum could be made monospecific for the two L1 strains by absorption with cells of *L. pneumophila* serogroups 4 and 5. For further evaluation a set of 131 isolates was tested. In the first instance 125 of 131 strains were serogrouped by SAT, and the results were confirmed by testing with the corresponding conjugated antisera only in the DFT. Six strains did not react with one of the agglutinating sera in the SAT, but belonged to serogroup 1 according to the DFT. Again a new serogroup was expected. Antisera raised against these strains could be made monospecific by absorption with the cell suspensions of *L. pneumophila* serogroup 1 and L1. This new serogroup is provisionally designated Leeuwarden 2 (L2).

It can be concluded that isolates can be serogrouped by the SAT with rabbit sera raised by injecting heat-killed cells. Based on the serogroup-specific antigens, sera raised by injecting heat-killed cells are preferable to sera raised by injecting Formalin-killed cells.

LITERATURE CITED

1. **Collins, M. T., S.-N. Cho, N. Høiby, F. Espersen, L. Baek, and J. S. Reif.** 1983. Crossed immunoelectrophoretic analysis of *Legionella pneumophila* serogroup 1 antigens. Infect. Immun. **39:**1428–1440.
2. **Meenhorst, P. L., A. C. Reingold, G. L. Gorman, J. C. Feeley, B. J. van Kronenburg, C. L. M. Meyer, and R. van Furth.** 1983. *Legionella pneumophila* in guinea pigs exposed to aerosols of concentrated potable water from a hospital with nosocomial Legionnaires' disease. J. Infect. Dis. **147:**129–132.

Rapid Microagglutination Test and Early Diagnosis of Legionnaires Disease

T. G. HARRISON AND A. G. TAYLOR

Central Public Health Laboratory, Colindale, London NW9 5HT, England

We have described a simple and rapid microagglutination test (RMAT) which permits the testing of serum specimens for antibodies to *Legionella pneumophila* serogroup 1 in about 30 min (2). Recent results indicate that the RMAT has good specificity (3), and data presented in another paper at this symposium (A. G. Taylor, T. G. Harrison, A. Buré, and E. Dournon, this volume), using sera from cases proven by isolation or immunofluorescent demonstration of *L. pneumophila* in appropriate clinical specimens, show that the test is highly sensitive.

In the absence of suitable local facilities and expertise for the culture or demonstration of *L. pneumophila*, serology is relied upon for the diagnosis of Legionnaires disease (LD). This study was undertaken to investigate the earliest time in the illness when a diagnosis can be made by using the RMAT.

Due to the irregularity with which serum samples are taken from patients, it is not easy to investigate seroconversion in relation to time. However, in an attempt to obtain this information, sera from all patients diagnosed in this laboratory in 1981–1982 by the indirect fluorescent-antibody test using formolized yolk sac antigen (1, 4) were tested by the RMAT, provided that the dates of onset of illness and of hospital admission were known and that a serum specimen had been taken within 28 days of admission (131 cases). In this series seven patients acquired LD nosocomially; for this analysis their date of onset was plotted as day 0 of admission. All serum specimens from patients were included. Single specimens were used in 25 cases, paired sera in 70 cases, and multiple specimens (three or more) in 36 cases.

We have shown that 98% of healthy control subjects show titers of <8 in the RMAT (2), and hence a titer of ⩾8 is suggestive of LD. In this series 22 of 53 (41%) of specimens taken on either the day of admission (day 0) or the next day (day 1) were seropositive (titer of ⩾8). On days 2 and 3, 18 of 37 (48%), on days 4 and 5, 18 of 24 (75%), and on days 6 and 7, 22 of 27 (81%) were seropositive. Thus, the data presented (Fig. 1) indicate that in a patient with LD, a suggestive positive result is likely to be obtained with the RMAT at an early stage of the illness.

A fourfold rise of titer to ⩾16 or a single or standing titer of ⩾32 in this test is considered diagnostic for LD (3). In this series of patients, at least 2 serum specimens were taken within 14 days of hospital admission in 89 of 131 patients, and the titers of the first two specimens from each of these patients are shown in Fig. 2. The mean delay between hospital admission and the time that the first specimen was taken was 2.1 days (range, 0 to 12 days), and the time between taking the first and second specimens was 6.8 days (range, 1 to 12 days). Within 2 days of admission 25% of the 89 patients had titers of ⩾8, and 9% met the diagnostic criteria. By day 7, 60% had positive titers and 40% were diagnostic, and within the 14-day period 96% were positive and 88% were diagnostic. These results demonstrate that a diagnosis of LD can often be made early in the course of illness when paired sera are available. The results (Fig. 1 and 2) indicate that by taking a specimen on admission and then taking frequent additional specimens the interval between hospital admission and diagnosis could be reduced.

In summary, if appropriately timed serum specimens are obtained in the acute phase of illness, a positive result can be expected in the

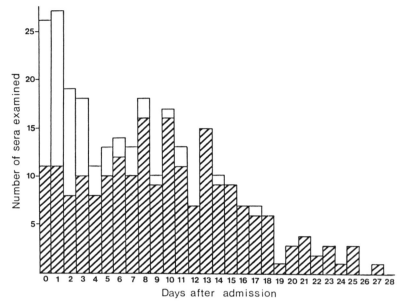

FIG. 1. Number of positive serum specimens taken in relation to stage of illness (day after hospital admission). ▨, Positive; □, negative.

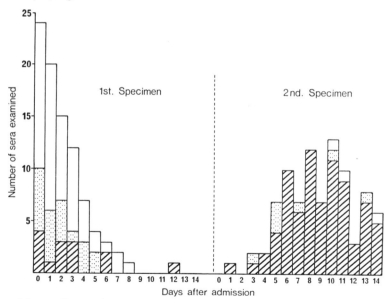

FIG. 2. Timing of first and second serum specimens in relation to stage of illness (day after hospital admission), and results obtained. ▨, Diagnostic; ▩, positive; □, negative.

RMAT. This test is specific and sensitive; therefore, such a positive result can be helpful in the clinical management of the patient.

LITERATURE CITED

1. **Harrison, T. G., and A. G. Taylor.** 1982. Diagnosis of *Legionella pneumophila* infection by means of formolised yolk-sac antigens. J. Clin. Pathol. **35:**211–214.
2. **Harrison, T. G., and A. G. Taylor.** 1982. A rapid microag-
glutination test for the diagnosis of *Legionella pneumophila* (serogroup 1) infection. J. Clin. Pathol. **35:**1028–1031.
3. **Taylor, A. G., and T. G. Harrison.** 1983. Serological tests for *Legionella pneumophila* (serogroup 1) infections. Zentralbl. Bakteriol. Hyg. Parasitenkd. Infektionskr. Abt. 1 Orig. Reihe A **255:**20–26.
4. **Taylor, A. G., T. G. Harrison, M. W. Dighero, and C. M. P. Bradstreet.** 1979. False positive reactions in the indirect fluorescent antibody test for Legionnaires' disease eliminated by use of formolised yolk-sac antigen. Ann. Intern. Med. **90:**686–689.

Diagnostic Potential of an Anti-*Legionella* Monoclonal Antibody

JEAN-GÉRARD GUILLET, JOHAN HOEBEKE, CUONG TRAM, FRANÇOISE PETITJEAN, AND A. D. STROSBERG

Laboratoire d'Immunologie Moléculaire, Institut Jacques Monod, Centre National de la Recherche Scientifique, Université Paris VII, F-75251 Paris Cedex 05, and Service d'Ecologie Bactérienne, Institut Pasteur, F-75724 Paris Cedex 15, France

Hybridomas were obtained by the fusion of splenocytes of BALB/c mice immunized intravenously with heat-killed *Legionella pneumophila* Philadelphia strain 1 and nonsecreting NS-1 myeloma cells (4). Ten clones were tested against legionellae of the other five serogroups and against the atypical strains NY-23 (ATCC 33279), TATLOCK (ATCC 33218), and WIGA (ATCC 33217). No cross-reaction could be shown with any of the antibodies.

Among the clones secreting anti-*Legionella* antibodies, two were selected for amplification by inducing ascites in BALB/c mice. The ascitic fluids of one of the clones (II-6-18) were further used for the characterization of the monoclonal antibody and for the study of its specificity and its diagnostic potential.

The anti-*Legionella* monoclonal antibody is of the gamma-3 isotype. Its stability was tested by frequent freezing and thawing during storage over several months at $-20°C$; there was no noticeable decrease in activity (2). The calibration of the ascitic fluid showed a half-maximal response at a dilution of 1/1,000. The fixation of the antibody could be inhibited up to 80% by a polyclonal rabbit anti-*Legionella* antiserum, suggesting that a common antigenic determinant is recognized by the rabbit and mouse antibodies.

The specificity of the ascitic fluid was tested against the other serotypes of *L. pneumophila*. No cross-reaction could be shown. Furthermore, the specificity was tested by enzyme-linked immunosorbent assay (5) and indirect immunofluorescence (6) against *Klebsiella pneumoniae*, *Mycoplasma pneumoniae*, *Mycobacterium tuberculosis*, *Pseudomonas thomasii*, *Haemophilus influenzae* type b, *Bacteroides thetaiotaomicron*, *Bacteroides fragilis*, and *Bacteroides vulgatus*, which are involved in pneumopathies or known to cross-react with polyclonal anti-*Legionella* antibodies (1, 3). No cross-reaction could be observed. As of now, the ascitic fluids have been tested against six cultures isolated from different clinical samples and shown to be positive for *Legionella* serotype 1 by indirect immunofluorescence. Of the six cultures, one reacted only weakly with the monoclonal antibody. A quantitative comparison in an enzyme-linked immunosorbent assay of the nonreacting strain and the Philadelphia strain 1 by using a rabbit polyclonal antiserum showed that the former was recognized at least 16 times more weakly than the latter.

Finally, the monoclonal antibody was used in a sensitivity test with decreasing amounts of bacteria. The test could detect fewer than 30,000 bacteria (Fig. 1). Preliminary tests on clinical samples (bronchoscopic samples and urine) suggest that the amount of antigen present in these types of samples is enough to be detectable.

In summary, we have developed an anti-*Legionella* serotype 1 monoclonal antibody available in large amounts and with the qualities required for its use in the quantitation of a specific bacterial antigen in clinical samples. This antibody would therefore be suited for the diagnosis of legionellosis in the early steps of the disease as well as for detecting the nature of the bacterial antigens responsible for the immune response.

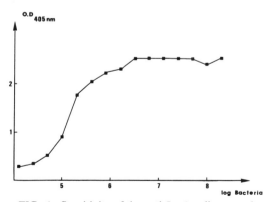

FIG. 1. Sensitivity of the anti-*Legionella* monoclonal antibody. The ascitic fluid containing the antibody was tested at a dilution of 1/100 against decreasing amounts of bacteria by using the enzyme-linked immunosorbent assay described in the text. The ordinate shows the optical density at 405 nm of the H_2O_2-ABTS product formed by horseradish peroxidase.

We thank B. Vray (Free University of Brussels, Belgium) and M. L. Guillou-Joly and C. Gibert (Hopital Bichat, Paris, France) for the *Legionella* cultures.

This work was supported by funds from the Centre National de la Recherche Scientifique, University of Paris VII, the Fondation de la Recherche Medicale Francaise, and the Association pour la Recherche sur le Cancer.

LITERATURE CITED

1. **Berdal, B. P., and J. Eng.** 1982. Coreactivity of *Legionella pneumophila* immune sera in the *Mycoplasma pneumoniae* complement fixation test. J. Clin. Microbiol. **16**:794–797.
2. **Cherry, W. B., B. Pittman, P. P. Harris, G. A. Hebert, B. M. Thomason, L. Thacker, and R. E. Weaver.** 1978. Detection of Legionnaires disease bacteria by direct immunofluorescence staining. J. Clin. Microbiol. **8**:329–338.
3. **Edelstein, P. H., R. M. McKinney, R. D. Meyer, M. A. C. Edelstein, C. J. Krause, and S. M. Finegold.** 1980. Immunological diagnosis of Legionnaires' disease: cross-reactions with anaerobic and microaerophilic organisms and infections caused by them. J. Infect. Dis. **141**:652–655.
4. **Guillet, J.-G., J. Hoebeke, C. Tram, S. Marullo, and A. D. Strosberg.** 1983. Characterization, serological specificity, and diagnostic possibilities of monoclonal antibodies against *Legionella pneumophila*. J. Clin. Microbiol. **18**:793–797.
5. **Tilton, R. C.** 1979. Legionnaires' disease antigen detected by an enzyme-linked immunosorbent assay. Ann. Intern. Med. **90**:697–698.
6. **Wilkinson, H. W., B. J. Fikes, and D. D. Cruce.** 1979. Indirect immunofluorescence test for serodiagnosis of Legionnaires disease: evidence for serogroup diversity of Legionnaires disease bacterial antigens and for multiple specificity of human antibodies. J. Clin. Microbiol. **9**:379–383.

Cross-Reacting Bacterial Antigens in a Radiometric Assay for Urinary *Legionella pneumophila* Antigen

RICHARD B. KOHLER, STEPHEN D. ALLEN, AND EDWARD R. WILSON

Indiana University School of Medicine, Indianapolis, Indiana 46202

Most patients with serogroup 1 Legionnaires disease excrete an antigen in their urine which can be detected by various immunoassays (3–6). Urine specimens from patients with a variety of other microbial infections do not contain cross-reactive antigens, and the assay appears to be highly specific. We have tested only a few specimens from patients with other *Legionella* infections, some of which have appeared to contain antigens. Our goal in this study was to determine whether our solid-phase radioimmunoassay, which uses immunoglobulin G (IgG) from a rabbit immunized with autoclaved whole serogroup 1 *Legionella pneumophila*, can detect antigens from cultures of other *Legionella* organisms and bacteria from other genera.

The bacterial strains tested were 6 strains of *L. pneumophila* serogroup 1; 1 strain each from *L. pneumophila* serogroups 2 through 6; 1 strain each of *Legionella bozemanii*, *Legionella dumoffii*, *Legionella gormanii*, *Legionella jor-*

danis, and *Legionella longbeachae* serogroups 1 and 2; 2 strains of *Legionella micdadei*; 37 strains of *Pseudomonas fluorescens*, including 2 strains previously reported to cross-react with serogroup 1 *L. pneumophila* direct fluorescent-antibody (DFA) reagents (1); 24 strains of *Pseudomonas alcaligenes*, including 1 strain previously reported to cross-react with serogroup 1 DFA reagents (1); 41 strains of other glucose-nonfermenting gram-negative bacilli representing 10 other species; 49 strains of *Enterobacteriaceae* representing 23 species; 14 strains of *Streptococcus pneumoniae*, including 1 of each of the pneumococcal vaccine serotypes; 4 strains of *Staphylococcus aureus*; 2 strains of *Staphylococcus epidermidis*; and 53 strains of *Bacteroides fragilis*, including 3 strains previously reported to cross-react with serogroup 1 DFA reagents (2).

Quantitated broth cultures of the bacteria were tested in a previously described direct

TABLE 1. Detection of bacterial antigens in tubes coated with specific IgG, specific F(ab')$_2$ fragments, or nonspecific IgG

Antigen	Reactivity (cpm)[a] in tubes coated with:			
	No coat[b]	Specific IgG	Specific F(ab')$_2$	Nonspecific IgG
L. pneumophila serogroup 1 (Bellingham 1)	244 ± 61	11,512 ± 389	10,389 ± 449	205 ± 31
P. fluorescens (CDC 93)	2,189 ± 898	13,381 ± 861	10,882 ± 378	2,245 ± 67
P. fluorescens (CDC EB)	3,789 ± 550	9,230 ± 1,336	8,536 ± 969	2,853 ± 363
Positive patient urine	742 ± 41	2,989 ± 142	2,011 ± 187	568 ± 15
S. aureus (Copenhagen)	1,033 ± 65	5,234 ± 397	846 ± 68	3,981 ± 601
S. aureus (Du-4916S)	496 ± 137	23,651 ± 5,610	694 ± 147	10,300 ± 977

[a] Counts per minute in solid-phase radioimmunoassay with tubes coated as indicated; results are expressed as mean ± standard deviation of triplicate tubes.

[b] Assays were carried out in uncoated tubes to determine the amount of nonspecific binding of the various antigens to the polystyrene tubes. This must be taken into account in interpreting the counts per minute in the remaining three columns.

TABLE 2. Reactions of various antisera with three *B. fragilis* isolates

Antiserum specificity	Lot no.	Fluorescence[a] with:			
		B. fragilis 78-189A	*B. fragilis* 78-235H	*B. fragilis* 79-84F	*L. pneumophila* Philadelphia 1
L. pneumophila serogroup 1	CDC80-0189	0	0	0	4+
L. pneumophila serogroup 1	CDC82-0073	3+[b]	3+[b]	3+[b]	4+
L. pneumophila serogroup 1	IU-LR2	0	0	0	4+
L. pneumophila poly A[c]	CDC82-0092	2+[b]	2+[b]	2+[b]	4+
L. pneumophila serogroup 4	CDC81-0068	2+[b]	2+[b]	2+[b]	0

[a] Values indicate the subjective degree of fluorescence.

[b] About 10% of bacterial population fluoresced.

[c] Poly A includes *L. pneumophila* serogroups 1 through 4.

(single-antibody sandwich) solid-phase radioimmunoassay for serogroup 1 *L. pneumophila* (3). The assay results were expressed as a binding ratio, which was the counts per minute of the bacterial culture in the assay divided by the counts per minute of uninoculated broth culture medium.

Five of the six serogroup 1 *L. pneumophila* isolates, the strains of *P. fluorescens* and *P. alcaligenes* previously reported to cross-react with serogroup 1 DFA reagents, and two of the four strains of *S. aureus* yielded binding ratios ranging from 54.87 to 208.96. The remaining serogroup 1 *L. pneumophila* strain, which grew relatively poorly in broth, the serogroup 2 through 6 *L. pneumophila* strains, the *L. dumoffii* strain, and one strain of *P. fluorescens* yielded binding ratios ranging from 2.31 to 4.80. The remaining bacterial strains, including the three strains of *B. fragilis* previously reported to cross-react with serogroup 1 *L. pneumophila* DFA reagents, yielded ratios of 1.72 or less.

Five bacterial strains and urine from a patient with Legionnaires disease were tested in the radioimmunoassay, using tubes coated with either specific anti-serogroup 1 *L. pneumophila* IgG, nonspecific IgG, or specific F(ab')$_2$ fragments (Table 1). The two *S. aureus* isolates with high binding ratios were highly reactive in tubes coated with specific IgG as well as in tubes coated with nonspecific IgG. They were not reactive in tubes coated with specific F(ab')$_2$ fragments. By contrast, two *P. fluorescens* strains, a serogroup 1 *L. pneumophila* strain, and the urine specimen were reactive in the tubes coated with specific IgG or F(ab')$_2$ fragments but not in tubes coated with nonspecific IgG. These results indicate that the *S. aureus* reactivity is due to nonspecific binding to IgG, probably due to protein A.

A fluorescein isothiocyanate conjugate was made with the IgG used in the radioimmunoassay. This conjugate and one (lot no. CDC 80-0189) supplied by the Centers for Disease Control, Atlanta, Ga., reacted with a serogroup 1 *L. pneumophila* strain but not the three *B. fragilis* strains previously reported to cross-react with serogroup 1 DFA reagents (Table 2). However, another serogroup 1 *L. pneumophila* DFA conjugate (lot no. CDC 82-0073), a polyvalent *L. pneumophila* DFA conjugate (lot no. CDC 82-0092), and a serogroup 4 *L. pneumophila* conjugate (lot no. CDC 81-0068) produced 2+ to 3+ fluorescence with the three *B. fragilis* isolates. Thus, the cross-reactive *B. fragilis* antigen appears to differ from the serogroup 1 antigen(s) as well as from the urinary antigen of *L. pneumophila*.

The cross-reactivity demonstrated by *L. pneumophila* serogroups 2 through 6 and by *L. dumoffii* suggests that the current assay might detect antigens in urine from patients with these infections if the cross-reactive antigens are excreted in sufficient concentrations. Urine from patients with *S. aureus* infections has not produced positive results when tested in the radioimmunoassay (3, 5). However, this study indicates that the potential for *S. aureus* assay reactivity exists and that the use of F(ab')$_2$-coated assay tubes would eliminate the cross-reactivity. Cross-reactivity also appears to be possible from a small proportion of strains of *P. fluorescens* and *P. alcaligenes*; however, infections with these organisms, particularly pneumonia, are rare. Finally, the previously reported cross-reactivity of a few *B. fragilis* strains should not pose problems in *L. pneumophila* antigen detection systems if appropriate antisera are selected.

LITERATURE CITED

1. **Broome, C. V., W. B. Cherry, W. C. Winn, Jr., and B. R. MacPherson.** 1979. Rapid diagnosis of Legionnaires' disease by direct immunofluorescent staining. Ann. Intern. Med. **90:**1–4.

2. **Edelstein, P. H., R. M. McKinney, R. D. Meyer, M. A. C. Edelstein, C. J. Krause, and S. M. Finegold.** 1980. Immunologic diagnosis of Legionnaires' disease: cross reaction with anaerobic and microaerophilic organisms and infections caused by them. J. Infect. Dis. **141:**652–655.

3. **Kohler, R. B., S. E. Zimmerman, E. Wilson, S. D. Allen,**

P. H. Edelstein, L. J. Wheat, and A. White. 1981. Rapid radioimmunoassay diagnosis of Legionnaires' disease: detection and partial characterization of urinary antigen. Ann. Intern. Med. **94**:601–605.
4. **Sathapatayavongs, B., R. B. Kohler, L. J. Wheat, A. White, and W. C. Winn, Jr.** 1983. Rapid diagnosis of Legionnaires' disease by latex agglutination. Am. Rev. Respir. Dis. **127**:559–562.

5. **Sathapatayavongs, B., R. B. Kohler, L. J. Wheat, A. White, W. C. Winn, Jr., J. C. Girod, and P. H. Edelstein.** 1982. Rapid diagnosis of Legionnaires' disease by urinary antigen detection: comparison of ELISA and radioimmunoassay. Am. J. Med. **72**:576–582.
6. **Tang, P. W., D. DeSavigny, and S. Toma.** 1982. Detection of *Legionella* antigenuria by reverse passive agglutination. J. Clin. Microbiol. **15**:998–1000.

Legionella pneumophila Serogroup 1 Flagellar Antigen in a Passive Hemagglutination Test to Detect Antibodies to Other *Legionella* Species

FRANK G. RODGERS AND TONY LAVERICK

Department of Microbiology and PHLS Laboratory, University Hospital, Nottingham, England

The indirect immunofluorescent-antibody (IFA) technique is the standard test for the detection of human antibody to *Legionella* species (3, 6). However, this procedure is time consuming, needs experience in interpretation of the immunofluorescence, and requires a panel of several species-specific antigens to detect antibody to the increasing number of members of the *Legionellaceae*. We describe and give preliminary results for a passive hemagglutination test (PHA) using *Legionella pneumophila* serogroup 1 flagellar antigen to detect serum antibody to a number of different *L. pneumophila* serogroups and other *Legionella* species with a single antigen preparation.

A modification of the indirect hemagglutination technique (5) was developed, using chick erythrocytes (CRBC). After washing six times in phosphate-buffered saline (PBS), 10 ml of a 10% (vol/vol) CRBC suspension was incubated with an equal volume of 1% (vol/vol) glutaraldehyde in PBS at 4°C for 30 min with gentle stirring. After fixation, the cells were washed three times in PBS by centrifugation and suspended at 10% in PBS containing 0.1% (wt/vol) sodium azide. The glutaraldehyde-fixed CRBC were tanned by mixing 10 ml of the 10% suspension with 10 ml of 0.005% (wt/vol) tannic acid for 30 min at 37°C. The tanned, fixed chick erythrocytes (T-GA-CRBC) were washed three times in buffer and suspended at 10% in PBS.

A clinical isolate of *L. pneumophila* serogroup 1, strain Nottingham N7, previously found to produce abundant flagella (4) was used. This had been passaged twice only on low-sodium chloride legionella blood agar (1) before use. Inoculated plates were incubated at 37°C in a humid atmosphere for 36 h, and the organisms were harvested into PBS. The flagella were stripped from the bacteria by repeated forcing through a 26-gauge syringe needle in a safety cabinet. Separation was by differential centrifugation at 7,000 × *g* for 20 min to pellet organisms and at 180,000 × *g* for flagella. The flagella were washed three times in PBS, and purity was checked by IFA and electron microscopy (Fig. 1). A 10% suspension of T-GA-CRBC was mixed with an equal volume of the flagellar antigen and incubated at 37°C for 30 min with occasional shaking. The sensitized cells were washed three times in PBS containing 0.25% (wt/vol) bovine serum albumin and finally suspended in the same solution to yield a stable 0.5% stock suspension.

A range of human and animal sera were diluted under code at 1:16 in PBS, and 25-μl volumes were added to U-bottomed microdilution plates. Aliquots (25 μl) of unsensitized (control) and flagella-sensitized (test) T-GA-CRBC were added, the plates were shaken, and the mixtures were allowed to stand for 90 min at room temperature before being read. Positive reactions were taken as those sera showing a hemagglutination pattern as compared with the control wells showing buttons. A total of 43 human sera were tested, of which 27 gave IFA titers of <1:16 and 16 gave positive titers of between 1:64 and 1:512 to *L. pneumophila* serogroup 1. In addition, 18 rabbit sera were tested, 15 of which were hyperimmune antisera raised against a number of different strains of *L. pneumophila* (serogroups 1 through 6) and other *Legionella* species. The remaining three sera included an antiserum to *L. pneumophila* serogroups 1 through 6 (polyvalent) and two control sera, a preimmunization normal rabbit serum and a polyvalent *Escherichia coli* agglutinating serum, both negative by IFA. As a positive control, hyperimmune anti-flagella serum was raised in New Zealand White rabbits by multiple subcutaneous inoculation of purified flagellar antigen in Freund complete adjuvant. This was followed 5 weeks later by an intramuscular booster dose into each thigh.

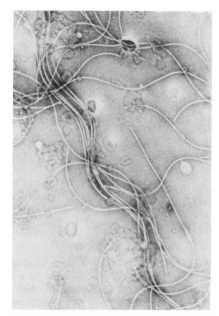

FIG. 1. Electron micrograph of flagella from *L. pneumophila* serogroup 1, strain N7, negatively stained with 1% phosphotungstic acid (pH 6.7). Magnification, ×20,000.

TABLE 1. Detection of *Legionella* antibodies in rabbit sera by PHA with *L. pneumophila* serogroup 1 flagellar antigen

Species	Sero-group	Strain	PHA result Test	PHA result Con-trol
L. pneumophila	1	Pontiac	+	−
	1	RH	+	+[a]
	1	Heysham 2	+	+[a]
	1	N7 (anti-fla-gella)	+	−
	2	Togus	+	−
	3	Blooming-ton 2	+	−
	4	Los Angeles 1	+	−
	5	Cambridge 2	+	−
	6	Oxford 1	+	−
	Polyva-lent (1–6)		+	−
L. bozemanii		WIGA	+	−
L. dumoffii		NY-23	+	−
L. gormanii		LS-13	−	−
L. longbeachae	1	LB-4	−	−
	2	Tucker 1	+	+[a]
L. jordanis		BL 540	−	−
L. micdadei		PPA	+	−
Normal rabbit serum			−	−
Escherichia coli (polyvalent 1) serum			−	−

[a] Chick cell agglutinins.

Of the 27 IFA-negative sera, 19 gave negative results in the PHA test and 8 had chick cell agglutinins, as shown by hemagglutination of the unsensitized control cells. With the 16 IFA-positive sera, 8 were PHA positive and 8 had chick cell agglutinins. The results of tests on the rabbit sera are shown in Table 1.

Work is in progress to further modify the test for use with turkey erythrocytes to reduce cross-adsorption procedures to overcome the problems of chick cell agglutinins which were present in 40% of the human sera and in three rabbit sera. The removal of these agglutinins proved variable. A reduction in the concentration of tannic acid to 0.0002% may decrease nonspecific agglutinations (R. Powell, personal communication). Immunological and biochemical assays have shown a similarity between the flagella of *L. pneumophila* serogroups 1, 2, and 3 (2). Our preliminary results suggest a broad common antigenicity for the flagella of all of the serogroups of *L. pneumophila* and those of many other species of *Legionella*, although sera to serogroups 7 and 8, *L. wadsworthii*, and *L. oakridgensis* remain to be tested. Failure to detect antibodies in the rabbit sera raised against *L. gormanii*, *L. jordanis*, and *L. longbeachae* was interesting. This may reflect a genuine diversity in antigenic structure for the flagella of some legionellae or a lack of flagella on those

antigen preparations used to produce the antisera. The presence of nonspecific inhibitors in the three negative rabbit sera cannot at present be excluded. As a rapid, simplified screening method, the procedure has the advantage of detecting antibodies to a number of legionellae with one antigen preparation used as a single test. The role of *Legionella* flagella in the pathogenesis of disease and the antibodies they induce in the outcome of legionellosis remain to be established.

LITERATURE CITED

1. **Dennis, P. J., J. A. Taylor, and G. I. Barrow.** 1981. Phosphate buffered, low sodium chloride blood agar medium for *Legionella pneumophila*. Lancet ii:636.
2. **Elliott, J. A., and W. Johnson.** 1981. Immunological and biochemical relationships among flagella isolated from *Legionella pneumophila* serogroups 1, 2, and 3. Infect. Immun. 33:602–610.

3. **Harrison, T. G., and A. G. Taylor.** 1982. Diagnosis of *Legionella pneumophila* infections by means of formolised yolk sac antigens. J. Clin. Pathol. **35**:211–214.

4. **Rodgers, F. G., P. W. Greaves, A. D. Macrae, and M. J. Lewis.** 1980. Electron microscopic evidence for flagella and pili on *Legionella pneumophila*. J. Clin. Pathol. **33**:1184–1188.

5. **Sawada, T., R. B. Rimler, and K. R. Rhoades.** 1982. Indirect hemagglutination test that uses glutaraldehyde-fixed sheep erythrocytes sensitized with extract antigens for detection of *Pasteurella* antibody. J. Clin. Microbiol. **15**:752–756.

6. **Wilkinson, H. W., B. J. Fikes, and D. D. Cruce.** 1979. Indirect immunofluorescence test for serodiagnosis of Legionnaires disease: evidence for serogroup diversity of Legionnaires disease bacterial antigens and for multiple specificity of human antibodies. J. Clin. Microbiol. **9**:379–383.

Stability of Three *Legionella pneumophila* Serogroup 1 Antigenic Preparations

M. NOWICKI, N. BORNSTEIN, AND J. FLEURETTE

Centre National de Référence des Légionelloses, Faculté de Médecine Alexis Carrel, 69372 Lyon Cedex 2, France

Indirect immunofluorescence is the established reference method for the serodiagnosis of *Legionella* infection. Two techniques are used for antigen preparation; in one, antigen (Ag A) is obtained from a culture on agar medium and heat inactivated, and in the other, the antigen (Ag B) is prepared from bacteria developed in an embryonated hen yolk sac and formolized. In a number of studies, the sensitivity and specificity of the antigens have been investigated. We have tested their stability under various conditions of preservation duration and temperature.

Three antigens, A, B, and C, of *Legionella pneumophila* serogroup 1 were prepared from strain Philadelphia 1 obtained from the Centers for Disease Control, Atlanta, Ga. Ag A was prepared on charcoal-yeast extract agar according to the recommendations of the Centers for Disease Control (3). Ag B was prepared on hen yolk sac according to the recommendations of the Public Health Laboratory, United Kingdom (2). As this corresponds to the antigen used in our laboratory for serodiagnosis, standardization of the preparation was checked by testing four different batches prepared under similar conditions at different times. The preparation of Ag C stock suspension differed slightly from that of Ag A: it was carried out in phosphate-buffered saline (pH 7.2) containing 3% yolk sac rather than distilled water, but dilutions for use were carried out in phosphate-buffered saline without supplementation with 0.5% yolk sac. The sterility of each preparation was checked by culturing for 2 weeks at 35°C on charcoal-yeast extract agar, 2 weeks on blood agar under aerobic and anaerobic conditions, and 20 days in Sabouraud medium at 20°C.

Indirect immunofluorescence, which is the reference method, was used to compare the three antigens; initial serum titers were determined by using indirect immunofluorescence and Ag B. A total of 174 sera were thus tested. Specificity was evaluated for 60 sera from blood donors and for 42 sera from patients with chlamydia ($n = 4$), mycoplasma ($n = 14$), or candida ($n = 24$) infections. Reproducibility was estimated for 71 sera from patients with *L. pneumophila* infection diagnosed serologically; a number of them had been confirmed by strain isolation. Of these 71 sera, 12 had a titer of 32, 11 had a titer of 64, 13 had a titer of 128, 17 had a titer of 256, 11 had a titer of 512, and 7 had a titer of 1,024.

Each type of antigen and each serum were tested three times under similar conditions; 213 tests were carried out for Ag A, for each batch of Ag B, and for Ag C. Sensitivity was evaluated by testing sera with low titers (32 to 64). Stability was tested by using the accelerated degradation method of Jerne and Perry (1). Each antigen preparation was divided into three equal parts, each remaining for 1 week, 3 weeks, 8 weeks, and 5 months at temperatures of −20, 4, 20, 37, and 56°C. The same rabbit immune serum was selected as a reference for each test (titer, 8.192), and each test was carried out three times. The results obtained were expressed as dilution variations compared with the reference titer. Arithmetic means of variations as well as standard deviations (SD) were calculated. Stability was recorded as positive when the mean plus 2 standard deviations was less than or equal to 1 dilution.

Reproducibility was almost perfect for all three antigens. No 2-dilution variations were observed, and 1-dilution variations were rare: 16% for Ag A (SD = 0.4), 3% for Ag B (SD = 0.17), and 16% for Ag C (SD = 0.4). No significant differences were demonstrated between the four Ag B preparations; the 1-dilution variations ranged from 1.9 to 4.2% (0.14 ≤ SD ≤ 0.2).

Specificity was excellent for the three types of antigens; the titers remained less than 16 for all

TABLE 1. Stability of the three antigens under various conditions of preservation duration and temperature

Preservation temp (°C)	Antigen	Stability after: 1 week Mean[a]	SD	3 weeks Mean	SD	8 weeks Mean	SD	22 weeks Mean	SD
−20	A	0	0			0	0	1.3	0.5
	B	0	0			0	0	0	0
	C	0	0			0	0	0.2	0.4
4	A	0	0	0.2	0.4	0.55	0.7	1.4	0.5
	B	0	0			0	0	0	0
	C	0	0	0	0	0.1	0.3	0.2	0.4
20	A	0.4	0.5	1.4	0.5	2.3	0.4	2.6	0.5
	B	0	0			0	0	1.6	0.6
	C	0.2	0.4	0.2	0.4	0.8	0.4	1.3	0.5
37	A	0.8	0.4	1.7	0.4	2.5	0.5	3.3	0.4
	B	0.03	0.2			0.1	0.3	1.8	0.5
	C	0.3	0.5	0.6	0.5	0.8	0.4	2.1	0.5
56	A	3.4	0.5	5.1	2.8				
	B	1.2	0.4			4.8	1.7		
	C	1.9	0.7	2.6	0.7				

[a] Arithmetic mean of variations.

of the sera from blood donors and from patients with chlamydia, mycoplasma, or candida infections.

The sensitivity of the various preparations tested on sera with low titers was good, since the results are similar to those obtained for all sera. One-dilution variations were observed for 22% of sera with Ag A (SD = 0.42), for 2% with Ag B (SD = 0.25), and for 7% with Ag C (SD = 0.26).

Antigen stability in relation to duration and temperature of preservation varied according to the type of preparation (Table 1). Optimal preservation was obtained at −20 and 4°C for all three preparations. Ag A was altered after 5 months at −20°C and 8 weeks at 4°C. At 20°C, Ag B remained stable after 8 weeks and Ag C after 3 weeks. At 37°C, Ag B was stable after 8 weeks, whereas the other antigens were altered. At 56°C, all of the antigens were modified.

The three antigen preparations were identical in reproducibility, specificity, and sensitivity. However, they differed in stability according to duration and temperature of preservation. Under normal preservation conditions, i.e., between −20 and 4°C, stability is satisfactory for all three antigens, lasting 4 weeks for Ag A and 5 months for Ag B and Ag C. Ag B had excellent stability, but its preparation is more complicated. On the other hand, preparation of Ag A is easier, but its preservation should be improved by a simple procedure, the addition of yolk sac in stock suspension (Ag C).

LITERATURE CITED

1. Jerne, N. K., and W. L. M. Perry. 1956. The stability of biological standards. Bull. W.H.O. 14:167–182.
2. Taylor, A. G., T. G. Harrison, and M. W. Dighero. 1979. False positive reactions in the indirect fluorescent antibody test for Legionnaires' disease eliminated by use of formolised yolk-sac antigen. Ann. Intern. Med. 90:686–689.
3. Wilkinson, H. W., B. J. Fikes, and D. D. Cruce. 1979. Indirect immunofluorescence test for serodiagnosis of Legionnaires disease: evidence for serogroup diversity of Legionnaires disease bacterial antigens and for multiple specificity of human antibodies. J. Clin. Microbiol. 9:379–383.

Immune Response of Cynomolgus Monkeys to Vaccines of *Legionella pneumophila* Serogroups 1 to 6: Antigenic Relatedness to Human Immunoglobulins

BONNIE J. BRAKE, HAZEL W. WILKINSON, ROBERT GELOK, AND ARTHUR MARKOVITS

Division of Bacterial Diseases, Center for Infectious Diseases, Centers for Disease Control, Atlanta, Georgia 30333, and BioDx, Inc., Denville, New Jersey 07834

Since the original outbreak of Legionnaires disease in Philadelphia, Pa., the indirect immunofluorescence assay (IFA) has been the most widely used diagnostic test to detect antibodies against *Legionella* antigens (2, 3). Like any serological test, the IFA requires positive control sera to ensure that the reagents are performing satisfactorily. Sufficient quantities of human sera have been difficult to obtain for this purpose, especially since the number of species and serogroups of *Legionella* has increased. An alternative serum source is the nonhuman primate, as immunoglobulins of human and nonhuman primates cross-react to various degrees depending on the extent to which they are phylogenetically related (1). The purpose of this study was to determine whether antisera from cynomolgus monkeys (*Macaca fascicularis*) that were immunized with vaccines to *Legionella pneumophila* serogroups 1 to 6 could be used instead of human sera as positive controls for the *Legionella* IFA.

Each monkey was immunized with a single footpad injection of heat-killed vaccine and was bled once a week for 12 weeks. These sera were compared with sera from two culture-documented cases of *Legionella* pneumonia, one due to *L. pneumophila* serogroup 1 infection and the other due to serogroup 6 infection. Ouchterlony analysis with human heavy-chain-specific immunoglobulin G (IgG), IgM, and IgA and human light-chain-specific κ and λ antisera showed reactions of partial identity for the human and monkey IgG, IgM, and IgA. Similarly, identical immunoelectrophoretic precipitation patterns were detected when the monkey and human sera were allowed to react with anti-human IgG, IgM, and IgA. Precipitation of partially purified *L. pneumophila* serogroup 6 anti-

gen occurred in an area of the gel that corresponded to the IgG-IgM mobility zone. Precipitin bands were not detectable in either analysis with κ or λ antiserum.

To determine whether the degree of cross-reactivity demonstrated with Ouchterlony and immunoelectrophoretic analyses was sufficient for the IFA, titers were determined for each monkey serum specimen against antigens of *L. pneumophila* serogroup 1 to 6 with anti-human fluorescein isothiocyanate-labeled conjugate. Homologous titers exceeded 16,000 within 4 to 6 weeks after immunization (Table 1). As shown previously for human sera (3, 4), the monkey sera also showed titers against heterologous serogroup antigens, although the latter titers were invariably lower than the homologous titers. Additional tests with anti-human IgG, IgM, and IgA conjugates showed higher peak titers with the IgG than with the IgM conjugate and revealed no anti-*Legionella* IgA reactivity. The fact that the immune response detected in immunized monkeys was primarily IgG, whereas all three immunoglobulin classes were detected in sera from the humans convalescing from *Legionella* pneumonia, may be attributed to different routes of inoculation or infection, viable versus killed cells, different extent of cross-reactivity among the immunoglobulin classes, differences in antigen degradation and persistence, individual variation, or, most likely, a combination of all of these factors. The polyvalent conjugate used in the reference IFA reacts with all three classes; therefore, the monkey antisera are adequate daily controls in the standard test. They have been stable for 2 years after dilution in normal human serum to give IFA titers of 256 to 512.

In summary, this study indicates that sera

TABLE 1. Peak *L. pneumophila* IFA titers in cynomolgus monkey antisera

Antigen serogroup	Peak titer with antiserum serogroup:					
	1	2	3	4	5	6
1	32,768	128	128	64	64	128
2	128	16,384	4,096	1,024	256	2,048
3	128	4,096	131,072	1,024	128	2,048
4	2,048	2,048	4,096	32,768	2,048	2,048
5	64	64	<64	128	131,072	256
6	128	2,048	2,048	4,096	256	32,768

from immunized cynomolgus monkeys can be used instead of human sera as positive controls for the *Legionella* IFA, and it is possible that sera from cynomolgus monkeys immunized with other agents can be used in serological assays for other diseases for which human sera are difficult to obtain.

LITERATURE CITED

1. **Mohagheghpour, N., and C. A. Leone.** 1969. An immunological study of the relationships of non-human primates to man. Comp. Biochem. Physiol. **31:**437–452.
2. **Wilkinson, H. W.** 1980. Immune response to *Legionella pneumophila*, p. 500–503. *In* N. R. Rose and H. Friedman (ed.), Manual of clinical immunology, 2nd ed. American Society for Microbiology, Washington, D.C.
3. **Wilkinson, H. W., B. J. Fikes, and D. D. Cruce.** 1979. Indirect immunofluorescence test for serodiagnosis of Legionnaires disease: evidence for serogroup diversity of Legionnaires disease bacterial antigens and for multiple specificity of human antibodies. J. Clin. Microbiol. **9:**379–383.
4. **Wilkinson, H. W., A. L. Reingold, B. J. Brake, D. J. McGiboney, G. W. Gorman, and C. V. Broome.** 1983. Reactivity of serum from patients with suspected legionellosis against 29 antigens of Legionellaceae and *Legionella*-like organisms by indirect immunofluorescence assay. J. Infect. Dis. **147:**23–31.

ROUND TABLE DISCUSSION

Clinical Features and Laboratory Diagnosis of *Legionella* Infections

Moderator: P. H. EDELSTEIN

Participants: H. N. Beaty, R. J. Fallon, J. Fleurette, A. Macrae, P. Meenhorst, A. W. Pasculle, A. Saito, S. Toma, and W. C. Winn, Jr.

Do the various *Legionella* species cause different clinical diseases? It was the feeling of members of the panel, as well as participants in the audience, that there were no major differentiating features among infections caused by the different *Legionella* species.

Is the host spectrum of *Legionella* disease species specific? Although there has been very limited experience with non-*L. pneumophila Legionella* infections, it is likely that the host spectrum is not species specific. In other words, the non-*L. pneumophila Legionella* infections may occur in non-immunosuppressed patients.

How common is extrapulmonary *Legionella* infection? Members of the panel and audience pointed out many examples of extrapulmonary *Legionella* infection. Many people felt that extrapulmonary infection will be recognized more commonly in the future.

Can *Legionella* infection be distinguished clinically or radiographically from other pneumonic diseases in similar hosts? The consensus was that in general, especially in sporadic cases, there were no distinguishing features that could be relied upon. It was the feeling of many people, however, that in an epidemic situation patients with Legionnaires disease could be distinguished from patients with other nosocomial infections.

What is the role of combination antimicrobial therapy in the treatment of Legionnaires disease? Very few people have had experience with combination chemotherapy, but for the most part it is reserved for treatment of severely ill patients.

What drugs can be used instead of erythromycin? Again, there was little experience with treatment of Legionnaires disease with single agents other than erythromycin. However, several people related anecdotal experience with the successful use of the combination of doxycycline and rifampin. There were also anecdotal reports of successful use of high-dosage cotrimoxazole (sulfamethoxazole-trimethoprim) therapy.

How specific and sensitive are the diagnostic tests being used? This question could not be directly answered because few people have been able to study these diagnostic tests in epidemic situations or in patients with culture-confirmed disease. Regardless, some participants in the discussion felt that the formolized yolk sac antigen for indirect fluorescent-antibody testing was more specific than was the heat-killed antigen. This was more of an impression than a conclusion based on reliable data. It was pointed out that the major gap in our knowledge regarding the utility of diagnostic testing is the lack of studies in epidemic situations or with culture-confirmed cases.

What is the best way to diagnose *Legionella* infections? This was not a question that could be easily answered, in part because of the use of different diagnostic methodologies in different laboratories. It seemed evident that culture diagnosis was often very successful, but there were certainly strong advocates of serological diagnosis.

What can the small laboratory do to aid the physician in diagnosing this disease? Rapid diagnostic methodologies such as urinary antigen detection may be helpful in the small laboratory. It is possible that other diagnostic methods such as culture should be performed by the small laboratory.

SUMMARY

Clinical Features and Laboratory Diagnosis

SYDNEY M. FINEGOLD

Wadsworth Veterans Administration Medical Center and UCLA School of Medicine, Los Angeles, California 90073

In summarizing the clinical features and laboratory diagnosis of Legionnaires disease, I will present a brief summary of papers and posters presented at the symposium and then discuss additional points or areas that I feel need investigation.

CLINICAL FEATURES

Clinical picture. Bouza and Rodriguez described the clinical features of Legionnaires disease as seen in 45 cases diagnosed by them in Spain. Of their cases, 45% had neurological involvement and 14% had cavitary pulmonary disease. They also described one possible case of bacterial endocarditis. Miller and MacFarlane described the characteristics of sporadic cases of Legionnaires disease seen by them in Nottingham, England. They noted that it may be difficult to distinguish Legionnaires disease from other community-acquired pneumonias, particularly pneumococcal pneumonia with bacteremia. They found that radiographic progression of Legionnaires pneumonia and subsequent slow clearing were common. Extrapulmonary complications were more common in severely ill patients. Four of fifteen patients with Legionnaires disease requiring assisted ventilation died. Six patients requiring dialysis all had total recovery. Mayaud et al. studied 21 patients admitted to a respiratory intensive-care unit in France, half of whom were immunocompromised and half not. They concluded that the prognosis correlated more with the speed with which treatment was instituted than with the immunological status of the patient. It was clear that immunocompromised patients might still do relatively well if treatment was begun very early. Pastoris and colleagues described arrhythmias and possible myocarditis in three children who had significant rises in titers to *Legionella pneumophila* but had no evidence of pneumonia or respiratory infection. This was a retrospective study. The authors now plan to study patients prospectively and to attempt to obtain bacteriological proof of the disease. During the Round Table on clinical features and laboratory diagnosis, Dournon mentioned one case of peritonitis involving *L. pneumophila* and noted that the organism was found in the stool by direct fluorescent-antibody (DFA) testing in four cases of Legionnaires disease. He also described a suggestive case of bacterial endocarditis.

Additional points of interest not reported on are as follows. (i) Some of the newly described species of *Legionella* have been reported only from environmental sources. Are they also involved in human disease and if not, why not? (ii) Additional study is needed to further define the characteristics of the disease caused by various species other than *L. pneumophila*. Or are there really no clinical differences in the disease produced by various species, as was suggested in the Round Table discussion? There are reports suggesting that there may be significant differences. For example, *Legionella micdadei* infection has been noted to produce nodular infiltrates, often with a prominent pleural component. Recently a case has been described with recurrent hemorrhagic pleural effusion. (iii) We should be on the lookout for additional evidence for bacterial endocarditis involving legionellae, particularly since bacteremia seems to be not uncommon and since there is a report of a patient with an infected hemodialysis shunt, in addition to the possible cases of endocarditis reported at this symposium and cited above. (iv) Further studies are necessary to determine the incidence of "formes frustes" of Legionnaires disease. How often will one find pneumonia on chest film in the absence of respiratory symptoms or signs, as in the three cases reported by Beaty in which the only symptom was diarrhea but in which chest films showed pneumonia that was not suspected?

Radiographic features. MacFarlane and colleagues compared the radiographic features of Legionnaires disease with those of other pneumonias. They found that Legionnaires disease may be difficult to distinguish from other pneumonias, particularly severe pneumococcal pneumonia, and that residual lesions in Legionnaires disease were relatively common. Twenty-nine percent of patients had linear streaking opacities suggesting the possibility of fibrosis. This was also found in 26% of patients with severe pneumococcal pneumonia. The question of appropriateness of therapy and its influence on these findings remains to be analyzed.

An additional point of interest relates to the question mentioned earlier as to whether pneumonia caused by Legionnaires disease is really indistinguishable from other pneumonias, both radiologically and clinically, as suggested by the discussion during the Round Table. While it is certainly true that one could never definitively diagnose Legionnaires disease pneumonia by clinical or radiographic criteria, it is my feeling that many cases of Legionnaires disease are very distinctive even in the nonepidemic setting. Important clues suggesting Legionnaires disease include multisystem involvement (central nervous system, gastrointestinal system, liver, kidney, heart, and skeletal muscle involvement in addition to pneumonia), the usual lack of sputum or, when small volumes of sputum are produced, a relative absence of host and bacterial cells in this material, temperature-pulse deficit, low serum phosphorus, and lack of response to beta-lactam and aminoglycoside agents.

Incidence of *Legionella* and *Legionella*-like organisms in pneumonia. Renner et al. studied the incidence of these organisms in fatal pneumonia. Their study was a retrospective analysis based on examination of lungs from 1,100 consecutive autopsies. *Legionella* infections were found in 83 subjects, and *Legionella*-like organisms were found in another 65. In all, there were more of these organisms in fatal pneumonia than there were fungi, acid-fast organisms, or *Pneumocystis carinii*.

Additional incidence data based on prospective studies of pneumonia of various severity are certainly needed. It would be desirable for this information to be based on well-documented cases, ideally with isolation of the organism.

Therapy and prognosis. As noted above, the study by Mayaud et al. noted a relatively good prognosis in immunocompromised patients when therapy was started early. Two studies by Ristuccia and colleagues described the in vitro activity of several drugs against legionellae and included data on bactericidal activity and synergism. How reliable this type of data is in predicting clinical response is not known. Of particular interest was a study reported by Baskerville et al. of guinea pigs infected with legionellae by the aerosol route. These workers noted that erythromycin led to good survival but that pathological lesions persisted, as did large numbers of organisms. Rifampin also eliminated the lesions and organisms.

Additional points of interest not reported on or not resolved are as follows. (i) Much more work is needed with regard to therapy. We need studies with immunomodulating drugs as they become available. We need to define better the effectiveness of co-trimoxazole, rifampin, and tetracyclines. Are there important differences among tetracyclines? What are the indications for combination therapy? Which combinations are most effective? Are they additive or synergistic? Are they ever antagonistic? We must keep in mind that there is some risk involved in using several drugs in an immunocompromised host; this would increase the risk of superinfection with resistant organisms such as fungi. Should therapy be different for different species? These points will all require prospective studies of a comparative nature in an outbreak situation. (ii) How often do true relapses occur and under what circumstances? How often is there a second infection with the same or another serotype or species? Is a different subtype involved if the serotype is the same? (iii) Many more data are needed regarding sequelae of Legionnaires disease, such as restrictive lung disease, permanent neurological damage, etc.

Miscellaneous. Information is needed on the following points. (i) Is there a carrier state and if so, where? How do we account for the placental isolate with no associated disease in mother or infant that was described some years ago? (ii) Why do apparently like strains sometimes produce severe pneumonia and sometimes produce Pontiac fever? Of particular interest is the report by Beaty and colleagues of two workers exposed when a cooling tower was accidentally turned on while they were working in it. One developed severe Legionnaires pneumonia, and the other developed Pontiac fever.

LABORATORY DIAGNOSIS

Direct demonstration of organisms in specimens. In the paper referred to above, Bouza and Rodriguez demonstrated *Legionella* organisms in the brain at autopsy in one patient and in cerebrospinal fluid in another patient on two occasions. Brown and colleagues noted, in a study of lung specimens, that indirect fluorescent-antibody (IFA) testing gave better results than DFA testing for detection of legionellae. The IFA procedure requires only one reagent, a panvalent antiserum. This permits screening for a large number of organisms in a single test. Reingold et al. presented data on the incidence of positive DFAs in various unselected and selected specimens submitted to the Centers for Disease Control. In the group of unselected specimens, 6% were positive for legionellae. Positivity rates, by type of specimen, were 14% for lung, 5% for expectorated sputum, and 3% for pleural fluid. Durham et al. found that commercially available reagents from sources other than the Centers for Disease Control were satisfactory only 56% of the time.

Additional points of interest not reported on

are as follows. (i) What is the optimum number of specimens of various types to study by the DFA procedure? (ii) Many more data are needed on the sensitivity and specificity of the DFA test, especially with regard to newer serogroups and species and pools of these. (iii) Can we concentrate respiratory tract secretions so as to increase the sensitivity of the DFA test?

Demonstration of antigen in urine and other specimens. Guillet et al. reported on the use of monoclonal antibody in an enzyme-linked immunosorbent assay to detect the organism in sputum and other materials. There was no comparison with the DFA on clinical samples, nor were there any specific data on the performance of their test. Kohler et al. studied cross-reactions between *L. pneumophila* and various organisms by a radioimmunoassay technique as used to detect antigen in the urine. They obtained strong cross-reactions with some strains of a few *Pseudomonas* species. Low-level cross-reactivity was noted between *L. pneumophila* serogroups 2 to 6 and *Legionella dumoffii*. Some protein A-related immunological nonspecific activity occurred with some strains of *Staphylococcus aureus*. Data reported previously by Kohler's group and mentioned by Kohler in the Round Table showed that the latex particle agglutination test for antigen in the urine is 80% as sensitive as the radioimmunoassay and enzyme immunoassay for detection of antigen in the urine.

Additional points of interest are as follows. (i) How often can *Legionella* antigen be detected in cerebrospinal fluid? (ii) Can *Legionella* endotoxin be demonstrated in the central nervous system? (iii) Would concentration of body fluids such as spinal fluid increase the yield of or permit earlier antigen detection? (iv) Commercial availability of the simple latex particle agglutination antigen detection technique, even with its somewhat reduced sensitivity, would be extremely helpful for early diagnosis of Legionnaires disease.

Culture of the organism. Keathley and Winn compared seven culture media for effectiveness, using clinical specimens (one sputum and eight lung samples). Media containing alpha-ketoglutarate gave more positive cultures, larger amounts of growth, and more rapid growth than did media without this substance. Media without charcoal were poor. It was encouraging that alpha-ketoglutarate-containing buffered charcoal-yeast extract medium from one commercial source was as good as that made in-house by these workers. Barbaree et al. found that albumin enhanced the growth of *L. micdadei, L. dumoffii,* and *L. bozemanii.* MacLeod and colleagues studied the ability of *Francisella* medium to support the growth of legionellae and vice versa, since both of these organisms are fastidious and require cysteine. They found that a single medium could be used to grow both types of organism. Pierre et al. studied the use of the protected bronchial brush to obtain specimens for culture. However, there were no specimens taken without the brush for comparative purposes. Groothuis and colleagues studied the use of monospecific sera for serogrouping of isolates by the microagglutination technique. This proved to be more specific than the fluorescing antisera used by the Centers for Disease Control for this purpose.

Additional points of interest are as follows. (i) What is the sensitivity of various culture procedures? Presumably the specificity is essentially 100%. There was mention of one positive throat culture, but this was in a patient with Legionnaires disease and may have represented contamination of the throat with sputum. (ii) Which of the available media should laboratories use routinely? (iii) What is the optimum number of specimens of various types to culture? (iv) Ways are needed to expedite the growth of *Legionella* spp., preferably to the point that organisms would routinely grow in 1 day from clinical specimens. (v) We need further improvements in selective media. (vi) Are egg yolk sac inoculation and intraperitoneal injection of guinea pigs of value in picking up species of *Legionella* other than *L. pneumophila*, as compared with cultural techniques? (vii) What is the incidence of bacteremia? We need a better blood culture medium, particularly since growth in the present one is very slow. (viii) What is the quantitative level of bacteremia (number of organisms per milliliter)? (ix) To what extent are various ingredients of selective media inhibitory to newer species of *Legionella*? For example, it is known that vancomycin is less inhibitory for *L. micdadei* than is cefamandole. (x) How often would "sterile" abscesses yield *Legionella* organisms if cultured for them? A recent report described legionellae in a perirectal abscess, along with a number of anaerobes. (xi) What is the sensitivity of various culture media and methods for various species of *Legionella*?

Demonstration of antibody. Durham et al., as noted above in connection with the DFA procedure, evaluated the reliability of commercially available reagents (from sources other than the Centers for Disease Control) and found them unsatisfactory. In the case of the IFA procedure, only 39% of these reagents were satisfactory. Nowicki and colleagues compared the stability of three antigen preparations used in the IFA procedure. The heat-inactivated antigen from growth on charcoal-yeast extract suspended in phosphate-buffered saline (pH 7.2) containing 3% yolk sac material was the most satisfactory in terms of stability and simplicity of preparation. Brake et al. found that they could produce satisfactory control sera in monkeys for serological tests for Legionnaires disease that detect immunoglobulin G. Kallings and Blomqvist studied the reliability of the IFA procedure in patients with pneumonia. Their data indicated the possibility of nonspecific stimulation of production of antibody to *L. pneumophila* in some patients with other types of infection. These, of course, could have represented dual infections or cross-reactions, as the authors noted. Carter and Carter found that determination of titers of immunoglobulin M and immunoglobulin G antibodies in the IFA procedure permitted early presumptive diagnosis in about 90% of patients studied. Data were not presented on just how early (from the time of onset of the disease) the diagnosis could be made presumptively. Guillet et al., in a study mentioned above, used monoclonal antibody in an enzyme-linked immunosorbent assay to detect antibody to *L. pneumophila*. As with the direct demonstration of the organism, no data or comparisons were given. Harrison and Taylor reported on a rapid microagglutination test for early diagnosis of Legionnaires disease. A total of 97% of control subjects had a titer of <8, 41% of patients had a titer of >8 by the day after admission to the hospital, and 75% had such a titer by day 5 of hospitalization (the time from onset of the disease is not given). Taking a second serum specimen early often permitted early detection of a greater-than-fourfold titer rise (40% were diagnosed by hospital day 7 and 88% by hospital day 14). Collins et al. correlated the microagglutination and IFA titers for 10 *Legionella* antigens among 273 pneumonia patients suspected of having Legionnaires disease. They concluded that a standardized microagglutination test provides a rapid, simple, reliable, and economical one-step test for serodiagnosis of Legionnaires disease. Taylor and colleagues compared the microagglutination test and IFA by using a formalinized yolk sac antigen and sera from 33 proven cases (21 proven by culture and 12 proven by DFA). The sensitivity of the two tests was comparable. The microagglutination test is semiautomatable and rapid and detects primarily immunoglobulin M. Rodgers and Laverick presented a preliminary report of a passive hemagglutination test with chicken erythrocytes and a flagellar *Legionella* antigen. There were some problems with specificity.

Additional points of interest are as follows. (i)

We have good data on the sensitivity and specificity of serological tests with regard to *L. pneumophila* serogroup 1. More correlation is needed with definitively diagnosed cases (especially cases diagnosed by culture) to determine the sensitivity and specificity of serological testing, particularly with newer species and serogroups and newer procedures. (ii) Are other methods of antigen preparation better than those in common use? (iii) The micro-IFA test saves reagents. Is it comparable in all ways to the standard IFA? (iv) What is the relative value of the enzyme-linked immunosorbent assay technique as compared with IFA and microagglutination? (v) To what extent may there be cross-reactivity between *Legionella* spp. and other agents of acute bacterial pneumonia?

General. (i) Can we standardize reagents in tests so that we can compare results more readily from one laboratory to another and from one country to another? (ii) More data are needed on the relative value, for both DFA and culture, of various respiratory tract specimens such as expectorated sputum, induced sputum, percutaneous lung aspirate, transtracheal aspirate, bronchial washings (obtained with and without the aid of a protected double-lumen catheter), bronchial brushings, transbronchial biopsy, and open lung biopsy. (iii) More studies are needed on the persistence of positive DFA tests and positive cultures during therapy and the significance, if any, of these findings in terms of the cure rate and incidence of sequelae. (iv) Can we find one or more genus-specific antigens which would eliminate the problem of multiple antigenic types with little cross-reactivity? If so, can we develop reagents of high specificity and high sensitivity with these?

Finally, there was an exciting but frustrating report by Kohne on a DNA probe which could detect as few as 400 *Legionella* organisms in studies done to date (and which theoretically could detect a single organism) and which is said to be capable of detecting all 23 species of *Legionella*. This probe could identify the specific species involved in a particular patient and quantitate the number of organisms. However, since a patent is being applied for, no methodology and very little data were given. Nevertheless, studies to date with this system done by Brenner have given favorable results. We all await the published data and general availability of this procedure, which should revolutionize not only the diagnosis of Legionnaires disease but that of infectious diseases in general.

MICROBIOLOGY

STATE OF THE ART LECTURE

Classification of Legionellae

DON J. BRENNER

Center for Infectious Diseases, Centers for Disease Control, Atlanta, Georgia 30333

The First International Symposium on Legionnaires Disease was held in November 1978. It was on that occasion that the formal classification of legionellae was initiated with the introduction of the family name *Legionellaceae*, the genus *Legionella*, and the single species *Legionella pneumophila* (6). *L. pneumophila* had been isolated as early as 1947 (19), and four serogroups of *L. pneumophila* had been encountered (22).

Substantial progress was made in the 4.5 years between the first symposium and the second, which was held in June 1983. There are now 8 described *L. pneumophila* serogroups (1, 3, 12, 22, 23) and 10 described species (5, 6, 10, 11, 15, 21, 24, 26; L. A. Herwaldt et al., submitted for publication), 1 of which, *Legionella longbeachae*, contains 2 serogroups (22). Work in progress indicates the existence of 13 additional *Legionella* species.

Legionellae are presumptively identified at the genus level by their ability to grow on buffered charcoal-yeast extract agar or on other media that contain cysteine and iron salts, but not on media that lack these ingredients. Confirmation of legionellae at the genus level is obtained biochemically and by a distinctive pattern of predominantly branched-chain fatty acids, but these methods are not sensitive enough for identification at the species level. Identification of species is done serologically. Strains that are negative in reactions with all *Legionella* antisera may represent new serogroups within an existing species or a new species. DNA hybridization is the only method by which one can unequivocally choose between these alternatives. DNA relatedness, as determined by hybridization, is the definitive means by which new *Legionella* species and serogroups are classified. All of the species and serogroups cited above were classified in my laboratory. In this "state-of-the-art" presentation, I will review our previous DNA relatedness studies, present our current work on additional *Legionella* species, and discuss these data with respect to the classification of legionellae.

The sources, isolation, and cultivation of strains have been presented for all of the described *Legionella* species and for the other species with which *L. pneumophila* was compared (2, 5–8, 10, 11, 15, 19, 21, 24, 26; Herwaldt et al., submitted for publication). The methods used for phenotypic characterization and for DNA hybridization as assayed on hydroxyapatite are also given in these references. Strains of *Legionella*-like organisms that have not yet been described in the literature were isolated by G. W. Gorman, Centers for Disease Control, or were sent to the Centers for Disease Control for identification and provided to us by W. F. Bibb. The DNA hybridization experiments were done by A. G. Steigerwalt, Centers for Disease Control.

So-called Legionnaires disease bacterium (LDB) strains were first isolated by methods used for rickettsiae (20). Conjecture that the LDB might be similar or identical to *Rochalimaea (Rickettsia) quintana* was dispelled by DNA rate-of-reassociation experiments that showed LDB to have a genome size of about 2.5 $\times$ 10^9 daltons, similar to those of many bacteria and twice that of *R. quintana* (8). LDB DNA had a guanine-plus-cytosine content of 39 mol% (8). Its DNA was tested for relatedness to a large number of bacteria, including those with similar phenotypic characteristics and a similar guanine-plus-cytosine content (6–8). It showed little, if any, total DNA relatedness to other bacteria (Table 1).

DNAs of LDB strains from outbreak-associated and sporadic cases of pneumonia (Legionnaires disease), from mild, self-limiting, acute febrile disease (Pontiac fever), and from one isolate obtained in 1947 were all 70% or more related and were part of a single species which was named *L. pneumophila* (6, 8, 19). These data and data confirming the existence of eight serogroups in *L. pneumophila* (1, 3, 6, 12, 22, 23) are shown in Table 2.

Legionella-like organisms were defined as "gram-negative bacteria that do not meet the definition of *L. pneumophila*, but will grow on at least one agar medium designed for the growth of *L. pneumophila* and fail to grow on other commonly used laboratory media" (5). Each newly isolated *Legionella*-like organism was

TABLE 1. DNA relatedness of LDB strain Philadelphia 1 to other bacteria[a]

Species	RBR[b] to LDB strain Philadelphia 1
Escherichia coli	3
Proteus mirabilis	2
Edwardsiella tarda..................	2
Yersinia enterocolitica	0
Yersinia intermedia	0
Yersinia frederiksenii................	4
Yersinia kristensenii	0
Yersinia pseudotuberculosis	0
Francisella tularensis	4
Vibrio cholerae.....................	0
Aeromonas hydrophila	0
Flavobacterium meningosepticum	2
Flavobacterium group IIB	3
Pasteurella multocida................	3
Pasteurella ureae...................	7
Pasteurella haemolytica..............	2
Pasteurella piscicida.................	1
Pasteurella bettii	1
Pasteurella BL group	2
Pasteurella sp. strain NCTC 10699	0
Actinobacillus seminus...............	0
Actinobacillus equuli................	0
Actinobacillus actinomycetumcomitans	1
Haemophilus influenzae	1
Haemophilus paragallinarum	1
Haemophilus parahaemolyticus	0
Haemophilus paraphrophilus	1
Cytophaga johnsonae.................	0
Cytophaga sp. strain CYPE	0
Cytophaga sp. strain CYH3	0
Cytophaga sp. strain FK17	0
Cytophaga sp. strain H72	0
Cytophaga sp. strain H106	0
Flexibacter canadensis...............	0
Microcyclus major..................	3
Rochalimaea quintana	5
Kingella kingii	17[c]

[a] Data are taken from references 6–8. Assays were done on hydroxyapatite. DNAs were incubated at 60°C.

[b] RBR, Relative binding index = (percent DNA bound to hydroxyapatite in heterologous reactions/percent DNA bound in homologous reactions) × 100.

[c] Considered insignificant relatedness for reasons given by Herwaldt et al. (in press).

tested for DNA relatedness to *L. pneumophila* to determine whether it represented a new *Legionella* species. As other species were described, *Legionella*-like organisms were tested for relatedness to all known species. As a result of these studies, nine additional *Legionella* species were described and classified between 1980 and 1983.

Legionella bozemanii was first isolated by Bozeman et al. in 1959 from embryonated hen eggs inoculated with brain tissue from a guinea pig that had been inoculated with lung tissue

from a patient who died of pneumonia (4). A second strain of *L. bozemanii* was reported by Thomason et al. in 1979 from a fatal case of pneumonia (29). The organism was classified and named by Brenner et al. (5). The third species, *Legionella micdadei*, was first isolated from embryonated eggs infected with tissue from a guinea pig that had been inoculated with human blood in 1943 (28) and was subsequently isolated from patients with pneumonia (14, 16, 25, 27). *L. micdadei* is commonly referred to as the Pittsburgh pneumonia agent (25), although the disease it causes is a pneumonia that is clinically indistinguishable from that caused by other legionellae. *L. micdadei* was characterized and named in 1980 (15). The fourth species, *Legionella dumoffii* (5), was isolated from a patient with pneumonia (18) and from water. The fifth species, *Legionella gormanii*, was first classified on the basis of a single strain isolated from soil on a creek bank (24). Serological evidence has presumptively implicated *L. gormanii* as a cause of human pneumonia, and additional strains have been isolated from the environment. *Legionella longbeachae*, the sixth species, was isolated from human lung tissue and transtracheal aspirates (21). A second sero-

TABLE 2. DNA relatedness among strains of *L. pneumophila*[a]

Source of unlabeled DNA[b]	RBR[c] to strain Philadelphia 1 at:	
	60°C	75°C
Philadelphia 1 (outbreak)	100	100
Philadelphia 2 (outbreak)	96	ND[d]
Philadelphia 3 (outbreak)	94	90
Philadelphia 4 (outbreak)	93	100
Pontiac 1 (outbreak, Pontiac fever)	82	77
Flint 1 (sporadic)	89	74
Flint 2 (sporadic)	98	ND
Knoxville 1 (sporadic)	87	82
Burlington 1 (outbreak)	86	93
Bellingham 1 (sporadic)	81	82
Albuquerque 1 (sporadic)	97	77
Berkeley 1 (sporadic)	97	ND
OLDA (1947 isolate)	90	81
Bloomington 1 (sporadic)	80	71
Bloomington 2 (serogroup 3)	84	81
Togus 1 (serogroup 2)	85	ND
Los Angeles 1 (serogroup 4)	71	52
Dallas 10 (serogroup 5)	73	81
Chicago 2 (serogroup 6)	79	81
SC-74-C5 (serogroup 7)	74	72
Concord 3 (serogroup 8)	93	ND

[a] Data are taken from references 1, 3, 6, 12, 22, and 23. Reactions were assayed on hydroxyapatite.

[b] All strains are *L. pneumophila*. The type strain of *L. pneumophila* is Philadelphia 1.

[c] RBR is defined in Table 1.

[d] ND, Not done.

group of *L. longbeachae* has been reported (2). The seventh and eighth species, *Legionella jordanis* (10) and *Legionella oakridgensis* (26) were isolated from water. *L. jordanis*, but not *L. oakridgensis*, was serologically implicated as the cause of a human case of pneumonia. The ninth species, *Legionella wadsworthii*, contains a single strain isolated from a patient with pneumonia (11). The only strain of the tenth species, *Legionella feeleii*, was isolated from an oil-water cooling mixture in an automobile assembly plant in Canada (Herwaldt et al., submitted for publication). Serological evidence implicated this strain as the causative agent in an outbreak of Pontiac fever that occurred in the plant.

The phenotypic properties of these species are given in Table 3. It is impossible to identify most of the described species solely on the basis of these reactions. All species grew on buffered charcoal-yeast extract agar, but not on media that lacked cysteine. All contained predominantly branched-chain fatty acids, failed to ferment D-glucose, gave positive tests for catalase and gelatin liquefaction, and gave negative tests for urea and nitrate reduction. Colonies of *L. bozemanii*, *L. dumoffii*, and *L. gormanii* produced a blue-white autofluorescence under long-wavelength UV light. All species except *L. micdadei* and *L. wadsworthii* produced a soluble brown pigment on tyrosine-containing agar. All species except *L. micdadei* gave positive (sometimes weak) reactions in chromogenic assays for beta-lactamase production. All species except *L. oakridgensis* were motile by means of polar or lateral flagella. *L. pneumophila* and possibly *L. feeleii* were the only species to hydrolyze hippurate. *L. oakridgensis* was the only species to give a negative alkaline phosphatase test at pH 8.5. Some species had different-colored colonies on a dye-containing medium (13, 30). *L. pneumophila* colonies were white-green, *L. micdadei* colonies were blue-gray, and colonies of *L. bozemanii*, *L. dumoffii*, and *L. gormanii* were green. In another study, *L. wadsworthii* colonies were green, compared with turquoise-green for *L. pneumophila* and blue for *L. micdadei* (11).

The guanine-plus-cytosine contents of described *Legionella* species were 39 to 45 mol% (6, 9, 11, 13, 19, 26; A. G. Steigerwalt, personal communication). With the exception of one *L. bozemanii* strain, relatedness of DNAs from strains of any single *Legionella* species was 75 to 100% (5, 6, 10, 11, 15, 21, 24, 26) (Table 4). One *L. bozemanii* strain originally showed 56% relatedness to the type strain of *L. bozemanii* (5). It was subsequently shown to be more than 75% related to the *L. bozemanii* type strain (A. G. Steigerwalt, personal communication). DNA relatedness among the described species of *Legionella* was 0 to 46% (5, 6, 10, 11, 15, 21, 24, 26) (Table 4). *L. micdadei*, *L. oakridgensis*, and *L. feeleii* were 10% or less related to all other species. *L. jordanis* was most related (10 to 20%) to *L. longbeachae* and *L. pneumophila*. *L. pneumophila* and *L. longbeachae* were both most related (12 to 25%) to the fluorescent species, *L. bozemanii*, *L. dumoffii*, and *L. gormanii*. These three fluorescent species were 12 to 46% interre-

TABLE 3. Phenotypic characteristics of legionellae[a]

Species	Browning of YE agar with tyrosine	Hippurate hydrolysis	Motility	Oxidase	Beta-lactamase	Autofluorescence	Color on dye-containing media	Alkaline phosphatase, pH 8.5
L. pneumophila	+	+	+	+ or +/−[b]	+	−	White-green	+
L. bozemanii	+	−	+	+/−	+/−	+	Green	+
L. micdadei	−	−	+	+	−	−	Blue-gray	+
L. dumoffii	+[c]	−	+	−	+	+	Green	+
L. gormanii	+	−	+	−	+	+	Green	+
L. longbeachae	+	−	+	+	+/−	−	ND[d]	+
L. jordanis	+	−	+	+	+	−	ND	+
L. oakridgensis	+	−	−	−	+[e]	−	ND	−
L. wadsworthii	−	−	+	−	+	−	Green	ND
L. feeleii	+[e]	+/−	+	−	−	−	ND	ND

[a] Data are taken from references 5, 6, 10, 11, 13, 15, 21, 24, and 26, and from Herwaldt et al. (submitted for publication). All species were positive in tests for growth on buffered charcoal-yeast extract agar (much of the data in the literature describes growth on charcoal-yeast extract agar; we assume that all species grow on both media) and required cysteine for growth (*L. oakridgensis* strains adapt to grow in the absence of cysteine but require cysteine when first isolated). All species were negative in tests for growth on blood agar, Gram reaction, nitrate reduction, urea, and acid production from carbohydrates. YE, Charcoal-treated yeast extract agar.
[b] +/−, Weakly or not always positive.
[c] A strain of *L. dumoffii* did not produce a diffusible brown pigment (P. H. Edelstein, personal communication).
[d] ND, Not done.
[e] Weak reaction.

TABLE 4. DNA relatedness among described *Legionella* species

Species	% Range of DNA relatedness in reciprocal 60°C reactions[a]									
	L. pneu-mophila	*L. boze-manii*	*L. mic-dadei*	*L. du-moffii*	*L. gor-manii*	*L. long-beachae*	*L. jor-danis*	*L. oakridg-ensis*	*L. wads-worthii*	*L. feel-eii*
L. pneu-mophila	75–100	2–25	0–5	6–22	1–20	2–8	10–19	8	6	2
L. boze-manii		56–77	0–6	15–32	12–27	17	6–8	8	30	3
L. micdadei			90–100	0–9	6–8	4	6–9	6	2	6
L. dumoffii				90	24–46	12	4–9	7	22	3
L. gormanii					—[b]	25	7–8	6	25	5
L. long-beachae						97–100	5–17	5	8	5
L. jordanis							93	7	2	3
L. oakridg-ensis								98–100	4	3
L. wads-worthii									—[b]	4
L. feeleii										—[b]

[a] All reactions between strains of each species pair are included, regardless of the species from which labeled DNA was obtained. All reactions were done at least twice and assayed on hydroxyapatite. Data are taken from references 5, 6, 10, 11, 15, 21, 24, and 26, and from Herwaldt et al. (in press).

[b] Only one known strain at the time of test.

lated. Although not fluorescent, *L. wadsworthii* was 20 to 30% related to the fluorescent species.

DNA relatedness studies now underway in our laboratory indicate the existence of another 13, as yet unnamed, species of *Legionella*. All of these organisms exhibit the phenotypic characteristics of legionellae. They are gram negative, grow on buffered charcoal-yeast agar, but not blood agar, require cysteine for growth, fail to reduce nitrates, and do not degrade urea or produce acid from carbohydrates. All are motile, are catalase positive, and liquify gelatin. One of them hydrolyzes hippurate, and six of them produce colonies that autofluoresce. Two of the six new autofluorescent species exhibit red rather than blue-white autofluorescence. One of these species contains both autofluorescent and nonautofluorescent strains. They all contain branched-chain cell wall fatty acids. Eleven of the new species were isolated from the environment, and two were isolated from patients with pneumonia.

Guanine-plus-cytosine contents of DNAs from 10 of the 13 new *Legionella* species were 38 to 43 mol%. One species had a guanine-plus-cytosine content of 46 mol%, and the remaining two species had guanine-plus-cytosine contents of 51 and 52 mol%. Selected DNA relatedness data for 12 of the new species are shown in Table 5. With the new species included, the range of DNA relatedness among legionellae was 0 to 67%.

Another group has proposed the creation of additional genera in the family *Legionellaceae* on the basis of DNA relatedness and phenotypic characteristics (9, 13, 30). If we group the legionellae on the basis of DNA relatedness, we have *L. pneumophila*, *L. jordanis*, *L. oakridgensis*, *L. feeleii*, species 14, and species 17 in groups of their own (less than 25% relatedness to all other species). *L. micdadei* and species 13 were 23% related and could be put into the same group or into two separate groups. *L. longbeachae*, species 11, and species 22 formed a relatedness group at 37%. Species 15 and 16 formed a relatedness group at 60%. Species 18 was 34 and 20% related to species 15 and species 16, respectively, and therefore might be included with species 15 and species 16 or put into a relatedness group of its own, as it showed 20% relatedness to species in several groups (Table 5). *L. bozemanii*, *L. dumoffii*, *L. gormanii*, *L. wadsworthii*, species 12, species 19, species 20, and species 21 form a relatedness group at 12 to 67% (mainly 35% or higher).

Relying strictly on this interpretation of DNA relatedness data, we would have some 10 to 12 genera in *Legionellaceae*. We know of no way to identify legionellae phenotypically other than by serology or by DNA hybridization, neither of which is available to clinical laboratories. Consideration of the phenotypic parameters of these "generic" relatedness groups brings up additional problems. Other than *L. pneumophila*, very few strains (fewer than 10, and more often 1 to 5) of *Legionella* species have been examined. Even with this limited sample, we have begun to see phenotypic differences and inconsistencies within *Legionella* species and among species grouped by DNA relatedness. There is an *L.*

TABLE 5. DNA relatedness of new *Legionella* species[a]

Source of unlabeled DNA	% DNA relatedness of labeled DNA from unnamed *Legionella* species:												
	10	11	12	13	14	15	16	17	18	19	20	21	22
L. pneumophila	2			5							21		
L. bozemanii			52							65	51	46	
L. micdadei			5	23	1				16	5	18		
L. dumoffii			25					9	19	35	57	48	
L. gormanii			39		11				19	40	47	40	
L. longbeachae		37				2	2	3	15				38
L. jordanis		3				10			20			10	7
L. oakridgensis									22		6		
L. wadsworthii							12		5	37	21	34	
L. feeleii	100												
Legionella 11		91–94								21		33	64
Legionella 12			84–90							21	56	53	25
Legionella 13				100[b]							56		
Legionella 14					100			17					
Legionella 15				4		100	62		34				
Legionella 16	18					59	100		20				
Legionella 17					14			100					
Legionella 18									100				
Legionella 19										100			
Legionella 20										28	94–99	54	
Legionella 21											67	100	
Legionella 22													100

[a] Selected values are given to demonstrate high and low levels of DNA relatedness. All data were obtained from 60°C reactions assayed on hydroxyapatite.

[b] One hundred percent is arbitrarily used for homologous reactions in species where only one strain exists. For species 13 the two known strains are 100% related. Ranges of relatedness are shown for other species with more than one known strain.

dumoffii strain that does not produce a brown pigment on tyrosine-containing agar (P. H. Edelstein, personal communication); one of five strains of species 12 does not autofluoresce; *L. wadsworthii* does not autofluoresce, but is highly related to the species that exhibit blue-white autofluorescence. These exceptions are probably understated and should be kept in mind by anyone attempting to use "literature reactions" for the characterization of legionellae. We need additional tests and much more data on existing tests before attempting to meaningfully describe legionellae biochemically. One step in this direction is the use of isoprenoid quinones to group *Legionella* species (17).

There is a good working definition of a genetic species (70% or higher relatedness in optimal DNA hybridization reactions, 6% or less divergence in related DNA sequences, and 55% or more relatedness in stringent DNA relatedness reactions), but no such genetic definition exists for a genus. It has been our belief that a genus should consist of a group of phenotypically similar species that must be dealt with together at the microbiology bench to differentiate one from another. If genetic and phenotypic similarity do not agree, then a phenotypic genus is better than a genotypic genus. Certainly legionellae form a good phenotypic genus. They are all isolated by the same methods, have similar biochemical profiles, cause the same disease, and are sensitive to the same antibiotics. For these reasons we have considered it premature and of little, if any, benefit to create additional genera in *Legionellaceae*. Our results with the new *Legionella* species support the wisdom of maintaining the single-genus concept for *Legionellaceae*.

LITERATURE CITED

1. **Bibb, W. F., P. M. Arnow, D. L. Dellinger, and S. R. Perryman.** 1983. Isolation and characterization of a seventh serogroup of *Legionella pneumophila*. J. Clin. Microbiol. 17:346–348.
2. **Bibb, W. F., R. J. Sorg, B. M. Thomason, M. Hicklin, A. G. Steigerwalt, D. J. Brenner, and M. R. Wolf.** 1981. Recognition of a second serogroup of *Legionella longbeachae*. J. Clin. Microbiol. 14:674–677.
3. **Bissett, M. L., J. O. Lee, and D. S. Lindquist.** 1983. A new serogroup of *Legionella pneumophila*: serogroup 8. J. Clin. Microbiol. 17:887–891.
4. **Bozeman, F. M., J. W. Humphries, and J. M. Campbell.** 1968. A new group of rickettsia-like agents recovered from guinea pigs. Acta Virol. (Prague) 12:87–93.
5. **Brenner, D. J., A. G. Steigerwalt, G. W. Gorman, R. E. Weaver, J. C. Feeley, L. G. Cordes, H. W. Wilkinson, C. Patton, B. M. Thomason, and K. R. Lewallen Sasseville.** 1980. *Legionella bozemanii* sp. nov. and *Legionella dumoffii* sp. nov.: classification of two additional species of *Legionella* associated with human pneumonia. Curr. Microbiol. 4:111–116.

6. Brenner, D. J., A. G. Steigerwalt, and J. E. McDade. 1979. Classification of the Legionnaires' disease bacterium: *Legionella pneumophila*, genus novum, species nova of the family *Legionellaceae*, familia nova. Ann. Int. Med. **90**:656–658.

7. Brenner, D. J., A. G. Steigerwalt, S. Pohl, H. Behrens, W. Mannheim, and R. E. Weaver. 1981. Lack of relatedness of *Legionella pneumophila* to *Cytophagaceae*, "*Pasteurellaceae*," and *Kingella*. Int. J. Syst. Bacteriol. **31**:89–90.

8. Brenner, D. J., A. G. Steigerwalt, R. E. Weaver, J. E. McDade, J. C. Feeley, and M. Mandel. 1978. Classification of the Legionnaires' disease bacterium: an interim report. Curr. Microbiol. **1**:71–75.

9. Brown, A., G. M. Garrity, and R. M. Vickers. 1981. *Fluoribacter dumoffii* (Brenner et al.) comb. nov. and *Fluoribacter gormanii* (Morris et al.) comb. nov. Int. J. Syst. Bacteriol. **31**:111–115.

10. Cherry, W. B., G. W. Gorman, L. H. Orrison, C. W. Moss, A. G. Steigerwalt, H. W. Wilkinson, S. E. Johnson, R. M. McKinney, and D. J. Brenner. 1982. *Legionella jordanis*: a new species of *Legionella* isolated from water and sewage. J. Clin. Microbiol. **15**:290–297.

11. Edelstein, P. H., D. J. Brenner, C. W. Moss, A. G. Steigerwalt, E. M. Francis, and W. L. George. 1982. *Legionella wadsworthii* species nova: a cause of human pneumonia. Ann. Int. Med. **97**:809–813.

12. England, A. C., III, R. M. McKinney, P. Skaliy, and G. W. Gorman. 1980. A fifth serogroup of *Legionella pneumophila*. Ann. Int. Med. **93**:58–59.

13. Garrity, G. M., A. Brown, and R. M. Vickers. 1980. *Tatlockia* and *Fluoribacter*: two new genera of organisms resembling *Legionella pneumophila*. Int. J. Syst. Bacteriol. **30**:609–614.

14. Hébert, G. A., C. W. Moss, L. K. McDougal, F. M. Bozeman, R. M. McKinney, and D. J. Brenner. 1980. The rickettsia-like organisms TATLOCK (1943) and HEBA (1959): bacteria phenotypically similar to but genetically distinct from *Legionella pneumophila* and the WIGA bacterium. Ann. Int. Med. **92**:45–52.

15. Hébert, G. A., A. G. Steigerwalt, and D. J. Brenner. 1980. *Legionella micdadei* species nova: classification of a third species of *Legionella* associated with human pneumonia. Curr. Microbiol. **3**:255–257.

16. Hébert, G. A., B. M. Thomason, P. P. Harris, M. D. Hicklin, and R. M. McKinney. 1980. "Pittsburgh pneumonia agent": a bacterium phenotypically similar to *Legionella pneumophila* and identical to the TATLOCK bacterium. Ann. Int. Med. **92**:53–54.

17. Karr, D. E., W. F. Bibb, and C. W. Moss. 1982. Isoprenoid quinones of the genus *Legionella*. J. Clin. Microbiol. **15**:1044–1047.

18. Lewallen, K. R., R. M. McKinney, D. J. Brenner, C. W. Moss, D. H. Dail, B. M. Thomason, and R. A. Bright. 1979. A newly identified bacterium phenotypically resembling, but genetically distinct from, *Legionella pneumophila*: an isolate in a case of pneumonia. Ann. Int. Med. **91**:831–834.

19. McDade, J. E., D. J. Brenner, and F. M. Bozeman. 1979. Legionnaires' disease bacterium isolated in 1947. Ann. Int. Med. **90**:659–661.

20. McDade, J. E., C. C. Shepard, D. W. Fraser, T. R. Tsai, M. A. Redus, W. R. Dowdle, and the Laboratory Investigation Team. 1977. Legionnaires' disease: isolation of a bacterium and demonstration of its role in other respiratory disease. N. Engl. J. Med. **297**:1197–1203.

21. McKinney, R. M., R. K. Porschen, P. H. Edelstein, M. L. Bissett, P. P. Harris, S. P. Bondell, A. G. Steigerwalt, R. E. Weaver, M. E. Ein, D. S. Lindquist, R. S. Kops, and D. J. Brenner. 1981. *Legionella longbeachae* sp. nov., another etiologic agent of human pneumonia. Ann. Int. Med. **94**:739–743.

22. McKinney, R. M., L. Thacker, P. P. Harris, K. R. Lewallen, G. A. Hébert, P. H. Edelstein, and B. M. Thomason. 1979. Four serogroups of Legionnaires' disease bacteria defined by immunofluorescence. Ann. Int. Med. **90**:621–624.

23. McKinney, R. M., H. W. Wilkinson, H. W. Sommers, B. J. Fikes, K. R. Sasseville, M. M. Yungbluth, and J. S. Wolf. 1980. *Legionella pneumophila* serogroup six: isolation from cases of legionellosis, identification by immunofluorescence staining, and immunological response to infection. J. Clin. Microbiol. **12**:395–401.

24. Morris, G. K., A. G. Steigerwalt, J. C. Feeley, E. S. Wong, W. T. Martin, C. M. Patton, and D. J. Brenner. 1980. *Legionella gormanii* sp. nov. J. Clin. Microbiol. **12**:718–721.

25. Myerowitz, R. L., A. W. Pasculle, J. M. Dowling, G. J. Pazin, M. Puerzer, R. B. Yee, C. R. Rinaldo, and T. R. Hakala. 1979. Opportunistic lung infection due to "Pittsburgh pneumonia agent." N. Engl. J. Med. **301**:953–958.

26. Orrison, L. H., W. B. Cherry, R. L. Tyndall, C. B. Fliermans, S. B. Gough, M. A. Lambert, W. F. Bibb, L. K. McDougal, and D. J. Brenner. 1983. *Legionella oakridgensis*: unusual new species isolated from cooling tower water. Appl. Environ. Microbiol. **45**:536–545.

27. Pasculle, A. W., R. L. Myerowitz, and C. R. Rinaldo. 1979. New bacterial agent of pneumonia isolated from renal transplant patients. Lancet **ii**:58–61.

28. Tatlock, H. 1944. A rickettsia-like organism recovered from guinea pigs. Proc. Soc. Exp. Biol. Med. **57**:95–99.

29. Thomason, B. M., P. P. Harris, M. D. Hicklin, J. A. Blackmon, C. W. Moss, and F. Matthews. 1979. *Legionella*-like bacterium related to WIGA in a fatal case of pneumonia. Ann. Int. Med. **91**:673–676.

30. Vickers, R. M., A. Brown, and G. M. Garrity. 1981. Dye-containing buffered charcoal-yeast extract medium for differentiation of members of the family *Legionellaceae*. J. Clin. Microbiol. **13**:380–382.

STATE OF THE ART LECTURE

Bacterial Physiology

PAUL HOFFMAN

Department of Microbiology and Immunology, University of Tennessee Center for Health Sciences, Memphis, Tennessee 38163

Modest advances in the area of physiology and metabolism of the legionellae have been made since the First International Symposium on Legionnaires Disease in 1978. Many problems still remain unresolved. While characterization of the various species of a new genus may not be as creative or glamorous as other endeavors of microbiology, this information is desperately needed for taxonomic and diagnostic purposes. Most of our knowledge of physiology and metabolism has been gained from studies of the type species, *Legionella pneumophila*. Now that there are 23 species (Brenner, this volume), the task of determining biochemical similarities and differences at the species level is even more difficult. In this chapter, I will attempt to summarize major aspects of the physiology and metabolism of the legionellae as they pertain to growth and metabolic activities. Since this is an overview, not all aspects of physiology and metabolism will be discussed, and I must emphasize that much more has been accomplished than can be discussed herein.

Morphology and ultrastructure. The legionellae are gram-negative rods (2 to 6 μm in length) which sometimes form filaments, depending on stage of growth and type of medium employed. The cell wall has been characterized and consists of an outer membrane, a thin peptidoglycan (PG) layer, and a cytoplasmic membrane. The peptidoglycan contains *m*-diaminopimelic acid and is believed to be highly branched. Amano and Williams found the peptidoglycan to be sensitive to lysozyme, but not to alkali, and that 80 to 90% of the *m*-diaminopimelic acid was cross-linked (1). The total cellular fatty acid content of legionellae has been characterized and was found to contain a high proportion of branched-chain fatty acids (19, 20). These include iso and anti-iso monounsaturated and saturated fatty acids as well as a C_{17} cyclopropane-terminated fatty acid. Mono- and dihydroxy fatty acids have also been detected in the legionellae (2 to 3% of the total fatty acid) and are likely bound to peptidoglycan (18). Mayberry proposed that these fatty acids may be β-hydroxyisomyristic acid and β-hydroxyarachidic

acid (18). Ketooctanoic acid, but not β-hydroxymyristic acid, has been detected in the outer membrane (8), but information on outer membrane structure, including protein composition and membrane function, must await further experimentation.

While flagella have been observed by light (flagella stain) and electron microscopy in all but one species, generally these organisms are only sluggishly motile to nonmotile in wet mounts.

A brown, melanin-like pigment was initially reported as characteristic of most legionellae, and its production occurs in late-log-phase cultures. Tyrosine enhances pigment production (2). Fluorescent pigments are produced by some species. These can be seen with light in the near-UV range, and the fluorescence is dull yellow or blue-white and may be species specific.

Nutrition. Amino acids are the primary sources of carbon and energy for the legionellae (10, 22, 25, 26, 29). These organisms are proteolytic, hydrolyze gelatin and serum proteins, and have chymotrypsin-like activity (3, 28). Several groups have clearly demonstrated that the legionellae can be successfully cultured in a chemically defined medium composed of amino acids (10, 22, 23, 25, 26). Purines, pyrimidines, vitamins, and other growth factors are not required. While there is some disagreement regarding the precise amino acid requirements, most workers agree that arginine, threonine, methionine, serine, isoleucine, leucine, valine, and cysteine are required. Proline, glutamate, and phenylalanine or tyrosine are helpful (10, 25, 26). Tesh and Miller reported that good growth of *L. pneumophila* could be obtained in medium containing the eight required amino acids plus glutamate (26). Glutathione, but not cystine, will substitute for cysteine (22).

Initial studies on the nutrition of the legionellae indicated that carbohydrates were not fermented. Pine et al. (22) later reported that a small amount of lactic acid was produced from D-glucose, but that less than 1% of the glucose was utilized. Since the legionellae are strict aerobes, oxidation of glucose rather than fermentation would be the more likely route. In this regard, Weiss et al (30) reported that D-glucose was slowly oxidized as evidenced by $^{14}CO_2$ release from whole cells. They concluded that glucose oxidation was very low relative to glutamate oxidation. Radiorespirometric studies by Weiss that were later confirmed by Tesh et al. (27) suggested that glucose was catabolized by the pentose pathway and by the Entner-Doudoroff pathway, but not by the Embden-Meyerhoff pathway (EMP). In my laboratory, we examined the kinetics of D-glucose uptake in legionellae and the effect of uncoupling agents on glucose transport. Using a membrane filter technique, we also observed low glucose uptake over a 30-min period, but results obtained at various carrier substrate concentrations suggested that glucose uptake was nonenzymatic and most likely due to diffusion. In view of our observations, I would suggest that conclusions based on radiotracer experiments with glucose be interpreted with caution. While there may be slight glucose metabolism by the legionellae, no one has been able to demonstrate that carbohydrates enhance either the growth rate or the growth yield. At this point, until someone can clearly demonstrate an advantage for inclusion of carbohydrates in culture media, they should be left out.

Reports that α-ketoglutarate (α-KG) and pyruvate stimulate or enhance growth are generally true in some media but not in others. There is no question that α-KG enhances growth in charcoal-yeast extract or buffered charcoal-yeast extract medium. However, in some chemically defined media, pyruvate and α-KG are inhibitory (25, 29). Pyruvate is a good energy source for legionellae, but it is also a key regulatory molecule which might be inhibitory at high concentrations. α-KG is not readily oxidized by whole cells, despite a very active α-KG dehydrogenase (see below), suggesting a transport problem. Keto acids react with cysteine as well as with peroxide, and so the stimulatory effect of these compounds may not necessarily be nutritional.

Trace metal requirements have been worked out by Reeves et al. (23), and a requirement for magnesium and potassium has been demonstrated. The legionellae require the following metals: Fe, Cu, Mg, Mn, Co, Ca, Mo, Ni, V, and Zn. Reeves et al. (24) recently reported the absence of ferric iron-binding siderophores in *Legionella* species grown under a variety of iron-limiting conditions. Phenolate and hydroxamate ferric iron-binding compounds were not detected by chemical or microbiological assays. Quinn and Weinberg (this volume) reported that the legionellae are unusually susceptible to ferric iron-binding compounds such as transferrin, phenanthroline, desferoxamine, and enterobactin. My group has found an iron reductase (5 nmol/min per mg of protein) in extracts of *L. pneumophila*, but this enzyme is common in aerobic microorganisms.

pH and temperature. Generally, the legionellae grow better if the pH is less than 7. Optimal growth in buffered charcoal-yeast extract medium occurs at pH 6.9, but many other media, including chemically defined media, reportedly have optimal pH values of between 6.0 and 6.5 (10, 23, 25).

Most clinical isolates exhibit optimal growth between 35 and 37°C. Environmental isolates show optimal growth around 30°C. Fliermans and others have reported isolation of legionellae from water in the range of 50 to 65°C (7).

Metabolic pathways. (i) Krebs cycle. The primary route of carbon assimilation and energy production in *L. pneumophila* is via the Krebs cycle (12). The primary energy sources are glutamate, serine, glutamine, and threonine (10, 22, 25, 26). Other amino acids are oxidized to a lesser extent. One assumes that the amino acids are translocated into the cells and deaminated. My group could not verify the presence of L-amino acid oxidases in *L. pneumophila*, but transaminases or deaminases are likely to be involved.

My group in collaboration with Leo Pine has shown that the Krebs cycle is complete in *L. pneumophila*. However, the glyoxylate bypass (isocitrate lyase and malate synthase) was absent or marginal in cell-free extracts of *L. pneumophila*. These activities were present in control organisms. The specific activities for the various Krebs cycle enzymes are presented in Table 1. High activities were observed for malate dehydrogenase, isocitrate dehydrogenase, α-KG dehydrogenase, and citrate synthase. Despite the high specific activity for isocitrate

TABLE 1. Activities of Krebs cycle enzymes in crude cell-free extracts of *L. pneumophila* at 37°C

Enzyme	Sp act (nmol/min per mg of protein)
Succinate dehydrogenase	30
Malate dehydrogenase (NAD)	100
Malate dehydrogenase (NADP)	40
α-KG dehydrogenase	19.8
Isocitrate dehydrogenase (NAD)	1.0
Isocitrate dehydrogenase (NADP) . . .	220
Pyruvate dehydrogenase	1.64
Aconitate hydratase	275
Fumarate hydratase	8.5
Citrate synthase	125.6
Isocitrate lyase	Tr
Malate synthase	Tr
Lactate dehydrogenase	0
NADH dehydrogenase (membrane vesicles)	199

dehydrogenase, whole cells did not oxidize this substrate at an appreciable rate. In general, Krebs cycle acids (with the exception of pyruvate) may be poorly transported by the legionellae. Whether this is due to a repressed permease, lack of a permease, or other reasons is unknown. Certainly more research in this area is needed.

(ii) Carbohydrate metabolism. While carbohydrates are not appreciably catabolized by the legionellae, we nevertheless were interested in the types of metabolic pathways which might be present and how these might function. Radiorespirometric studies with D-glucose suggested that the pentose and the Entner-Doudoroff pathways, but not the EMP, were functional in *L. pneumophila* (27, 30). My group has recently examined the catabolic and anabolic routes of carbohydrate metabolism in *L. pneumophila*. Since experience has taught us to be cautious of every positive or negative result, we have carefully chosen several other microorganisms to serve as controls for these assays. Standard assay procedures were used, and the results of this study are presented in Table 2. In support of glucose catabolism, we found glucokinase activity. This activity was not found in *Campylobacter jejuni*, which does not transport carbohydrates. There was no glucose dehydrogenase activity in *L. pneumophila*.

We found no evidence for the Entner-Doudoroff pathway in *L. pneumophila* strains Philadelphia 1 and Knoxville 1. The activity reported is regarded as the limit of detection by our system. Both *Escherichia coli* and *C. jejuni* were also negative for this pathway.

Both glucose 6-phosphate dehydrogenase and the CO_2-evolving 6-phosphogluconate dehydrogenase were present in *L. pneumophila*, indicating that the pentose cycle was present. We did not assay the transketolase and transaldolase activities, but we assume that these enzymes also are present.

Enzymes characteristic of the EMP were found in legionellae. The fructose 1,6-biphosphate aldolase was present, but relative to that of *E. coli* was of low activity. Additional key enzymes of glucose catabolism by the EMP (aside from the aldolase) included phosphoglucose isomerase and phosphofructokinase. Both of these enzymes exhibited low activities. Glyceraldehyde 3-phosphate dehydrogenase and the lower portion of the EMP appeared to be intact, including pyruvate kinase and pyruvate dehydrogenase (low activity). We could not demonstrate lactate dehydrogenase activity.

In comparison to the catabolic function of the EMP, the gluconeogenic anabolic enzymes were much more active. Two enzymes which link the Krebs cycle with gluconeogenesis (phospho-

TABLE 2. Enzymatic activities for various pathways of carbohydrate metabolism in *L. pneumophila*

Enzyme	Sp act (nmol/min per mg of protein)[a]
6-Phosphogluconate dehydratase + KDPG[b] aldolase	0.01 (155)[c]
Glucose hexokinase	4.8
Glucose dehydrogenase	0.01
Glucose 6-phosphate dehydrogenase . .	20
6-Phosphogluconate dehydrogenase . . .	2
Fructose 1,6-biphosphatase	6 (5.6)[d]
Phosphoglucose isomerase	3.3 (7,300)[d]
Fructose diphosphate aldolase	1.8 (40.4)[d]
Phosphofructokinase	0.67 (63)[d]
Glyceraldehyde 3-phosphate dehydrogenase	140.75
Pyruvate kinase	4.5 (1.2)[e]
Phosphoenolpyruvate carboxylase	7.0 (1.1)[e]
Phosphoenolpyruvate carboxykinase . .	8.4 (1.1)[e]
Pyruvate carboxylase	140 (0.87)[e]

[a] High-speed supernatants (100,000 × g) were used for these assays to eliminate problems with membrane-bound NADH dehydrogenase.
[b] KDPG, 2-Keto-3-deoxy-6-phosphogluconate.
[c] Value in parentheses is the activity of *Azotobacter vinelandii*, used as a control.
[d] Value in parentheses is the activity of *E. coli*, used as a control.
[e] Value in parentheses is the activity of *C. jejuni*, used as a control.

enolpyruvate carboxylase and phosphoenolpyruvate carboxykinase) were detected in cell-free extracts of legionellae. In addition to these enzymes, we found appreciable activity for fructose 1,6-biphosphatase. Based on the relative specific activities of these enzymes, we concluded that the major function of the EMP is biosynthetic rather than catabolic.

Respiratory metabolism. The electron transport chain of *L. pneumophila* is composed of cytochromes of the *c*, *b*, *a*, and *d* types (12, 15). The relative concentrations of the cytochromes in membrane vesicles of *L. pneumophila* are presented in Table 3. Generally, the cytochrome concentrations were much lower than observed for other aerobes such as members of *Pseudomonas*, *Campylobacter*, and *Azotobacter*. Karr et al. (14) have shown that the major quinone is ubiquinone. The terpene side chain of ubiquinone is of 10 to 14 carbons, and this appears to be a unique characteristic of the legionellae. The ubiquinones of other bacteria have terpene carbon chain lengths of fewer than 10 carbons. The respiratory chain is most likely branched and is terminated by several oxidases. Oxidases of the *c*, *a*, and *d* types have been confirmed in my laboratory from their photochemical action spectrum. The organization of the respiratory chain remains to be determined. Respiratory activities of membrane vesicles were similar to the specific activities reported in Table 1. NADH oxidase was very active, and the activity was consistent with the conclusion that the Krebs cycle is the major energy-yielding metabolic pathway (amino acid catabolism). Our preliminary studies suggested that legionellae are relatively resistant to cyanide, but highly sensitive to the respiratory inhibitor 2-heptyl-4-hydroxyquinoline-N-oxide (12). The K_i for cyanide was 10 mM.

Preliminary experiments aimed at determining the energy conservation efficiency of legionellae by proton stoichiometric methods suggested that the coupling of respiration to ATP synthesis may be low. Our determinations of H/O ratios indicated an efficiency of 1 ATP molecule per atom of oxygen consumed. This value is preliminary and might be slightly higher, but will not be as high as values reported for other aerobes (H/O ratios between 4 and 8).

OXYGEN TOXICITY AND MECHANISM OF ACTION OF CHARCOAL

Many fastidious pathogens have recently been shown to be sensitive to low levels of hydrogen peroxide and superoxide radicals. Legionellae are also sensitive to these agents. The excess cysteine present in media used for culture of legionellae may contribute to this toxicity. Peroxide and superoxide are generated by autooxidation of medium components (particularly in the presence of cysteine) or by photochemical oxidation mechanisms. Hoffman et al. (13) reported that yeast extract (YE) medium (without added charcoal or cysteine) rapidly autooxidized when exposed to fluorescent light and that hydrogen peroxide and superoxide radicals rapidly accumulated in this medium. Moreover, autoclaved YE medium was more susceptible to photooxidation than medium sterilized by filtration. This observation is consistent with reports that filtered medium is less toxic in the absence of added charcoal than autoclaved medium. Fluorescent light also accelerated the autooxidation of cysteine. Carlsson et al. (5) reported that cysteine oxidation generated hydrogen peroxide, which in turn was consumed in other oxidation reactions. In the presence of metals such as ferric iron salts, cysteine is spontaneously oxidized. Gardner and Jursinic (9) demonstrated that at pH values above 7.0, superoxide radicals were generated in this reaction. Cysteine-ferric iron complexes also accelerated the decomposition of lipid hydroperoxides by free-radical reactions involving oxygen (9). A similar mechanism might be involved in destruction of membrane fatty acids in legionellae, and accumulation of peroxide by photochemical mechanisms in the YE medium would likely be involved.

Challenge experiments have demonstrated that legionellae are killed by hydrogen peroxide at levels of 25 μM or greater in 1 h (13, 17). Photochemical oxidation of YE medium leads to accumulation of 50 μM peroxide in 3 h. The significance of these observations in regard to primary isolation of legionellae from clinical and environmental sources is that in most cases low numbers of organisms are plated onto media (without charcoal) already containing toxic levels of hydrogen peroxide. For these organisms, even lower levels of peroxide (<50 μM) would likely be toxic. Heavy inocula, as would be used from stock cultures in the laboratory, when streaked onto autoclaved YE medium lacking charcoal, will overcome the peroxide toxicity and grow along the initial streaks. Peroxide toxicity provides a reasonable explanation for why most common laboratory media, in the absence of added charcoal, fail to culture *L. pneumophila*, despite inclusion of ferric pyrophosphate and cysteine. The cellular targets for the peroxide toxicity remain to be determined, but will likely involve membranes and membrane functions.

ACES-buffered charcoal-yeast extract medium (21) and other peptone-based formulations containing charcoal have been shown to be superior for isolation and culture of the legionellae. Activated charcoal has been used in the culture of many fastidious pathogens; yet the mechanism of action of charcoal has never been clearly established. Charcoal will adsorb and thereby detoxify free fatty acids, as will starch. Yet, starch will not substitute for charcoal in autoclaved YE medium and is only a poor substitute in medium sterilized by filtration. In chemically defined medium containing no fatty acids, starch does not substitute for charcoal, suggesting that fatty acid toxicity plays a minor role (13). Activated charcoal is commonly used in air and water purification systems, although little information is available on mechanisms. Our group has demonstrated that charcoal prevents photochemical oxidation reactions in YE medium and rapidly decomposes hydrogen peroxide (13). Charcoal functions similarly to a combination of superoxide dismutase (SOD) and catalase (Table 4), although the charcoal might be acting as a trap for free radicals, thereby blocking additional free radical chain-propagating reactions. Additionally, the charcoal may adsorb reactant molecules and thus prevent reactions leading to the formation of superoxide and peroxide. Charcoal also prevented the light-accelerated oxidation of cysteine. Starch, bovine serum albumin, and boiled enzymes (catalase and SOD) had no effect on autooxidation or photochemical oxidation reactions. There may be additional beneficial mechanisms for action of charcoal, and

TABLE 3. Cytochrome levels in membrane vesicles of *L. pneumophila*[a]

Cytochrome	Concn (nmol/mg of protein)
c_{552}	0.19
b_{558}	0.235
b_{562}	0.175
a_{598}	0.074
d_{628}	0.091

[a] Cytochrome levels were determined from difference spectra of membrane vesicles.

certainly more research is needed.

The sensitivity of legionellae to free radicals and peroxide seems almost paradoxical when one considers that the primary cells infected by these bacteria are phagocytes (lung monocytes and polymorphonuclear leukocytes) capable of generating these reactant molecules. Locksley et al. (17) and others (16) have demonstrated that despite the acute sensitivity of *L. pneumophila* to oxygen-dependent microbicidal systems (myeloperoxidase, eosinophil peroxidase, and xanthine oxidase) in vitro, these systems are not activated during infection in vivo. How legionellae prevent activation of the various killing mechanisms of phagocytic cells is unknown, as is how legionellae acquire nutrients and grow intracellularly. These are important questions that deserve much more attention.

SOD, catalase, and peroxidase. The sensitivity of legionellae to reduced forms of oxygen may not be due to an absence of SOD, catalase, or peroxidase (12, 13). Leo Pine and I examined legionellae for the presence of these enzymes, as has Locksley et al. (17). Essentially, *L. pneumophila* is catalase negative, despite reports to the contrary. *L. pneumophila* does have an active peroxidase and appears capable of disposing of peroxides generated intracellularly. Most of the other *Legionella* species do have catalase, but most that have catalase lack peroxidase. Table 5 presents a survey of the respiratory protective enzymes found in the various species of *Legionella*; additional details are presented elsewhere in this volume. All *Legionella* species thus far tested have SOD in levels comparable to those reported for *E. coli* (12 to 30 U). *L. pneumophila* lacks a true catalase but has appreciable peroxidase activity. Moreover, when the peroxidase is partially purified, it has some catalase activity. The characteristics of this enzyme are very similar to those reported by Clairborn and Fridovich for a hydroperoxidase of *E. coli* (6). Legionellae can be grouped by the presence of the three enzymes (Table 5). Group 1 legionellae have catalase, peroxidase, and SOD; group 2 lacks peroxidase but has SOD and catalase; group 3 has peroxidase and SOD. We

are presently working on a rapid test to differentiate these groups based on the activities of catalase and peroxidase. The only other rapid test (besides serology) is hippurate hydrolysis, which distinguishes *L. pneumophila* from other legionellae (11).

The presence of catalase, SOD, and peroxidase in legionellae suggests that the toxic effects of exogenous hydrogen peroxide and superoxide may not be directed toward intracellular targets. Rather, it is more likely that the targets damaged by reduced forms of oxygen are external to and include the cytoplasmic membrane.

METABOLISM AND CHEMISTRY OF CYSTEINE

Cysteine is both a blessing and a curse for legionellae in that it is required for growth yet may contribute to the problem of oxygen toxicity inhibiting growth. Most auxotrophs for cysteine (*E. coli*, salmonellae, and bacilli) can use cystine and other sulfur-containing compounds, but this may not be true for legionellae. The apparent requirement for excess cysteine (2.5 μM), necessary for successful culture of legionellae, may be due to its unstable nature in

TABLE 4. Photochemical oxidation of YE medium by exposure to fluorescent light

YE medium and supplement(s)[a]	O_2 consumption (nmol/min per ml)	% O_2 regenerated	H_2O_2 decomposition[b] (nmol/min per ml)
YE (filtered)	1.0	0	<0.5
YE (autoclaved)	5.2 to 6.9	0	<0.5
YE + catalase	4.02	28	
YE + SOD	3.56	45	
YE + catalase + SOD	2.3	68	
YE + mannitol	4.6	2	
YE + histidine	7.2	0	
YE + boiled SOD + catalase	5.6	0	
YE + bovine serum albumin	5.06	0.3	
YE + starch	6.5	0	0
YE + Norit A charcoal	<0.3		6.44
YE + Norit SG charcoal	<0.3		5.92
Charcoal + phosphate buffer			18
CYE agar (control)	<0.1		23.6

[a] Except where indicated, the YE medium used in these experiments was autoclaved and did not contain cysteine and ferric pyrophosphate. The medium was illuminated with fluorescent light, and photooxidation was followed with an oxygen probe as detailed in reference 13.
[b] Hydrogen peroxide decomposition was determined by measuring oxygen evolution in media to which 1.5 mM hydrogen peroxide had been added.

TABLE 5. Catalase, peroxidase, and SOD activities in extracts of *Legionella* species[a]

Species	Group[b]	Cata-lase	Perox-idase	SOD
L. dumoffii	1	+	+	+
L. jordanis	2	+	−	+
L. wadsworthii	2	+	−	+
L. oakridgensis	2	+	−	+
L. bozemanii	2	+	−	+
L. micdadei	2	+	−	+
L. longbeachae	2	+	−	+
L. pneumophila (all serogroups)	3	−	+	+
L. gormanii	3	−	+	+

[a] Cells were grown on buffered charcoal-yeast extract agar, sonicated, and centrifuged at $39,000 \times g$ for 1 h. Extracts were passed through a Sephadex G-200 (fine) chromatographic column to separate the enzymes.
[b] See text.

complex medium. Essentially, cysteine is rapidly oxidized to cystine in YE medium, and ferric iron salts catalyze the oxidation. Charcoal does not prevent the autooxidation of cysteine. In broth, cysteine oxidizes at a rate of 18 to 28 nmol/min per ml, and when the level of cysteine reaches 0.5 μM, the oxidation decreases and levels off. The amount of cysteine remaining is sufficient to support good growth of the legionellae. Ferric pyrophosphate is rapidly reduced (5 to 9 nmol/min per ml) during the oxidation of cysteine. Since ferrous iron is soluble, and since legionellae produce no siderophores, cysteine autooxidation in the presence of ferric iron may make iron more available to the bacteria. Some of the cysteine which is oxidized may form compounds other than cystine. These might include cysteine sulfinic acid, sulfones, and adducts with keto acids and sugars. These latter forms may still be available to legionellae.

Several groups have demonstrated that cysteine is required as a nutrient, but incorporation into cellular material has only recently been demonstrated in my laboratory. Using [14C]cysteine, we examined the uptake of cysteine into whole cells of *L. pneumophila*. Cysteine was taken up by *L. pneumophila*, but at a lesser rate than observed for *E. coli* (Fig. 1). The biggest problem with cysteine uptake is that the substrate is constantly changing during the experiment. When the cysteine becomes oxidized, it is no longer translocated. On the other hand, *E. coli* translocates cystine. The rate of cysteine uptake was 2.5 pmol/min per mg (dry weight) for *L. pneumophila*, whereas for *E. coli* the rate was 10.4 pmol/min per mg (dry weight). Uptake was maximal in the presence of energy sources like serine or glutamate. Using [35S]cysteine, we demonstrated by autoradiographic techniques

that cysteine was incorporated into proteins.

In a recent examination of the biosynthetic pathway for cysteine in legionellae, we found that *L. pneumophila* lacks two key enzymes. These are serine transacetylase, which acetylates serine to *o*-acetylserine, and *o*-acetylserine sulfhydrylase, which adds sulfide to *o*-acetylserine to form cysteine and acetate. Both of these enzymes were present in extracts of *Legionella oakridgensis*, which does not require added cysteine for growth after primary isolation (20). The activity of these enzymes in *L. oakridgensis* was not affected by the presence of added cysteine in the medium. We have not examined the enzymes of assimilatory sulfate reduction in legionellae, so no information on deficiencies in this pathway is available.

GENETICS AND PLASMIDS

Plasmids have been found in several *Legionella* species and range in size from 23 to 80 megadaltons. Presently, there appears to be no correlation between virulence and presence of a plasmid. Brown et al. (4) have reported changes in surface antigens between plasmid-containing and plasmidless strains. The plasmids might prove useful in molecular genetic experimentation aimed at determining the genes involved in surface antigen expression as well as in the

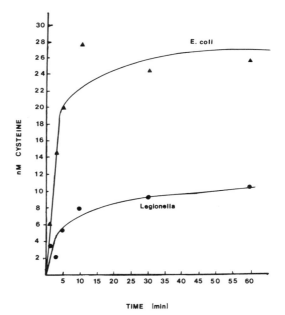

FIG. 1. [14C]cysteine uptake by *L. pneumophila*. [14C]cysteine uptake was determined with whole cells in 50 mM potassium phosphate buffer (pH 6.9) in the presence of 10 μM carrier cysteine. *E. coli* was used as a control. A membrane filter technique was used, and filters were placed in scintillation fluor and counted in a scintillation counter.

cloning of toxins and virulence factors. While many groups are in the process of developing genetic systems for legionellae, there is little to report on at the present time.

In regard to regulatory mechanisms at the genetic level in legionellae, we have probably seen some evidence of these already. In reviewing the differences in chemically defined media and reports of the specific amino acid requirements, the differences observed might well be a result of feedback regulation by one amino acid on the biosynthesis of other amino acids. The isoleucine-leucine-valine (Ilv) amino acid group required by legionellae for growth may be such an example. Cysteine is known to repress the expression of the *ilv* cluster in *E. coli*. Characterization of *Legionella* mutants with defects in regulatory loci might prove useful in resolving such questions.

While much new information on the physiology and metabolism of legionellae has appeared since the first symposium, much still remains to be accomplished. This is particularly true in the genetics area, in mechanisms of pathogenesis, and in the physiology of intracellular growth in phagocytic cells. It is hoped that the information presented in this overview will be helpful to those already in the field as well as stimulating to those casually interested in legionellae.

I thank members of my research group, Mark Keen, Tony Butler, and Denise Street, for permission to use their research results in this presentation and Leo Pine for helpful suggestions and critical reading of the manuscript.

LITERATURE CITED

1. **Amano, K.-I., and J. C. Williams.** 1983. Peptidoglycan of *Legionella pneumophila*: apparent resistance to lysozyme hydrolysis correlates with a high degree of peptide cross-linking. J. Bacteriol. **153**:520–526.
2. **Baine, W. B., and J. K. Rasheed.** 1979. Aromatic substrate specificity of browning by cultures of the Legionnaires' disease bacterium. Ann. Intern. Med. **90**:619–620.
3. **Berdal, B. P., O. Olsik, S. Myhre, and T. O. Omland.** 1982. Demonstration of extracellular chymotrypsin-like activity from various *Legionella* species. J. Clin. Microbiol. **16**:452–457.
4. **Brown, A., R. M. Vickers, E. M. Elder, M. Lema, and G. M. Garrity.** 1982. Plasmid and surface antigen markers of endemic and epidemic *Legionella pneumophila* strains. J. Clin. Microbiol. **16**:230–235.
5. **Carlsson, J., G. P. D. Granberg, C. K. Nyberg, and M. B. K. Edlund.** 1979. Bactericidal effect of cysteine exposed to atmospheric oxygen. Appl. Environ. Microbiol. **37**:383–390.
6. **Clairborn, A., and I. Fridovich.** 1979. Purification of the *o*-dianisidine peroxidase from *Escherichia coli*. J. Biol. Chem. **254**:4245–4252.
7. **Fliermans, C. B., R. J. Soracco, and D. H. Pope.** 1981. Measure of *Legionella pneumophila* activity *in situ*. Curr. Microbiol. **6**:89–94.
8. **Fliescher, A. R., S. Ito, B. J. Mansheim, and D. L. Kasper.** 1979. The cell envelope of the Legionnaires' disease bacterium. Morphologic and biochemical characteristics. Ann. Intern. Med. **90**:628–630.
9. **Gardner, H. W., and P. A. Jursinic.** 1981. Degradation of linoleic acid hydroperoxides by cysteine-FeCl$_2$ catalyst as

a model for similar biochemical reactions. 1. Study of oxygen requirement, catalyst and effect of pH. Biochim. Biophys. Acta **665**:100–112.
10. **George, J. R., L. Pine, M. W. Reeves, and W. K. Harrell.** 1980. Amino acid requirements of *Legionella pneumophila*. J. Clin. Microbiol. **11**:286–291.
11. **Hébert, G. A.** 1981. Hippurate hydrolysis by *Legionella pneumophila*. J. Clin. Microbiol. **13**:240–242.
12. **Hoffman, P. S., and L. Pine.** 1982. Respiratory physiology and cytochrome content of *Legionella pneumophila*. Curr. Microbiol. **7**:351–356.
13. **Hoffman, P. S., L. Pine, and S. Bell.** 1983. Production of superoxide and hydrogen peroxide in medium used to culture *Legionella pneumophila*: catalytic decomposition by charcoal. Appl. Environ. Microbiol. **45**:784–791.
14. **Karr, D. E., W. F. Bibb, and C. W. Moss.** 1982. Isoprenoid quinones of the genus *Legionella*. J. Clin. Microbiol. **15**:1044–1048.
15. **Kronick, P. L., and R. W. Gilpin.** 1980. Cytochrome spectra of *Legionella pneumophila*. Microbios Lett. **14**:59–63.
16. **Locher, J. E., R. L. Friedman, R. H. Bigley, and B. H. Iglewski.** 1983. Effect of oxygen-dependent antimicrobial systems on *Legionella pneumophila*. Infect. Immun. **39**:487–489.
17. **Locksley, R. M., R. F. Jacobs, C. B. Wilson, W. M. Weaver, and S. J. Klebanoff.** 1982. Susceptibility of *Legionella pneumophila* to oxygen-dependent microbicidal systems. J. Immunol. **129**:2192–2197.
18. **Mayberry, W. R.** 1981. Dihydroxy and monohydroxy fatty acids in *Legionella pneumophila*. J. Bacteriol. **147**:373–381.
19. **Moss, C. W., R. E. Weaver, S. B. Dees, and W. B. Cherry.** 1977. Cellular fatty acid composition of isolates from Legionnaires disease. J. Clin. Microbiol. **6**:140–143.
20. **Orrison, L. H., W. B. Cherry. R. L. Tyndall, C. B. Fliermans, S. B. Gough, M. A. Lambert, L. K. McDougal, W. F. Bibb, and D. J. Brenner.** 1983. *Legionella oakridgensis*: unusual new species isolated from cooling tower water. Appl. Environ. Microbiol. **45**:536–545.
21. **Pasculle, A. W., J. C. Feeley, R. J. Gibson, L. G. Cordes, R. L. Meyerowitz, C. M. Patton, G. W. Gorman, C. L. Carmack, J. W. Ezzell, and J. N. Dowling.** 1980. Pittsburgh pneumonia agent: direct isolation from human lung tissue. J. Infect. Dis. **141**:727–732.
22. **Pine, L., J. R. George, M. W. Reeves, and W. K. Harrell.** 1979. Development of a chemically defined liquid medium for growth of *Legionella pneumophila*. J. Clin. Microbiol. **9**:615–626.
23. **Reeves, M. W., L. Pine, S. H. Hutner, J. R. George, and W. K. Harrell.** 1981. Metal requirements of *Legionella pneumophila*. J. Clin. Microbiol. **13**:688–695.
24. **Reeves, M. W., L. Pine, J. B. Neilands, and A. Balows.** 1983. Absence of siderophore activity in *Legionella* species grown in iron-deficient media. J. Bacteriol. **154**:324–329.
25. **Ristroph, J. D., K. W. Hedlund, and S. Gowda.** 1981. Chemically defined medium for *Legionella pneumophila* growth. J. Clin. Microbiol. **11**:19–21.
26. **Tesh, M. T., and R. D. Miller.** 1981. Amino acid requirements for *Legionella pneumophila* growth. J. Clin. Microbiol. **13**:865–869.
27. **Tesh, M. J., S. A. Morse, and R. D. Miller.** 1981. Intermediary metabolism in *Legionella pneumophila*: utilization of amino acids and other compounds as energy sources. J. Bacteriol. **154**:1104–1109.
28. **Thompson, M. R., R. D. Miller, and B. H. Iglewski.** 1981. In vitro production of an extracellular protease by *Legionella pneumophila*. Infect. Immun. **34**:299–302.
29. **Warren, W. J., and R. D. Miller.** 1979. Growth of Legionnaires disease bacterium (*Legionella pneumophila*) in chemically defined medium. J. Clin. Microbiol. **10**:50–55.
30. **Weiss, E., M. G. Peacock, and J. C. Williams.** 1980. Glucose and glutamate metabolism of *Legionella pneumophila*. Curr. Microbiol. **4**:1–6.

Growth of *Legionella pneumophila* in Continuous Culture and Its Sensitivity to Inactivation by Chlorine Dioxide

JAMES D. BERG, JOHN C. HOFF, PAUL V. ROBERTS, AND ABDUL MATIN

Stanford University, Stanford, California 94305, and U.S. Environmental Protection Agency, Cincinnati, Ohio 45268

Substantial differences in the sensitivity of bacteria to inactivation by disinfectants have been attributed to the conditions experienced by the organisms immediately before treatment (2, 5). The physiology, growth rate, and gross morphology of bacteria and the ensuing responses to environmental stresses (e.g., disinfectants) can be altered significantly by changes in growth parameters such as nutrient concentration and composition, oxygen tension, and temperature (9, 10). These parameters cannot be adequately controlled with conventional batch culture techniques, in which the environmental conditions change drastically during growth, yielding populations of unknown phenotype. These problems are compounded because laboratory cultures, which can be more sensitive by orders of magnitude than naturally occurring populations (4), have usually been used in disinfection studies.

Several *Legionella* species isolated from a diverse set of aquatic environments (8) have been reported to exhibit differing sensitivities to disinfectants (6, 15). Also, there is some controversy concerning the growth requirements of legionellae, especially with respect to L-cysteine, iron(III), and oxygen (7, 11, 12, 14). Therefore, the objectives of this study were to utilize the continuous culture approach to develop additional information on the growth requirements of legionellae and to evaluate the effects of various growth conditions on the sensitivity of the cultures to disinfectants.

The medium of Ristroph et al. (12) was modified for use in the chemostat. The modified medium contains, per liter, yeast extract (Difco Laboratories), 10 g; L-cysteine, 0.2 g; and starch, 0.25 g. The pH was adjusted to 6.9 with 1 N KOH, and the medium was filter sterilized through a Seitz S3 filter with nitrogen. Nitrogen was bubbled continuously through the 10-liter reservoir to maintain a reducing environment. It was necessary to remove iron(III) from the medium to minimize the catalytic oxidation of cysteine that can occur under forced aeration. With the methods of Roberts and Rouser (13) and Alexander (1), cysteine at a concentration of 0.4 g liter^{-1} in the presence of iron(III) was undetectable after 11 min. Based on the reaction stoichiometry described by Taylor et al. (16) for the reaction of cysteine and oxygen (4:1), the cysteine concentration was estimated to approach zero in less than 60 min under typical continuous culture conditions in the presence of iron(III). The removal of iron(III) minimized the problem of cysteine oxidation. Moreover, the apparent increased stability of the L-cysteine in this medium permitted a yield approximately 3.5 times greater than could be achieved in batch culture by using the conventional yeast extract medium supplemented with L-cysteine and iron-(III) as ferric pyrophosphate. The addition of the starch enhanced growth, presumably because of its role as an absorbant of toxic substances that may be produced at the surface of the immersion heater in the chemostat vessel. Ristroph et al. (12) suggested that fatty acids which inhibit legionellae can be produced during heating of yeast extract.

The chemostat inoculum was obtained by streaking a stock suspension, previously stored at $-70°C$, onto a charcoal-yeast extract agar slant which was then incubated for 48 to 72 h at 37°C. The growth on the slant was then washed into a 250-ml flask containing 100 ml of chemostat medium (described above) and incubated in a shaking water bath at 37°C for 48 h. The liquid culture was transferred into the 2-liter chemostat vessel, which was purged with nitrogen. The culture was then diluted with medium to an optical density at 660 nm of ≥ 0.15. Once an optical density of about 0.40 was achieved (after 4 to 6 days), positive aeration and nutrient pumping was started. Steady-state cultures can be harvested in 2 to 5 days, depending on the dilution rate (D, expressed as hours^{-1}).

Legionella pneumophila was grown successfully under a variety of growth conditions, using batch and continuous culture. Two dilution rates ($D = 0.06$ h^{-1} and $D = 0.15$ h^{-1}) and two temperatures (37 and 44°C) were used. $D = 0.06$ h^{-1} represents the slow growth rate, and $D = 0.15$ h^{-1} represents an intermediate rate. The 37°C temperature was the optimum, whereas 44°C represents a more harsh environment thought to exist in thermal effluents and hot tap water systems. Aeration utilized ambient air or a mixture of 95% air and 5% CO_2. The flow rates of each gas ranged from 20 to 150 liter min^{-1},

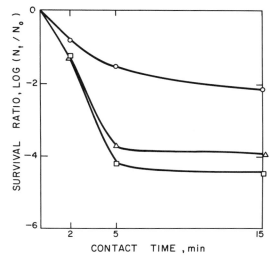

FIG. 1. Effect of growth environment on sensitivity of *L. pneumophila* to a dose of 0.75 mg of ClO_2 liter^{-1}. Symbols: ○, chemostat cultures grown at $D =$ 0.06 h^{-1} and 37°C, using air at 20 liters min^{-1}; △, batch cultures grown at 37°C on a shaker, harvested at 36 h; □, batch cultures grown at 37°C on a shaker, harvested at 24 h. The medium for the 24-h batch cultures was different from the other two in that it contained ferric pyrophosphate (0.25 g liter^{-1}) and additional L-cysteine (0.4 g liter^{-1}).

yielding 1 to 38% saturation of dissolved oxygen.

Steady-state optical density values ranged from 0.64 to 1.20, depending on the conditions used. High catalase activity was observed under all of the growth conditions. Pigment production, observed in the stationary-phase batch cultures in the iron-containing media, was not noted under chemostat conditions. A requirement for a titrant to maintain the pH at 6.9 was dependent on the form of aeration. Air (95%) plus CO_2 (5%) required the addition of base (0.5 N KOH), whereas without CO_2 (5%), acid (0.5 N HCl) was required (e.g., at $D = 0.06$ h^{-1} and 37°C, 2 mM acid h^{-1} was consumed).

The sensitivity of the chemostat populations to chlorine dioxide (ClO_2) was investigated in triplicate experiments using populations grown under three different conditions. The results of the inactivation experiments are shown in Fig. 1. *Legionella* cultures at a particular steady state were harvested, centrifuged, and then suspended in phosphate buffer (pH 7) to yield 10^7 CFU ml^{-1}. A dose of 0.75 mg of ClO_2 liter^{-1} was used in each experiment. The experiments were done in triplicate. The more slowly grown cells ($D =$ 0.06 h^{-1}) from the chemostat were about 2 orders of magnitude more resistant than either of the batch-grown populations. These data are

analogous to the data obtained in previous experiments using *Escherichia coli* (3). The two batch-grown populations differed with respect to the presence of iron(III) in the medium and the concentration of L-cysteine and exhibited no significant differences in sensitivity. Inactivation experiments using cultures grown under other conditions are in progress.

Comparing these sensitivity data with those obtained for *E. coli* (3), the legionellae in all three cases were more resistant in two respects. First, the initial inactivation rate for legionellae was much slower than that for *E. coli*. Second, the extent of kill ($t = 15$ min) was much less for legionellae.

LITERATURE CITED

1. **Alexander, N. M.** 1958. Spectrophotometric assay for sulfhydryl groups using *N*-ethylmaleimide. Anal. Chem. **30:**1292–1294.
2. **Berg, J. D., E. M. Aieta, and P. V. Roberts.** 1979. Effectiveness of chlorine dioxide as a wastewater disinfectant, p. 61–71. *In* A. D. Venosa (ed.), Progress in wastewater technology. U.S. Environmental Protection Agency, Cincinnati, Ohio.
3. **Berg, J. D., A. Matin, and P. V. Roberts.** 1982. Effect of antecedent growth conditions on sensitivity of *Escherichia coli* to chlorine dioxide. Appl. Environ. Microbiol. **44:**814–819.
4. **Berg, J. D., A. Matin, and P. V. Roberts.** 1983. Growth of disinfection-resistant bacteria and simulation of natural aquatic environments in the chemostat, p. 1137–1147. *In* R. L. Jolley (ed.), Water chlorination: environmental impact and health effects, vol. 4, book 2. Ann Arbor Science, Ann Arbor, Mich.
5. **Carson, L. A., M. S. Favero, W. W. Bond, and N. J. Petersen.** 1972. Factors affecting the comparative resistance of naturally occurring and subcultured *Pseudomonas aeruginosa* to disinfectants. Appl. Microbiol. **23:**863–869.
6. **Edelstein, P. H., R. E. Whittaker, R. L. Kreiling, and C. L. Howell.** 1983. Efficacy of ozone in eradication of *Legionella pneumophila* from hospital plumbing fixtures. Appl. Environ. Microbiol. **44:**1330–1334.
7. **Feeley, J. C., G. W. Gorman, R. E. Weaver, D. C. Mackel, and H. W. Smith.** 1978. Primary isolation medium for Legionnaires disease bacterium. J. Clin. Microbiol. **8:**320–325.
8. **Fliermans, C. B., W. B. Cherry, L. H. Orrison, S. J. Smith, D. L. Tison, and D. H. Pope.** 1981. Ecological distribution of *Legionella pneumophila*. Appl. Environ. Microbiol. **41:**9–16.
9. **Gill, C. O., and J. R. Suisted.** 1978. The effects of temperature and growth rate on the proportion of unsaturated fatty acids in bacterial lipids. J. Gen. Microbiol. **104:**31–36.
10. **Matin, A., A. Grootjins, and H. Hogenhuis.** 1976. Influence of dilution rate on enzymes of intermediary metabolism in fresh water bacteria grown in continuous culture. J. Gen. Microbiol. **105:**187–197.
11. **Orrison, L. H., W. B. Cherry, R. L. Tyndall, C. B. Fliermans, S. B. Gough, M. A. Lambert, L. K. McDougal, W. F. Bibb, and D. J. Brenner.** 1983. *Legionella oakridgensis:* unusual new species isolated from cooling tower water. Appl. Environ. Microbiol. **45:**536–545.
12. **Ristroph, J. D., K. W. Hedlund, and R. G. Allen.** 1980. Liquid medium for growth of *Legionella pneumophila*. J. Clin. Microbiol. **11:**19–21.
13. **Roberts, E., and G. Rouser.** 1958. Spectrophotometric assay for reaction of *N*-ethylmaleimide with sulfhydryl groups. Anal. Chem. **30:**1291–1292.
14. **Saito, A., R. D. Rolfe, P. H. Edelstein, and S. M. Finegold.**

gold. 1981. Comparison of liquid growth media for *Legionella pneumophila*. J. Clin. Microbiol. **14**:623–627.

15. **Skaliy, P., T. A. Thompson, G. W. Gorman, G. K. Morris, H. V. McEachern, and D. C. Mackel.** 1980. Laboratory studies of disinfectants against *Legionella pneu-*

mophila. Appl. Environ. Microbiol. **40**:697–700.

16. **Taylor, J. E., J. F. Yan, and J. Wang.** 1966. The iron (III)-catalyzed oxidation of cysteine by molecular oxygen in the aqueous phase. An example of a two-thirds order reaction. J. Am. Chem. Soc. **88**:1663–1667.

Loss of Virulence of *Legionella pneumophila* Serogroup 1 with Conversion of Cells to Long Filamentous Rods

N. BORNSTEIN, M. NOWICKI, AND J. FLEURETTE

Centre National de Référence des Légionelloses, Faculté de Médecine Alexis Carrel, 69372 Lyon Cedex 2, France

Legionella pneumophila serogroup 1 grows as a short bacillus (0.3 to 2 μm) in guinea pig spleen and embryonated egg yolk sacs, whereas it becomes long and filamentous (2 to 10 μm) after subculture on agar medium. It has been demonstrated that it also loses part of its virulence (2), which decreases more or less according to the type of agar medium used; the loss is more important when buffered charcoal-yeast extract (BCYE) agar rather than charcoal-yeast extract (CYE) agar is used (1, 3).

To obtain short, nonfilamentous forms of *Legionella* in agar media, we carried out a series of tests in an attempt to modify these media; addition of various concentrations of iron, amino acids, casein, sucrose, and embryonated egg or embryo tissue extracts all gave negative results. On the other hand, short forms occurred on a medium (SCYE agar) identical in composition to CYE agar except for containing, in the place of yeast extract, the residue from its predialysis in distilled water (porosity, 2.5 nm). It seemed interesting to investigate a possible correlation between the morphology and virulence of legionellae. This was done by examining different growth conditions. All tests were carried out with one strain of *L. pneumophila* serogroup 1 (Philadelphia 1), obtained from the Centers for Disease Control. Inocula were prepared from ground, infected guinea pig spleen and embryonated egg yolk sacs.

For each culture the size of bacteria was determined by Gram (two tests) and direct fluorescent-antibody (four tests) staining and by using a Zeiss micrometric microscope measurement. One hundred readings were made for each test (10 fields per slide). Lengths of bacilli (L) were divided into four classes ($0.5 \leq L \leq 2$ μm, $2 < L \leq 5$ μm, $5 < L \leq 10$ μm, and $L > 10$ μm); these lengths were measured after culture on yolk sacs and one or two subcultures on CYE, BCYE, and SCYE agar media. In addition, colonies from BCYE and CYE media were subcultured on SCYE agar.

Measurements of the 50% lethal dose (LD_{50})

were used to evaluate virulence on 7-day embryonated hen eggs, according to the method of Reed and Muench (4). Two identical assays were carried out with standard inocula from yolk sac and from CYE, BCYE, and SCYE agars subcultured once only, and a third test was done under the same conditions after 10 subcultures on each of the agar media.

Very similar results were obtained for the six measurements, and their means are shown in Table 1. Yolk sacs yielded the shortest forms (69% with $0.5 \leq L \leq 2$ μm), with an average length of 1.9 μm. Conversely, BCYE medium gave the longest forms (48% with $L > 10$ μm), with an average length of 9.35 μm. Of all the agar media, SCYE produced the shortest forms, with an average length of 3 μm, but failed to give filamentous rods. No significant modification in length was observed between first and second subcultures, regardless of the agar medium used. Passage from BCYE and CYE media to SCYE medium produced short forms equivalent to those observed after the first subculture on SCYE medium. Similar LD_{50} values were obtained for the three tests, the means of which are given in Table 1. The most virulent inoculum was that obtained from growth on yolk sacs ($LD_{50} = 10^{1.36}$ bacteria per ml). The least virulent inocula were those produced after culture on BCYE and CYE media, with LD_{50}s of $10^{4.79}$ and $10^{4.42}$ bacteria per ml, respectively (Table 1). Of all the agar media, SCYE yielded the most virulent inoculum ($LD_{50} = 10^{2.66}$ bacteria per ml). A positive and significant correlation existed between the length of bacilli and the LD_{50} logarithm ($r = 0.96$, $0.02 < P < 0.05$).

In conclusion, our results confirm that legionellae lose most of their virulence when passaged from live to agar media (CYE, BCYE), with, in parallel, a morphological modification (conversion into long filamentous rods). A slight modification in the CYE medium preparation (predialysis of yeast extract) produced bacilli with a morphology close to that observed in live tissues and a virulence greater than that obtained when

TABLE 1. Length and virulence of *L. pneumophila* serogroup 1 organisms under different culture conditions

Medium	% of bacteria with the following length (μm):				Mean length (μm) of bacteria	LD$_{50}$
	$0.5 \leqslant L \leqslant 2$	$2 < L \leqslant 5$	$5 < L \leqslant 10$	$L > 10$		
Yolk sac	69.5	30.5			1.94	$10^{1.36}$
BCYE						
1[a]		15.83	41	41		
2[b]		11.5	39.33	49.17		
Mean		13.49	40.17	46.33	9.3	$10^{4.79}$
CYE						
1	1	36.17	40.17	22.67		
2	2.33	38.67	36.67	22.33		
Mean	1.67	37.42	38.42	22.5	7.02	$10^{4.42}$
SCYE						
1	35.33	60.83	3.83			
2	31.83	64.83	3.33			
3[c]	31.5	64.67	3.63			
3[d]	27.83	68.67	4.1			
Mean	31.62	64.70	3.67		2.94	$10^{2.66}$

[a] First subculture on agar media from egg yolk sacs and ground spleen.
[b] Second subculture from agar medium to identical agar medium.
[c] Third subculture from BCYE medium to SCYE medium.
[d] Third subculture from CYE medium to SCYE medium.

using the other agar media. Work is in progress to investigate the factors associated with formation of short bacilli. The reason why the length and virulence are correlated remains to be established.

LITERATURE CITED

1. **Feeley, J. C., R. J. Gibson, G. W. Gorman, N. C. Langford, J. K. Mashead, D. C. Mackel, and W. B. Baine.** 1979. Charcoal-yeast extract agar: primary isolation medium for *Legionella pneumophila*. J. Clin. Microbiol. **10:**437–441.

2. **McDade, J. E., and C. C. Shepard.** 1979. Virulent to avirulent conversion of Legionnaires' disease bacterium (*Legionella pneumophila*). Its effect on isolation techniques. J. Infect. Dis. **139:**707–711.

3. **Pasculle, A. W., J. C. Feeley, R. J. Gibson, L. G. Cordes, R. L. Myerowitz, C. M. Patton, G. W. Gorman, C. L. Carmack, Y. W. Lezzell, and Y. N. Dowling.** 1980. Pittsburgh pneumonia agent: direct isolation from human lung tissue. J. Infect. Dis. **141:**727–731.

4. **Reed, L. Y., and H. Muench.** 1938. A simple method of estimating fifty percent endpoints. Am. J. Hyg. **27:**493–497.

Carbohydrate Profiling of Some *Legionellaceae* by Capillary Gas Chromatography-Mass Spectrometry

ALVIN FOX, PAULINE Y. LAU, ARNOLD BROWN, STEPHEN L. MORGAN, Z.-T. ZHU,† MICHAEL LEMA, AND MICHAEL WALLA

Department of Microbiology and Immunology and Department of Medicine, School of Medicine, University of South Carolina, and Department of Chemistry, University of South Carolina, Columbia, South Carolina 29208, and Research Service, William Jennings Bryan Dorn Veterans Administration Hospital, Columbia, South Carolina 29201

Garrity et al. have proposed that the first five species of *Legionellaceae* be divided into three genera: *Legionella*, *Tatlockia*, and *Fluoribacter* (6). Genetic and biochemical data, ubiquinone patterns, fatty acid analysis, peptide profiles, and antigenic characterization provide strong evidence to support this contention (2, 3, 6, 7a, 8). The data presented here that species of *Legionella*, *Tatlockia*, and *Fluoribacter* can be

distinguished by their carbohydrate profiles provide additional evidence for the proposed taxonomic division (A. Fox, P. Y. Lau, A. Brown, S. L. Morgan, Z.-T. Zhu, M. Lema, and M. Walla, submitted for publication).

The bacterial strains used in this study were incubated for 3 days on buffered charcoal-yeast extract agar in air at 37°C, and the confluent bacterial growth was suspended in distilled water, killed by heating in flowing steam at 100°C, washed in distilled water, and lyophilized. The neutral and amino sugar content of bacteria was determined by our modification of the alditol

† Present address: Department of Chemistry, Shanxi University, Tiayuan, People's Republic of China.

acetate procedure, employing capillary gas chromatography and gas chromatography-mass spectrometry (4, 5, 7).

The *Fluoribacter* species were distinguished by the presence of moderate amounts of both fucose and rhamnose and by the presence of two isomeric amino dideoxy hexoses (X1 and X2). *Fluoribacter (Legionella) dumoffii* strain NY-23 could be distinguished from *Fluoribacter (Legionella) bozemanae* and *Fluoribacter (Legionella) gormanii* by the absence of X2. *Tatlockia (Legionella) micdadei* was distinguished by the presence of large amounts of fucose and rhamnose and the absence of both X1 and X2. *Legionella pneumophila* was distinguished by the absence of fucose and the presence of X1. The eight most prominent sugars (rhamnose, fucose, ribose, mannose, X1, X2, muramic acid, and glucosamine) were quantitated. A measure of the dissimilarity between these carbohydrate profiles was obtained by treating each chromatogram as a point in an eight-dimensional space and by calculating the Euclidean distance

$$d_{ij} = \left[\sum_{k=1}^{8} (X_{ik} - X_{jk})^2 \right]^{1/2}$$

between each pair of patterns, Xi and Xj. Figure 1 shows a nonlinear map (9, 10) representing the chromatograms of *Legionellaceae*. This two-dimensional representation of the eight-dimensional data was obtained by adjusting the set of two-dimensional coordinates to minimize an error function involving the difference between the Euclidean distances in the eight-dimensional space and the distances between the points in the two-dimensional space. The nonlinear map is thus the two-dimensional representation best preserving the relationships and structure of the original data. Points that are close to one another on the map represent carbohydrate profiles that are similar to one another; points that are distant from one another represent profiles that are quite different. Replicate analyses were performed for all but one strain, and these replicate points are connected by line segments on the map. The three areas on the map were outlined to define the regions where *Tatlockia, Legionella*, and *Fluoribacter* patterns are found. The greater variability encountered with the *Fluoribacter* samples, illustrated in the nonlinear map, may be because the *Fluoribacter* isolates analyzed represented three different species. In the case of *Legionella*, the tight clustering in the map is consistent with the presence of a single species. One of the *Tatlockia* strains (PPA-GL) appears on the map to be more similar to the *Legionella* strains than to the other *Tatlockia* strains; however, when the original data are examined, the carbohydrate profile of this strain is easily differentiated from those of legionellae by the presence of fucose and the absence of X1, which clearly indicate that it belongs to *Tatlockia*.

Deoxyribose, a component of DNA, was not detected in any strains of *Legionellaceae* studied, probably because of the harsh hydrolysis conditions used. 2-Keto-3-deoxyoctulosonic acid was also not detected in the *Legionellaceae*; however, under the same hydrolysis conditions we also did not detect it in lipopolysaccharide preparations, as 2-keto-3-deoxyoctulosonic acid is destroyed by heating in strong mineral acids. Others have detected 2-keto-3-deoxyoctulosonic acid in *L. pneumophila* (11). Interestingly, we also did not detect heptose in the *Legionellaceae*, although we did detect this sugar in *Salmonella typhimurium* and *Pasteurella multocida* lipopolysaccharides. While it is possible that heptoses could have been present below the detection limits of our methodology, it is also possible that this sugar is not present in the lipopolysaccharide of *Legionellaceae*.

Others have characterized the peptidoglycan of *L. pneumophila* by amino acid analysis (1), but the presence of peptidoglycan in *Tatlockia* and *Fluoribacter* species has not been previously reported. The glycan backbone of peptidoglycan is a polymer consisting of alternating units

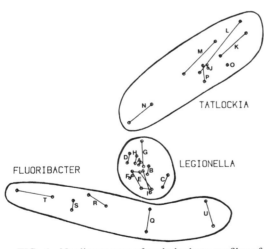

FIG. 1. Nonlinear map of carbohydrate profiles of five species of *Legionellaceae*. Replicates are joined by lines. The strains are identified on the map as follows. *L. pneumophila*: A, Philadelphia 1; B, Knoxville 1; C, Pontiac 1; D, SCH; E, Togus 1; F, Bloomington 2; G, Los Angeles 1; H, 684; I, Houston. *T. micdadei*: J, PPA-EK; K, PPA-PGH-12; L, PPA-JC; M, PPA-ML; N, PPA-GL; O, PPA-CAR; P, TATLOCK. *F. dumoffii*: Q, NY-23. *F. bozemanae*: R, WIGA; S, MI-15. Unclassified: T, E-327F. *F. gormanii*: U, LS-13.

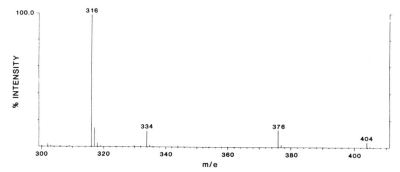

FIG. 2. Methane chemical ionization mass spectrum of X1.

of muramic acid and glucosamine. We have detected the presence of both muramic acid and glucosamine in *Legionella, Tatlockia,* and *Fluoribacter* strains. However, the glucosamine/muramic acid ratio was significantly higher than 1:1 in most isolates, suggesting the presence of an additional glucosamine-containing macromolecule.

The peaks labeled X1 and X2 have been identified as alditol acetates of amino dideoxy hexoses by electron impact and methane chemical ionization mass spectrometry. Figure 2 shows a chemical ionization mass spectrum of peak X1, for which a molecular weight of 375 is proposed based on interpreting the observed ion at 376 as the addition of a proton (M + 1 peak) and the ion at 404 as the addition of C_2H_5 (M + 29 peak) from the methane reagent gas. The loss of acetic acid (mass 60) and ketene (mass 42) from the M + 1 ion is observed at masses 316 and 334, respectively. The electron impact and methane chemical ionization spectra of component X2 are very similar to their X1 counterparts, indicating that X1 and X2 are isomers. In addition to the identification of these two rather unusual sugars, matching retention times and mass spectra compared with standards were obtained for the following neutral and amino sugars found in the hydrolysates of the *Legionellaceae*: rhamnose, fucose, ribose, mannose, glucose, muramic acid, and glucosamine.

Although strains of *Legionella, Tatlockia,* and *Fluoribacter* are similar in a number of characteristics, important differences are apparent not only by DNA homology, but also by a growing number of analytic techniques. The carbohydrate profiles of these bacteria support the taxonomic distinction between these genera. In addition, carbohydrate analysis by gas chromatography-mass spectrometry is a relatively simple yet powerful technique for the identification and characterization of bacteria.

This work was supported by Public Health Service grants 27135 and EY04715 from the National Institutes of Health, by the Veterans Administration Medical Research Service, and by the National Science Foundation EPSCOR Program.

LITERATURE CITED

1. **Amano, K.-I., and J. C. Williams.** 1983. Peptidoglycan of *Legionella pneumophila*: apparent resistance to lysozyme hydrolysis correlates with a high degree of peptide cross-linking. J. Bacteriol. **153**:520–526.

2. **Brown, A., G. M. Garrity, and R. M. Vickers.** 1981. *Fluoribacter dumoffii* (Brenner et al.) comb. nov. and *Fluoribacter gormanii* (Morris et al.) comb. nov. Int. J. Syst. Bacteriol. **31**:111–115.

3. **Collins, M. T., F. Espersen, N. Høiby, S.-N. Cho, A. Friis-Møller, and J. S. Reif.** 1983. Cross-reactions between *Legionella pneumophila* (serogroup 1) and twenty-eight other bacterial species, including other members of the family *Legionellaceae*. Infect. Immun. **39**:1441–1456.

4. **Fox, A., J. R. Hudson, S. L. Morgan, Z.-T. Zhu, and P. Lau.** 1983. Capillary gas chromatographic analysis of alditol acetates of neutral and amino sugars in bacterial cell walls. J. Chromatogr. **256**:429–438.

5. **Fox, A., J. H. Schwab, and T. Cochran.** 1980. Muramic acid detection in mammalian tissues by gas-liquid chromatography-mass spectrometry. Infect. Immun. **29**:526–531.

6. **Garrity, G. M., A. Brown, and R. M. Vickers.** 1980. *Tatlockia* and *Fluoribacter*: two new genera of organisms resembling *Legionella pneumophila*. Int. J. Syst. Bacteriol. **30**:609–614.

7. **Hudson, J. R., S. L. Morgan, and A. Fox.** 1982. High-resolution glass capillary columns for the separation of alditol acetates of neutral and amino sugars. J. High Resolut. Chromatogr. Chromatogr. Commun. **5**:285–290.

7a. **Karr, D. E., W. F. Bibb, and C. W. Moss.** 1982. Isoprenoid quinones of the genus *Legionella*. J. Clin. Microbiol. **15**:1044–1048.

8. **Lema, M., and A. Brown.** 1983. Electrophoretic characterization of soluble protein extracts of *Legionella pneumophila* and other members of the family *Legionellaceae*. J. Clin. Microbiol. **17**:1132–1140.

9. **Morgan, S. L., and C. A. Jacques.** 1982. Characterization of simple carbohydrate structure by glass capillary pyrolysis gas chromatography and cluster analysis. Anal. Chem. **54**:741–747.

10. **Sammon, J. W., Jr.** 1969. A non-linear mapping for data structure analysis. IEEE Trans. Comput. **C-18**:401–409.

11. **Wong, K. H., C. W. Moss, D. H. Hochstein, R. J. Arko, and W. O. Schalla.** 1979. Endotoxicity of the Legionnaires' disease bacterium. Ann. Intern. Med. **90**:624–627.

Taxonomic Evaluation of Amino Acid Metabolism by *Legionella* Species

MARTIN J. FRANZUS, GEORGIA B. MALCOLM, AND LEO PINE

Division of Bacterial Diseases, Center for Infectious Diseases, Centers for Disease Control, Atlanta, Georgia 30333

Amino acids are reported to be the sole source of energy and carbon for the growth of *Legionella* species (5, 9). Because the nine described species are distinguished primarily on the basis of DNA homology and serological reactivities, it appeared that a practical phenotypic differentiation might be made on the basis of specific amino acid metabolism. We elected to study the utilization of single amino acids by washed cell suspensions because such tests are more rapid and less complex than similar tests with growing cultures. We have compared the aerobic formation of high pH as indicated by phenol red and the anaerobic reduction of methylene blue by cell suspensions of *Legionella* species given single amino acids as substrates.

All strains were obtained from the Centers for Disease Control, Atlanta, Ga. All cultures or test samples were checked for contamination by direct or microscopic observation or by streak plates on blood or Trypticase (BBL Microbiology Systems) soy agar plates. We examined a total of 41 strains of *Legionella pneumophila* (eight serogroups), 17 strains of the remaining eight *Legionella* species, 24 *Legionella*-like organisms (LLOs), and, for comparative purposes, 13 strains of gram-negative bacteria, including species of *Pseudomonas, Escherichia, Alcaligenes, Flavobacterium, Bordetella*, and *Achromobacter*. Each strain was grown on three slants of buffered charcoal-yeast extract agar (4) until confluent growth was reached (approximately 72 h, maximum). The cultures were stored at 5°C and used within 72 h. Aseptic procedures were used through all manipulations. The cells were washed from three slants and centrifuged, and the pellet was suspended in 3.0 ml of distilled water. To each well of a 24-well flat-bottomed cell culture plate (3.5-ml capacity per well) was added 0.1 ml of cell suspension, 1 ml of 2.0 mM MOPS (morpholinepropanesulfonic acid) buffer (pH 6.3) with 25 μg of Tween 80, trace metals (6), and 1 of 21 amino acids at 30 μM. Each plate also had three wells, used as controls, to which all constituents were added except the amino acid. After 24 h, 4 drops of 0.03% phenol red was added to each well, the color was rated 0 (yellow), 1 (orange), 2 (pink), or 3 (very pink), and the plate was returned to the shaker for an additional 24 h for a final reading. For anaerobic tests, the cell suspensions were similar, but 1.5 ml of each was placed in a Becton Dickinson Red Dot 2.0-ml Vacutainer tube to which was added 0.1 ml of 0.02% methylene blue solution. The red rubber vaccine stopper was reinserted, and the tube was evacuated with the use of a 21-gauge needle attached to a vacuum pump. Tubes were incubated at 35°C for 24 h; readings of the relative reduction of methylene blue were recorded several times within this period.

Aerobically, no single species of *Legionella* could be delineated from another species by its amino acid metabolic profile (Table 1); similarly, no serogroup within the species *L. pneumophila* could be so identified. The reactions observed with the LLOs were essentially identical to those with *L. pneumophila*. However, these potential legionellae differed considerably from the non-*Legionella* gram-negative strains, which showed much more rapid and stronger alkaline reactions, particularly with the aspartic, pyruvic, and succinyl coenzyme A groups of amino acids (Table 1).

Of the 21 L-amino acids tested, the legionellae did not elicit any noticeable change within 24 h with ornithine, threonine, glycine, cystine, isoleucine, valine, methionine, leucine, tryptophan, phenylalanine, or tyrosine; at 48 h the reactions with these amino acids were, at best, only minimal. Of the remaining amino acids, glutamic acid was the most rapidly used, and strong reactions closely followed with aspartic acid, asparagine, lysine, alanine, and serine. Proline always gave an initial acid reaction which became alkaline by 48 h. Histidine showed strong permanent acid reactions with all of the legionellae but became alkaline with most of the non-*Legionella* gram-negative strains.

Anaerobically, reduction of the methylene blue progressed from the bottom of the vial to within 3 to 4 mm of the surface; the vials were rated according to the height of the uncolored column and not by the rates of reduction. The anaerobic utilization of the amino acids did not serve to characterize specifically any species or serogroup of *Legionella* (Table 2). Although there appeared to be marked differences between certain species (e.g., *L. pneumophila, L. gormanii*, and *L. longbeachae* in utilization of glycine, cystine, phenylalanine, and tyrosine), these differences are interpreted as being quantitative rather than qualitative. *L. pneumophila* strains and the LLOs, as a group, showed slow-

TABLE 1. Aerobic utilization of single amino acids by washed cell suspensions of *Legionella* species as indicated by phenol red color change

Species	No. of strains	Group I						Group II		Group III					Group IV			Group V			Group VI		Control[b]		
		Gln	Glu	Arg	Pro	Orn	His	Asp	Asn	Ala	Ser	Thr	Gly	Cys	Ile	Val	Met	Lys	Leu	Try	Phe	Tyr	1	2	3
L. pneumophila	41	2	3	2	2	0	A[c]	2	2	2	2	1	1	0	1	1	1	2	1	0	0	0	0	0	0
L. gormanii	1	2	2	2	2	0	A	2	1	2	2	2	0	0	0	0	0	0	0	0	0	0	0	1	0
L. dumoffii	3	1	3	1	2	0	A	2	1	2	2	0	1	0	1	0	1	1	1	1	0	0	0	0	0
L. bozemanii	1	1	3	2	2	1	A	2	2	2	2	1	1	0	1	1	1	2	1	1	1	0	1	0	0
L. jordanis	3	2	3	2	2	0	A	2	2	2	2	1	1	0	1	0	1	1	0	0	0	0	1	0	0
L. micdadei	2	1	1	1	2	0	A	1	1	2	1	0	1	0	1	0	0	2	0	0	0	0	0	0	0
L. longbeachae	5	2	3	1	2	0	A	1	1	2	2	0	1	0	2	0	0	1	1	1	1	1	0	0	0
L. wadsworthii	1	2	3	1	1	0	A	2	1	2	2	1	1	0	1	0	1	1	1	1	1	1	0	0	0
L. oakridgensis	1	1	3	1	2	1	A	2	2	3	3	1	1	0	1	1	1	0	0	0	1	1	0	0	0
LLOs	24	2	3	1	2	1	A	2	1	2	2	0	0	0	1	0	0	0	0	0	0	0	0	1	0
Non-*Legionella*, gram negative	13	2	3	2	2	2	2	3	3	3	2	2	3	1	2	1	2	1	1	1	0	0	1	0	0

[a] Reactions were read at 48 h with a subjective rating of 0 (yellow), 1 (orange), 2 (pink), or 3 (very pink). Amino acid metabolic groups are as given by Lehninger (3): I, glutamate; II, asparate; III, pyruvate; IV, succinyl coenzyme A; V, acetyl coenzyme A; VI, fumarate.

[b] Three control wells having the buffer-cell suspension with no added amino acids were present in each 24-well plate.

[c] A, Acid reaction.

TABLE 2. Anaerobic reduction of methylene blue by washed cell suspensions of *Legionella* species using single amino acids as substrates

Species	No. of strains	Group I						Group II		Group III					Group IV			Group V			Group VI		Control[b]		
		Gln	Glu	Arg	Pro	Orn	His	Asp	Asn	Ala	Ser	Thr	Gly	Cys	Ile	Val	Met	Lys	Leu	Try	Phe	Tyr	1	2	3
L. pneumophila	44	2	4	2	2	2	3	3	2	1	2	2	0	0	2	2	1	3	2	0	0	0	0	0	0
L. gormanii	1	4	4	3	4	2	3	4	4	3	3	3	3	3	4	4	3	3	4	2	4	3	3	3	3
L. dumoffii	2	3	4	2	4	2	3	3	3	3	3	2	3	0	2	3	2	2	2	0	2	0	0	0	0
L. bozemanii	1	3	4	2	3	2	2	3	3	4	2	3	0	2	4	3	3	2	2	2	2	3	0	0	0
L. jordanis	2	4	4	3	4	3	3	4	4	4	4	4	2	1	3	2	3	3	3	3	2	3	0	0	0
L. micdadei	2	3	3	1	3	4	3	2	2	2	3	3	2	3	2	1	1	1	2	2	3	0	0	0	0
L. longbeachae	5	4	4	3	4	3	3	4	2	4	3	4	3	2	3	4	2	3	3	2	3	2	0	0	0
L. wadsworthii	1	3	3	2	3	2	1	3	3	3	3	2	0	0	2	2	1	3	2	2	0	0	0	0	0
L. oakridgensis	1	3	4	3	3	2	3	4	4	3	3	3	2	2	2	2	2	3	3	2	2	2	3	2	2
LLOs	6	3	4	3	2	1	4	4	4	1	2	2	1	1	2	1	2	1	2	1	1	0	0	0	0
Non-*Legionella*, gram negative	12	4	4	3	4	2	4	4	4	3	3	4	4	4	4	3	3	3	4	3	4	2	2	3	2

[a] Reactions were read at 24 h, with ratings 0 to 4 given for the relative amount of each vial (bottom to top) having methylene blue totally reduced. Amino acid metabolic groups are as given by Lehninger (3): I, glutamate; II, asparate; III, pyruvate; IV, succinyl coenzyme A; V, acetyl coenzyme A; VI, fumarate.

[b] Three control vials having the buffer-cell suspension with no added amino acids were present in each test.

er and less complete reduction and were less metabolically active than the remaining species. Similarly, *L. dumoffii*, *L. wadsworthii*, and *L. micdadei* were of low activity. Strong endogenous reactions were observed with non-*Legionella* gram-negative species, *L. oakridgensis*, and *L. gormanii*. All strains of *L. pneumophila* showed strong utilization of glutamate, histidine, aspartate, and lysine. All other *Legionella* species showed strong reductions with glutamine, glutamate, arginine, proline, aspartate, asparagine, alanine, serine, and threonine. Ornithine, glycine, cystine, isoleucine, methionine, lysine, leucine, tryptophan, phenylalanine, and tyrosine were poorly used by all species of *Legionella*.

As originally described for washed suspensions of *L. pneumophila* by Weiss et al. (10), glutamate was the most actively metabolized substrate; glutamate, glutamine, glucose, succinate, and acetate were tested. However, *Legionella* species very rapidly metabolize glutamate, glutamine, arginine, proline, histidine, aspartate, asparagine, alanine, serine, and lysine aerobically; these amino acids plus ornithine, threonine, leucine, and isoleucine are utilized anaerobically. With the exceptions of arginine, threonine, and serine, the amino acids used aerobically are not the amino acids required for growth (2, 7, 8), and it may be assumed that these aerobically metabolized amino acids represent primary potential sources of carbon and energy. Those additional amino acids used for active reduction of methylene blue also appear to be available to the cell and may function as secondary carbon sources. Tyrosine and phenylalanine, which have been shown to be used for production of brown pigments by growing cells (1), were not actively used by any of the cell suspensions under our conditions, and no pigment was produced. However, these amino acids were used aerobically by other genera with pigment production.

The genus *Legionella* shows a pattern of metabolized amino acids which is different from that observed with non-*Legionella*, gram-negative species of *Pseudomonas*, *Escherichia*, *Alcaligenes*, *Flavobacterium*, and *Achromobacter*. These genera can show a high endogenous metabolism plus strong qualitative and quantitative differences from *Legionella* in the metabolism of histidine, tyrosine, tryptophan, phenylalanine, methionine, leucine, glycine, and cystine.

LITERATURE CITED

1. **Baine, W. B., and J. K. Rasheed.** 1979. Aromatic substrate specificity of browning by cultures of the Legionnaires' disease bacterium. Ann. Intern. Med. **90:**619–620.
2. **George, J. R., L. Pine, M. W. Reeves, and W. K. Harrell.** 1980. Amino acid requirements of *Legionella pneumophila*. J. Clin. Microbiol. **11:**286–291.
3. **Lehninger, A. L.** 1975. Biochemistry. Worth Publishers, Inc., New York.
4. **Pasculle, A. W., J. C. Feeley, R. J. Gibson, L. G. Cordes, R. L. Meyerowitz, C. M. Patton, G. W. Gorman, C. L. Carmack, J. W. Ezzell, and J. N. Dowling.** 1980. Pittsburgh pneumonia agent: direct isolation from human lung tissue. J. Infect. Dis. **141:**727–732.
5. **Pine, L., J. R. George, M. W. Reeves, and W. K. Harrell.** 1979. Development of a chemically defined liquid medium for growth of *Legionella pneumophila*. J. Clin. Microbiol. **9:**615–626.
6. **Reeves, M. W., L. Pine, S. H. Hutner, J. R. George, and W. K. Harrell.** 1981. Metal requirements of *Legionella pneumophila*. J. Clin. Microbiol. **13:**688–695.
7. **Ristroph, J. D., K. W. Hedlund, and S. Gowda.** 1981. Chemically defined medium for *Legionella pneumophila* growth. J. Clin. Microbiol. **13:**115–119.
8. **Tesh, M. J., and R. D. Miller.** 1981. Amino acid requirements for *Legionella pneumophila* growth. J. Clin. Microbiol. **13:**865–869.
9. **Warren, W. J., and R. D. Miller.** 1979. Growth of Legionnaires disease bacterium (*Legionella pneumophila*) in chemically defined medium. J. Clin. Microbiol. **10:**50–55.
10. **Weiss, E., M. G. Peacock, and J. C. Williams.** 1980. Glucose and glutamate metabolism of *Legionella pneumophila*. Curr. Microbiol. **4:**1–6.

Susceptibility of *Legionella pneumophila* to Iron-Binding Agents

FREDERICK D. QUINN AND EUGENE D. WEINBERG

Indiana University, Bloomington, Indiana 47405

The experiments in the present study were designed to determine the effect of iron-binding compounds on the growth of *Legionella pneumophila* Bloomington 2 (ATCC 33155). Results obtained with *L. pneumophila* were compared with those for *Legionella jordanis* BL-540, *Escherichia coli* K-12 and W1485, and a clinical isolate and a laboratory strain of *Salmonella typhimurium*.

Cultures were grown in buffered yeast extract broth which contained 18 to 24 μM iron. Various amounts of either 1,10-phenanthroline, apotransferrin, apoconalbumin, deferoxamine (Desferal; Ciba Pharmaceutical Co., Summit, N.J.), or enterobactin were included in the medium 1 h before addition of bacterial inoculum. Cultures were shaken at 41, 37, or 30°C at 200 rpm for 24 to 36 h. Doubling times at 41, 37, and 30°C were

2.8, 3.2, and 3.9 h, respectively.

The minimum inhibitory concentration of 1,10-phenanthroline in buffered yeast extract broth for *L. pneumophila* at 41, 37, and 30°C was 20, 28, and 34 μM, respectively. At ≥34 μM, 1,10-phenanthroline killed 99.99% of *L. pneumophila* cells within 30 h. At 41°C, the bactericidal action of the chelator was more rapid than at 37 or 30°C (Fig. 1).

The concentrations of 1,10-phenanthroline necessary to inhibit growth at 41, 37, and 30°C were the same for *L. pneumophila* and *L. jordanis*. However, these amounts were much lower than the 1,000 μM necessary for the inhibition of growth of *E. coli* K-12 and W1485 or the 1,600 μM required to inhibit either strain of *S. typhimurium*.

The minimum inhibitory concentration for transferrin or conalbumin against *L. pneumophila* or *L. jordanis* was 20 μM, regardless of temperature (Fig. 2). The strains of *E. coli* and *S. typhimurium* required ≥12-fold-higher concentrations of transferrin or conalbumin for inhibition at the three temperatures.

Deferoxamine, a hydroxamate siderophore, inhibited *L. pneumophila* at 50 μM at all three temperatures. Enterobactin, a catechol siderophore, was similarly active at 110 μM. At concentrations as high as 500 μM, these compounds had no effect on the growth of *E. coli* or *S.*

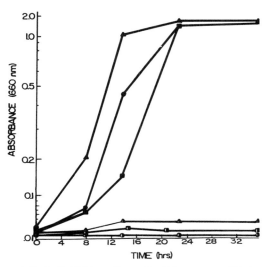

FIG. 2. Inhibition of growth of *L. pneumophila* by transferrin at 30, 37, and 41°C. *L. pneumophila* was grown in buffered yeast extract broth containing 0 μM transferrin at 41°C (▲), 37°C (●), and 30°C (■), and containing 20 μM transferrin at 41°C (△), 37°C (○), and 30°C (□).

typhimurium at any of the three temperatures.

Ferric pyrophosphate and ferric chloride were equally effective in suppressing the inhibitory action of 1,10-phenanthroline. At 30 and 37°C, between 10 and 40 μM was required to suppress the bactericidal action of 34 μM 1,10-phenanthroline. At both temperatures, 10 μM ferric pyrophosphate allowed growth after a 16-h lag phase, whereas 40 μM gave growth curves indistinguishable from those of 1,10-phenanthroline-free controls. However, iron compounds at concentrations as high as 100 μM were unable to suppress the bactericidal action of 1,10-phenanthroline at 41°C.

Ferric pyrophosphate at 10 μM suppressed the inhibition by transferrin and conalbumin of *L. pneumophila* at each of the three temperatures. Ferric pyrophosphate also suppressed the inhibitory effects of each siderophore. The iron salt at 10 μM was able to reverse completely the inhibitory effects of enterobactin. Ferric pyrophosphate at 50 μM was required to completely suppress deferoxamine inhibition of *L. pneumophila* cultures. Similar results were obtained with *L. jordanis*.

As a soil and water organism and as, occasionally, a mammalian pathogen, how can *L. pneumophila* survive and grow in diverse environments despite the relative ease with which it can be deprived of iron? Possibly the organism possesses a novel type of iron-chelating system or can utilize siderophores other than enterobactin

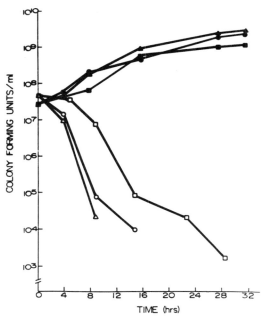

FIG. 1. *L. pneumophila* grown in buffered yeast extract broth containing 0 μM 1,10-phenanthroline at 41°C (▲), 37°C (●), and 30°C (■), and containing 40 μM 1,10-phenanthroline at 41°C (△), 37°C (○), and 30°C (□).

or deferoxamine (5, 9). This organism does not necessarily need more iron than other bacteria, but apparently has an unusual method for obtaining the metal from soil, water, and host niches.

At fever temperatures, the growth of many gram-negative bacterial pathogens is inhibited unless they are provided with extra iron or exogenous siderophores (3, 6). In the present study, 1,10-phenanthroline suppressed iron acquisition by *Legionella* species more efficiently at 41°C, but host defense iron-binding proteins were equally active at each of the temperatures employed.

Growth of *L. pneumophila* in human hosts may require increased plasma iron levels. The disease caused by *L. pneumophila* is especially prevalent in middle-aged and elderly males (8). Such persons may have a tendency to develop iron overload due to excessive absorption of iron from the diet or, in cases of alcoholism or liver disease, increased release of stored iron from the breakdown of hepatic parenchymal cells (4, 7). Accordingly, plasma iron levels, especially in middle-aged and elderly male patients, should be monitored. If elevated, hypoferremic procedures should be contemplated (1, 2). Differences in plasma iron levels between patients and resistant individuals may be a factor in the lack of communicability of the disease. The organism seems to be an opportunistic pathogen with an extremely narrow pathogenic niche. If iron levels, which may be an important part of the niche, can be controlled, the disease may be prevented or more easily managed. If use of iron withholding is shown to be safe and efficacious in animal studies, the procedure might become useful as an adjunct to antibiotic therapy.

LITERATURE CITED

1. **Davis, W. D., Jr., and W. R. Arrowsmith.** 1953. The treatment of hemochromatosis by massive venesection. Ann. Intern. Med. **39**:723–724.
2. **Finch, C. A.** 1949. Iron metabolism in hemochromatosis. J. Clin. Invest. **28**:780–781.
3. **Garibaldi, J. A.** 1972. Influence of temperature on the biosynthesis of iron transport compounds by *Salmonella typhimurium.* J. Bacteriol. **110**:262–265.
4. **Hershko, C.** 1977. Storage iron regulation. Prog. Hematol. **10**:105–148.
5. **Murphy, T. P., D. R. S. Lean, and C. Nalewajko.** 1976. Blue-green algae: their excretion of iron-selective chelators enables them to dominate other algae. Science **192**:900–902.
6. **Perry, R. D., and E. D. Weinberg.** 1973. Effect of iron deprivation and temperature upon bacterial survival. Microbios **8**:129–135.
7. **Powell, L. W.** 1966. Normal human iron storage and its relation to ethanol consumption. Aust. Ann. Med. **15**:110–115.
8. **Storch, G., W. B. Baine, and D. W. Fraser.** 1979. Sporadic community-acquired Legionnaires' disease in the United States. A case control study. Ann. Intern. Med. **90**:596–600.
9. **Tison, D. L., D. H. Pope, W. B. Cherry, and C. B. Fliermans.** 1980. Growth of *Legionella pneumophila* in association with blue-green algae (cyanobacteria). Appl. Environ. Microbiol. **42**:768–772.

Composition and Structure of Dihydroxy, Monohydroxy, and Unsaturated Nonhydroxy Fatty Acids of *Legionella* Species

WILLIAM R. MAYBERRY

Department of Microbiology, Quillen-Dishner College of Medicine, East Tennessee State University, Johnson City, Tennessee 37614

In a previously reported study (5), five strains of *Legionella pneumophila* were shown to contain, in addition to the nonhydroxy fatty acids reported by other workers, low but significant levels (ca. 5 mol% of total fatty acid) of 3-hydroxy acids, predominantly iso-C_{14} (i14h) and normal-C_{20} (n20h). Additionally, these strains contained 1 to 3 mol% of a novel class of 2,3-dihydroxy fatty acids, predominantly iso-C_{14} (i14h$_2$). The hydroxylated fatty acids were non-extractable in chloroform-methanol at 2:1 (wall associated, bound) and were alkali stable and acid labile (amide linked), in contrast to the non-hydroxylated fatty acids (comprising 90 to 95 mol% of total fatty acid), which were extractable (lipid associated) and alkali labile (ester linked).

In the current study, 23 strains, representing eight of the nine currently named species of *Legionella*, including most of the known serogroups of these species, were examined for nonhydroxy, monohydroxy, and dihydroxy fatty acid content and composition. In addition, the 2,3-dihydroxy fatty acids were examined for stereochemistry (*erythro* versus *threo* configuration), and the unsaturated nonhydroxy fatty acids were examined for stereochemistry (*cis* versus *trans* configuration) as well as for number and position of unsaturation.

Strains studied included *L. pneumophila* Philadelphia 1, Knoxville 1, Togus 1, Bloomington 2, Los Angeles 1, Dallas 1, Chicago 2, Chicago 8, and Concord 3; *L. micdadei* TATLOCK, HEBA, and PI-1; *L. bozemanii* WIGA and MI-15; *L. dumoffii* NY-23 and TEX-KL; *L. long-*

beachae Long Beach 4, Tucker 1, and Atlanta 5; *L. jordanis* BL-540 and ABB-9; *L. gormanii* LS-13; and *L. wadsworthii* Wadsworth 81-716A. Fatty acids of the different classes were acquired by three methods: (i) direct acid hydrolysis of wet cell pellets, followed by extraction, esterification with acid methanol, and separation of methyl esters into nonhydroxy, monohydroxy, and dihydroxy ester classes by thin-layer chromatography; (ii) extraction of wet cell pellets with chloroform-methanol (2:1, vol/vol) with subsequent release of nonhydroxy fatty acids from the lipid extract by either alkaline or acid methanolysis, and release of hydroxylated fatty acids from the defatted cell residue by acid hydrolysis, followed by esterification and separation into monohydroxy and dihydroxy ester classes by thin-layer chromatography; and (iii) release of ester-linked (nonhydroxy) fatty acids from wet cell pellets by mild alkaline methanolysis, and release of amide-linked (hydroxylated) fatty acids from the partially deacylated cell residues by acid hydrolysis, followed by esterification and separation into monohydroxy and dihydroxy ester classes by thin-layer chromatography.

Model compounds (commercially available monounsaturated fatty acids of known position of unsaturation and *cis/trans* configuration) and the unsaturated nonhydroxy fatty acids of the *Legionella* species were esterified and converted to dihydroxy compounds by either *cis* hydroxylation with Woodward's reagent (4) or *trans* hydroxylation with Fenton's reagent (1). The dihydroxy esters thus formed, as well as the naturally occurring 2,3-dihydroxy acids from certain *Legionella* species, were analyzed by thin-layer chromatography on plates of Silica Gel H (Merck) impregnated with boric acid, under conditions which permitted separation of *erythro* and *threo* diols (6).

All fatty acid methyl ester classes were analyzed by gas-liquid chromatography on a Hewlett-Packard 5840A instrument equipped with flame ionization detectors, using a 50-m fused silica capillary column of OV-1. When indicated, gas chromatography-mass spectrometry was carried out on a Finnegan 4000 instrument operated in the electron impact ionization mode, using a 25-m fused silica capillary column of DB-5. Methyl esters of nonhydroxy fatty acids were analyzed without further derivatization, those of monohydroxy fatty acids were analyzed as the trimethylsilyl, trifluoroacetyl, and acetate derivatives, and those of dihydroxy fatty acids were analyzed as the trimethylsilyl, trifluoroacetyl, acetate, isopropylidene, and *n*-butylboronyl derivatives.

The nonhydroxy fatty acid profiles of the test strains were in good agreement with those reported for the various species by other workers (2, 3, 7, 8), thus validating the identity of the strains under investigation. All strains studied showed low but significant levels (5 to 10 mol% of total fatty acid, depending on the species) of monohydroxy fatty acids.

The strains were readily divisible into two groups: (i) those in which i14h was relatively high (>10 mol% of monohydroxy acid content), with low n14h, including all strains of *L. pneumophila, L. micdadei,* and *L. jordanis,* and (ii) those in which i14h was low (<3 mol% of monohydroxy acid content), with relatively high n14h, including all strains of *L. bozemanii, L. dumoffii, L. gormanii, L. longbeachae,* and *L. wadsworthii.* Each of the species showed a characteristic quantitative profile of monohydroxy fatty acids, permitting rapid, unequivocal separation.

In only two species, *L. pneumophila* and *L. micdadei,* were detectable amounts of 2,3-dihydroxy fatty acids found, at levels approximately 1 to 3 mol% of the total fatty acid content. In strains of *L. pneumophila,* the $i14h_2$ component was predominant, with low to trace levels of $n14h_2$ and anteiso-$15h_2$ ($a15h_2$). In strains of *L. micdadei,* the $i14h_2$ and $a15h_2$ components as well as the $n14h_2$ component were well represented. Trace levels of $n12h_2$ and $n13h_2$ were found in these species as well. The three methods of analysis yielded essentially similar results.

Table 1 summarizes the arrangement of the *Legionella* species into "similarity groups" on the basis of the three classes of cellular fatty acid. This table also indicates the utility of rank-ordering the major components in each class as an aid to species identification.

Gas-liquid chromatography and gas chromatography-mass spectrometry results indicate that the monohydroxy fatty acids are predominantly of the 3-hydroxy family and are saturated. Trace levels of 2-hydroxy acids are also indicated. Straight-chain acids with both even and odd numbers of carbon atoms are found. Among the branched-chain compounds, the acids with an even number of carbon atoms are predominantly iso branched, while the compounds with an odd number of carbon atoms are predominantly anteiso branched. These patterns reflect those of the nonhydroxy fatty acids.

Thin-layer chromatography results indicate that the naturally occurring 2,3-dihydroxy acids are of the *erythro* configuration. The unsaturated nonhydroxy fatty acids yield *erythro* diols on *cis* hydroxylation and *threo* diols on *trans* hydroxylation. Gas chromatography-mass spectrometry analysis of the derived diols yields fragments characteristic of 9,10-dihydroxylation. Thus, the unsaturated nonhydroxy fatty

TABLE 1. Similarity groupings of *Legionella* species by cellular fatty acid classes

Fatty acid type	Species	Fatty acids
Nonhydroxy	Iso-C$_{16}$ highest	
	L. pneumophila	i16:0 > a15:0 > a17:0 > n16:1 > i14:0
	L. longbeachae	i16:0 > a15:0 ≃ n16:1 > 17cyc ≥ a17:0
	Anteiso-C$_{15}$ highest	
	L. micdadei	a15:0 > a17:0 > i16:0 > n16:1 ≃ n16:0
	L. jordanis	a15:0 > a17:0 > i16:0 > n16:1 ≃ 17cyc
	L. dumoffii	a15:0 > a17:0 > i16:0 ≃ 17cyc > n16:1
	L. wadsworthii	a15:0 > a17:0 > n16:1 > i16:0 > 17cyc
	L. bozemanii	a15:0 > i16:0 > 17cyc ≃ a17:0 > n16:1 ≃ n16:0
	L. gormanii	a15:0 > i16:0 > a17:0 > n16:1 ≥ n16:0
Monohydroxy	Significant iso-C$_{14}$	
	L. pneumophila	i14h > n20h
	L. micdadei	i14h > a15h ≥ n20h ≃ a23h > n22h > a21h > n21h
	L. jordanis	a15h > i14h ≥ n20h > a19h > a21h
	Significant normal-C$_{14}$	
	L. dumoffii	n14h > n20h > n18h > n16h
	L. bozemanii	n14h ≃ n18h ≃ n20h > n16h
	L. wadsworthii	n18h ≃ n16h ≃ n14h > a19h
	L. gormanii	n14h > i16h ≃ n20h > n16h
	L. longbeachae	i16h > n20h > n14h ≃ n18h
2,3-Dihydroxy	*L. pneumophila*[a]	br14h$_2$/n14h$_2$/br15h$_2$ (~15:1:tr)
	L. micdadei[a]	br14h$_2$/n14h$_2$/br15h$_2$ (~7:2:5)

[a] Traces of n12h$_2$ and n13h$_2$ were found. 2,3-Dihydroxy acids were not detected in *L. bozemanii*, *L. dumoffii*, *L. gormanii*, *L. jordanis*, *L. longbeachae*, or *L. wadsworthii*.

acids are of the *cis*-9,10-monounsaturated (oleic) acid family, which is not unusual in aerobic procaryotes.

I thank the Biological Products Division, Leo Pine, and H. W. Wilkinson, all of the Centers for Disease Control, for provision of strains. I especially thank H. W. Wilkinson for periodic serological verification of strains, and K. J. Mayberry-Carson, East Tennessee State University College of Medicine, for maintenance of cultures, and, along with Randall J. Rogers, for excellent technical assistance. Manuscript preparation by Paula J. Williams is also gratefully acknowledged.

LITERATURE CITED

1. Buehler, C. A., and D. E. Pearson. 1970. Survey of organic synthesis, p. 220–221. Wiley-Interscience, New York.
2. Edelstein, P. H., D. J. Brenner, C. W. Moss, A. G. Steigerwalt, E. M. Francis, and W. L. George. 1982. *Legionella wadsworthii* species nova: a cause of human pneumonia. Ann. Intern. Med. 97:809–813.
3. Finnerty, W. R., R. A. Makula, and J. C. Feeley. 1979. Cellular lipids of the Legionnaires' disease bacterium. Ann. Intern. Med. 90:631–634.
4. Gunstone, F. D., and L. J. Morris. 1957. Fatty acids. V. Applications of the Woodward *cis*-hydroxylation procedure to long-chain olefinic compounds. J. Chem. Soc. (London) 1957:487–490.
5. Mayberry, W. R. 1981. Dihydroxy and monohydroxy fatty acids in *Legionella pneumophila*. J. Bacteriol. 147:373–381.
6. Morris, L. J. 1967. Separation of higher fatty acid isomers and vinylogues by thin-layer chromatography. Chem. Ind. (London) 1967:1238–1240.
7. Moss, C. W., D. W. Karr, and S. B. Dees. 1981. Cellular fatty acid composition of *Legionella longbeachae* sp. nov. J. Clin. Microbiol. 14:692–694.
8. Orrison, L. H., W. B. Cherry, R. L. Tyndall, C. B. Fliermans, S. B. Gough, M. A. Lambert, L. K. McDougal, W. F. Bibb, and D. J. Brenner. 1983. *Legionella oakridgensis*: unusual new species isolated from cooling tower water. Appl. Environ. Microbiol. 45:536–545.

Taxonomic Delineation of *Legionella* Species by the Presence and Relative Molecular Weights of Catalase and Peroxidase

LEO PINE, PAUL S. HOFFMAN, GEORGIA B. MALCOLM, AND ROBERT F. BENSON

Division of Bacterial Diseases and Biological Products Program, Center for Infectious Diseases, Centers for Disease Control, Atlanta, Georgia, 30333, and Microbiology Section, Department of Biology, Memphis State University, Memphis, Tennessee, 39152

Although *Legionella pneumophila* was described originally as being catalase positive, our observations of strain Philadelphia 1 showed little or no live cell catalase activity when it was grown on diverse media (6, 10). Subsequent studies with cell-free systems showed very little or no catalase activity in several strains of *L. pneumophila*, although extracts from other species of *Legionella* indicated the presence of strong catalase (7). Because reduced-oxygen-scavenging enzymes may play an important role in the growth of legionellae, we examined 44 strains for the presence of catalase, peroxidase, and superoxide dismutase (SOD).

All strains were maintained on buffered charcoal-yeast extract agar (BYCE) and were grown on this medium, Ristroph yeast extract agar, Feeley-Gorman agar, fortified Mueller-Hinton agar, or synthetic media with or without agar for experimental purposes. A simple reaction chamber was constructed with an attached Swinney filter to determine the live-cell catalase activity by using the assay of Beers and Sizer (1). Peroxidase was determined with o-dianisidine (12), and SOD was quantitatively determined with pyrogallol (9); the peak SOD activities of purified fractions were determined by the procedure of Winterbourn et al. (11). We prepared cell-free extracts by using a French press or sonifier. Clear extracts, obtained by centrifugation at 39,000 × g for 1 h, were analyzed directly or purified chromatographically by gel filtration through Sephadex G-200 (fine). Relative molecular weights were determined with thyroglobulin, ferritin, catalase, aldolase, bovine albumin, horseradish peroxidase, bovine erythrocyte SOD, and cytochrome c as molecular weight standards.

For *L. pneumophila*, including serogroups 1 to 8, the observed first-order rate constants (k) (5) for 1 ml of cell suspension (absorbance = 2.0 at 660 nm and a 1.8-cm path length) in a 20-ml reaction volume ranged from 0.000 to 0.002 for 19 strains, from 0.002 to 0.003 for 7 strains, and from 0.004 to 0.008 for 3 strains. Strains of *Legionella gormanii*, *Legionella wadsworthii*, and *Legionella oakridgensis* had values of 0.001 to 0.004. The highest live-cell catalase activities were found for *Legionella jordanis*, *Legionella longbeachae*, *Legionella micdadei*, and *Legionella bozemanii*, with k values of 0.010 to 0.035.

Values obtained with two strains of *Escherichia coli* and one strain of *Pseudomonas aeruginosa* were 0.004, 0.032, and 0.076 to 0.143, respectively. Two strains of *Legionella dumoffii* (NY-23 and TEX-KL) were particularly inconsistent and showed k values ranging from 0.000 to

TABLE 1. Catalase, peroxidase, and SOD activities of 39,000 × g supernatants of sonic extracts of *Legionella* species

Strain[a]	Protein (mg/ml)	U/mg of protein		
		Catalase[b]	Peroxidase[b] ($\times 10^2$)	SOD[c]
L. pneumophila				
Philadelphia 1 (1)	6.4	0.0	0.2	13.4
Knoxville 1 (1)	3.6	2.5	2.7	9.1
Bellingham 1 (1)	20.2	2.8	2.3	13.6
Burlington 1 (1)	29.2	0.3	0.9	18.3
Atlanta 4 (2)	15.3	0.8	0.6	18.9
Bloomington 2 (3)	28.6	3.6	1.4	14.7
Baltimore 1 (4)	28.0	0.9	1.2	10.8
Los Angeles 1 (4)	14.0	1.6	1.1	15.8
Cambridge (5)	8.7	0.7	1.1	9.3
Dallas 17 (5)	23.3	1.4	0.4	12.0
Chicago 2 (6)	14.3	0.9	0.5	14.8
L. bozemanii				
WIGA	1.1	9.6	0.0	16.0
L. dumoffii				
TEX-KL	1.4	17.5	3.8	8.2
NY-23	1.8	7.8	2.0	29.3
L. gormanii				
LS-13	11.2	2.3	1.2	18.8
L. jordanis				
BL-540	8.5	7.2	0.0	8.4
L. longbeachae				
LB-4 (1)	3.0	7.5	0.0	22.5
Tucker 1 (2)	12.6	17.2	0.0	24.7
L. micdadei				
TATLOCK	2.3	16.6	0.0	30.5
HEBA	2.1	16.7	0.0	21.7

[a] Numbers given in parentheses designate the serogroup.
[b] International units (12).
[c] A unit is that amount of enzyme giving 50% inhibition of the spontaneous oxidation of pyrogallol under test conditions (9).

0.010. Differences between the low catalase producers ($k \leq 0.003$) and high catalase producers ($k \geq 0.010$) were easily discerned by a catalase test in which 1 drop of cell suspension mixed with 1 drop of 3% H_2O_2 on a glass slide was covered with a cover slip and observed for the streaming of oxygen bubbles.

Because we observed variation in catalase reactivity for the strains of *L. dumoffii* and *P. aeruginosa*, we tested various media for their effect on the catalase content of selected strains. Two strains each of *E. coli* and *Pseudomonas*, when grown on diverse agar media containing added cysteine (BCYE, *Actinomyces* agar, Mueller-Hinton agar, and Feeley-Gorman agar), showed two- to fivefold decreases in catalase activity compared with that observed when they were grown on blood agar base with no cysteine. Of the media tested, BCYE gave the poorest catalase production by these species.

Of the media which supported the growth of legionellae, BCYE gave not only the best growth but also the highest catalase activity. Strains grown in the chemically defined *Legionella* broth or agar, fortified Mueller-Hinton agar, Ristroph yeast extract agar, or Feeley-Gorman agar were virtually without catalase activity. However, certain anomolies were apparent; cells of *L. dumoffii* grown on BCYE sometimes showed low live-cell catalase activities but had high activities in their cell-free extracts.

We analyzed the cell-free sonic extracts of 20 strains grown on BCYE for catalase, peroxidase, and SOD. Certain catalase and peroxidase activities were rapidly inactivated by high concentrations of H_2O_2, and competitive reactions occurred between catalase and peroxidase. With adjusted substrate concentrations and modified procedures, we were able to delineate the specific activities of each enzyme (Table 1). Eleven strains of *L. pneumophila* and a single strain of *L. gormanii* showed a very low specific activity

TABLE 2. Relative molecular weights of catalase, peroxidase, and SOD of *Legionella* species as determined with Sephadex G-200

Species	Strain	M_r		
		Catalase[a]	Peroxidase[b]	SOD
L. dumoffii	NY-23	274,000	161,000	29,000
	TEX-KL	274,000	161,000	29,000
L. jordanis	BL-540	304,000	—	34,000
	ABB-9	288,000	—	34,000
L. wadsworthii	81-716A	304,000	—	34,000
L. oakridgensis	Oak Ridge 10	246,000	—	28,000
L. bozemanii	WIGA	199,000	—	33,000
	D-62	199,000	—	33,000
L. micdadei	TATLOCK	152,000	—	29,000
	HEBA	130,000	—	29,000
L. longbeachae	Tucker 1	130,000	—	31,000
	Long Beach 4	169,000	—	36,000
L. gormanii	LS-13	—	137,000	33,000
L. pneumophila	Philadelphia 1	—	137,000	25,000
	Knoxville 1	—	137,000	33,000
	Bellingham 1	—	144,000	33,000
	Togus 1	—	123,000	29,000
	Bloomington 1	—	137,000	38,000
	Los Angeles 1	—	152,000	34,000
	Dallas 17	—	144,000	36,000
E. coli	O111:B4	—	304,000	34,000
	K-12 Su65-42	—	304,000	36,000
P. aeruginosa	6045	179,000	111,000	43,000

[a] No catalase activity described exhibited peroxidase activity; —, no activity.
[b] All peroxidase peak activities demonstrated concomitant catalase activity.

of catalase and a high peroxidase activity. The two strains of *L. dumoffii*, however, showed a strong catalase content, and their peroxidase activities were comparable to that of *L. pneumophila*. All other species had no peroxidase but had catalase at an average ninefold-greater specific activity than that observed for *L. pneumophila* strains. SOD was present in all strains but in concentrations which did not characterize the species.

Because competing reactions occur between peroxidase and catalase (2, 3), the enzymes were separated by using Sephadex G-200; relative molecular weights (M_r) were also determined (Table 2). As based upon the presence or absence of catalase or peroxidase and their M_r, *L. pneumophila* and *L. gormanii* were distinguished from the other species, having only a peroxidase ($M_r = 139,000$) with minor catalase-like activity. *L. dumoffii* was characterized by having both a catalase ($M_r = 274,000$) with no peroxidase activity and a peroxidase ($M_r = 161,000$) with minor catalase-like activity. All other species had only a catalase, which had no peroxidase-like activity. The catalases of *L. jordanis* and *L. wadsworthii* had a mean M_r of 299,000, that of *L. oakridgensis* was 246,000, and that of *L. bozemanii* was 199,000. The mean M_r of the catalases of *L. micdadei* and *L. longbeachae* was noticeably lower than in the above-named species, being 145,000.

Claiborne and Fridovich (2, 3) described two hydroperoxidases of *E. coli* B, a catalase (312,000 daltons) having virtually no peroxidase activity and a peroxidase (337,000 daltons) with strong catalase-like activity. The hydroperoxidases observed in our two control strains of *E. coli* had an M_r of 304,000. Values for the SOD of *E. coli* B are reported as 39,500 for Mn-SOD (8) and 38,700 for Fe-SOD (13). An iron-containing SOD ($M_r = 42,000$) has been described for *Bacteroides fragilis* (4). M_r values for the SOD of the *Legionella* strains ranged from 29,000 to 38,000 when the most rapidly eluting peak values observed were used for the determination.

Our results suggest the presence of novel catalases, peroxidases, and perhaps SODs within the genus *Legionella*. The presence of a peroxidase, having a reactivity similar to that of the hydroperoxidase I of Claiborne and Fridovich (2) but of much lower M_r, in *L. pneumophila* and *L. gormanii*, the presence of both catalase and peroxidase in *L. dumoffii*, and the presence of a very low-molecular-weight catalase in *L. micdadei* and *L. longbeachae* would appear to clearly separate these species from the remaining four species, *L. jordanis*, *L. wadsworthii*, *L. oakridgensis*, and *L. bozemanii*, which have high-molecular-weight catalase only.

LITERATURE CITED

1. **Beers, R. F., Jr., and I. W. Sizer.** 1952. A spectrophotometric method for measuring the breakdown of hydrogen peroxide by catalase. J. Biol. Chem. **195**:133–140.
2. **Claiborne, A., and I. Fridovich.** 1979. Purification of the *o*-dianisidineperoxidase from *Escherichia coli* B. J. Biol. Chem. **254**:4245–4252.
3. **Claiborne, A., and I. Fridovich.** 1979. Purification and characterization of hydroperoxidase II of *Escherichia coli* B. J. Biol. Chem. **254**:11664–11667.
4. **Gregory, E. M., and C. H. Dapper.** 1983. Isolation of iron-containing superoxide dismutase from *Bacteroides fragilis*: reconstitution as a Mn-containing enzyme. Arch. Biochem. Biophys. **220**:293–300.
5. **Hebert, D.** 1955. Catalase from bacteria (*Micrococcus lysodeikticus*). Methods Enzymol. **2**:784–788.
6. **Hoffman, P. S., and L. Pine.** 1982. Respiratory physiological cytochrome content of *Legionella pneumophila*. Curr. Microbiol. **1**:351–356.
7. **Hoffman, P. S., L. Pine, and S. Bell.** 1983. Production of superoxide and hydrogen peroxide in medium used to culture *Legionella pneumophila*: catalytic decomposition by charcoal. Appl. Environ. Microbiol. **45**:784–791.
8. **Keile, B. B., Jr., J. M. McCord, and I. Fridovich.** 1970. Superoxide dismutase from *Escherichia coli* B. J. Biol. Chem. **245**:6176–6181.
9. **Marklund, S., and G. Marklund.** 1974. Involvement of the superoxide anion radical in the autoxidation of pyrogallol and a convenient assay for superoxide dismutase. Eur. J. Biochem. **47**:469–474.
10. **Pine, L., J. R. George, M. W. Reeves, and W. K. Harrell.** 1979. Development of a chemically defined liquid medium for growth of *Legionella pneumophila*. J. Clin. Microbiol. **9**:615–626.
11. **Winterbourn, C. C., R. E. Hawkins, M. Brian, and R. W. Carrell.** 1975. The estimation of red cell superoxide dismutase activity. J. Lab. Clin. Med. **85**:337–341.
12. **Worthington Biochemical Corp.** 1972. Worthington enzyme manual, p. 43–45. Worthington Biochemical Corp., Freehold, N.J.
13. **Yost, F. J., Jr., and I. Fridovich.** 1973. An iron-containing superoxide dismutase from *Escherichia coli*. J. Biol. Chem. **248**:4905–4908.

Proteolytic Action in Actively Growing Legionellae

HANS E. MÜLLER

Staatliches Medizinaluntersuchungsamt Braunschweig, D-3300 Braunschweig, West Germany

The proteolytic action of non-proliferating cells of *Legionella pneumophila* on human serum was recently described (2). The induction of proteolytic enzymes in growing *L. pneumophila* and some other *Legionella* species was subsequently studied. For this purpose, a plate assay was developed to investigate the enzymatic activity of legionellae on human serum proteins under in vivo-like conditions. The following bacterial strains were investigated: *L. pneumophila* serogroups 1 through 6, *L. bozemanii*, *L. dumoffii*, *L. gormanii*, and *L. micdadei* (ATCC strains 33152, 33153, 33154, 33155, 33156, 33204, 33215, 33216, 33217, 33218, 33279, 33297, and Berlin 1). The bacteria were plated on ACES-buffered, sterile, filtered, yeast extract (BYE) agar containing about 50% sterile human serum or on BYE agar containing about 50% sterile human serum plus 10% aprotinin (20,000 U/ml), an inhibitor of proteases.

The growing bacteria release their soluble products, e.g., proteolytic enzymes which have an effect on human serum proteins, in a manner similar to that in an infected organism. The action of bacterial enzymes was studied by immunoelectrophoresis with monospecific rabbit antisera against 24 different serum proteins. The procedure used is summarized in Fig. 1. The concentrations of some serum proteins, i.e., prealbumin, α_1-lipoprotein, α_1-antitrypsin, haptoglobin, α_2-macroglobulin, β-lipoprotein, hemopexin, immunoglobulin A, and immunoglobulin G, with and without action of *L. pneumophila* enzymes, were determined by using commercial immunodiffusion plates. The wells of the immunodiffusion plates were filled with 5 µl of BYE agar containing human serum. After a diffusion time of 48 h, the diameters of the precipitates were measured. The concentrations of proteins were determined by means of a reference curve. The antisera used in immunoelectrophoresis and the results of immunoelectrophoresis and of the immunodiffusion assay are listed in Table 1.

All 24 serum proteins tested were altered enzymatically in the presence of legionellae, as could be demonstrated by immunoelectrophoretic patterns. The shifts were anodic and, in some proteins, cathodic. Furthermore, some protein bands disappeared. This strongly suggests that the shifts were affected by changes of electrical charges of the proteins by loss of terminal amino acids or peptides. Different en-

zymes, e.g., the extracellular protease described and characterized by Thompson et al. (5) but also one or more other hydrolases (3, 4, 6), may have been responsible for these phenomena. There was no substantial difference in activity of different *Legionella* species and their individual strains. However, *L. pneumophila* serogroup 1 strains seemed to be somewhat more active, and species other than *L. pneumophila* showed less proteolytic activity, which confirmed the findings of Berdal and Fossum (1). Gelatin liquefaction was positive in all species tested.

The fact that aprotinin showed no inhibition of the observed proteolysis may have been due to a lack of action of aminopeptidases on human proteins, because aprotinin inhibits their enzymatic activity (3).

There are important differences between the proteolytic action of non-proliferating cells in the absence of induction of proteinases (by growth on agar containing appropriate substrates) and of bacteria grown on BYE agar containing human serum. The results indicated that proteolytic enzymes are induced by their substrates during growth of the cells. The extent of enzyme inducibility and the degree of enzyme action on human serum proteins and other human proteins may determine the virulence of individual *Legionella* strains.

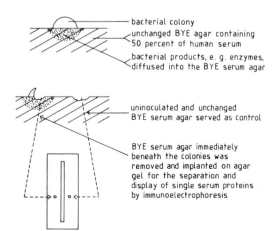

FIG. 1. Method used for studying the action of bacterial proteolytic enzymes on human serum or plasma proteins under in vivo-like conditions.

TABLE 1. Proteolytic action of legionellae on human serum proteins, investigated by immunoelectrophoresis and by a quantitative immunodiffusion plate assay

Serum protein	Degradation of proteins by cells[a]			
	With human serum			Without serum (immunoelectrophoresis)
	Immunoelec-trophoresis	Immunodiffusion plate assay		
		Without aprotinin	With aprotinin	
Prealbumin	AS	0%	5–6%	—
Albumin	D	ND	ND	—
α_1-Lipoprotein	D, AS	0–28%	100%	—
α_1-Acid glycoprotein	AS	ND	ND	CS
α_1-Antitrypsin	AS	12–14%	4–11%	—
α_1-Antichymotrypsin	D, AS	ND	ND	D, CS
α_1-β-Glycoprotein	D	ND	ND	—
Inter-α-trypsin inhibitor	D	ND	ND	—
Haptoglobin	D, AS	15–21%	23–44%	—
Ceruloplasmin	D	ND	ND	—
Gc-Globulin	D	ND	ND	—
α_2-Macroglobulin	D, AS	16%	8%	—
α_2-HS-Glycoprotein	D	ND	ND	—
α_2-PA-Glycoprotein	D	ND	ND	ND
α_2-Zn-Glycoprotein	D, AS	ND	ND	—
β-Lipoprotein	D, AS	16–28%	40–52%	AS
Transferrin	D, AS	ND	ND	—
β_{1C}/β_{1A}-Globulin	D, AS	ND	ND	—
β_{1E}-Globulin	D, AS	ND	ND	CS
Hemopexin	D, AS	71%	71%	—
β_2-Glycoprotein I	D	ND	ND	D
Immunoglobulin A	D, AS	26–50%	35–46%	—
Immunoglobulin M	D, AS	ND	ND	—
Immunoglobulin G	D, AS	46–81%	63–100%	—

[a] Cells were grown on BYE agar with or without human serum. Abbreviations: AS, anodic shift of band; CS, cathodic shift of band; D, degradation of band; ND, not done; —, no reaction.

LITERATURE CITED

1. Berdal, B. P., and K. Fossum. 1982. Occurrence and immunogenicity of proteinases from Legionella species. Eur. J. Clin. Microbiol. 1:7–11.
2. Müller, H. E. 1980. Proteolytic action of Legionella pneumophila on human serum proteins. Infect. Immun. 27:51–53.
3. Müller, H. E. 1981. Enzymatic profile of Legionella pneumophila. J. Clin. Microbiol. 13:423–426.
4. Nolte, F. S., G. E. Hollick, and R. G. Robertson. 1982. Enzymatic activities of Legionella pneumophila and Legionella-like organisms. J. Clin. Microbiol. 15:175–177.
5. Thompson, M. R., R. D. Miller, and B. H. Iglewski. 1981. In vitro production of an extracellular protease by Legionella pneumophila. Infect. Immun. 34:299–302.
6. Thorpe, T. C., and R. D. Miller. 1981. Extracellular enzymes of Legionella pneumophila. Infect. Immun. 33:632–635.

Polyacrylamide Gel Electrophoresis in the Characterization of Unusual *Legionella pneumophila* Strains Isolated from Patients in a San Francisco Hospital

MARJORIE BISSETT, GENEVIEVE NYGAARD, JADE LEE, JANICE LOPEZ, and M. CLAIRE MELANEPHY

Microbial Diseases Laboratory, California Department of Health Services, Berkeley, California 94704, and University of California Medical Center, San Francisco, California 94143

Specific identification of legionellae is based currently on results of the direct fluorescent-antibody test (DFA). This test requires the use of specific fluorescein-conjugated antibodies. To date, no common specific antigen for legionellae has been described. Therefore, at least 17 different conjugates are required to identify the known species and serogroups of the legionellae. In this report, we describe the characterization of strains of *Legionella pneumophila* that react weakly with some, but not all, conjugates prepared to different strains of serogroup 1. The described strains were encountered in nosocomial infections that occurred at the University of California Medical Center hospital in San Francisco. The epidemiology of this outbreak is reported elsewhere.

In October 1981, one of us (M.C.M.) isolated *Legionella* organisms from two kidney transplant patients. These strains reacted very weakly or not at all with conjugates prepared to the Knoxville 1 strain of *L. pneumophila* serogroup 1. The Centers for Disease Control laboratory reported these strains as "unidentified *Legionella*-like" organisms. The strains, designated by the Centers for Disease Control as SF 9 and SF 10, were studied further at the California Microbial Diseases Laboratory. In June and July 1982, two additional strains of *Legionella*-like organisms were isolated from kidney transplant patients at the hospital. These isolates, designated SF 13 and SF 14, reacted similarly to SF 9 and SF 10 with our Knoxville conjugate, but reacted in DFA tests at the Centers for Disease Control at the 4+ level with a lot of *L. pneumophila* serogroup 1 conjugate prepared by combining Knoxville 1 and Bellingham 1 antisera. All four SF strains reacted strongly (4+) in DFA tests using conjugate prepared against the SF 9 strain.

Biochemical, DFA cross-absorption, gel diffusion, gas-liquid chromatography, and polyacrylamide gel electrophoresis (PAGE) assays were performed to characterize these strains. Biochemical and gas-liquid chromatography tests of the SF strains produced results that were similar to reactions obtained with other *L. pneumophila* strains. DFA tests using conjugates to *L. pneumophila* serogroups 2 through 8, *Legionella micdadei*, *Legionella bozemanii*, *Legionella gormanii*, *Legionella dumoffii*, *Legionella longbeachae* serogroups 1 and 2, and *Legionella jordanis* were negative. Conjugates were prepared to each of the four SF strains. Table 1 shows the titers obtained with the homologous organisms and with heterologous *L. pneumophila* serogroup 1 named strains. Cross-absorption tests (1) indicated that the four strains contained antigens distinct from those of the named strains of *L. pneumophila* serogroup 1 as well as antigens that cross-reacted with some serogroup 1 strains, namely, Bellingham 1 and OLDA (Table 1). The major cross-reacting antibodies could be absorbed from the SF strain conjugates without significantly lowering the titer to the homologous organism or to the other SF organisms (Table 1). Gel diffusion tests (not shown) indicated that these four strains had a common specific precipitin band not present in soluble antigen extracts (1) of *L. pneumophila* serogroup 1 named strains (Knoxville 1, Philadelphia 2, Bel-

TABLE 1. Relationship of SF strains and named strains of *L. pneumophila* serogroup 1 as determined by DFA staining

Conjugate	Antigen	Antibody titer[a] of unabsorbed conjugate	Antibody titer[a] of conjugate absorbed with:		
			SF 9	SF 14	Bellingham 1
SF 9	SF 9	320	<10	<10	320
	SF 10	320	<10	<10	160
	SF 13	640	<10	<10	160
	SF 14	160	<10	<10	160
	Bellingham 1	160	40	40	<10
	OLDA	20	10	10	<10
	Knoxville 1	<10	—[b]	—	—
SF 14	SF 9	320	<10	<10	160
	SF 10	160	<10	<10	160
	SF 13	320	<10	<10	160
	SF 14	320	<10	<10	160
	Bellingham 1	20	<10	<10	<10
	OLDA	160	<10	20	40
	Knoxville 1	<10	—	—	—
Bellingham 1	SF 9	<10	—	—	—
	SF 10	<10	—	—	—
	SF 13	<10	—	—	—
	SF 14	20	20	10	<10
	Bellingham 1	80	80	80	<10
	OLDA	40	80	40	<10
	Knoxville 1	<10	—	—	—

[a] Titer represents the reciprocal of the conjugate dilution at which staining intensity of the majority of organisms was at least 3+ to 4+.
[b] Antigen not tested with absorbed conjugate because titer with unabsorbed conjugate was <10.

lingham 1, Pontiac 1, and OLDA), *L. pneumo-phila* serogroups 2 through 8, or other species of legionellae. These strains were further characterized by sodium dodecyl sulfate (SDS)-pore gradient PAGE assay utilizing vertical slabs (14 cm by 16 cm by 1.5 mm) of polyacrylamide and *N*,*N*′-diallyltartardiamide in a gradient of 13.5% (14.2% T, 4.8% C) to 27% (28.4% T, 4.8% C) acrylamide (Fig. 1). The gels were run in a discontinuous buffer system by a modification of the method of Laemmli (2, 3). Buffer in the resolving gels was at pH 8.5; the stacking gel was buffered at pH 7.0. Bromphenol blue (0.4%) was added to the stacking gel as a tracking dye. Equal volumes of the soluble antigen extract sample and disruption mixture (0.25 ml of 1 M Tris [pH 7.0], 0.25 ml of sucrose [60%, wt/vol], 0.25 ml of β-mercaptoethanol, 1.62 ml of distilled water, and 1.0 ml of 10% SDS solution) were heated for 5 min at 95°C. The sample inoculum was 20 μl, containing 40 to 50 μg of protein. The gels were electrophoresed in Tris buffer (pH 8.5) at 5°C and 40 mA constant

current until the tracking dye line reached the resolving gel line and then at 150 V constant voltage until the tracking dye line was approximately 1 cm from the bottom of the gel. Gels were removed from the plates, fixed, stained with Coomassie blue R-250 stain, destained, and dried. A low-molecular-weight standard (Bio-Rad Laboratories) was tested with each gel.

The patterns obtained in tests of the soluble antigen extracts of the SF strains, several named strains of *L. pneumophila* serogroup 1, and serogroups 2 through 4 are shown in Fig. 1. The SF strains, SF 9, SF 10, and SF 13, gave identical patterns. The pattern for SF 14 differed from the other SF strains; SF 14 apparently had lost or altered some of the peptides present in the other SF strains. Recognizable differences existed between patterns seen with the named strains within serogroup 1 and serogroups 2 through 4. Different patterns were also observed when pore gradient PAGE was run with other serogroups and species of legionellae. Results of the diffusion-in-gel and cross-absorption tests

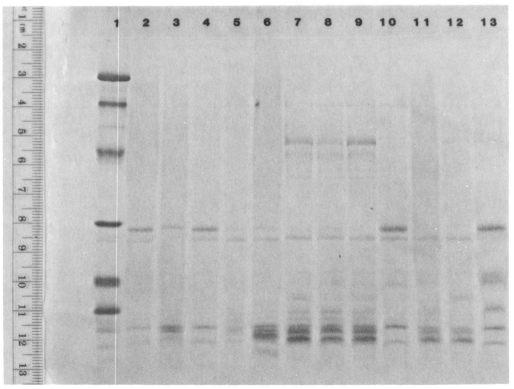

FIG. 1. SDS-pore gradient polyacrylamide gel (13.5% to 27% acrylamide) of soluble antigen extracts of various *L. pneumophila* strains. Lanes: 1, low-molecular-weight standard (Bio-Rad Laboratories); 2, serogroup 1, Knoxville 1; 3, serogroup 1, Philadelphia 2; 4, serogroup 1, Bellingham 1; 5, serogroup 1, OLDA; 6, serogroup 1, Pontiac 1; 7, SF 9; 8, SF 10; 9, SF 13; 10, SF 14; 11, serogroup 2, Togus 1; 12, serogroup 3, Bloomington 2; 13, serogroup 4, Los Angeles 1.

indicated that SF 14 is more similar to the other SF strains than to the named serogroup 1 strains and other serogroups and species tested. The presence of unique antigens in the SF strains, indicated by results of DFA cross-absorption, gel diffusion, and SDS-pore gradient PAGE assays, offers strong evidence that these strains represent either a serological variant of serogroup 1 or a new serogroup of *L. pneumophila*.

The diagnosis of Legionnaires disease has been advanced by the development of media that permit earlier isolation of the agent, but the DFA test still provides the most rapid diagnosis in many cases. DFA assays are limited by their inability to identify previously unrecognized variants, species, and serogroups that are undetected by available conjugates. Although DFA and gel diffusion were useful in characterizing the strains in this study, SDS-pore gradient PAGE made a strong contribution to the detection of similarities and differences among the studied strains and representative legionellae strains. Pore gradient PAGE can provide a useful tool for the characterization of strains in epidemiological studies and can contribute to the improvement of reagents for use in conventional diagnostic procedures.

LITERATURE CITED

1. **Bissett, M. L., J. O. Lee, and D. S. Lindquist.** 1983. New serogroup of *Legionella pneumophila*, serogroup 8. J. Clin. Microbiol. **17:**887–891.
2. **Chrambach, A., T. M. Jovin, P. J. Svendsen, and D. Rodbard.** 1976. Analytical and preparative polyacrylamide gel electrophoresis, p. 27–144. *In* N. Catsimpoolas (ed.), Methods of protein separation. Plenum Press, New York.
3. **Laemmli, U. K.** 1970. Cleavage of structural proteins during the assembly of the head of bacteriophage T4. Nature (London) **227:**680–685.

Isolation and Partial Chemical Characterization of Membrane-Bound Fluorescent Compounds from *Legionella gormanii* and *Legionella dumoffii*

WILLIAM JOHNSON AND KENNETH RINEHART

Department of Microbiology, University of Iowa, Iowa City, Iowa 52242, and School of Chemical Sciences, University of Illinois, Urbana, Illinois 61801

Legionella dumoffii was first isolated from water from a cooling tower (1), and *Legionella gormanii* was isolated from soil collected from a creek bank (2). Although these two organisms differ in their major cell surface antigens and can be separated by DNA relatedness, they have similar phenotypic characteristics, one of which is a characteristic blue-white fluorescence when exposed to long-wavelength UV radiation (2). The objective of this study was to determine whether the fluorescent compounds produced by *L. gormanii* (LS-13) were identical to those produced by *L. dumoffii* (NY-23) or whether each organism produced fluorescent compounds with unique chemical and structural features.

Preliminary experiments showed that the fluorescent compounds were bound to the cell membrane and could not be extracted with aqueous solvents. An extraction procedure using methylene chloride-acetic acid (99:1, vol/vol) was used for the initial extraction. Approximately 20 g of cells was suspended in 100 ml of distilled water. An equal volume of acidified methylene chloride was added, and the mixture was stirred at room temperature overnight. The mixture was then placed in a separatory funnel, and the organic phase containing the fluorescent compounds was removed and evaporated to dryness. The residue was dissolved in methylene chloride, applied to a silica gel thin-layer chromatography (TLC) plate, and developed with a methylene chloride-heptane-acetic acid (90:10:1, vol/vol) solvent mixture. The results are shown in Fig. 1. The extracts from both *L. gormanii* and *L. dumoffii* contained at least eight distinct fluorescent bands when exposed to long-wavelength UV radiation. The R_f values of the eight fluorescent compounds of *L. gormanii* were identical to the R_f values of the eight fluorescent compounds of *L. dumoffii*.

Two of the fluorescent compounds (FP-1 and FP-2) from both organisms were eluted from the silica TLC plates with acidified methylene chloride, evaporated to dryness, and redissolved in methylene chloride. The purity of the compounds was analyzed by high-pressure liquid chromatography, using a C-18 reversed-phase column (4 by 250 mm) and an 80 to 100% methanol gradient. The FP-1 compound from both organisms eluted as two peaks with retention times of 12.26 and 12.94 min. Each peak, when rechromatographed, eluted as two peaks with retention times identical to those of the original sample. These results suggested that the FP-1 compounds existed in two forms in equilibrium with each other. Analytical TLC of the FP-

1 compound isolated from both organisms also suggested that the FP-1 fraction contained two forms of the compound. The FP-1 compound isolated from the high-pressure liquid chromatography column contained two fluorescent bands with R_f values of 0.8 and 0.3 (Fig. 2, lanes A and B). The FP-1 compound from both organisms was converted to the monoacetate form by reaction with acetic anhydride. Conversion of the FP-1 compound of both organisms to the monoacetate form resulted in a single fluorescent band with an R_f of 0.3 when analyzed by TLC (Fig. 2, lanes C and D).

The FP-1 compounds from both *L. gormanii* and *L. dumoffii* had identical UV-visible spectra, with absorption maxima at 393, 377, 338, 324, 273, and 240 nm. The compounds also had identical fluorescence spectra, with excitation maxima at 390, 375, 338, 324, 273, and 238 nm and an emission maximum at 464 nm. Partial chemical characterization of the FP-1 compound isolated from *L. gormanii* was done by high-resolution fluid desorption mass spectral analysis and proton and ^{13}C nuclear magnetic reso-

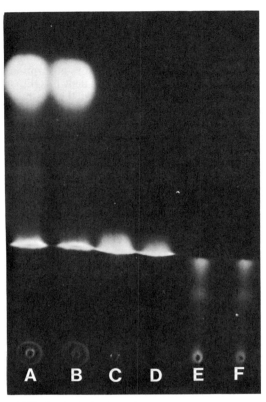

FIG. 2. Analytical TLC of purified fluorescent compounds isolated from *L. gormanii* (LG) and *L. dumoffii* (LD). Lanes: A, LG FP-1; B, LD FP-1; C, LG FP-1 monoacetate; D, LD FP-1 monoacetate; E, LG FP-2; F, LD FP-2.

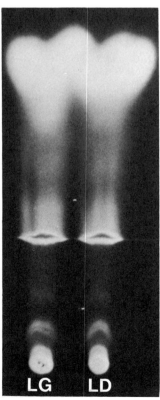

FIG. 1. Analytical TLC of acidified methylene chloride extracts from *L. gormanii* (LG) and *L. dumoffii* (LD). The TLC plate was illuminated with long-wavelength UV irradiation.

nance. Based on these studies, we propose a molecular weight of 290.0943 and a chemical formula of $C_{19}H_{14}O_3$ for the FP-1 compound. Preliminary data on the chemical structure of the FP-1 compound suggest that it consists of two unsaturated six-carbon rings with a conjugated diene system, *trans*-substituted.

The FP-2 compound isolated from both *L. gormanii* and *L. dumoffii* eluted from the high-pressure liquid chromatography column as two peaks with retention times of 3.24 and 3.68 min. When analyzed by TLC, the FP-2 fraction from both organisms contained two fluorescent bands with R_f values of 0.18 and 0.29 (Fig. 2, lanes E and F). Each fluorescent band, when eluted and rechromatographed, gave two fluorescent bands with R_f values of 0.18 and 0.29, suggesting that the FP-2 compound existed in two forms in equilibrium with each other.

The FP-2 compounds from both organisms had identical UV-visible spectra, with absorption maxima at 325, 246, and 215 nm. The compounds also had identical fluorescence spectra, with excitation maxima at 327, 248, and 217 nm and an emission maximum at 400 nm. Structural studies of this compound have not been completed.

The results of these studies suggest that *L. gormanii* and *L. dumoffii* both produce at least eight fluorescent compounds. Two of the fluorescent compounds produced by each of these organisms, FP-1 and FP-2, are identical as determined by TLC, high-pressure liquid chromatography, and UV-visible and fluorescence spectroscopy. Studies on the biosynthesis and function of these unique fluorescent compounds are in progress.

This investigation was supported by Public Health Service grant AI-15807 from the National Institute of Allergy and Infectious Diseases.

LITERATURE CITED

1. Brenner, D. J., A. G. Steigerwalt, G. W. Gorman, R. E. Weaver, J. C. Feeley, L. G. Cordes, H. W. Wilkinson, C. Patton, B. M. Thomason, and K. R. L. Sasseville. 1980. *Legionella bozemanii* sp. nov. and *Legionella dumoffii* sp. nov.: classification of two additional species of *Legionella* associated with human pneumonia. Curr. Microbiol. 4:111–116.
2. Morris, G. K., A. Steigerwalt, J. C. Feeley, E. S. Wong, W. T. Martin, C. M. Patton, and D. J. Brenner. 1980. *Legionella gormanii* sp. nov. J. Clin. Microbiol.12:718–721.

Comparison of the Rapid Tube Test, Paper Chromatography, and Gas-Liquid Chromatography for Detection of Hippurate Hydrolysis by *Legionella* Species

MICHAEL L. TOWNS, JAMES M. BARBAREE, JAMES C. FEELEY, AND EDWARD O. HILL

Emory University School of Medicine, Atlanta, Georgia 30322, and Centers for Disease Control, Atlanta, Georgia 30333

The hydrolysis of sodium hippurate has been used for a number of years to aid in the identification of bacterial species (1). The test determines the ability of an organism to hydrolyze sodium hippurate to form two products, glycine and benzoic acid.

The ability of *Legionella* species to hydrolyze sodium hippurate has previously been determined by a rapid tube test (RTT) in which ninhydrin is used to detect the presence of glycine. Using this test, Hébert (6) showed that almost all strains of the six then-known *L. pneumophila* serogroups hydrolyzed hippurate. Two of ten strains of serogroup 4 tested were unable to hydrolyze hippurate. *L. bozemanii, L. micdadei,* and *L. dumoffii* strains were negative. In addition, two environmental isolates, LS-13 and BL-540, which are now known as *L. gormanii* and *L. jordanis,* respectively, were negative.

There are two potential problems with the RTT. First, weak reactions are difficult to interpret. Second, since ninhydrin reacts with any alpha amino group, there is the potential problem of obtaining a false-positive reaction, i.e., a reaction with an amino group other than that of glycine.

As part of our effort to expand the application of substrate utilization to characterize legionellae, we investigated the use of specific identification of the end products of hippurate hydrolysis to detect any differences between serogroups, strains, or species.

The *Legionella* species investigated in this study and strain identifications are given in Table 1. *Streptococcus agalactiae* was included as a positive control. Samples were analyzed by circular paper chromatography to confirm the presence or absence of glycine (2) and to ascertain whether false-positive reactions occurred with the RTT.

Benzoic acid was detected in the test suspension by gas-liquid chromatography (GLC) as previously described (3, 7) except that a Fisher 2400 series gas chromatograph was used. In all three studies thermal conductivity detectors were employed. The test medium was 0.5% sodium hippurate in buffered yeast extract broth (BYEB) with alpha-ketoglutarate. The 48- to 72-h growth from a buffered charcoal-yeast extract agar plate was suspended in BYEB to a density equivalent to a no. 3.0 McFarland standard, and 0.2 ml of this suspension was used to inoculate 10 ml of the test medium. Samples of 1 ml were analyzed for benzoic acid after 24, 48, 72, and 96 h. Detection of benzoic acid within 96 h was considered a positive test. Tests included an uninoculated control and a positive control. The only *L. pneumophila* serogroups in which benzoic acid was detected were groups 2, 3, and 5. However, these serogroups did not consistently produce a benzoic acid peak. *L. pneumophila* serogroups 1, 4, and 6, as well as the strains of the other *Legionella* species, failed to produce detectable benzoic acid even after 96 h.

In addition to the GLC procedure, three other techniques were used for the detection of hippurate hydrolysis. In these studies, strains were grown on buffered charcoal-yeast extract agar plates for 72 h and then inoculated into a 1% sodium hippurate solution in distilled water. The analyses were performed in parallel after 18 to 20 h of incubation at 35°C.

The MA colorimetric test has been shown to detect as little as 0.1 μg of benzoic acid per ml (4). When 1% uranyl acetate solution is shaken with colorless saturated rhodamine B in benzene, the benzene layer turns red if benzoic acid is present. All samples tested were negative by the MA procedure, thus confirming most of the GLC results. However, three strains of *L. pneumophila* were weakly positive by GLC.

The RTT was performed as described by Hébert (6). After incubation, 0.2 ml of a 3.5% ninhydrin in acetone solution was added to each tube and reincubated for 10 min. Positive tests showed a dark purple-blue color, while negative tests showed no color change or a faint tinge of purple. All strains of the six *L. pneumophila* serogroups were positive except one serogroup 4 strain, thus confirming Hébert's earlier work.

Circular paper chromatography was performed as described previously (2). The solvent consisted of a mixture of 120 ml of *n*-butanol, 30 ml of glacial acetic acid, and 50 ml of distilled water. After the incubation period, 0.1 ml of the inoculum was added to 0.9 ml of acetone for each test sample. The amino acid standards were prepared by making a 10% concentration in 1% sodium hippurate and then a 1:10 dilution in acetone, giving a final amino acid concentration of 1%. The samples and standards, in 50-μl amounts, were concentrated on the stationary phase, and the system was allowed to run to completion (about 90 min). The R_f values of the samples were compared with those of the amino acid standards. The results showed that for each test strain positive in the RTT, a compound which gave the same R_f value, color, and staining properties of glycine was identified. This evidence highly suggested, although did not unquestionably prove, that glycine was present.

Since we still could not account for the absence of benzoic acid in many samples, we speculated that in the presence of benzoic acid some legionellae might have further utilized this compound through β-ketoadipate or a similar metabolic pathway. We investigated this possibility by adding 0.5, 1.0, 2.0, and 4.0 mmol of benzoic acid to BYEB with alpha-ketoglutarate. A 50-ml Erlenmeyer flask with 10 ml of BYEB-benzoic acid was inoculated with 0.1 ml of a *L.*

TABLE 1. Determination of hippurate hydrolysis by detection of glycine or benzoic acid

Strain	Test result[a]			
	Glycine		Benzoic acid	
	RTT	CPC[b]	MA	GLC
L. pneumophila SG[c] 1, Philadelphia 1 and Pontiac 1	+	+	−	−
L. pneumophila SG 2, Maine	+	+	−	±
L. pneumophila SG 2, Atlanta	+	+	−	−
L. pneumophila SG 3, BL-433	+	+	−	±
L. pneumophila SG 3, SC2-C5	+	+	−	−
L. pneumophila SG 4, Rockless	−	−	−	−
L. pneumophila SG 4, PF-159C-C1	+	+	−	−
L. pneumophila SG 5, Dallas 4	+	+	−	±
L. pneumophila SG 5, Dallas 17	+	+	−	−
L. pneumophila SG 6, Chicago 2 and PT 41C-C1	+	+	−	−
L. micdadei Pi-12 and TATLOCK	−	−	−	−
L. dumoffii TEX-KL and NY-23	−	−	−	−
L. bozemanii MI-15 and WIGA	−	−	−	−
L. jordanis BL-540 and ABB-9	−	−	−	−
L. longbeachae SG 1, Los Angeles 24 and Long Beach 4	−	−	−	−
L. longbeachae SG 2, Tucker	−	−	−	−
L. gormanii LS-13	−	−	−	−
Streptococcus agalactiae Emory (positive control)	+	+	+	+
Uninoculated control (negative control)	−	−	−	−
Benzoic acid (1 mmol/ml)[d]			+	+

[a] GLC was for detection of benzoic acid in BYEB. All other assays were performed in 1% sodium hippurate (wt/vol) in distilled water.

[b] CPC, Circular paper chromatography test for amino acids.

[c] Serogroup.

[d] Concentration in water for MA test and in BYEB for GLC test.

pneumophila serogroup 1 cell suspension adjusted to a 0.5 McFarland standard. One-milliliter samples were removed at 24, 48, 72, and 96 h and tested for benzoic acid as in the GLC procedure. If *L. pneumophila* utilized benzoic acid, there would be a disappearance or decrease in the benzoic acid peak when compared with the uninoculated broths. The results showed that even after 96 h, the benzoic acid peak remained unchanged.

In summary, we confirmed by using the circular paper chromatography technique to detect the presence of glycine that *L. pneumophila* strains (excluding the one strain of serogroup 4 tested) hydrolyzed hippurate. Because we were unable to detect the presence of benzoic acid being produced by most of the *Legionella* strains, we recommend that the GLC and MA tests not be used to determine whether *Legionella* strains hydrolyze hippurate. We do not understand why benzoic acid is not produced in detectable amounts; we suggest that further research is needed to determine whether hippuratase acts intracellularly in *L. pneumophila* without release from the cell, as in some streptococci (5).

LITERATURE CITED

1. **Ayers, S. H., and P. Rupp.** 1922. Differentiation of hemolytic streptococci from human and bovine sources by the hydrolysis of sodium hippurate. J. Infect. Dis. **30:**388–399.
2. **Block, R. J., E. L. Durrum, and G. Zweig.** 1955. A manual of paper chromatography and paper electrophoresis, p. 27–30. Academic Press, Inc., New York.
3. **Dezfulian, M., and V. R. Dowell, Jr.** 1980. Cultural and physiological characteristics and antimicrobial susceptibility of *Clostridium botulinum* isolates from foodborne and infant botulism cases. J. Clin. Microbiol. **11:**604–609.
4. **Edberg, S. C., and S. Samuels.** 1976. Rapid colorimetric test for the determination of hippurate hydrolysis of group B *Streptococcus.* J. Clin. Microbiol. **3:**49–50.
5. **Ferrieri, P., L. W. Wannamaker, and J. A. Nelson.** 1973. Localization and characterization of hippuratase activity of group B streptococci. Infect. Immun. **7:**747–752.
6. **Hébert, G. A.** 1981. Hippurate hydrolysis of *Legionella pneumophila.* J. Clin. Microbiol. **13:**240–242.
7. **Kodaka, H., G. L. Lombard, and V. R. Dowell, Jr.** 1982. Gas-liquid chromatography technique for detection of hippurate hydrolysis and conversion of fumarate to succinate by microorganisms. J. Clin. Microbiol. **26:**962–964.

Cytolytic Exotoxin and Phospholipase C Activity in *Legionella* Species

WILLIAM B. BAINE

The University of Texas Health Science Center at Dallas, Dallas, Texas 75235

The histopathology of pneumonia caused by *Legionella pneumophila* is characterized by a cytoclastic picture with necrosis of mononuclear and polymorphonuclear cells in the alveolar inflammatory exudate. The pathogenesis of this lesion has not been elucidated. One possible mechanism could be release of an extracellular bacterial cytotoxin acting on host cell membranes.

My co-workers and I have previously reported that cultures of *L. pneumophila* lyse guinea pig, horse, rabbit, and sheep erythrocytes in agar medium, whereas human erythrocytes are less susceptible. Erythrocytes from guinea pigs, in which the membrane lecithin content is particularly high (4), are most readily hemolyzed by *L. pneumophila* (1).

Since dog erythrocyte membranes are also particularly rich in lecithin (4), representative strains of *L. pneumophila, Legionella bozemanii, Legionella micdadei, Legionella dumoffii, Legionella longbeachae,* and *Legionella jordanis* were cultured on agar medium containing defibrinated dog blood as well as on guinea pig, human, rabbit, and sheep blood agar. Plates were examined daily for bacterial growth and hemolysis. All *Legionella* species showed hemolytic activity within 3 days, but *L. micdadei* was weakly hemolytic at best. Dog erythrocytes

showed the greatest susceptibility to lysis (Table 1).

Unconcentrated broth culture filtrates of *L. pneumophila* were not hemolytic, but cytolytic activity was detectable by a radial hemolysis assay in supernatants of *L. pneumophila* broth cultures after the fluid was concentrated by ultrafiltration through a membrane with a nominal molecular weight cutoff of 10,000.

In several bacterial species hemolysis is associated with bacterial phospholipase C (lecithinase), an enzyme that hydrolyzes phosphatidyl choline to phosphorylcholine and diglyceride. We have reported that cultures of *L. pneumophila* on agar with 5% egg yolk produce clouding of the medium resembling that seen with lecithinase-positive species of *Clostridium* (1).

Representative strains of *Legionella* species were screened for phospholipase C activity on egg yolk agar. Plates were examined daily for bacterial growth and early clearing of the opaque medium, iridescence, and delayed formation of a cloudy precipitate, which are suggestive of possible protease, lipase, and lecithinase activity, respectively. *L. micdadei* showed no evidence of presumptive lecithinase activity. Clouding was most pronounced around growth of *L. bozemanii.*

The chromogenic compound *p*-nitrophenylphosphorylcholine, a water-soluble analog of lecithin (3), was used as a substrate to quantify phospholipase C activity in legionellae. Hydrolysis of this compound by phospholipase C yields phosphorylcholine and yellow p-nitrophenol. Washed bacterial cells were incubated overnight in a shaker bath at 37°C with 20 mM p-nitrophenylphosphorylcholine in 5 mM Tris buffer (pH 7.2) containing 5 mM $CaCl_2$. The cells were then removed by centrifugation, and p-nitrophenol released into the supernatant was measured colorimetrically. A standard curve was prepared with serial dilutions of p-nitrophenol. Controls included buffer alone, bacteria alone, and the substrate alone. All seven species appeared to produce phospholipase C (Table 2), although enzymatic activity was greatest with *L. bozemanii*, the species with the strongest lecithinase reaction on egg yolk.

An agar medium containing 20 mM p-nitrophenylphosphorylcholine was evaluated as a screening assay for bacterial phospholipase C that might be used to detect enzymatically defective or hyperproducing mutants of legionellae. *L. bozemanii* turned the agar bright yellow. Gradual but definite yellowing of the medium was also observed with cultures of *L. pneumophila* but not with those of the five other species tested.

L. pneumophila Philadelphia 1 was grown for 48 h at 37°C in 4.2-liter batches of a supplemented dialysate of yeast extract. Cell-free broth culture supernatants were concentrated by using ultrafiltration membranes with a nominal molecular weight cutoff of 10,000. Phospholipase C activity was assayed by hydrolysis of p-nitrophenylphosphorylcholine. Enzyme activity was present in the concentrated retentate but not in the ultrafiltrate of the broth culture supernatant. The retentate was dialyzed against 0.01 M Tris (pH 7.4) and brought to 70% saturation with

TABLE 2. Hydrolysis of p-nitrophenylphosphoryl-choline by *Legionella* spp.

Species	p-Nitrophenol liberated[a]
L. pneumophila	7.62 ± 0.36
L. bozemanii	29.2 ± 0.5
L. micdadei	3.71 ± 0.10
L. dumoffii	2.79 ± 0.14
L. gormanii	7.51 ± 0.00
L. longbeachae	6.06 ± 0.10
L. jordanis	3.71 ± 0.07

[a] Expressed as nanomoles per absorbance unit (590 nm) of bacteria per milliliter. The mean and range of duplicate samples are shown.

ammonium sulfate. The precipitate was dissolved in buffer, dialyzed to remove residual ammonium sulfate, and passed over a Sephadex G-100 column. Phospholipase C activity appeared in the void volume.

Erythrocytes provide a convenient target for the detection of cytotoxins that disrupt cell membranes. Although hemolytic anemia is not seen in *Legionella* pneumonia, the anatomic site of infection is probably critical in determining the extent and location of damage produced by a cytolytic bacterial toxin. Profound intravascular hemolysis produced by the hemolytic alpha-toxin of *Clostridium perfringens* accompanies septicemia caused by that organism. However, *C. perfringens* pneumonia is instead characterized by necrosis of lung tissue without hemolysis (2).

Seven distinct species of *Legionella* possess cytolytic activity that can be detected by using dog erythrocytes as convenient target cells. One or more cytolytic exotoxins may participate in the pathogenesis of the pulmonary lesions of *Legionella* pneumonia.

Phospholipase C activity, analogous to that of clostridial alpha-toxin, also appears to be common in members of the genus *Legionella*. *L. pneumophila* releases extracellular phospholipase C that has an apparent molecular weight of greater than 100,000. Studies to assess the role of phospholipase C in the hemolytic activity of *Legionella* spp. are in progress.

TABLE 1. Hemolysis by cultures of *Legionella* spp. on modified buffered yeast extract agar with 5% defibrinated blood from five mammalian species

Species	Hemolysis[a] with erythrocytes from:				
	Dogs	Guinea pigs	Humans	Rabbits	Sheep
L. pneumophila	+ + + +	+	+	+ +	+
L. bozemanii	+ + +	+ + +	+	+ + + +	+
L. micdadei	+ +	−	−	−	−
L. dumoffii	+ + + +	+	±	+ +	+
L. gormanii	+ + + +	+	±	+ +	+
L. longbeachae	+ + +	+ +	+	+ + + +	+ +
L. jordanis	+ +	+	±	+	+

[a] Evaluated after incubation for 3 days at 37°C.

This work was supported by Biomedical Research Support grant 2 SO7 RR05426-20 from the National Institutes of Health.

I thank Edward K. Davis for his capable technical assistance and Robert S. Munford III for his valued criticism and suggestions.

LITERATURE CITED

1. **Baine, W. B., J. K. Rasheed, D. C. Mackel, C. A. Bopp, J. G. Wells, and A. F. Kaufmann.** 1979. Exotoxin activity associated with the Legionnaires disease bacterium. J. Clin. Microbiol. 9:453–456.
2. **Bayer, A. S., S. C. Nelson, J. E. Galpin, A. W. Chow, and**

L. B. Guze. 1975. Necrotizing pneumonia and empyema due to *Clostridium perfringens*. Am. J. Med. **59**:851–856.

3. Kurioka, S., and M. Matsuda. 1976. Phospholipase C assay using *p*-nitrophenylphosphorylcholine together with sorbitol and its application to studying the metal and detergent requirement of the enzyme. Anal. Biochem. **75**:281–289.

4. Turner, J. C., H. M. Anderson, and C. P. Gandal. 1958. Species differences in red blood cell phosphatides separated by column and paper chromatography. Biochim. Biophys. Acta **30**:130–134.

Assay of Extracellular *Legionella bozemanii* Hemolysin by a Rapid Microtiter Method

CAROLYN F. FRISCH AND WILLIAM B. BAINE

The University of Texas Health Science Center at Dallas, Dallas, Texas 75235

Legionella bozemanii is a fastidious, aerobic, gram-negative bacillus that can cause pneumonia associated with freshwater immersion (2). When grown on buffered yeast extract-blood agar, *L. bozemanii* produces zones of hemolysis around areas of heavy growth (W. B. Baine, this volume). A membrane-active cytolytic exotoxin responsible for hemolytic activity could play a role in the pathogenesis of infection with this bacterium, as necrosis of inflammatory cells in lung lesions is part of the pathology of this disease (3). A rapid method of detecting hemolytic activity of broth culture filtrates of *L. bozemanii* was pursued to facilitate preliminary attempts to isolate and purify the factor or factors responsible for the hemolysis observed.

Concentrated supernatant was prepared by culturing the WIGA strain of *L. bozemanii* in dialysate of 1% yeast extract supplemented with 1% ACES [*N*-(2-acetamido)-2-aminoethanesulfonic acid], 0.04% L-cysteine-hydrochloride, and 0.025% ferric pyrophosphate (pH 6.9) for 41 to 48 h in a Microferm fermentor at 37°C. Supernatant was harvested by pelleting cells at $10,400 \times g$ for 10 min and then filtering through a Pellicon filtration unit with a nominal molecular weight cutoff of 10,000. The filtrate was centrifuged again to remove residual cells and filter sterilized through a 0.45-μm Nalgene filter unit before concentration by membrane filtration or vacuum dialysis against 0.1 M Tris (pH 7.6). No difference was observed between activities of concentrate prepared by membrane filtration and vacuum dialysis.

Several variables were considered which might affect lysis of the erythrocytes: osmolality, temperature, divalent cation concentration, sublytic concentrations of detergents, and pH (1). Dog erythrocytes were chosen as substrate for their susceptibility to hemolysis by *Legionella* spp. (Baine, this volume). These variables were tested in a microtiter assay by adding test samples in 50-μl aliquots to 96-well V-bottom microtiter plates. Either 50 μl of erythrocytes suspended to 5% in test buffer or 5 μl of stock solutions of divalent cation, detergent, or both and 45 μl of erythrocytes suspended to 5.5% in test buffer were added. Plates were incubated for 1 h at 37°C and then for 1 h at 4°C or as otherwise specified and then centrifuged at $450 \times g$ for 6 min. Hemolysis was determined by inspection, using redness of buffer and lessening of the erythrocyte pellet as indicators. Hemolysis was scored as − (no hemolysis) to + + + + (complete hemolysis).

Osmolalities of reagents involved in the microtiter assay were measured by freezing-point depression, and calculation of final osmotic strength per well with Tris buffer in a 1:1 ratio with sample reagents gave a range of between 255 and 262 mOsm per well. These values are above the 225 mOsm level, below which hemolysis of erythrocytes in the microtiter assay due to hypotonicity is observed. Therefore, hemolytic activity in wells containing concentrated *L. bozemanii* supernatant was not attributable to hypotonic lysis.

Four different incubation conditions were tested: 1-h hot (37°C) and 1-h cold (4°C) incubation; 2-h hot and 2-h cold incubation; 3-h hot and overnight cold incubation; or overnight cold incubation. One hour each of hot and cold incubation gave the best results most rapidly, permitting some hemolysis in wells containing concentrated supernatant but not in wells containing control medium.

The effect of divalent cation concentration was tested by titration in 0.1 M Tris buffer (pH 7.2 or 7.65) with 10 to 0.25 mM $CaCl_2$, $ZnCl_2$, or both. $CaCl_2$ gave the best results.

Complete hemolysis of concentrated supernatant wells was still not achieved, so a panel of detergents was tested for their effects in sublytic concentrations. Detergent solutions in various strengths were added to the samples and brought to the desired final concentrations by the addition of erythrocytes in Tris buffer. Evaluation of sodium dodecyl sulfate, Triton X-100, and Tween 80 in this manner showed that only in 0.5% Tween 80 did erythrocytes in control wells

TABLE 1. Effect of detergent and CaCl₂ concentration on hemolytic activity of concentrated *L. bozemanii* supernatant

CaCl$_2$ concn (mM)	Hemolytic activity[a] with:					
	0.0075% Triton X-100		0.002% SDS[b]		0.5% Tween 80	
	Control	Supernatant	Control	Supernatant	Control	Supernatant
10	NT[c]	NT	NT	NT	−	−
5	NT	NT	NT	NT	−	+
1	+	−	+	−	−	+++
0.5	+	−	+	−	−	+++
0.25	NT	NT	NT	NT	+	+++
0	+	+	+	+	+	+++

[a] Assayed at pH 7.2 or 7.65. See text for scoring of activity.
[b] SDS, Sodium dodecyl sulfate.
[c] NT, Not tested.

remain intact, whereas considerable hemolysis was observed in wells with concentrated supernatant (Table 1).

The pH of the Tris buffer was tested in increments of 0.1 pH unit from pH 7.0 to 8.0. Optimal hemolytic activity of the concentrated *L. bozemanii* supernatant was observed with 0.7% Tween 80 at pH 7.5, 0.6% Tween 80 at pH 7.6, and 0.5% Tween 80 at pH 7.7. Optimal conditions of 0.5 mM CaCl₂, 0.6% Tween 80, and 0.1 M Tris (pH 7.6) were chosen for the remaining experiments.

Attempts to increase the yield of the hemolysin were pursued. Cell lysates were obtained from bacterial cultures by agitation with glass beads or sonication. Negligible hemolysis was observed in microtiter assays of cell lysate preparations, indicating that most of the available hemolysin was an extracellular product. Complete hemolysis was obtained, however, when the concentration of *L. bozemanii* supernatant was increased from 233× to 649×.

TABLE 2. Purification by gradient ammonium sulfate precipitation

Sample	Total protein (mg)	Hemolytic activity[a]
Control medium (2,692×)	0.07	−
Concentrated supernatant	3.47	+
Pellet of ultracentrifuged concentrate	0.29	−
Supernatant of ultracentrifuged concentrate	2.35	+
30% Ammonium sulfate cut	0.73	++++
40% Ammonium sulfate cut	0.19	−
50% Ammonium sulfate cut	0.45	−
60% Ammonium sulfate cut	0.37	−
70% Ammonium sulfate cut	0.52	−
70% Ammonium sulfate supernatant	0.15	−

[a] See text for scorings.

Preliminary efforts were then made to characterize the *L. bozemanii* hemolysin. Concentrated *L. bozemanii* supernatant was subjected to ultracentrifugation, and the resulting supernatant proteins were precipitated in 70% saturated ammonium sulfate. The resultant fractions were separated over a Sephadex G-100 column. Complete hemolysis in the microtiter assay was observed only in wells containing the void volume fraction, indicating that the hemolytic activity is recovered as monomers or aggregates of greater than 100,000 molecular weight. Concentrated *L. bozemanii* supernatant was also separated by precipitation with increasing concentrations of ammonium sulfate after ultracentrifugation. A gradient of ammonium sulfate cuts from 30 to 70% saturation in 10% increments was made. Only precipitate from the 30% ammonium sulfate fraction had any hemolytic activity, and it completely lysed the erythrocytes in the microtiter assay. This property of the hemolysin should aid in its isolation and purification, as the majority of proteins are not precipitable at 30% ammonium sulfate saturation (Table 2).

Hemolytic activity of concentrated *L. bozemanii* supernatant, the Sephadex G-100 void volume fraction, and the 30% ammonium sulfate cut in the microtiter assay was destroyed by heating in a boiling water bath for 5 min. Hemolytic activity of these samples was greatly inhibited by the addition of 100 μg of phosphatidylcholine to each microtiter well.

This work was supported by Biomedical Research Support grant 2 SO7 RR05426-20 from the National Institutes of Health.

We thank Edward K. Davis for his indispensable technical assistance and Robert S. Munford III, Paul A. Gulig, and Susan A. Jones for their critical observations, technical advice, and patience. We are also grateful to Jonathan W. Uhr for his encouragement.

LITERATURE CITED

1. **Arbuthnott, J. P.** 1982. Bacterial cytolysins (membrane-damaging toxins), p. 107–129. *In* P. Cohen and S. Van Heyningen (ed.), Molecular action of toxins and viruses.

Elsevier Biomedical Press, Amsterdam.
2. **Brenner, D. J., A. G. Steigerwalt, G. W. Gorman, R. E. Weaver, J. C. Feeley, L. G. Cordes, H. W. Wilkinson, C. Patton, B. M. Thomason, and K. R. Lewallen Sasseville.** 1980. *Legionella bozemanii* sp. nov. and *Legionella dumof-*

fii sp. nov.: classification of two additional species of *Legionella* associated with human pneumonia. Curr. Microbiol. **4**:111–116.
3. **Winn, W. C., Jr., and R. L. Myerowitz.** 1981. The pathology of the legionella pneumonias. Hum. Pathol. **12**:401–422.

Characterization of an Extracellular Hemolysin from *Legionella pneumophila*

THURMAN C. THORPE AND RICHARD D. MILLER

Department of Microbiology and Immunology, University of Louisville School of Medicine Health Sciences Center, Louisville, Kentucky 40292

Hemolytic activity by *Legionella pneumophila* has been reported by Baine and co-workers (1, 2) although detailed descriptions of the hemolysin were not presented. The importance of hemolysins (i.e., hemolytic toxins) in the pathogenicity of other bacteria is well documented, and the hemolysin of *L. pneumophila* may have similar toxic properties. The purpose of the present investigation was to document the presence of this hemolysin and to characterize its activity.

The growth medium (BSYE blood agar) used to detect hemolytic activity contained yeast extract (1.0%), ACES [*N*-(2-acetamido)-2-aminoethanesulfonic acid] buffer (1.0%), corn starch (0.5%), and agar (1.7%) and was supplemented with L-cysteine (0.04%) and ferric pyrophosphate (0.025%). Fresh, defibrinated guinea pig erythrocytes were then added to the basal medium to achieve a 5.0% final concentration. Zones of beta-hemolysis on this medium were typically observed by 24 h around areas of heavy growth, although hemolysis around single colonies was not detectable until 2 days after appearance of the colony.

A comparison of the hemolytic activity of different species of *Legionella* was performed by spot inoculating on BSYE blood agar and incubating at 37°C in 2.5% CO_2. The diameter of the zones of hemolysis were then measured after 24 and 48 h. Differences in the rate of hemolysis between species were detected (perhaps related to the growth rates). However, by 48 h all species of *Legionella* examined (*L. pneumophila*, *L. micdadei*, *L. bozemanii*, *L. dumoffii*, *L. gormanii*, *L. longbeachae*, and *L. jordanis*) produced zones of similar size, with the exception of *L. micdadei*, which had no detectable hemolytic activity even after extended incubation beyond 48 h.

To determine whether hemolysin production could be related to the virulence of *L. pneumophila*, we examined a number of environmental, human, animal-passaged, and extensively laboratory-adapted strains. No significant differences in the zones of hemolysis were noted between any of these strains. These results do not preclude a role for the hemolysin in pathogenicity, but merely indicate that this property is not solely responsible for the virulent-to-avirulent conversion observed during in vitro culture.

In addition to hemolysis with guinea pig erythrocytes, comparable hemolysis by *L. pneumophila* was detected on BSYE agar containing rabbit erythrocytes. Good hemolysis was also observed with human erythrocytes after 48 h, although the rate of initial hemolysis was slow. Hemolytic activity against bovine and sheep erythrocytes was significantly less pronounced.

Further characterization of the hemolysin was carried out on culture supernatants prepared from late-exponential-phase cultures of *L. pneumophila* Knoxville 1 grown in a biphasic medium composed of yeast extract broth (1.0%) and buffered charcoal-yeast extract agar. Culture supernatants were then concentrated by (i) dialysis against distilled water followed by lyophilization, (ii) a collodion bag vacuum dialysis apparatus, or (iii) precipitation with 80% ammonium sulfate.

Hemolytic activity was measured by using a tube assay, a plate assay, or both. The tube assay consisted of 3 ml of a 1.0% suspension of defibrinated guinea pig erythrocytes in either 10 mM phosphate buffer (pH 7.2) containing 0.9% NaCl or 10 mM Tris-hydrochloride buffer (pH 7.2) containing 0.9% NaCl. After addition of 150 μl of test supernatant, the tubes were incubated at 37°C. Hemolytic activity was measured by release of hemoglobin or by a decrease in the turbidity of the erythrocyte suspension. The plate assay consisted of 5.0% defibrinated guinea pig erythrocytes in 10 mM Tris-hydrochloride buffer with 0.9% NaCl and 1.7% agar. Wells were punched in the agar, and 50 μl of the test supernatants was added. Zones of beta-hemolysis were measured after 24 h of incubation at 37°C.

The culture supernatant concentrated 10-fold by collodion bag vacuum dialysis showed good

TABLE 1. Characterization of the hemolysin

Treatment	Activity[a]
Standard osmotic dialysis[b] retentate ($\geq$12,000 MW)	+
Collodion bag vacuum dialysis[c] retentate ($\geq$25,000 MW)	+
80% $(NH_4)_2SO_4$ precipitate	+
100°C, 10 min	−
Pronase digestion[d]	−

[a] Hemolytic activity was determined based on zones of hemolysis on 5% guinea pig erythrocyte agar or hemolysis of a 1% suspension of guinea pig erythrocytes (or both).

[b] Overnight dialysis against distilled water. MW, Molecular weight.

[c] Dialyzed against distilled water.

[d] Sample was incubated with pronase (400 µg/ml) for 3 h at 37°C.

hemolytic activity on the plate assay, and the supernatant could be diluted to at least 1:8 and still retain detectable activity. Initial characterization of the hemolysin in the tube assay indicated that the reaction took place at 22 and 42°C, but not 4°C. Hemolytic activity in the supernatants was stable at 4 or −70°C for more than 2 months. Additional characteristics of the hemolysin are summarized in Table 1. Standard osmotic dialysis and collodion bag vacuum dialysis indicated a molecular weight of $\geq$25,000. The protein nature of the hemolysin was evidenced by its precipitation with 80% ammonium sulfate, its destruction by boiling for 10 min, and its inactivation after pronase digestion.

Due to the large number of extracellular products produced by this bacterium (5), it was important to distinguish between the hemolysin and other enzymes (i.e., protease or lipase) or toxins. Separation of hemolysin, protease, and lipase activities in the supernatants was carried out by electrophoresis on nondenaturing 12% polyacrylamide gels. After the run, lanes were cut and either stained for protein with Coomassie blue or placed on the surface of agar plates containing the desired substrate. Numerous protein bands were apparent at the top of the gel (Fig. 1). Protease and lipase activity corresponded to protein bands with an R_f of 0.42 and 0.66, respectively. Hemolytic activity was somewhat more diffuse but appeared to be localized near or on the dye front.

Thus, we have demonstrated that the hemolysin appears to be a heat-labile protein with a molecular weight of $\geq$25,000 that is active on human, guinea pig, rabbit, sheep, and bovine

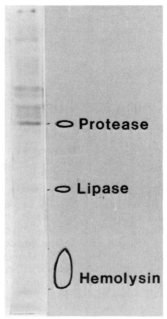

FIG. 1. Polyacrylamide gel electrophoresis of collodion bag-concentrated *L. pneumophila* culture supernatant proteins, with the corresponding location of protease, lipase, and hemolytic activity.

erythrocytes. Hemolytic activity can be clearly distinguished from the protease and lipase activities, and the size and heat stability properties would eliminate the cytotoxin described by Friedman et al. (3) and the cell-associated toxin reported by Hedlund (4). In addition, no phospholipase C activity could be detected in any of our supernatants.

The mechanism of action of this hemolysin and its role in the pathogenesis of legionellosis are currently under investigation.

LITERATURE CITED

1. **Baine, W. B., J. K. Rasheed, H. W. Maca, and A. F. Kaufmann.** 1979. Hemolytic activity of plasma and urine from rabbits experimentally infected with *Legionella pneumophila*. Rev. Infect. Dis. **1**:912–917.
2. **Baine, W. B., J. K. Rasheed, D. C. Mackel, C. A. Bopp, J. G. Wells, and A. F. Kaufmann.** 1979. Exotoxin activity associated with the Legionnaires disease bacterium. J. Clin. Microbiol. **9**:453–456.
3. **Friedman, R. L., B. H. Iglewski, and R. D. Miller.** 1980. Identification of a cytotoxin produced by *Legionella pneumophila*. Infect. Immun. **29**:271–274.
4. **Hedlund, K. W.** 1981. *Legionella* toxin. Pharmacol. Ther. **15**:123–130.
5. **Thorpe, T. C., and R. D. Miller.** 1981. Extracellular enzymes of *Legionella pneumophila*. Infect. Immun. **33**:632–635.

Bactericidal Activity of Five Antibiotics Against *Legionella pneumophila* Under Conditions Simulating Sepsis

PATRICIA A. RISTUCCIA, ANGELA M. RISTUCCIA, AND BURKE A. CUNHA

Infectious Disease Division, Department of Medicine, Nassau Hospital, Mineola, New York 11501

Antimicrobial agents which have been shown to effectively inhibit *Legionella pneumophila* in vitro have not consistently proven to be effective in vivo (2). It was suggested that the discrepancies may be due to the rate of killing or degree of antibiotic penetration into areas containing the organism (1). This investigation was designed to compare the bactericidal activity of five antimicrobial agents against *L. pneumophila* under in vitro conditions simulating sepsis. The bactericidal activity of rifampin, erythromycin, doxycycline, minocycline, and tetracycline for *L. pneumophila* ATCC 33152 were determined by using the Abbott MS-2, in which in vivo conditions of sepsis were simulated. Therapeutically achievable concentrations of each drug were used in buffered yeast extract broth (BYEB) with or without human serum (inactivated) at normal pH (7.4) and acidotic pH (7.2) and at 35 and 38°C to determine the effect of acidotic conditions on the activity of the drugs. The investigation was divided into three phases. The first phase was a comparison of the bactericidal activity of each agent against *L. pneumophila* in BYEB or in heat-inactivated pooled human serum diluted with BYEB (HS) at normal pH (7.4) at 35°C. The second phase was a comparison of the bactericidal activity of each agent with these media under acidotic conditions (pH 7.2) at 35°C. The third phase was a repetition of phases 1 and 2 with an increase in temperature (38°C). The minimal bactericidal concentrations (MBCs) were those producing 99.9% killing within 35 h.

The most active drug was rifampin, followed by erythromycin, minocycline, doxycycline, and tetracycline. The activity of rifampin and erythromycin was greater in HS than in BYEB, with a mean threefold and twofold decrease, respectively, in the MBC in HS. Minocycline showed no difference in activity with any change in medium or temperature. Both doxycycline and tetracycline were less active in HS than in BYEB, with a twofold increase in the MBC when tested in HS. The most rapidly bactericidal drugs for *L. pneumophila* were rifampin and erythromycin. Of the tetracyclines tested, doxycycline and minocycline had considerably greater activity than conventional tetracycline. The bactericidal activity of rifampin and erythromycin was greater in HS than in BYEB; the differences between the MBCs in BYEB and HS were threefold and twofold, respectively. The MBCs for doxycycline and tetracycline were two times higher in HS than in BYEB. Minocycline activity remained equivalent under all conditions. Rifampin and erythromycin were both rapidly bactericidal in vitro at readily achievable therapeutic concentrations under conditions simulating sepsis. The tetracyclines were bactericidal at concentrations above those therapeutically achievable. Therefore, if tetracyclines are used to treat *L. pneumophila* infections, higher-than-usual doses should be employed to assure achieving bactericidal concentrations.

LITERATURE CITED

1. **Dowling, J. N., S. W. Robbin, and A. W. Pasculle.** 1982. Bactericidal activity of antibiotics against *Legionella micdadei* (Pittsburgh pneumonia agent). Antimicrob. Agents Chemother. 22:272–276.
2. **Fraser, D. W., I. K. Wachsmuth, C. Bopp, J. C. Feeley, and T. F. Tsai.** 1978. Antibiotic treatment of guinea-pig infected with agents of Legionnaires' disease. Lancet i:175–178.

In Vitro Activity of Antimicrobial Agents Against *Legionella pneumophila* and Type Strains of Four Other *Legionella* Species

PATRICIA A. RISTUCCIA, BURKE A. CUNHA, AND ANGELA M. RISTUCCIA

Infectious Disease Division, Department of Medicine, Nassau Hospital, Mineola, New York 11501

Antimicrobial susceptibility testing of *Legionella* species has been previously reported with use of both buffered charcoal-yeast extract agar and buffered yeast extract broth (BYEB) to determine whether there was any effect of the media on the minimal inhibitory concentrations (MICs) of antimicrobial agents against these organisms (1–5). The comparison of activity of various antimicrobial agents against legionellae under conditions simulating sepsis in patients has not been undertaken. The present study was designed to determine and compare the anti-*Legionella* activity of five antimicrobial agents, alone and in combination, under normal laboratory testing conditions and under conditions simulating sepsis in vitro. The *Legionella* strains used were *L. pneumophila* ATCC 33152, *L. dumoffii* ATCC 33279, *L. micdadei* ATCC 33204, *L. gormanii* ATCC 33297, and *L. bozemanii* ATCC 33217.

The investigation was divided into three main phases. The first phase was a comparison of susceptibilities of each organism in BYEB or in heat-inactivated pooled human serum diluted with BYEB at normal pH (pH 7.4) at 35°C. The second phase was a comparison of susceptibilities in these media under acidotic conditions (pH 7.2) at 35°C. The third phase was a comparison of the first two phases (media at pH 7.2 and 7.4) under increased temperature (38°C).

The MIC of each antibiotic for each organism was taken as the lowest concentration of antibiotic for which the isolates showed no growth or a barely visible haze. Combinations of drugs were considered synergistic when the fractional inhibitory concentration, i.e., the MIC of drug in combination divided by the MIC of drug acting alone, was less than or equal to 1.0. Any antibiotic combination with a fractional inhibitory concentration of between 1 and 2 was considered indifferent, and any with a fractional inhibitory concentration greater than 2 was considered antagonistic.

The in vitro activities of rifampin, erythromycin, tetracycline, doxycycline, and minocycline, alone and in combination, were compared. All drugs showed activity at therapeutically achievable concentrations. Rifampin was the most active against all strains tested, with a mean MIC of 0.03 µg/ml. Erythromycin was the next most active, with an MIC of 0.2 µg/ml against all strains except those of *L. bozemanii* and *L. micdadei* (mean MIC, 0.6 and 0.8 µg/ml, respec-

tively). The most active tetracycline was doxycycline, with a mean MIC of 1.1 µg/ml, followed by minocycline, with a mean MIC of 1.4 µg/ml for all strains tested. Tetracycline was the least active of all drugs used, with a mean MIC of 2.9 µg/ml. MICs tended to be lower in broth than in serum.

The MICs for each antimicrobial agent against the *Legionella* organisms tested were lower in BYEB than in heat-inactivated pooled human serum diluted with BYEB. The activity of the antimicrobial agents studied was not affected by the changes in media pH or temperature at which the experiment was performed. The most synergistic combination was determined to be that of rifampin and erythromycin, which was synergistic for all strains tested.

Synergy testing of rifampin plus a tetracycline also showed synergy in all cases. The combinations of rifampin-minocycline and rifampin-doxycycline were the most synergistic, and rifampin-tetracycline was less synergistic. Rifampin and erythromycin, the recommended drugs of choice for therapy of Legionnaires disease, were the most active antibiotics, alone or in combination, against all strains of *Legionella* under all in vitro conditions investigated.

LITERATURE CITED

1. **Dowling, J. D., R. S. Weyant, and W. A. Pasculle.** 1982. Bactericidal activity of antibiotics against *Legionella micdadei* (Pittsburgh pneumonia agent). Antimicrob. Agents Chemother. **22:**272–276.
2. **Edelstein, P. H., and R. D. Meyer.** 1980. Susceptibility of *Legionella pneumophila* to twenty antimicrobial agents. Antimicrob. Agents Chemother. **18:**403–408.
3. **Saravalatz, L. D., D. J. Pohlod, and E. L. Quinn.** 1979. In vitro susceptibility of *Legionella pneumophila* serogroups I–IV. J. Infect. Dis. **40:**251.
4. **Thornsberry, C., C. H. Baker, and L. A. Kirven.** 1978. In vitro activity of antimicrobial agents on Legionnaires' disease bacterium. Antimicrob. Agents Chemother. **13:**78–80.
5. **Thornsberry, C., and L. K. McDougal.** 1982. In vitro susceptibility of *Legionella* species to antimicrobial agents, p. 24–26. *In* P. Periti and G. G. Grassi (ed.), Current chemotherapy and immunotherapy, vol. 1. American Society for Microbiology, Washington, D.C.

Inhibition and Hydrolysis Studies of Beta-Lactamases Found in *Legionella* Species: Antimicrobial Activity of New Macrolides on Legionellae

RONALD N. JONES, LINDA K. McDOUGAL, AND CLYDE THORNSBERRY

Kaiser Regional Laboratory, Clackamas, Oregon 97015, and Centers for Disease Control, Atlanta, Georgia 30333

The clinical significance of *Legionella pneumophila* has been well recognized since 1976, and the use of the macrolides as drugs of choice

was recognized shortly thereafter (4). Thornsberry and Kirven (12) and Fu and Neu (5) have described the presence of a beta-lactamase produced by *L. pneumophila*. However, other *Legionella* species have also been reported to produce enzymes capable of hydrolyzing some beta-lactam substrates. This resistance mechanism may contribute to the relative ineffectiveness of beta-lactam antimicrobial agents against *Legionella* infections, although the intracellular position of the pathogen appears to be a more significant factor (1). In this report we present additional information about the beta-lactamases found among the *Legionella* species. These investigations include beta-lactamase substrate hydrolysis profiles, beta-lactamase inhibition studies, combined drug interaction (synergy) analyses, and in vitro activity studies of new beta-lactam compounds, beta-lactamase inhibitors, and new potentially usable macrolide drugs (2, 6).

The methods used to study in vitro activity were previously described: agar dilution with buffered charcoal-yeast extract agar (pH 6.9) and an inoculum of 3×10^5 CFU/ml at 35°C (without increased CO_2), with endpoints read at 48 and 72 h (2, 3, 11). The methods for beta-lactamase hydrolysis and inhibition tests were those of Jones et al. (7, 9), utilizing a scanning spectrophotometer and a centrifugal-fast analyzer, respectively. Synergy studies were interpreted by criteria described by Jones and Packer (8). The drugs were obtained from Hoechst-Roussel Pharmaceuticals Inc., Eli Lilly & Co., Bristol Laboratories, Glaxo Inc., Pfizer Inc., Beecham Laboratories, Schering Corp., and Dow Chemical Co. The *Legionella* strains were obtained from the Centers for Disease Control and includ-

ed six serogroups of *L. pneumophila* (21 strains), *L. bozemanii* (3 strains), *L. dumoffii* (2 strains), *L. micdadei* (2 strains), *L. longbeachae* (2 strains), and 1 strain each of *L. gormanii*, *L. jordanis*, and *L. oakridgensis*. *Escherichia coli* ATCC 25922 was included as a control.

Eighteen antimicrobial agents or drug combinations were tested by the buffered charcoal-yeast extract agar dilution method against representative strains of eight *Legionella* species (Table 1). Rifampin remained the most active antimicrobial agent for all the species tested (11). The new macrolides, especially RU28965, were found to have equal or slightly superior activity when compared with erythromycin. The activity of ampicillin and amoxicillin against these strains was enhanced when those drugs were combined with beta-lactamase inhibitors. Among these inhibitors, clavulanate had the most potent anti-*Legionella* activity and was comparable to erythromycin. The activity of the combinations was superior to the in vitro erythromycin activity. The cephalosporins that were tested generally had relatively higher mean minimum inhibitory concentrations (MICs), but cefamandole, previously shown to be hydrolyzed by *Legionella* beta-lactamases, had the highest MICs (5, 12). Of the "third-generation" cephalosporins, cefotaxime and moxalactam were more active than cefoperazone.

Synergy, or drug interaction, studies were performed with combinations of amoxicillin and clavulanate (Augmentin; Beecham Laboratories), ampicillin and sulbactam (sultamicillin), and two ratios of BL-P2013 and amoxicillin (data

TABLE 1. Activities of rifampin, macrolides, ampicillin and analogs, beta-lactamase inhibitors and combinations, and cephalosporins against 33 strains of *Legionella*, representing eight species

Antimicrobial agent	Geometric mean MIC (µg/ml) against:					
	L. pneumophila (n = 21)	*L. bozemanii* (n = 3)	*L. dumoffii* (n = 2)	*L. micdadei* (n = 2)	*L. longbeachae* (n = 2)	Other species (n = 3)[a]
Rifampin	0.03	0.05	0.09	0.05	0.25	0.07
Erythromycin	0.7	1.0	1.3	0.8	0.8	1.7
RU28965	0.4	0.5	0.4	0.3	0.4	0.7
RU29065	0.7	0.8	0.8	0.6	0.5	1.7
RU29702	1.5	1.7	1.5	1.3	1.0	3.0
Ampicillin	2.1	0.3	12	0.1	2.5	2.2
Amoxicillin	2.3	0.3	12	0.1	3.0	1.8
BL-P2013	1.9	2.0	10	0.8	8.0	8.0
Clavulanate	0.7	0.5	1.0	0.3	0.2	1.2
Sulbactam	2.4	1.0	12	0.3	2.5	5.3
Augmentin	0.4	0.1	2.2	0.04	0.8	0.4
Sultamicillin	0.4	0.2	5.0	0.05	0.8	0.8
BL-P–amox[b]	0.5	0.3	5.0	0.05	2.5	1.0
Cefamandole	28	6.7	32	4	16	16
Cefoperazone	12	0.8	12	0.6	8.0	4.7
Cefotaxime	1.6	≤0.06	4.0	≤0.06	4.0	2.1
Moxalactam	7.1	0.8	2.0	0.6	0.8	0.6

[a] Includes a single strain of *L. gormanii*, *L. jordanis*, and *L. oakridgensis*.
[b] BL-P2013 and amoxicillin in fixed ratio of 1:2.

not shown). The 21 *L. pneumophila* strains from six serogroups showed remarkable in vitro synergism when ampicillin or amoxicillin were combined with the inhibitors. The rates of partial or complete synergy were: Augmentin, 95.2%; sultamicillin, 100%; BL-P2013–amoxicillin in a ratio of 2:1, 100%; and BL-P2013–amoxicillin in a ratio of 1:2, 90.5%. Lower rates of synergy were found among the other seven *Legionella* species (12 strains), and the rate was 75% for all combinations. For all species the synergy rates ranged from 90.9% for sultamicillin and BL-P2013–amoxicillin (2:1) to 84.8% for BL-P2013–amoxicillin (1:2). These rates are slightly higher than those suggested in studies of clavulanate by Pohlod et al. (10). The most likely interpretation of these findings is that the first antibiotic is protected by the enzyme inhibitor, but this does not rule out a different mechanism such as synergistic interaction at the cell wall target site.

Beta-lactamase hydrolysis studies required extended monitoring of incubation and critical correction for the autohydrolysis rates of the most enzyme-labile substrates, e.g., nitrocefin and cefamandole. Six *Legionella* species showed spectrophotometer-detectable cephalosporinase activity (nitrocefin substrate, 0.04 to 0.57 μM/min). *L. oakridgensis* and *L. micdadei* showed no evidence of beta-lactamase activity (<0.01 μM/min). Only *L. gormanii* showed significant detectable penicillinase activity against the penicillin and ampicillin substrates (Table 2). The level of beta-lactamase production and rate of hydrolysis were comparable to those measured in the *E. coli* ATCC 25922 quality control strain. This strain has a chromosomally mediated beta-lactamase yet is considered susceptible to nearly all gram-negative active beta-lactams, including ampicillin (MIC, 0.5 μg/ml) and "first-generation" cephalosporins. Comparisons of cefamandole hydrolysis rates and MICs failed to reveal a significant correlation (Table 2). Thus,

we believe the beta-lactamases of legionellae contribute little to the clinical failures of beta-lactam drugs against Legionnaires disease.

The beta-lactamase inhibitors, clavulanate, sulbactam, and BL-P2013, proved to be good inhibitors of the enzymes produced by *L. pneumophila*, *L. gormanii*, *L. dumoffii*, and *L. longbeachae*. The inhibition of the nitrocefin substrate was comparable to that of SCH29482 (a penem) and dicloxacillin, but markedly superior to new broad-spectrum cephalosporins such as cefotaxime, cefodizime, and ceftazidime. Very similar beta-lactamase inhibition profiles were found for all of the legionellae studied, e.g., BL-P2013 = SCH29482 = dicloxacillin > sulbactam > clavulanate. The beta-lactamase produced by *Legionella* species appears unique in the combined substrate hydrolysis profiles and inhibition patterns. Although this enzyme was previously thought to most resemble a type I cephalosporinase (5), these data show that the beta-lactamase affinity profiles are most similar to that of type III, TEM beta-lactamase, but the substrate hydrolysis patterns appear unique. These results are reproducible between strains within species and between representatives of beta-lactamase-producing legionellae.

In conclusion, these in vitro data suggest that beta-lactamase inhibitors interact favorably with ampicillin or amoxicillin to produce synergistic inhibition. The clinical efficacy of these drug combinations has yet to be proven. The investigated macrolides are excellent against all legionellae; RU28965, which had MICs more than twofold lower (0.3 to 0.7 μg/ml) than those of erythromycin (0.7 to 1.7 μg/ml), was especially good. These drugs and other new macrolides (2) may be acceptable alternatives to erythromycin because of their increased gram-positive spectrum or improved pharmacokinetic qualities (2, 6). The beta-lactamases of legionellae are unique, are of low potency, are inhibited by

TABLE 2. Beta-lactamase hydrolysis of beta-lactam substrates by *Legionella* enzymes[a]

Source of beta-lactamase	Hydrolysis rate (μM/min)[b]			
	Nitrocefin	Cefamandole[c]	Ampicillin	Penicillin G
L. gormanii LS-13	0.57	0.84 (16)	0.42 (4.0)	0.09
L. bozemanii WIGA	0.52	0.27 (8.0)		
L. pneumophila Philadelphia 1	0.78	<0.01 (16)		
L. dumoffii NY-23	0.45	<0.01 (32)		
L. longbeachae Long Beach 4	0.11	<0.01 (16)		
L. jordanis BL540	0.04	<0.01 (16)		
L. oakridgensis Oak Ridge 16	<0.01	<0.10 (16)		
L. micdadei TATLOCK	<0.01	<0.01 (8.0)		
E. coli ATCC 25922	0.14	0.03 (0.5)		

[a] Hydrolysis was measured by sensitive spectrophotometric methods for a minimum of 1 h.
[b] Data are corrected for substrate degradation rates and maximum possible instrument drift during analysis (0.22 μM/min for nitrocefin and 0.16 μM/min for cefamandole or ampicillin). Numbers in parentheses are MICs.
[c] Significant hydrolysis has been detected by bioassay systems (12).

clavulanate and sulfones, and probably do not contribute significantly to beta-lactam resistance. Since beta-lactams do not enter leukocytes or pulmonary macrophages, it is unlikely that beta-lactamase inhibitors do, but this should be investigated in vitro and in an animal model.

We are grateful for the technical support of Harold Wilson and the word processing of Barbara Beardsley.

LITERATURE CITED

1. **Bacheson, M. A., H. M. Friedman, and C. E. Benson.** 1981. Antimicrobial susceptibility of intracellular *Legionella pneumophila.* Antimicrob. Agents Chemother. **20:**691–692.
2. **Edelstein, P. H., K. A. Pasiecznik, V. K. Vasui, and R. D. Meyer.** 1982. Susceptibility of *Legionella* spp. to mycinamicin I and II and other macrolide antibiotics: effects of media composition and origin of organisms. Antimicrob. Agents Chemother. **22:**90–93.
3. **Feeley, J. C., R. J. Gibson, G. W. Gorman, N. C. Langford, J. K. Rasheed, D. C. Mackel, and W. B. Baine.** 1979. Charcoal-yeast extract agar: primary isolation medium for *Legionella pneumophila.* J. Clin. Microbiol. **10:**437–441.
4. **Fraser, D. W., T. R. Theodore, W. O. Orenstein, W. E. Parkin, H. J. Beecham, R. G. Sharrar, J. Harris, G. F. Mallison, S. M. Martin, J. E. McDade, C. C. Shepard, P. S. Brachman, and the Field Investigation Team.** 1977. Legionnaires' disease: description of an epidemic of pneumonia. N. Engl. J. Med. **297:**1189–1197.
5. **Fu, K. P., and H. C. Neu.** 1979. Inactivation of beta-lactam antibiotics by *Legionella pneumophila.* Antimicrob. Agents Chemother. **16:**561–564.
6. **Jones, R. N., A. L. Barry, and C. Thornsberry.** 1983. In vitro evaluation of three new macrolide antimicrobial agents, RU28965, RU29065, and RU29702, and comparisons with other orally administered drugs. Antimicrob. Agents Chemother. **24:**209–215.
7. **Jones, R. N., A. L. Barry, C. Thornsberry, and H. W. Wilson.** 1981. In vitro antimicrobial activity evaluation of cefodizime (HR221), a new semisynthetic cephalosporin. Antimicrob. Agents Chemother. **20:**760–768.
8. **Jones, R. N., and R. R. Packer.** 1982. Antimicrobial activity of amikacin combinations against *Enterobacteriaceae* moderately susceptible to third-generation cephalosporins. Antimicrob. Agents Chemother. **22:**985–998.
9. **Jones, R. N., H. W. Wilson, and W. J. Novick.** 1982. In vitro evaluation of pyridine-2-azo-*p*-dimethylaniline cephalosporin, a new diagnostic chromogenic reagent, and comparison with nitrocefin, cephacetrile, and other beta-lactam compounds. J. Clin. Microbiol. **15:**677–683.
10. **Pohlod, D. J., L. D. Saravolatz, E. L. Quinn, and M. M. Somerville.** 1980. Effect of clavulanic acid on minimal inhibitory concentrations of 16 antimicrobial agents tested against *Legionella pneumophila.* Antimicrob. Agents Chemother. **18:**353–354.
11. **Thornsberry, C., C. N. Baker, and L. A. Kirven.** 1978. In vitro activity of antimicrobial agents on Legionnaires disease bacterium. Antimicrob. Agents Chemother. **13:**78–80.
12. **Thornsberry, C., and L. A. Kirven.** 1978. Beta-lactamase of Legionnaires' bacterium. Curr. Microbiol. **1:**51–54.

Action of Ampicillin and Erythromycin on the Growth and Morphology of *Legionella pneumophila*

FRANK G. RODGERS AND TOM S. ELLIOTT

Department of Microbiology and PHLS Laboratory, University Hospital, Nottingham, England

A number of antibiotics are active against *Legionella* species. Both ampicillin and erythromycin have been used successfully to treat clinical legionellosis. In this investigation, the in vitro morphological response of *Legionella pneumophila* serogroup 1, strain Nottingham N7, and of *L. pneumophila* serogroup 3, strain Bloomington 2, to either ampicillin or erythromycin was studied by negative-stain, thin-section, and scanning electron microscopy. Organisms were grown in enriched broth (3), and at 24 h of incubation, corresponding to the logarithmic phase of growth, antibiotics were added to give final concentrations of 10 μg/ml. Samples were removed at various time intervals for electron microscopy, and viable counts (as CFU per milliliter) were made in duplicate on each sample by using 10-fold dilutions inoculated onto enriched blood agar (1). Colony counts were read daily for 5 days to evaluate regrowth after antibiotic exposure. Minimum inhibitory concentrations for both strains were assayed by plate incorporation techniques. The minimum inhibi-

tory concentration for ampicillin was 0.5 μg/ml, and that for erythromycin was 0.4 μg/ml.

Growth kinetics for both strains were similar, with a mean generation time of 2 h which was maintained for approximately 48 h. Each strain attained viable counts of 10^{10} to 10^{11} CFU/ml after 60 h. Electron microscopy of these organisms demonstrated that they had the typical appearance of *Legionella* cells (2). The cells had a relatively smooth outer surface by scanning electron microscopy and negative staining, and the outer and inner membranes of the cell wall were seen in thin sections. Division and separation of the bacteria were also evident and occurred at the middle of cells in a symmetrical pinching fashion. Nuclear filaments together with ribosomes were dispersed throughout the bacterial cytoplasm.

Exposure of both strains to ampicillin at 10 μg/ml for 6 h induced changes in the shape of organisms, with grossly abnormal cells and disturbed division. Bacterial collapse was evident by scanning electron microscopy (Fig. 1a). Dis-

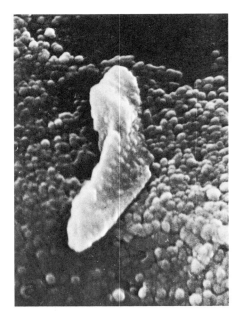

crete membranous vesicles, randomly distributed, appeared associated with the surface membranes. After 24 h of treatment with ampicillin, these membranous vesicles enlarged and the cell wall was broken in several sites, involving both

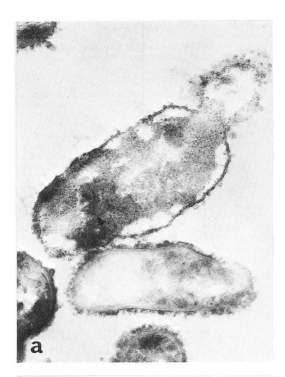

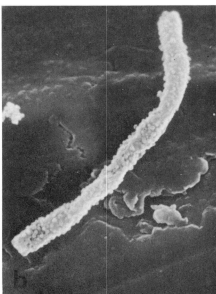

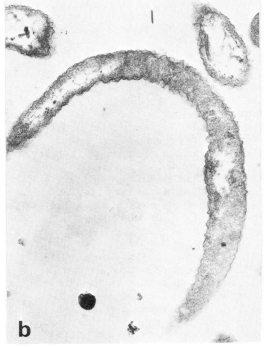

FIG. 1. Scanning electron microscopy of *L. pneumophila* morphologically damaged by exposure to antibiotics at 10 μg/ml. (a) Serogroup 3 organism after 24 h of ampicillin treatment. Note collapse of the bacterium. Magnification, ×24,000. (b) Serogroup 1 organism after 6 h of exposure to erythromycin. Note fine surface lesions and elongation of the bacterium. Magnification, ×12,000.

FIG. 2. Thin sections of *L. pneumophila* morphologically damaged by exposure to antibiotics at 10 μg/ml. (a) Serogroup 3 organisms after 6 h of ampicillin treatment. Note cell wall lesions and bacterial lysis. Magnification, ×35,600. (b) Serogroup 1 organisms after 6 h of exposure to erythromycin. Note convoluted cell membranes, bacterial elongation, and cytoplasmic swelling. Magnification, ×23,140.

the inner and outer membranes with extrusion of cytoplasmic contents (Fig. 2a). The membranous vesicles and lysis points did not coincide. A few spheroplasts were also present, some grossly enlarged and lysed. Occasional cells with apparently normal morphology were evident after extended ampicillin treatment. A few spherical minicells of 0.15-μm diameter with normal cell wall structure were found. The corresponding viable counts decreased dramatically in the presence of ampicillin, with a 4 to 5 $\log_{10}$ reduction from approximately 10^8 CFU/ml occurring after 6 h of exposure. The time required for regrowth of organisms after removal of ampicillin was directly proportional to the time of exposure to the antibiotics.

Exposure of both strains to erythromycin at 10 μg/ml induced less microbial damage, with more normal cells compared with results of exposure to ampicillin. After 6 h of exposure, cell division was disturbed and no pinching or separation sites were seen. In addition, the cell walls were modified, the outer membranes were covered with small vesicle-like lesions (Fig. 1b), and filamentous organisms predominated (Fig. 2b). By thin section, the number and staining density of cytoplasmic ribosomes increased markedly. On exposure to erythromycin for 24 h bacterial lysis occurred, but as with ampicillin-treated cultures, cells with normal morphology persisted. No spheroplasts were found, and lysed cells had discrete breakage points in the cell walls. Viable counts for both strains decreased after exposure to erythromycin; however, this was not as pronounced as with ampicillin. These findings confirmed electron microscopy studies in which lysis was more evident with ampicillin.

After removal of antibiotics by either dilution or centrifugation, regrowth studies showed that viable organisms regrew faster and to higher numbers after erythromycin treatment than after ampicillin treatment. The time required to regrow was directly related to the time of ampicillin treatment. However, increasing exposure to erythromycin had little effect.

These in vitro studies showed that both antibiotics resulted in lysis of *L. pneumophila*, but that the effect of ampicillin on the growth and morphology of the bacteria was more disruptive than that of erythromycin. With both antibiotics normal cells persisted, whereas with ampicillin minicells also occurred. The lesions produced by each antibiotic were distinct. *Legionella* organisms act as facultative intracellular pathogens (2; Rodgers and Oldham, this volume); it is therefore possible that differences between the activity of antibiotics on bacterial replication in vitro and in infected cells in vivo may be pronounced. Indeed, the apparent efficacy of erythromycin in clinical disease may be due more to the concentration of the antibiotic achieved within macrophages and its action intracellularly than to the potency of the drug per se. It is tempting to speculate that the 'persisters' and the minicells partly account for the prolonged nature of some cases of legionellosis, even in the presence of apparently adequate antimicrobial therapy.

LITERATURE CITED

1. **Dennis, P. J., J. A. Taylor, and G. I. Barrow.** 1981. Phosphate buffered, low sodium chloride blood agar medium for *Legionella pneumophila*. Lancet **ii**:636.
2. **Rodgers, F. G.** 1979. Ultrastructure of *Legionella pneumophila*. J. Clin. Pathol. **32**:1195–1202.
3. **Rodgers, F. G., P. W. Greaves, A. D. Macrae, and M. J. Lewis.** 1980. Electron microscopic evidence for flagella and pili on *Legionella pneumophila*. J. Clin. Pathol. **33**:1184–1188.

Construction of a *Legionella pneumophila* Gene Bank

N. CARY ENGLEBERG, DAVID J. DRUTZ, AND BARRY I. EISENSTEIN

Division of Infectious Diseases, Department of Medicine, and Department of Microbiology, University of Texas Health Science Center, San Antonio, Texas 78284

A full understanding of the pathogenic processes and host response during infection with *Legionella pneumophila* will depend ultimately upon the isolation and purification of individual components of the complex cellular structure. Using recombinant DNA technology, we have begun to isolate some of these components at the genetic level. We constructed a "bank" of the *L. pneumophila* genome by cloning sized, random chromosomal fragments into an *Escherichia coli* plasmid vector. As a first step in selecting potentially important genes, we screened our bank with pooled rabbit antisera to detect clones that express *L. pneumophila* antigens.

For all experiments we used *L. pneumophila* serogroup 1 strain 130b, which was animal passaged and then passed only once on buffered charcoal-yeast extract agar before use. After organisms were harvested from plates, chromosomal DNA was extracted and purified by using a modification of the method of Nakamura et al. (4).

L. pneumophila DNA was partially restricted with Sau3A endonuclease and fractionated on a 5 to 40% sucrose density gradient. Fragments of 2.5 to 7.0 megadaltons were isolated and inserted into the phosphatase-treated BamHI site of vector pBR322. Hybrid plasmids were used to transform CaCl₂-treated E. coli K-12 strain HB101 (recA hsdR hsdM). Transformants were selected on media containing ampicillin. All of 40 randomly selected ampicillin-resistant (Ampʳ) transformants were sensitive to tetracycline, indicating inactivation of the tetracycline resistance gene of pBR322.

Antisera were prepared by injecting New Zealand rabbits subcutaneously with suspensions of heat-killed or Formalin-killed L. pneumophila

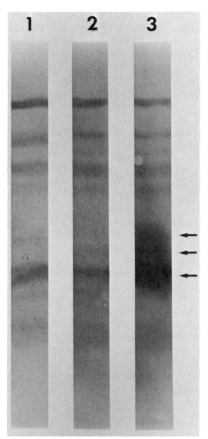

FIG. 2. Western blot of a clone (no. 11) strongly reactive by FBEIA. Lane 1, clone 11 plus preimmune sera; lane 2, E. coli HB101(pBR322) plus immune sera; lane 3, clone 11 plus immune sera.

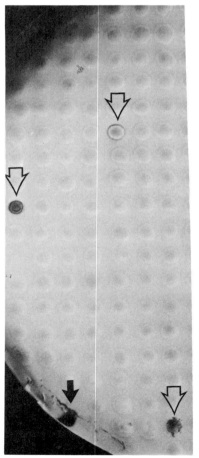

FIG. 1. Section of a nitrocellulose filter processed by FBEIA. The positive control (Formalin-killed L. pneumophila) is indicated by solid arrow; positive clones are indicated by open arrows.

biweekly for 2 months. Sera collected at least 3 weeks after the last injection were tested for specific activity by using an L. pneumophila whole cell enzyme-linked immunosorbent assay. All preimmune sera had titers of <1:10; immune sera had titers of 1:1,280 and ≥1:10,240 for heat- and Formalin-killed cellular antigens, respectively. Immune sera were absorbed extensively with E. coli whole cells and E. coli sonic extract without significantly affecting their activity in the enzyme-linked immunosorbent assay.

A total of 2,559 Ampʳ transformants were screened for L. pneumophila by using a filter-binding enzyme immunoassay (FBEIA). Transformant colonies (~300) and negative control colonies [E. coli HB101(pBR322)] were transferred from agar plates to detergent-free nitrocellulose paper (Millipore Corp., Bedford, Mass.). Positive controls (suspensions of Formalin-killed L. pneumophila) were spotted directly onto each filter. Colonies were lysed in situ by the method of Meyer et al. (3). Briefly, filters were sequentially overlaid (colony side up) for 5 min each on a series of Whatman 3MM paper disks saturated, respectively, with 0.1 N NaOH, 1.5 M Tris-hydrochloride (pH 7.4), 0.3 M NaCl–0.03 M sodium citrate, and 70% ethanol and then dried by baking at 60°C in a vacuum for 2 h. Dry filters were incubated in 3% gelatin in Tris-buffered saline (TBS) for 2 h at 22°C to block nonspecific protein binding and then transferred to pooled rabbit antisera in TBS (1:800 dilution) and incubated with gentle rotation at 22°C overnight. After extensive washings with TBS, the filters were treated with peroxidase-conjugated goat anti-rabbit immunoglobulin G (1:3,000 dilution) for 2 h and, after additional, extensive TBS washes, were bathed in a color development solution (0.05% 4-chloro-1-naphthol [Bio-Rad Laboratories, Richmond, Calif.], 0.015% H_2O_2, and 20% methanol in TBS). Colonies expressing L. pneumophila antigens and positive controls developed a dark blue color within minutes (Fig. 1).

A total of 77 (30%) of the 2,559 transformant colonies we screened by FBEIA were positive; 31 (1.2%) were strongly positive, and 46 (1.8%) were weakly positive.

To confirm the presence of L. pneumophila antigen in FBEIA-positive colonies, we analyzed one strongly reactive clone (no. 11) by Western blotting (5). Stationary-phase cells were lysed by boiling in sample buffer containing 2% sodium dodecyl sulfate for 2 min and separated by sodium dodecyl sulfate-polyacrylamide gel electrophoresis (10% acrylamide). After electrotransfer to nitrocellulose paper in a Trans-blot cell (Bio-Rad), antigenic bands were visualized by using rabbit sera, peroxidase-conjugated goat anti-rabbit immunoglobulin G, and color development reagent as described above for the FBEIA.

Unique, strongly reactive bands were present in clone 11 lysate exposed to rabbit immune sera (Fig. 2, lane 3); these bands were not present in controls [E. coli HB101(pBR322) plus immune sera and clone 11 plus preimmune serum].

We conclude that our gene bank contains several E. coli clones that express L. pneumophila antigens. Using the formula of Clarke and Carbon (2) and an estimate of the L. pneumophila genome size (1), we calculate that any average-sized (~1,000 kilobase pairs) strain of L. pneumophila has an 85 to 99% probability of being included in our initial bank of ~2,500 clones. We expect to find expression of many of the major L. pneumophila antigens represented among the 77 FBEIA-positive clones. By using novel methods of selection, we may be able to identify other clones that express nonimmunogenic proteins or polypeptides of importance in pathogenesis.

LITERATURE CITED

1. Brenner, D. J., A. G. Steigerwalt, R. E. Weaver, J. E. McDade, J. C. Feeley, and M. Mandel. 1978. Classification of the Legionnaires' disease bacterium: an interim report. Curr. Microbiol. 1:71–75.
2. Clarke, L., and J. Carbon. 1976. A colony bank containing synthetic Col E1 hybrid plasmids representative of the entire E. coli genome. Cell 9:91–99.
3. Meyer, T. F., N. Mlawer, and M. So. 1982. Pilus expression in Neisseria gonorrhoeae involves chromosomal rearrangement. Cell 30:45–52.
4. Nakamura, K., P. M. Pirtle, and M. Inouye. 1979. Homology of the gene coding for outer membrane lipoprotein within various gram-negative bacteria. J. Bacteriol. 137:595–604.
5. Towbin, H., J. Staehelin, and J. Gordon. 1979. Electrophoretic transfer of proteins from polyacrylamide gels to nitrocellulose sheets: procedure and some applications. Proc. Natl. Acad. Sci. U.S.A. 76:4350–4354.

Nucleic Acid Probe Specific for Members of the Genus *Legionella*

DAVID E. KOHNE,† ARNOLD G. STEIGERWALT, AND DON J. BRENNER

Division of Bacterial Diseases, Center for Infectious Diseases, Centers for Disease Control, Atlanta, Georgia 30333

A radioactive nucleic acid probe has been developed that is specific only for members of the genus *Legionella*. Nucleic acid hybridization experiments demonstrated that the probe hybridized to greater than 50% with nucleic acids from diverse members of the genus *Legionella* and did not hybridize significantly with nucleic acids from mammals, *Saccharomyces cerevisiae*, and a wide variety of bacterial strains (Table 1). The probe hybridized well even with *Legionella* species, such as *L. pneumophila* and *L. micdadei*, which show very little total DNA relatedness. Preliminary data with a less refined probe indicated that all of the 23 described and unnamed *Legionella* species can be detected by this probe and distinguished from all other bacteria listed in Table 1.

† 11575 Sorrento Valley Road, Suite 203, San Diego, CA 92121.

TABLE 1. Hybridization of *Legionella*-specific probe with nucleic acids from widely different sources

Unlabeled nucleic acid source	Normalized % of probe hybridization	Unlabeled nucleic acid source	Normalized % of probe hybridization
Controls		*Flavobacterium gleum*	<1
No nucleic acid	<1	*Flavobacterium meningo-*	
Mock nucleic acid isolation	<1	*septicum*	<1
L. pneumophila-infected		*Flavobacterium spiritovorum* ...	<1
hamster tissue .... 100 (actual %, 81)		*Flavobacterium thalpophilum* ..	<1
		Flexibacter canadensis	<1
Legionellaceae		*Proteus mirabilis*	<1
L. bozemanii (WIGA)	>50	*Proteus vulgaris*	<1
L. dumoffii (TEX-KL)	>50	*Providencia alcalifaciens*	<1
L. gormanii (LS-13)	>50	*Providencia rettgeri*	<1
L. jordanis (BL-540)..........	50	*Providencia stuartii*	<1
L. longbeachae (Long Beach 4)	>50	*Pseudomonas alcaligenes*	<1
L. micdadei (HEBA)	>50	*Vibrio cholerae*	<1
L. oakridgensis (OR-10)	>50	*Mycoplasma hominis*	<1
L. pneumophila (Philadelphia 1)	100	*Mycoplasma hyorhinis*.........	<1
Unnamed *Legionella* species 16	>50	*Mycoplasma salivarium*.......	<1
Unnamed *Legionella* species 17	>50	*Acholeplasma laidlawii*	<1
Unnamed *Legionella* species 18	>50	*Spiroplasma* SMCA, cornstunt,	
		honey bee, or cactus	<1
Other microbial species			
Aeromonas hydrophila........	<1	Yeast	
Bacillus subtilis	<1	*Saccharomyces cerevisiae*	<1
Campylobacter jejuni	<1		
Cytophaga johnsonae	<1	Mammals	
Escherichia coli..............	<1	Human, hamster, or mouse	<1
Flavobacterium breve	<1		

The specificity of this probe makes it possible to detect and quantitate the presence of *Legionella* species, even in the presence of large numbers of nonrelated bacterial or mammalian cells. Liver from a hamster infected with *L. pneumophila* was assayed for the presence and number of legionellae by using the specific probe with well-established nucleic acid hybridization procedures. The liver had previously been assayed by a microbiological growth test, which indicated that about 10^7 *Legionella* organisms/g were present in the infected liver. Nucleic acid hybridization analysis indicated 1×10^8 to about 2×10^8 *Legionella* organisms/g of liver. This result suggests that the plating efficiency in the growth test is about 5 to 10%, that some 90% of the organisms present in the sample were nonviable, or both.

The specific probe allows the highly sensitive and rapid detection of *Legionella* organisms even in the presence of large numbers of mammalian cells. In an assay that took less than 1 day, the presence of about 400 *Legionella* organisms, which were mixed with 0.4 mg of liver (about 2×10^5 cells), was easily detected with the probe. We anticipate that both the sensitivity and the speed of detection can be substantially increased.

SUMMARY

Microbiology of *Legionella*

A. BALOWS

Center for Infectious Diseases, Centers for Disease Control, Atlanta, Georgia 30333

Rather than adhere to the program format in which papers and posters were presented at the symposium, I elected to organize this summary into seven of the areas covered in which bacteria are usually studied. This was done to provide the reader with the highlights of the presentations and the discussions and to suggest areas where research is still needed.

Bacterial growth. Berg et al. developed a complex medium in which *Legionella pneumophila* cells were grown in an aerated continuous-culture chemostat. *L. pneumophila* grew quite well during vigorous agitation; the only growth-limiting factor appeared to be the labile nature of L-cysteine in the presence of an iron catalyst. This approach appears to be especially useful in studying the physiology of a large culture mass, the quantitation of growth-limiting factors, and the action of a large number of germicides on actively multiplying legionellae.

Morphology. Bornstein et al. described work that dealt with the loss or retention of virulence in laboratory-maintained strains of legionellae. When predialyzed yeast extract was incorporated in the "standard" charcoal-yeast extract agar, the resulting *L. pneumophila* cells were rod shaped and 3.0 μm or less long. The virulence of these agar-grown cells, using embryonated hen eggs as the assay for virulence, showed a 50% lethal dose closely approximating that of control *L. pneumophila* cells that were maintained continuously in yolk sacs. No, or very few, filamentous forms were cultivated on the modified charcoal-yeast extract agar. The authors believe that there is significant correlation between cell length and virulence, with the bacillary form of 3.0 μm or less in length being more likely to be virulent than are the longer and filamentous forms. It is interesting to note that McDade in 1978 observed the production of filamentous forms by *L. pneumophila* in yolk sac but made no quantitation or assessment of virulence at that time.

Taxonomy and nomenclature. Brenner stated that his laboratory is now working with 23 defined species, 10 of which are named, and that more species are yet to come. All of the named and unnamed species studied are closely related phenotypically, and 12 of the 23 species have been implicated in human disease. DNA relatedness is the principal tool for species designation, supported by antigen analysis, cell wall fatty acid profiles obtained by gas-liquid chromatography, and a variety of differential reactions·and characteristics. A source of confusion regarding names at the genus level has developed after the proposal of Brown et al. that two generic designations, *Tatlockia* and *Fluoribacter*, be added to *Legionella*. I have neither the wisdom nor the desire to attempt to resolve this potential dilemma. Suffice it to say that similar situations have occurred in the past. As long as the proposed genus or species designations follow the existing rules of nomenclature, the only apparent way to resolve the issue is through common usage by the scientific community. Time will tell whether we are better served with the single current genus, as one school holds, and 23 or more species, or with three genera that subsequently may have to be increased to 12 or more if one applies the same rule of thumb for generic DNA relatedness as has been advocated by the other school.

Fox et al. reported the establishment of carbohydrate profiles to elucidate taxonomic interrelationships. They used gas chromatography and mass spectrometry for carbohydrate analyses of each of the serogroups (1 through 6) of *L. pneumophila* and initiated analyses of *L. micdadei*. The carbohydrate peaks were identified, and the authors indicate that by extending this approach to include similar analyses of several strains of each species and serogroups within a species it should be possible to shed light on the taxonomic characterization of *Legionella* species.

Physiology and metabolism. Hoffman presented a complete picture of the current knowledge on the physiology and metabolism of legionellae. The cell wall is composed of diaminopimelic acid, typical of gram-negative bacilli, but shows an unusually high cross-linkage of 80 to 90%. Why this occurs is not yet understood. The *Legionella* species that have been studied are strict aerobes and have an absolute requirement for cysteine. Cystine substitutes very poorly for cysteine. Pine et al. have reported that glutathione could substitute for cysteine; this may require confirmation.

Legionellae require some 18 amino acids as energy sources. Some if not all of the amino acids are utilized through oxidative pathways. Carbohydrates provide very little in the way of energy, as less than 2% of glucose added to a defined medium is conserved. Through generally predictable pathways, legionellae derive energy for metabolic activities; i.e., the Krebs cycle has been reported to occur in *L. pneumophila*, glucose is synthesized from amino acids as would be expected, and the metabolism of pyruvate is not a major source of energy. The possibility of amino acid metabolism following species or serogroup definition was explored by Franzus et al., but no clear differences were observed.

The earlier report by Reeves et al. on the lack of siderophore production was confirmed and amplified by the elegant work of Quinn and Weinberg on the susceptibility of *L. pneumophila* and *L. jordanis* to iron-binding agents. They studied five different purified chelating agents, using the two *Legionella* species growing in a suitable broth medium; both species were quite susceptible to these natural siderophores. Reversal of the inhibition by any of the five chelating agents was accomplished by adding 10 to 50 μmol of ferric pyrophosphate to the broth medium. However, when legionellae were grown at 41°C (well within the optimum growth range), 1,000 μmol of ferric pyrophosphate could not suppress the growth inhibition effect of phenanthroline.

Mayberry analyzed two or more strains of seven different species and presented data on the composition and structure of the dihydroxy, monohydroxy, and unsaturated monohydroxy fatty acids. Previous work by Moss et al. indicated that these fatty acids were present in a range of 5 to 10 mol%. Mayberry showed that dihydroxy fatty acids are found only in *L. pneumophila* and *L. micdadei*. The monohydroxy fatty acid profiles made it possible to divide the seven species tested into two groups, one with about 3 mol% and a second with about 10 mol%.

Pine et al. examined 44 strains of different *Legionella* species for the presence of the scavenging enzymes catalase, peroxidase, and superoxide dismutase. None of the species examined could be differentiated by superoxide dismutase. *L. pneumophila* and *L. gormanii* produced little or no catalase but did produce readily detectable amounts of peroxidase. Conversely, *L. micdadei*, *L. bozemanii*, *L. jordanis*, and *L. longbeachae* showed high catalase but no peroxidase activity. Expanding this data base to include more strains of each species and more of the recognized and emerging species may make presumptive identification of species less complex by testing for the presence of these scavenging enzymes.

Miller et al. were able to cultivate strains of five different *Legionella* species on buffered charcoal-yeast extract agar containing 50% human serum. Inducible proteolytic enzymes were present in each of the species tested, and many of the specific proteins acted upon by the proteolytic enzymes were identified by electrophoresis with monospecific rabbit antisera against 23 serum proteins. This proteolysis was not influenced by the addition of aprotinin. Miller suggested that extensive proteolytic degradation of important human proteins may be a virulence factor of legionellae.

Characterization of legionellae. It was inevitable that laboratory scientists would detect isolates of legionellae that did not conform to what previously had been described in the characterization of the family, genus, and species. Thus we learned that new serogroups have been added to *L. pneumophila*, and eight are now recognized, with several others yet to be antigenically differentiated.

Bissett, in describing her work that points to a new serogroup, made the point that the exact antibody composition of a direct immunofluorescent conjugate needs to be on the label and known by the user. By using two "different" serogroup 1 *L. pneumophila* conjugates, Bissett et al. were able to demonstrate a new serovariant, one that has been implicated in 13 cases of disease, 10 of which were nosocomial. It is safe to predict that new serogroups and types within groups will continue to be described.

The need for additional antigen analyses of the cell and its structural components was given strong support by the interesting report by Rodgers in which a passive hemagglutination test using *L. pneumophila* flagella as antigen was developed in his laboratory. The flagellar antigen was prepared by using the method described by Rodgers and Laverick in which *L. pneumophila* was grown on a nutritionally complete blood agar base; the flagella were harvested and partially characterized for use as an antigen. If this hemagglutination test can be standardized, it may be a useful diagnostic and epidemiological tool.

Johnson and Reinhart described the chemical characterization of two different fluorescent compounds from the cell membrane of *L. gormanii* and *L. dumoffii*. One compound was hydrophobic, and the authors described the UV-visible spectrum absorption maxima, spectroscopy, and mass spectrum for determination of the molecular weight. A chemical formula of $C_{19}H_{14}O_3$ was established, with no nitrogen, sulfur, or iron detected. The other fluorescent compound was water soluble and gave different values for UV absorption maxima and fluorescent spectroscope emission.

Brenner covered briefly a number of characteristics that are useful in species differentiation. Some of these are autofluorescence, beta-lactamase production, motility, alkaline phosphatase production, and hippurate hydrolysis. The last test is useful for separating L. pneumophila from the other recognized species.

Towns presented data from a study on hippurate hydrolysis in which several strains of five species in addition to L. pneumophila were tested, and end product determination was done by paper chromatography and gas-liquid chromatography. This report confirmed the usefulness of the hippurate hydrolysis test and also indicated that one, and possibly a second, strain of L. pneumophila gave negative results, a finding previously reported by Hebert.

Three presentations dealt with the hemolytic activity of legionellae. Miller described an extracellular hemolysin in six Legionella species. This hemolytic activity was best demonstrated on a modified charcoal-yeast extract agar in which 0.5% corn starch was substituted for the charcoal, and 5% sheep erythrocytes (RBC) was added. The hemolysin was inactivated by heating at 100°C and by pronase. Preliminary polyacrylamide gel electrophoresis indicated that the hemolysin was not associated with lipase or protease activity. However, no in vivo studies or studies with tissue cell-grown legionellae were done.

Baine, using dog RBC as the target cells, demonstrated the elaboration of a cytolytic exotoxin produced by seven species of Legionella. The dog RBC membranes are rich in lecithin and were lysed when incorporated into an agar medium inoculated with any of the species tested. The elaboration of phospholipase C or lecithinase was established by an appropriate assay. Guinea pig and rabbit RBC were also lysed by most of the species tested, but human RBC showed little or no lysis.

Frisch and Baine used the supernatant fluid from L. bozemanii grown in a broth medium and demonstrated the lysis of freshly washed dog RBC after 1 h at 37°C and 1 h at 4°C. This hemolytic activity was inactivated by heating at 100°C and inhibited by phosphatidylcholine. These last two hemolysins appear to be different in molecular weight and mechanism of RBC lysis.

Cytotoxic activity was observed by Rodgers and Oldham, who grew L. pneumophila in HEp-2, MRC-5, and Vero tissue cell lines. With intracellular growth of the legionellae they noted fatty degeneration of the mitochondria and dilation of the endoplasmic reticulum of the cell lines.

In vitro antibiotic susceptibility studies. Ristuccia et al. tested five species of Legionella for minimal inhibitory concentrations of rifampin, erythromycin, and three tetracyclines, using the Abbott MS-2 instrument with buffered yeast extract broth and human serum as diluents. Consistent with previous reports, rifampin was most active, followed by erythromycin and doxycycline. Alteration of the pH from 7.2 to 7.4 and of the temperature of incubation from 35 to 38°C resulted in no differences. In vitro synergy was best observed with the combination of erythromycin plus rifampin at readily achievable in vitro serum levels.

Jones et al. reported on the hydrolysis of beta-lactam antibiotics with beta-lactamase-producing legionellae and inhibition of this hydrolysis by beta-lactamase inhibitors. The authors used nitrocefin for the hydrolysis and its inhibition, as these reactions can be followed spectrophotometrically. Their data indicated that five species of Legionella produce rather small amounts of beta-lactamases that "seem most similar" to the TEM-1 beta-lactamase but do have unique substrate profiles.

The most interesting finding was with the synergy studies using beta-lactamase inhibitors such as sulbactam. For example, mean minimal inhibitory concentrations against 21 strains of L. pneumophila were 2.1 µg/ml for ampicillin and 2.2 µg/ml for sulbactam when the drugs were tested singly, but when ampicillin and sulbactam were combined the mean minimal inhibitory concentration was 0.38 µg/ml. The clinical implications of these findings are unknown.

Rodgers and Elliott examined the effects of ampicillin and erythromycin on the morphology and regrowth ability of L. pneumophila by using thin-section negative staining and scanning electron microscopy. Cell lysis was evident, as was the production of long filamentous forms and organisms with small surface lesions. Viable cell counts showed that ampicillin was more bactericidal than erythromycin, and regrowth to normal cell division and morphology was more rapid in cells exposed to erythromycin.

Molecular biology. Engleberg et al. isolated cellular proteins at the generic level by using recombinant DNA techniques. They inserted fragments of L. pneumophila genome into an Escherichia coli vector. The E. coli then synthesized the L. pneumophila antigens. This led to the preparation of a plasmid gene bank of L. pneumophila that contained many of the genes earmarked for future study. Several of these genes were cloned, and the clones showed some potentially unique antigens by the Western blot technique. These findings were interpreted as a means of identifying, at the molecular level, virulence factors that ultimately could be identified by specifically prepared antisera.

Kohne presented an enticing and tantalizing narrative on the development of a DNA probe that was specific for legionellae at either the genus or species level. He presented limited data showing that the probe detected each of the 23 named and unnamed species of *Legionella* but did not detect any of about 40 species of other bacteria and mycoplasmas. Regrettably, no data were presented on the makeup of the probe because of pending patent applications.

Future needs. Space does not permit a detailed listing of all of the future research needs on the bacteriology of the legionellae. Three major areas are as follows. (i) A systematic characterization of the recognized species by careful study of as many strains as possible of each species is needed to establish a reliable and statistically valid data base for biochemical reactions; production of enzymes, hemolysins, and other virulence factors; cell structure and composition; and antigen analysis. (The small peptide containing six amino acid residues that has been described by Iglewski and shown to exert a variety of destructive actions on monocytes and polymorphonucleocytes is an example of a fascinating finding that must be pursued.) (ii) Extensive studies on the physiology of legionellae with a focus on in vivo metabolic activities are critical if we are to understand the role of essential nutrients, such as iron, and how these requirements are met when legionellae invade leukocytes, amoebae, and various tissue cell lines. (iii) Continued and in-depth studies on the molecular biology and related aspects of legionellae should be pursued. The initial reports indicate that these organisms can be studied with existing technology, and the questions to be answered are many.

PATHOLOGY AND PATHOPHYSIOLOGY

STATE OF THE ART LECTURE

Pathology and Pathogenesis of Legionella Infections

WASHINGTON C. WINN, JR.

Department of Pathology, University of Vermont College of Medicine, Burlington, Vermont 05405

In the 5 years since the First International Symposium on Legionnaires Disease, we have learned much about the pathology and pathogenesis of the infections produced by this new group of human pathogens. It is difficult to define precisely the boundaries for this presentation, which clearly overlaps considerations of bacterial structure and function on the one hand and immunological defense mechanisms of the infected host on the other. Likewise, clinical studies of human pathophysiology and epidemiological investigation of bacterial transmission impinge directly on our understanding of the pathogenesis of these infections, but are best considered in other sections of this Symposium.

The major areas of inquiry into bacterial pathogenesis, however, can be categorized chronologically and, admittedly somewhat arbitrarily, by the degree to which the investigations have been completed. The initial areas of investigation included the pathological anatomy, physiology, and biochemistry of human infection; transmission of infection to humans; and potential bacterial virulence mechanisms, including structural and biochemical factors. Recent areas of investigation have been the interactions of legionellae and defense mechanisms in vitro and in vivo and the development of experimental models of infection in vivo. Areas for future effort include evaluation of potential virulence mechanisms in vivo; study of mediators of inflammation, including complement; evaluation of pathological physiology in vivo, including mechanisms of tissue damage; evaluation of effects of immunity on defense mechanisms; study of residua of infection, including fibrosis; and study of interactions between antibacterial chemotherapy and host defense mechanisms.

A substantial base of knowledge has been accumulated about human infection. Additional information may accrue, but the difficulty of accomplishing sequential observations of complicated physiological events in humans will probably limit these investigations severely. Rather than review the mass of data on human infections and bacterial virulence factors, I will concentrate on the development of experimental models of infection in laboratory animals, with which some of the difficult questions on patho-

genesis may be investigated in the future.

The first experimental model of legionella infection may, in fact, have been the acute respiratory disease that was produced in epidemiologists from the Centers for Disease Control who investigated the outbreak of Pontiac fever in 1968 (10). Exposure of guinea pigs to environmental samples from this epidemic was the first investigation in which a respiratory route of infection was employed and the first investigation in which animals were not used solely for the isolation of the etiological agent (15). Meenhorst and associates have used guinea pigs similarly to demonstrate the presence of infectious *Legionella pneumophila* in potable water that had been obtained during an outbreak of legionella pneumonia (17).

After the recognition of a bacterial etiology both for Pontiac fever and for Legionnaires disease, the pathological response to intraperitoneal inoculation of bacteria was described (6). A purulent peritonitis developed regularly; hepatic and splenic abscesses were produced, and transdiaphragmatic extension of the inflammation produced a purulent pleural exudate. Pneumonia also resulted, but the inflammatory process was concentrated in the interstitium and resembled neither the lesions that were produced after respiratory infection of experimental animals (8, 14) nor the inflammation in cases of Legionnaires disease in humans (21).

The peripheral location and pleural base of many inflammatory nodules in clinical Legionnaires disease has raised the possibility of septic pulmonary emboli in some cases. The dissimilarity of the experimental postbacteremic pneumonia from the peripheral nodules of human disease and the reproduction of the nodular lesions by respiratory inoculation of animals mitigate against a vascular portal of entry into the lung in human Legionnaires disease. The absence of emboli in macroscopically visible pulmonary arteries and the absence of extrapulmonary foci of inflammation in most human cases also suggest another pathogenetic mechanism for the pleural-based, nodular lesions.

A model legionella infection in which bacteria were placed into diffusion chambers that had been inserted subcutaneously has also been described (1). Such a system may be useful for studying components of the inflammatory reaction and for analysis of chemoattractants, but it is not a model of natural disease.

The experimental models of legionella pneumonia in which the bacteria have been intro duced by the respiratory route are summarized in Table 1. All of the reports describe serogroup 1 *L. pneumophila* pneumonia, except for a brief description of the lesions that were produced in guinea pigs after intratracheal instillation of *Legionella micdadei*. Bacteria have been introduced by intranasal and intratracheal instillation and by exposure of animals to an aerosolized cloud of bacteria. Aerosol exposure has been accomplished both in systems that expose the whole animal to the bacterial mist and in systems that provide varying degrees of restriction of the inoculum to the snouts of the animals. One can construct a list of pros and cons for each method. Intranasal and intratracheal instillation provide a route of infection that is analogous to aspiration of oropharyngeal contents. Although this mechanism of infection has not been documented definitively, it has not been excluded as an important pathogenetic mechanism, especially when disease is acquired nosocomially by compromised hosts. The intranasal route suffers from the serious disadvantage that it is difficult to quantitate the bacterial inoculum, which may be significantly different from animal to animal. Intratracheal inoculation necessitates the introduction of an artificial bolus of fluid into the lower respiratory tract of animals that have already been compromised by obligatory general anesthesia. Normal resident defense mechanisms are undoubtedly disordered by this insult, and evaluation of early inflammatory events is impossible.

No evidence of animal-to-animal transmission of *L. pneumophila* in experimentally infected guinea pigs has been found (13). It is easier to minimize the biohazard when animals are inoculated by the intratracheal or intranasal routes than when the exposure is accomplished with concentrated bacterial aerosols, especially of large volume. Intratracheal or intranasal inoculation may be accomplished, therefore, in existing facilities that contain laminar-flow biohazard hoods.

Exposure of animals to an infected aerosol, however, has obvious advantages over other pulmonary routes. The inoculum delivered to the lower respiratory tract can be estimated by mathematical calculation (2, 4, 18) or can be determined experimentally by quantitative culture of the total lung volume immediately after completion of the aerosol (7). The inoculum that is delivered to various areas of a snout-only exposure chamber is uniform, so that legitimate comparisons can be made between individual animals in each exposure (7). Whole-body exposure chambers have the theoretical advantage of mimicking the inoculum that one might encounter when walking through contaminated drift from a cooling tower but have the theoretical

TABLE 1. Animal models of legionella pneumonia after respiratory exposure

Animal	Strain	Route of inoculation	Serology	Bacteriology	Antigen detection	Histology	Clinical	Reference
			Means of diagnosis					
Guinea pig	Hartley outbred	Intranasal, intratracheal	−	+	+	+	+	14
		Intratracheal	+	+	+	+	+	20
		Intratracheal (*L. micdadei*)	−	+	−	+	+	19
		Whole-body aerosol	−	+	−	+	+	15
			+	+	−	−	+	4
			+	+	+	+	+	17
			+	+	−	−	+	18
		Snout-only aerosol	+	+	+	+	+	7
			+	+	+	+	+	2, 3, 8
Rat	Sprague-Dawley outbred	Intratracheal	+	+	+	+	+	20
	Lewis	Snout-only aerosol	+	+	+	+	+	7
Mouse	Porton	Snout-only aerosol	+	+	−	+	+	8
Marmoset		Snout-only aerosol	+	+	+	+	+	2, 3, 8
Monkey	Rhesus	Snout-only aerosol	+	+	+	+	+	2, 8
	Cynomolgus	Snout-only aerosol	+	+	+	+	+	8

disadvantage that bacteria are deposited onto a large surface area of body fur as well as into the respiratory tract. Practically, it is easier to achieve the delivery of large inocula into the lower respiratory tract with a snout-only system. The volumes of contaminated fluid, and the resulting biohazards, can therefore be reduced without sacrificing sensitivity of the system. The major disadvantage of snout-only aerosol systems is that animals must be physically restrained during the exposure.

Guinea pigs were used successfully in the initial aerosol experiments with condensate from the evaporative condenser in Pontiac, Mich., and this species was a mainstay of diagnostic efforts for several years. Various other rodent and primate species have also been infected, however (Table 1). With one notable exception, it has been possible to produce pneumonia in each species. It is unfortunate that the limited numbers of murine strains that have been investigated to date (8) have shown little or no evidence of acute inflammation in the airspaces after exposure. The fact that persistence of bacteria in the lungs of these animals could be documented at least 5 days after exposure emphasizes that data should be interpreted cautiously if criteria for infection and disease exclude histological documentation of pneumonia.

The clinical and pathological disease that has been produced by various investigators appears similar qualitatively, although several different strains of serogroup 1 *L. pneumophila* and diverse methods for preparing and delivering the inoculum have been used. It is probable that there are quantitative differences, which may be explained on the basis of differences in virulence of the bacterial strains that have been employed or differences in susceptibility of outbred animals from differing sources. Even *L. micdadei*, which appears to require considerably higher inocula to produce an infection that is equivalent to *L. pneumophila*-induced pneumonia, elicits a very similar inflammatory response (19), as one would expect from studies of human disease (21).

Fortunately, the model infections that are produced in experimental animals mimic human disease rather well, at least in most pathological aspects and in the rather limited clinical parameters that can be applied to rodents. Guinea pigs are moderately susceptible to infection. The 50% lethal dose after aerosol administration of bacteria is between 10^3 and 10^4 bacteria (7, 8); after intratracheal inoculation the lethal dose appears to be between 10^5 and 10^6 bacteria (20). This level of susceptibility permits adjustment of the lethality of experimental infection and does not preclude generation of mildly to moderately severe pneumonia with minimal mortality. Ex-

tensive pneumonia has also been produced in a small number of marmosets (3). Rats and monkeys, on the other hand, are more resistant to extensive pneumonia and lethal disease. They are not resistant to infection, as determined by seroconversion or intrapulmonary growth of legionellae (7, 20), and histological pneumonia can be produced even with small doses of inhaled bacteria (7). These species, therefore, provide useful models in which to investigate the effects of deletion of specific host defenses.

After introduction of legionellae into the respiratory tract, a multifocal inflammatory process occurs in both lungs. Small lesions coalesce to form macroscopically visible nodular inflammatory infiltrates, similar to those described in human cases. In severely affected animals the entire lung volume may be consolidated. There is perhaps a greater tendency for the macroscopic lesions to concentrate centrally around the major bronchi after intratracheal instillation than after aerosol deposition of bacteria (20); whatever the method by which the inoculum is presented to the lung, however, the inflammatory process develops in the distal airspaces and not in the larger bronchi. The inflammatory lesions that develop after inhalation of an infected aerosol are randomly, but not necessarily uniformly, distributed through the pulmonary parenchyma. In preliminary studies of a quantitative morphometric technique for assessing the extent of pneumonia, the mean percent lung area consolidated in a single histological section from each lung of 20 guinea pigs was identical, but considerable differences existed between the representative right and left lung samples from individual animals (7).

The inflammatory infiltrate in the distal airspaces consists of a mixture of polymorphonuclear neutrophilic leukocytes (PMNLs) and macrophages in proportions that are similar to those in representative human cases. The influx of cells into the airspaces has been studied by sequential bronchoalveolar lavage of aerosol-infected guinea pigs (G. S. Davis, W. C. Winn, Jr., D. W. Gump, and H. N. Beaty, J. Infect. Dis., in press). The initial inflammatory influx consists of PMNLs and occurs as early as 12 h after exposure. The speed with which PMNLs are recruited into the airspaces is related directly to the bacterial inoculum that is deposited into the airspaces initially. A net increase in the numbers of macrophages within the airspaces is not detected until 48 to 72 h after exposure. Other components of the inflammatory response appear simultaneously. Exudation of albumin into the airspaces has been quantitated, and fibrin is evident in histological sections. Considerable exudation of erythrocytes into the alveolar spaces may also occur, although the lesions are not frankly hemorrhagic. Karyorrhexis and karyolysis of the inflammatory exudate are evident, producing a histopathological reaction that has been called leukocytoclastic by analogy to a cutaneous vasculitis, which is also characterized by fragmentation of the inflammatory cells. The extremes of this phenomenon that are observed in some human cases have not been described for the experimental animals. Hyaline membranes are seen occasionally. Leukocytic infiltration of septal blood vessels occurs, but, as in the human lesions, is not prominent and does not appear to be the primary pathological lesion.

By 7 to 10 days after exposure to an infectious aerosol, most PMNLs have been cleared from the airspaces (Davis et al., in press). The remaining inflammatory cells are predominantly macrophages, but increased numbers of lymphocytes have also been identified in the stages of healing (3). Histologically the resolving lesions often assume a distinctly nodular outline similar to that in the healing phase of human legionella pneumonia. An intense collection of PMNLs is concentrated in the center of the nodules, and the periphery is populated by macrophages. In human lesions there is clearly necrosis of pulmonary parenchyma; septal destruction may be

present also in guinea pigs. Early fibroblastic proliferation may be observed in the alveoli and in the alveolar septa (3, 19), but extensive studies of the late lesions have not been reported.

Extrapulmonary inflammatory lesions have been clearly documented in human legionella pneumonia (21), but they are unusual. Splenic necrosis occurs with high frequency in guinea pigs that have been inoculated intratracheally or intranasally (14, 20) and has also been documented after aerosol infection in some instances (7) but not in others (3). The reason for these divergent results is not clear but may be related to differences in bacterial or animal strains. In keeping with the less extensive pulmonary disease that develops in rats and rhesus monkeys, extrapulmonary inflammatory lesions are infrequent or absent in these species (3, 20). In this respect the infection in these systems approximates human disease more closely than does that in guinea pigs.

Possibilities abound for use of these animal models to explore questions of pathogenesis, and we are just beginning to see the harvest from the time-consuming developmental work. Muller et al. have investigated the role of noncellular defenses against infection (18). Davis and co-

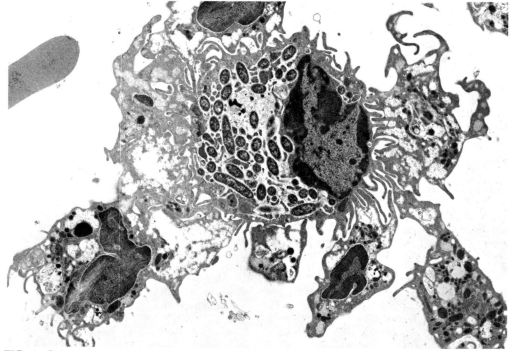

FIG. 1. Guinea pig alveolar macrophage, obtained by bronchoalveolar lavage 24 h after infection with *L. pneumophila*, containing large numbers of bacilli within an enlarged membrane-bound space. PMNLs cluster around the macrophage, and the processes of the two cell types interdigitate. Lead citrate- and uranyl acetate-stained ultrathin section. Magnification, ×9,500.

workers have examined the role of mucociliary clearance mechanisms in the early response to infection (7).

I would like to concentrate, however, on the information that these models have already provided about the non-immunological cellular inflammatory response to a primary infection. This area of investigation provides instructive examples of the way in which animal models on the one hand have buttressed in vitro observations with human cells and solidified our appreciation of the role of the alveolar macrophage and on the other hand have raised questions about the interpretation of in vitro results with human PMNLs.

Early investigations of the pathology of human legionella pneumonia emphasized the alveolar macrophage heavily and appropriately (5). Subsequent in vitro investigations of the interaction of *L. pneumophila* and macrophages from a variety of sites and animal species have confirmed the importance of the macrophage in this infection. The elegant experiments of Horwitz and Silverstein with human peripheral blood monocytes are exemplary (12). This topic will be covered in more detail later in the Symposium. It will be discussed only briefly here, because the animal models have provided information

that is confirmatory rather than revelatory. It is important, nevertheless, to know that the cell-related association of legionellae and macrophages can be documented in a developing infection in vivo (14; Davis et al., in press) (Fig. 1), that exponential bacterial growth occurs in the lung at a time when the only important cellular defense available is the alveolar macrophage (7), and that an association of viable legionellae with this cell type can be demonstrated in lavaged airspace contents (Davis et al., in press).

The second major inflammatory cell type, the PMNL, was clearly underemphasized in the initial pathologic reports (21). Neutrophils were as numerous as macrophages in the exudate and in many cases were the predominant inflammatory cell (11, 21). Subsequent histochemical studies with the Leder stain for neutrophilic granules (16) suggested that the difficulty of distinguishing neutrophilic precursors from mononuclear phagocytes in a necrotizing exudate may be partially responsible for the underemphasis of the PMNL. In several cases from which the inflammatory exudate had been classified as mixed PMNLs and macrophages initially, the vast majority of the inflammatory cells stained as PMNLs when the tissue sections were treated by the Leder method (Fig. 2). The limita-

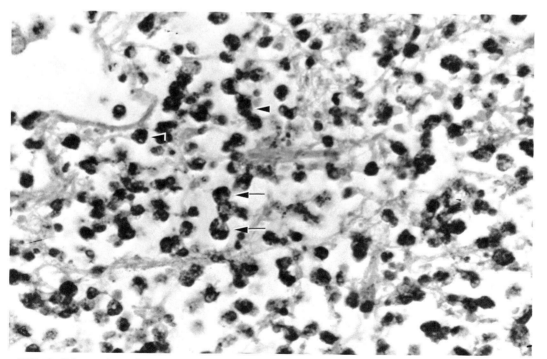

FIG. 2. Polymorphonuclear neutrophils participating in the inflammatory exudate of a patient who died of Legionnaires disease. Cells with multiply segmented nuclei (arrowheads) are easy to identify as PMNLs. This Leder stain identifies many of the remaining cells, which were difficult to characterize in hematoxylin-eosin-stained sections, as PMNLs (arrows). Magnification, ×950.

tions of human material, however, precluded sequential study and analysis of the functional role that these cells were playing.

In vitro studies with normal human PMNLs have actually suggested that this cell type might be an ineffective defense mechanism, although the unimpeded growth that was obtained after infection of monocytes did not occur. A very careful study by Friedman and co-workers of the effects of purified legionella toxin on the function of normal human PMNLs (9) documented the inhibition of bactericidal activity and the respiratory metabolic burst when toxin-treated PMNLs were exposed to a reference strain of *Escherichia coli*. The doses of toxin that were employed in the study did not affect cell viability and did not interfere with the attachment and subsequent uptake of the *E. coli*.

Horwitz and Silverstein demonstrated in carefully controlled studies that an egg yolk-passaged strain of virulent *L. pneumophila* was phagocytized by normal human PMNLs only in the presence of specific antibody and complement. No bactericidal activity developed in the absence of these substances, and the maximal loss in bacterial viability was less than 10-fold (12). The deficiency in bacterial killing was probably not due to the production of intracellular legionella toxin, because *E. coli* was effectively killed by the PMNLs when the cells were dually infected with both pathogens.

This gloomy picture of woefully inadequate non-immunological cellular defense mechanisms is brightened somewhat by data that have been accumulated from the animal models of legionella infection. Careful sequential observation of bacterial growth and inflammatory cell recruitment into the airspaces suggested that the first diminution in the rate of bacterial growth within the lung coincided with the arrival of PMNLs in the airspaces (Davis et al., in press). When relatively large bacterial inocula were used, the predominant inflammatory cell in the airspaces was the PMNL within 24 h after exposure to an aerosol, a time when a major contribution from immunologically specific defenses is unlikely. By this time the viable bacteria were highly cell associated; by direct immunofluorescence staining, legionellae were found in both macrophages and PMNLs. When the inflammatory exudate was separated on discontinuous Percoll density gradients, however, viable bacteria were preferentially associated with macrophages rather than PMNLs.

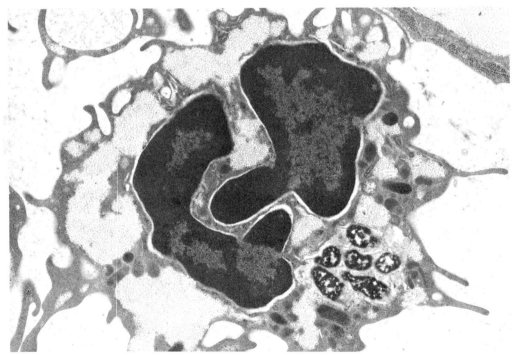

FIG. 3. Six bacterial profiles present within membrane-bound spaces in the PMNLs of a guinea pig infected with *L. pneumophila* 24 h earlier. The bacteria appear morphologically intact. Large glycogen lakes are present within the cytoplasm of the phagocyte. Lead citrate- and uranyl acetate-stained ultrathin section. Magnification, ×22,000.

Electron microscopic evaluation of the lavaged airspace exudate supported the conclusions that PMNLs were effectively phagocytizing legionellae very early in the course of the infection and that bactericidal activity was probably being generated by these PMNLs. Almost all of the inflammatory cells that contained large numbers of intact legionellae were macrophages or cells that were too degenerate to identify with certainty. In contrast, although intact bacteria were identified within PMNLs, the majority of the legionellae that had been phagocytized by this cell type appeared damaged morphologically (Fig. 3). This dichotomy in the morphological appearance of intracellular bacteria within phagocytes has been described for ultrathin sections from the lungs of guinea pigs that had been infected intratracheally (14). A dramatic clustering of PMNLs around a heavily parasitized macrophage can be seen in Fig. 1, although intracellular bacteria are not evident within the PMNLs.

Such divergent, although not necessarily discrepant, clues as to the role of the PMNL in legionella infection indicate the necessity for continued exploration of this question. The importance of employing all possible approaches to the resolution of experimental questions is clear. The relationships among several different approaches to the central questions of pathogenetic mechanisms are illustrated schematically in Fig. 4. Clinical and pathological evaluation of human cases is extremely important, because these tools provide the most direct observations on the infected host of greatest interest and importance. The arrows that connect these modalities to the central question are small, however, because of the limitations on human studies that have been mentioned previously. In vitro investigations with human cells and reagents provide important approaches to the problems, again because of our admittedly anthropocentric orientation. There is a block to this powerful

method, however, because the conditions that must be selected for the in vitro model are, of necessity, arbitrary and may not always reflect the complex interrelationships that exist in vivo. Animal models provide an opportunity to bridge the gap, because the infection can be studied sequentially from its earliest stages and sophisticated methods of study can be brought to bear. The block to the in vivo model approach derives from the very real danger that species-specific differences in host-parasite interactions may vitiate transfer of information from animal to human. Selective use of primate models may help to alleviate some of this concern. An additional approach that may be of use, however, is the use of animal cells in vitro in a correlative fashion with in vitro human studies on the one hand and whole-animal investigations on the other. Particularly when the problems are thorny and the preliminary information is confusing or contradictory, a multidisciplinary and multifactorial approach may be not only useful but essential.

Many important problems remain to be addressed. With the impressive groundwork that has been laid in the 5 years since the first legionella symposium, progress should be steady toward elucidation of some very complex pathogenetic mechanisms that will advance our understanding of host-parasite interactions in general.

LITERATURE CITED

1. Arko, R. J., K. H. Wong, and J. C. Feeley. 1979. Immunologic factors affecting the in-vivo and in-vitro survival of the Legionnaires' disease bacterium. Ann. Intern. Med. 90:680–683.
2. Baskerville, A., M. Broster, R. B. Fitzgeorge, P. Hambleton, and P. J. Dennis. 1981. Experimental transmission of Legionnaires' disease by exposure to aerosols of Legionella pneumophila. Lancet ii:1389–1390.
3. Baskerville, A., R. B. Fitzgeorge, M. Broster, and P. Hambleton. 1983. Histopathology of experimental Legionnaires' disease in guinea pigs, rhesus monkeys and marmosets. J. Pathol. 139:349–362.
4. Berendt, R. F., H. W. Young, R. G. Allen, and G. L. Knutsen. 1980. Dose-response of guinea pigs experimentally infected with aerosols of Legionella pneumophila. J. Infect. Dis. 141:186–192.
5. Chandler, F. W., J. A. Blackmon, M. D. Hicklin, R. M. Cole, and C. S. Callaway. 1979. Ultrastructure of the agent of Legionnaires' disease in the human lung. Am. J. Clin. Pathol. 71:43–50.
6. Chandler, F. W., J. E. McDade, M. D. Hicklin, J. A. Blackmon, B. M. Thomason, and E. P. Ewing, Jr. 1979. Pathologic findings in guinea pigs inoculated intraperitoneally with the Legionnaires' disease bacterium. Ann. Intern. Med. 90:671–675.
7. Davis, G. S., W. C. Winn, Jr., D. W. Gump, J. E. Craighead, and H. N. Beaty. 1982. Legionnaires' pneumonia after aerosol exposure in guinea pigs and rats. Am. Rev. Respir. Dis. 126:1050–1057.
8. Fitzgeorge, R. B., A. Baskerville, M. Broster, P. Hambleton, and P. J. Dennis. 1983. Aerosol infection of animals with strains of Legionella pneumophila of different virulence: comparison with intraperitoneal and intranasal

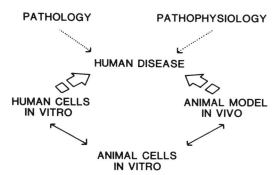

FIG. 4. Diagrammatic representation of the experimental approach to pathogenesis of legionella pneumonia.

routes of infection. J. Hyg. **90:**81–90.

9. **Friedman, R. L., J. E. Lochner, R. H. Bigley, and B. H. Iglewski.** 1982. The effects of *Legionella pneumophila* toxin on oxidative processes and bacterial killing of human polymorphonuclear leukocytes. J. Infect. Dis. **146:**328–334.

10. **Glick, T. H., M. B. Gregg, B. Berman, G. Mallison, W. W. Rhodes, Jr., and I. Kassanoff.** 1978. Pontiac fever. An epidemic of unknown etiology in a health department. 1. Clinical and epidemiologic aspects. Am. J. Epidemiol. **107:**149–160.

11. **Hernandez, F. J., B. D. Kirby, T. M. Stanley, and P. H. Edelstein.** 1980. Legionnaires' disease. Postmortem pathologic findings of 20 cases. Am. J. Clin. Pathol. **73:**488–495.

12. **Horwitz, M. A., and S. C. Silverstein.** 1981. Interaction of the Legionnaires' disease bacterium (*Legionella pneumophila*) with human phagocytes. 1. *L. pneumophila* resists killing by polymorphonuclear leukocytes, antibody, and complement. J. Exp. Med. **153:**386–397.

13. **Katz, S. M., W. A. Habib, J. M. Hammel, and P. Nash.** 1982. Lack of airborne spread of infection by *Legionella pneumophila* among guinea pigs. Infect. Immun. **38:**620–622.

14. **Katz, S. M., and S. Hashemi.** 1982. Electron microscopic examination of the inflammatory response to *Legionella pneumophila* in guinea pigs. Lab. Invest. **46:**24–32.

15. **Kaufmann, A. F., J. E. McDade, C. M. Patton, J. V. Bennett, P. Skaliy, J. C. Feeley, D. C. Anderson, M. E. Potter, V. F. Newhouse, M. B. Gregg, and P. S. Brachman.** 1980. Pontiac fever: isolation of the etiologic agent (*Legionella pneumophila*) and demonstration of its mode of transmission. Am. J. Epidemiol. **114:**337–347.

16. **Leder, L. D.** 1964. Uber die selektive fermentcytochemische Darstellung von neutrophilen myeloischen Zellen und Gewebsmastzellen im Paraffinschnitt. Klin. Wochensch. **42:**553.

17. **Meenhorst, P. L., A. L. Reingold, G. W. Gorman, J. C. Feeley, B. J. van Cronenburg, C. L. M. Meyer, and R. van Furth.** 1983. Legionella pneumonia in guinea pigs exposed to aerosols of concentrated potable water from a hospital with nosocomial Legionnaires' disease. J. Infect. Dis. **147:**129–132.

18. **Muller, D., M. L. Edwards, and D. W. Smith.** 1983. Changes in iron and transferrin levels and body temperature in experimental airborne legionellosis. J. Infect. Dis. **147:**302–307.

19. **Pasculle, A. W.** 1981. Experimental studies of Pittsburgh pneumonia agent, p. 169–172. *In* D. Schlessinger (ed.), Microbiology—1981. American Society for Microbiology, Washington, D.C.

20. **Winn, W. C., Jr., G. S. Davis, D. W. Gump, J. E. Craighead, and H. N. Beaty.** 1982. Legionnaires' pneumonia after intratracheal inoculation of guinea pigs and rats. Lab. Invest. **47:**568–578.

21. **Winn, W. C., Jr., and R. L. Myerowitz.** 1981. The pathology of the *Legionella* pneumonias. A review of 74 cases and the literature. Hum. Pathol. **12:**401–422.

Comparison of the Virulence of the Philadelphia and Pontiac Isolates of *Legionella pneumophila*

ROBIN E. HUEBNER, PAUL W. REESER, AND DONALD W. SMITH

Department of Medical Microbiology, University of Wisconsin, Madison, Wisconsin 53706

Infection with *Legionella pneumophila* may lead to Legionnaires disease, a rapidly progressing pneumonia with a fatality rate of 20%, or to Pontiac fever, a self-limiting febrile illness. It is unclear why *L. pneumophila* causes two such distinct syndromes; however, several hypotheses have been proposed to explain the different forms seen. One hypothesis is that isolates obtained from outbreaks of either form of disease differ in virulence, i.e., in their ability to multiply and disseminate in vivo.

This hypothesis was examined by comparing the virulence of the Philadelphia and Pontiac isolates of *L. pneumophila* in a guinea pig model of experimental airborne legionellosis. The study was divided into two parts; the first experiment examined the virulence as measured by the 50% lethal dose (LD_{50}) and the FID_{50}, the dose required to induce fever in 50% of the animals. Groups of guinea pigs were exposed under conditions leading to the inhalation and retention of between 10 and 5,000 CFU per animal. The body temperature of each animal was recorded daily, as were any deaths that occurred. Defining fever as a body temperature of >39.0°C, at the lowest infection doses (10 CFU of the Pontiac isolate and 20 CFU of the Philadelphia isolate), four of six guinea pigs in each treatment group showed a febrile response (Table 1). Based on these data, the FID_{50} is <10 CFU per animal for the Pontiac isolate and <20 CFU per animal for the Philadelphia isolate. Animals exposed at the highest concentration to Formalin-killed organisms of either isolate were afebrile, as were those exposed to placebo. With regard to the lethal dose of *L. pneumophila*, inhalation of 5,000 CFU of either the Pontiac or Philadelphia isolate led to the deaths of three of six and four of six animals, respectively. Therefore, the LD_{50} for both isolates is in the range of 500 to 5,000 CFU (Table 1).

The second experiment evaluated the virulence of the two isolates by their ability to multiply at the site of implantation in the lungs and by their ability to disseminate to and multiply in the spleen. Animals were infected via the respiratory route under conditions leading to the inhalation and retention of approximately 100 CFU per animal of organisms from either the Pontiac or Philadelphia outbreaks. Body temperatures were measured daily, and at several intervals postinfection, groups of animals were killed and their lungs and spleens were excised aseptically and homogenized. The homogenates were plated on buffered charcoal-yeast extract agar for measurement of the number of recoverable legionellae. The results show that both isolates multiplied at the site of implantation in the lung, reaching maximum numbers by day 5, and were then cleared from the lung at the same rate (Fig. 1). The Philadelphia and Pontiac isolates both disseminated from the lungs to the spleen, where they appeared in relatively high numbers as early as day 2. By day 8 the Pontiac isolate was no longer detectable, and only about 10 CFU of the Philadelphia isolate were recovered from the spleens of single animals on days 8 and 11. An analysis of variance of the lung data and a two-tailed *t* test of the spleen data showed

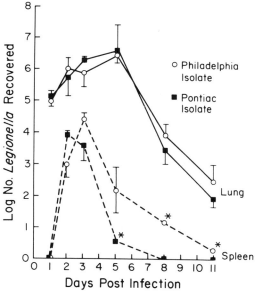

FIG. 1. Change with time in the number of *L. pneumophila* isolated from the lungs and spleens of guinea pigs infected by the respiratory route with 100 CFU of either the Philadelphia or Pontiac isolate. Bars show the standard error of the mean. (∗) Counts from one animal only. The other animals showed no recoverable *L. pneumophila*.

TABLE 1. LD$_{50}$ and FID$_{50}$ of Philadelphia and Pontiac isolates of *L. pneumophila* as determined in a guinea pig model of experimental airborne legionellosis

Treatment	Challenge level (CFU)	No. of febrile animals[a]	No. of deaths[a]
Philadelphia isolate	20	4	0
	250	6	0
	1,000	6	0
	5,000	6	4
Pontiac isolate	10	4	0
	15	6	0
	500	6	0
	5,000	6	3
Philadelphia isolate, Formalin-killed		0	0
Pontiac isolate, Formalin-killed		0	0
Placebo		0	0

[a] Six animals per group.

no significant differences between animals infected with either isolate. The peak febrile response of the animals exposed to either the Philadelphia or Pontiac isolates was also the same, coinciding with the organisms reaching their maximum levels in the lung and spleen.

By all the parameters measured, the virulence of the Philadelphia and Pontiac isolates was the same. Therefore, using a guinea pig model of experimental airborne legionellosis, we were unable to obtain evidence supporting the hypothesis that differences between Legionnaires disease and Pontiac fever are due to differences in the in vivo properties of the Pontiac and Philadelphia isolates.

Results in animal models, however, may not mirror the situation in humans. For example, even with the lowest infection doses we were unable to produce in guinea pigs an illness analogous to human Pontiac fever. It may be that guinea pigs are too susceptible to *L. pneumophila* and therefore not a good animal to use in studying the differences between the Philadelphia and Pontiac isolates. Further studies, perhaps developing new animal models for Pontiac fever, may have to be investigated before the differences between Legionnaires disease and Pontiac fever can be explained.

Febrile Illness Produced in Guinea Pigs by Oral Inoculation of *Legionella pneumophila*

SHEILA MORIBER KATZ AND JOSEPH P. MATUS

Department of Pathology, Hahnemann University School of Medicine, Philadelphia, Pennsylvania 19102

The natural habitat of *Legionella pneumophila* is water, and the microbe has been cultured from a variety of aqueous environments, including streams, ponds, cooling towers, and potable water systems (7, 11, 12). Although evidence indicates that *L. pneumophila* is transmitted to humans from aqueous environmental sources by inhalation (3, 4), it is possible that illness can be acquired via an oral route.

Our study, designed to evaluate oral inoculation of *L. pneumophila*, consisted of five groups of guinea pigs. In group 1, *L. pneumophila* serogroup 1 (range of dose, 10^3 to 10^8 organisms) was inoculated into the drinking water of 56 guinea pigs. In group 2, *L. pneumophila* (range of dose, 10^3 to 10^{10} organisms) was inoculated via gastric intubation into 59 guinea pigs. In group 3, heat-killed *L. pneumophila* (range of dose, 10^3 to 10^9 organisms) was inoculated into the drinking water of 19 guinea pigs. In group 4, *L. pneumophila* (dose, 10^6 organisms) was inoculated intraperitoneally into 24 guinea pigs; this group was the positive control that confirmed virulence. In group 5, the negative control, 27 guinea pigs were either intubated gastrically with phosphate-buffered saline or given drinking water without *L. pneumophila*. Methods of study were measurement of body temperature, observation for diarrhea and cachexia, and postmortem examination including light microscopy, culture for *L. pneumophila* on buffered charcoal-yeast extract agar, direct immunofluorescent-antibody test for *L. pneumophila*, and indirect immunofluorescent-antibody test for circulating antibodies to *L. pneumophila*.

A total of 29 of 56 (52%) guinea pigs of group 1, 37 of 59 (63%) of group 2, and 20 of 24 (83%) of group 4 had a fever ($\geq 39.4°C$), compared with 0 of 19 (0%) of group 3 and 1 of 27 (4%) of group 5. The febrile response in groups 1 and 2 was short lived, usually 3 days or less. By contrast, animals given *L. pneumophila* by intraperitoneal injection (group 4) typically exhibited fever for more than 3 days. There were no fatalities in groups 1, 2, 3, and 5, compared with 15 of 24 (63%) guinea pigs of group 4. Pneumonitis and splenitis were consistent features of groups 1, 2, and 4. The pneumonitis of groups 1 and 2 was

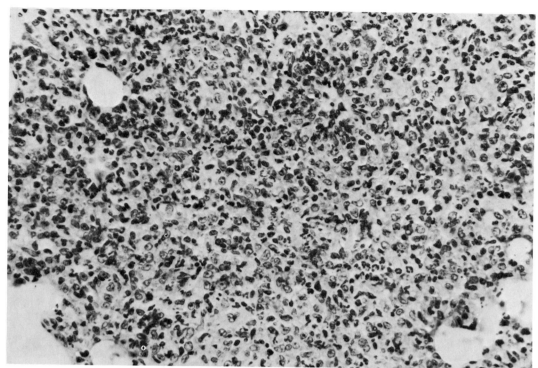

FIG. 1. Photomicrograph of lung obtained at autopsy from a febrile guinea pig given *L. pneumophila* in drinking water. Note the consolidation produced by the conspicuous inflammatory infiltrate. Most of the inflammatory cells are macrophages, but there are some neutrophils (hematoxylin and eosin stain; original magnification, ×400).

typically mild, predominantly focal, interstitial, and mainly composed of macrophages. Some lesions, however, did involve alveolar lumens, and neutrophils were occasionally present. Neither gross nor microscopic evidence of aspiration was seen. By contrast, guinea pigs of group 4 exhibited extensive consolidation and a florid inflammatory exudate of neutrophils and macrophages. Splenitis, featuring congestion, sinus histiocytosis, scattered neutrophils, and rare microabscesses, was conspicuous in group 4 and slight but present in groups 1 and 2. Inflammation of the small and large intestine rarely occurred in the guinea pigs inoculated by gastric intubation (group 2). The gastrointestinal lesions were sparse and microscopic, consisting of sparse, focal mucosal infiltration by neutrophils and macrophages. A total of 4 of 29 (14%) guinea pigs of group 1 and 5 of 15 (33%) guinea pigs of group 4 seroconverted to *L. pneumophila*, compared with 0 of 7 (0%) guinea pigs of group 3 and 0 of 10 (0%) guinea pigs of group 5. In groups 1 and 2 combined, *L. pneumophila* was isolated from samples of lung and spleen of 5 of 47 guinea pigs, and the organism was visualized by direct immunofluorescent-antibody test from samples

of lung and spleen of 7 of 47 guinea pigs. Stool samples from 24 of 37 (65%) guinea pigs in group 1 and from 16 of 35 (46%) guinea pigs in group 2 exhibited staining for *L. pneumophila* by direct immunofluorescence 1 week after inoculation. Preinoculation stool samples and samples from the negative control group (group 5) were consistently negative for *L. pneumophila* by direct fluorescent-antibody test thus minimizing but not completely excluding the possibility of a cross-reacting fecal bacterium.

In a small-scale experiment similar to the one detailed herein—the only other study to date of oral inoculation of *L. pneumophila* in animals—Conner and Gilbert (2) added *L. pneumophila* to sterile distilled drinking water (10^4 to 10^6 bacteria per ml) of three guinea pigs. Each animal ingested between 150 to 200 ml of water per day, and after 7 days, no clinical signs of illness were detected. None of the animals seroconverted, but postmortem examination was not performed, and there was no mention made of the body temperature of test animals. Stools of some of the test animals, however, were positive for *L. pneumophila* by direct fluorescent-antibody test for at least 4 days after the water

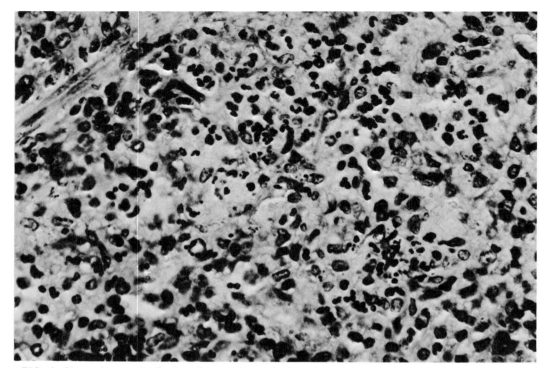

FIG. 2. Photomicrograph of spleen from a guinea pig given *L. pneumophila* by gastric intubation. Note the macrophages and numerous neutrophils (hematoxylin and eosin stain; original magnification, ×400).

containing *L. pneumophila* was discontinued.

Approximately one-third to one-half of patients who have Legionnaires disease develop gastrointestinal symptoms (1, 10, 14), and gastrointestinal illness is often a major clinical manifestation of the disease (9, 13). Peritonitis (6) and abscess of the appendix (8) have been linked to infection by *L. pneumophila*, and one investigation suggests that aspiration of gastric contents might produce Legionnaires disease (5). Thus, the passage of *L. pneumophila* through the gastrointestinal tract might account for the gastrointestinal manifestations of legionellosis.

The self-limited febrile illness that we have produced in guinea pigs by gastrointestinal administration of *L. pneumophila* constitutes a new animal model of legionellosis. In some respects, the model is reminiscent of Pontiac fever. Both are characterized by development of a mild form of febrile illness shortly after exposure to *L. pneumophila*. The contribution, if any, of aerosolization of *L. pneumophila* in guinea pigs that drank the water containing *L. pneumophila* toward development of illness remains to be elucidated. Perhaps a similar illness might be produced in animals in the wild.

Thus, we produced a self-limited febrile illness in guinea pigs by oral inoculation of *L. pneumophila*. This animal model might further

unravel the epidemiology and pathobiology of legionellosis.

LITERATURE CITED

1. **Beaty, H. N., A. A. Miller, C. V. Broome, S. Goings, and C. A. Phillips.** 1978. Legionnaires' disease in Vermont, May to October, 1977. J. Am. Med. Assoc. **240:**127–131.
2. **Conner, R. W., and D. N. Gilbert.** 1979. *Legionella pneumophila* in drinking water of guinea pigs. Ann. Intern. Med. **91:**323.
3. **Cordes, L. G., D. W. Fraser, P. Skaliy, C. A. Perlino, W. R. Elsea, G. F. Mallison, and P. S. Hayes.** 1980. Legionnaires' disease outbreak at an Atlanta, Georgia, country club: evidence for spread from an evaporative condenser. Am. J. Epidemiol. **111:**425–431.
4. **Dondero, T. J., Jr., R. C. Rendtorff, G. F. Mallison, R. M. Weeks, J. S. Levy, E. W. Wong, and W. Schaffner.** 1980. An outbreak of Legionnaires' disease associated with a contaminated air-conditioning cooling tower. N. Engl. J. Med. **302:**365–370.
5. **Dournon, E., A. Bure, N. Desplaces, M. F. Carette, and C. H. Mayaud.** 1982. Legionnaires' disease related to gastric lavage with tap water. Lancet **i:**797–798.
6. **Dournon, E., A. Bure, J. L. Kemeny, J. L. Pourriat, and P. Valeyre.** 1982. *Legionella pneumophila* peritonitis. Lancet **i:**1363.
7. **Fliermans, C. B., W. B. Cherry, L. H. Orrison, S. J. Smith, D. L. Tison, and D. H. Pope.** 1981. Ecological distribution of *Legionella pneumophila*. Appl. Environ. Microbiol. **41:**9–16.
8. **Holt, P.** 1981. Legionnaires' disease and abscess of appendix. Br. Med. J. **282:**1035–1036.
9. **Kirby, B. D., K. M. Synder, R. D. Meyer, and S. M. Finegold.** 1978. Legionnaires' disease: clinical features of

24 cases. Ann. Intern. Med. **89:**297–309.

10. **Meenhorst, P. L., J. W. M. van der Meer, and J. Borst.** 1979. Sporadic cases of Legionnaires' disease in the Netherlands. Ann. Intern. Med. **90:**529–532.

11. **Stout, J., V. L. Yu, R. M. Vickers, J. Zuravleff, M. Best, A. Brown, R. Vee, and R. Wadowski.** 1982. Ubiquitousness of *Legionella pneumophila* in the water supply of a hospital with endemic Legionnaires' disease. N. Engl. J. Med. **306:**466–468.

12. **Stout, J., V. L. Yu, R. M. Vickers, and J. Shonnard.** 1982. Potable water supply as the hospital reservoir for Pittsburgh pneumonia agent. Lancet **i:**471–472.

13. **Swartz, M. N.** 1979. Clinical aspects of Legionnaires' disease. Ann. Intern. Med. **90:**492–495.

14. **Tsai, T. F., D. R. Finn, B. D. Plikaytis, W. McCauley, S. M. Martin, and D. W. Fraser.** 1979. Legionnaires' disease: clinical features of the epidemic in Philadelphia. Ann. Intern. Med. **90:**509–517.

Bull's-Eye Lesion of *Legionella pneumophila*

BARBARA S. LOWRY, GLEN D. COTTRILL, AND ROBERT S. THOMPSON

United States Army Medical Research Institute of Infectious Diseases, Frederick, Maryland 21701

The pathophysiology of human pulmonary legionellosis has been poorly documented to date, since tissue for gross and microscopic examination typically has been obtained from cases of advanced disease. Consequently, to better understand the disease, animal models have been utilized. Berendt et al. (2) in 1980 first reported the use of a guinea pig experimentally infected with aerosols of *Legionella pneumophila*, but no description of diseased tissue was given.

Recent reports of gross and microscopic observations (1, 3, 4) have not described the gross lesion resembling a bull's-eye which we commonly encounter on day 6 ± 1 in guinea pigs exposed to sublethal aerosolized doses of *L. pneumophila*; our guinea pigs exposed to lethal doses of *L. pneumophila* died before day 6.

Male Hartley strain guinea pigs (Hilltop) weighing 250 to 350 g were housed in stainless steel cages and given free access to guinea pig pellets and water. *L. pneumophila* Philadelphia 1, originally obtained from the Centers for Disease Control, Atlanta, Ga., was grown on a modification of charcoal-yeast extract agar and harvested in tryptose saline. Portions stored at −70°C showed no significant loss in viability. Aerosol infective doses were obtained by diluting stock cultures with tryptose saline.

Fluorescence studies were conducted with *L. pneumophila* polyvalent (groups 1, 2, 3, and 4) fluorescein isothiocyanate-labeled rabbit globulin (Centers for Disease Control). Fresh tissue was fixed in neutral buffered Formalin, processed, and paraffin embedded for fluorescence and light microscopy studies. Gross photographs of the fresh autopsy tissue were recorded on Kodak Ecktachrome and Polaroid film. The fresh pleural surfaces were documented first, and then cut surfaces were exposed for photographic recording and comparison.

A Collison atomizer and Henderson-type aerosol mixing and transit tube attached to a plastic exposure box housing the guinea pigs were used for exposure to *L. pneumophila*. All-glass impingers (7) were used for aerosol sampling. The estimation of the concentration of *L. pneumophila* in the sampler for each exposure dose was obtained by spreading diluted impinger fluid on charcoal-yeast extract agar plates, with CFU counts made after appropriate incubation. The inhalation dose was based on an inhalation rate of 0.15 liter/min (5), and it was assumed that retention was approximately 50% of the inhaled dose, from the developed data of Hatch and Ross (6).

Guinea pigs were exposed to a predetermined infective dose of aerosolized *L. pneumophila*. The calculated number of retained viable organisms in the three runs ranged from 280 to 21,000 CFU. The severity of the disease paralleled the dose of infective organisms; with low retained numbers of *L. pneumophila*, few and small foci of pneumonia were encountered, and as the dose increased, the number of individual pneumonic foci and the maximal size of the lesions also increased. No deaths from *L. pneumophila* were observed.

Counting the infective day as day zero, guinea pigs were sacrificed by halothane beginning on day 2. Immediately after being sacrificed the animals were autopsied, fresh tissue was photographed, and tissues were fixed.

The earliest gross lung lesion was the subpleural petechia, appearing 1 to 2 days postexposure. These petechiae gradually enlarged in a rounded or geographic fashion. At day 5 to 6, 24 to 48 h after the height of the fever, a glistening, white, mildly raised center was frequently superimposed upon the lesion's erythematous background which demarcated it from the surrounding normal, pink, pleural surface of the lung, giving a bull's-eye appearance (Fig. 1). Zonal demarcation within this lesion was usually not identified after day 7, except in rare cases where the appearance of fresh petechiae suggested continued infection of the lungs. The bull's-eye lesion faded to a dull whitish-tan by

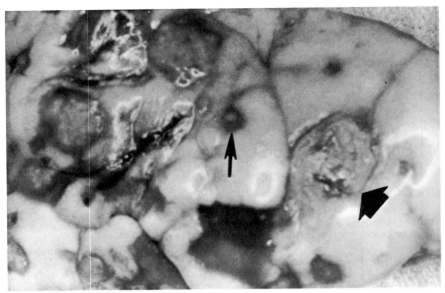

FIG. 1. Gross pleural surface of lung at day 6. Thin arrow points to bull's-eye lesion; thick arrow designates more advanced pneumonic focus.

day 7 to 8, occasionally becoming depressed below the level of the surrounding pleura, and then slowly resolved.

L. pneumophila was identified in the bull's-eye lesions by the direct fluorescent-antibody technique and culture of the ground infected lung on charcoal-yeast extract medium.

Hemotoxylin and eosin stains of tissue sections showed the bull's-eye lesions to be small or often enlarging areas of subpleural consolidation with the influx of macrophages and neutrophils varying in concentration from focus to focus, although increased numbers of neutrophils were more often associated with the bull's-eye, suggesting an acute abscess.

Edema fluid and precipitated fibrin were prominent. We noted that rather than actual hemorrhage, congested septae, both within and surrounding the area of consolidation, were responsible for the erythematous ring of the bull's-eye lesion.

Some subpleural lesions at day 6 ± 1 failed to resemble a "good" bull's-eye, but appeared erythematous or graded into tan. Microscopic examination of these lesions showed consolidation without either the concentration of neutrophils described above or the accompanying degree of edema.

Although lesions were diffusely scattered in the lungs, the bull's-eye lesion was only identified from the pleural aspect. From the appearance of the first petechiae, the majority of lesions were found associated with or immediately beneath the pleural surfaces, although the reason for this remains obscure. Perhaps the congested and edematous pneumonic lesions tend to expand toward the distensible pleural surface.

We thank Richard F. Berendt for his technical support in aerosolization of L. pneumophila and James C. Hardy for the superior quality of the tissue slides.

LITERATURE CITED

1. Baskerville, A., R. B. Fitzgeorge, M. Broster, P. Hambleton, and P. J. Dennis. 1981. Experimental transmission of Legionnaires' disease by exposure to aerosols of Legionella pneumophila. Lancet ii:1389–1390.
2. Berendt, R. F., H. W. Young, R. G. Allen, and G. L. Knutsen. 1980. Dose response of guinea pigs experimentally infected with aerosols of Legionella pneumophila. J. Infect. Dis. 141:186–192.
3. Davis, G. S., W. C. Winn, Jr., D. W. Gump, J. E. Craighead, and H. N. Beaty. 1982. Legionnaires' pneumonia after aerosol exposure in guinea pigs and rats. Am. Rev. Respir. Dis. 128:1050–1057.
4. Davis, G. S., W. C. Winn, Jr., D. W. Gump, J. M. Craighead, and H. N. Beaty. 1983. Legionnaires' pneumonia in guinea pigs and rats produced by aerosol exposure. Chest 83 (Suppl.):15–16.
5. Guyton, A. C. 1947. Measurement of the respiratory volumes of laboratory animals. Am. J. Physiol. 150:70–77.
6. Hatch, T. F., and P. G. Ross. 1946. Pulmonary deposition and retention of inhaled aerosols, p. 50. Academic Press Inc., New York.
7. Henderson, D. W. 1952. An apparatus for the study of airborne infection. J. Hyg. 50:53–68.

Quantitative Morphometry as a Tool for Analysis of Experimental Pneumonia

WASHINGTON WINN, JR., GERALD DAVIS, JON DURDA, AND HARRY BEATY

Departments of Pathology and Medicine, University of Vermont College of Medicine, Burlington, Vermont 05405

Experimental models of infectious diseases provide a means for evaluation of pathophysiological or therapeutic variables that cannot be easily studied in human disease. It is important to have objective parameters for assessment of potential differences between experimental groups, preferably measurements that can be examined with statistical techniques. Common endpoints that have been used are the inocula that produce infection (ID_{50}) or lethal disease (LD_{50}) in 50% of animals, but these techniques lack precision in detection of small differences between groups.

We have explored the possibility that morphometric techniques might provide a quantitative parameter for assessing the severity of experimental pneumonia. For this purpose we used a model of *Legionella pneumophila* pneumonia that is produced by exposure of guinea pigs and rats to a bacterial aerosol that is limited to the snout (1).

For the morphometric studies, 36 Hartley strain guinea pigs (Charles River Breeding Laboratories, Quebec, Canada; 200 to 300 g) were exposed to an aerosol that contained 2×10^2 to 5×10^5 CFU of serogroup 1 *L. pneumophila* (Burlington 1 strain). The initial inoculum was determined by quantitative culture of the lungs from at least two animals immediately after completion of the aerosol. The details of preparation of inoculum, generation of the aerosol, and evaluation of animals have been reported previously (1, 2). Guinea pigs that appeared healthy and had become afebrile 1 week after exposure were considered survivors. Three animals that were sacrificed before 7 days were excluded from analysis. For preliminary analysis of morphometric techniques, two additional guinea pigs (initial dose was not recorded for one animal and was 5×10^5 CFU for the second) and three Lewis rats (Charles River Breeding Laboratories; 150 to 200 g; initial dose, 75 CFU) were used.

At the time of planned sacrifice 7 or 11 days after infection, or at death, the lungs were fixed with cold 10% formaldehyde solution that was instilled through the trachea. After fixation, the lungs were cut at 2-mm thickness in a plastic mold. Hematoxylin-eosin-stained sections were prepared from paraffin-embedded tissue. The image of the lung sections was then projected onto paper, and the lesions were outlined. Image analysis was performed with an Optomax image analysis system that included an Apple II microcomputer and an Apple graphics tablet. The ratio of area of consolidated lung to total area of the section was determined.

Technical variability. Three sections from guinea pigs with moderately severe disease were examined on five separate occasions over a period of several days. The analyses were done by a single investigator (J.D.) without reference to the identifying number of the section. The results of the replicate determinations are depicted graphically in Fig. 1. The mean area of pneumonia in the three sections ranged from 29.2 to 54.3%. The standard deviation for the five observations ranged from 1.0 to 3.3%. Thus the technical reproducibility of the measurements appeared acceptable, at least if a single individual performed the measurements.

Variation within an animal. As a preliminary assessment of variance within the lungs of individual animals, the extent of consolidation in histological sections from consecutive paraffin blocks that encompassed the entire lung was determined for one guinea pig with extensive pneumonia, three guinea pigs with moderately severe disease, and three rats that had minimal consolidation. The standard deviation and coefficient of variation were small in the animal with virtually total consolidation of the lung. The absolute value of the standard deviation was highest for those animals with moderately severe disease, but the relative variability (as reflected in the coefficient of variation) was also high for the rats with minimal pneumonia.

Despite the variability in extent of pneumonia from section to section, the data can be analyzed statistically if the values are normally distributed. To examine this possibility, the values for percent lung area consolidated in 37 sections from three animals with moderately severe disease were pooled and displayed graphically. The scatter of the data around the mean approximated a normal distribution.

Variation from animal to animal. The variance of morphometric data from different animals was similar to the variation in sections from a single animal, but quantitatively greater. The coefficient of variation was small for an exposure which resulted in extensive, lethal pneumonia and high for experiments in which less extensive disease was produced. Once again the absolute

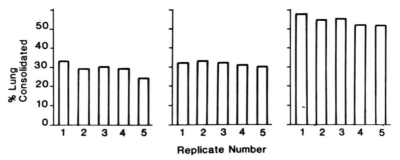

Replicate Number

FIG. 1. Replicate morphometric measurements by a single investigator. A single section from each of three guinea pigs with moderately severe disease was tested over a period of several days. The standard deviation in percent lung area consolidated ranged from 1.0 to 3.3%.

standard deviation was highest in experiments that resulted in moderately severe disease. Because of this interanimal variation, the statistical benefit derived from examining multiple sections from an individual animal is minimal.

Relationship of inoculum to extent of pneumonia. Thirty-three guinea pigs that had died or survived for at least 7 days after exposure to infectious aerosols (2×10^2 to 5×10^5 CFU) in four separate experiments were examined. The relationship between percent lung area consolidated and the $\log_{10}$ CFU was linear over this dose range. The percent lung area consolidated ranged from $17.9 \pm 5.7\%$ (mean $\pm$ standard error of the mean) for the lowest dose to $93.6 \pm 2.0\%$ for the highest dose (Table 1).

Relationship of extent of pneumonia to outcome of infection. No fatalities were observed with the lowest inoculum, whereas the highest inoculum produced uniformly fatal infection. At the two intermediate doses, the lethality of the infection was approximately 60%. The relationship of

TABLE 1. Relationship of pneumonia to initial bacterial inoculum

Expt no.	Initial dose (CFU)	No. of animals	% Pneumonia (mean)	No. of deaths/ no. exposed
1	2×10^2	4	18	0/4
2	4×10^3	12	48	7/12
3	2×10^4	5	62	3/5
4	5×10^5	12	94	12/12

inoculum to extent of pneumonia and outcome of infection was clear. In an outbred animal like the guinea pig, however, variations in host defenses may also determine outcome if the inoculum is not overwhelming. It would be of interest to know whether animals surviving an exposure that produced lethal disease had less pneumonia than their lethally infected counterparts, or whether they were able to avoid pathological effects of the infection entirely. After exposure to the two intermediate inocula (Table 1, experiments 2 and 3), the percent lung area consolidated was $23.2 \pm 7.4\%$ (mean $\pm$ standard error of the mean; $n = 7$) in surviving animals and $71.9 \pm 8.0\%$ ($n = 10$) in fatally infected animals ($P < 0.01$).

Quantitative morphometry provides a useful tool for evaluating experimental pneumonia. This parameter is amenable to statistical analysis; with knowledge of variance of the measurements, it is possible to calculate the numbers of animals that will be required to demonstrate differences of stated magnitude between experimental groups.

LITERATURE CITED

1. **Davis, G. S., W. C. Winn, Jr., D. W. Gump, J. E. Craighead, and H. N. Beaty.** 1982. Legionnaires' pneumonia after aerosol exposure in guinea pigs and rats. Am. Rev. Respir. Dis. **126:**1050–1057.
2. **Winn, W. C., Jr., G. S. Davis, D. W. Gump, J. E. Craighead, and H. N. Beaty.** 1982. Legionnaires' pneumonia after intratracheal inoculation of guinea pigs and rats. Lab. Invest. **47:**568–578.

Pathological and Bacteriological Findings After Aerosol *Legionella pneumophila* Infection of Susceptible, Convalescent, and Antibiotic-Treated Animals

A. BASKERVILLE, R. B. FITZGEORGE, D. H. GIBSON, J. W. CONLAN, L. A. E. ASHWORTH, AND A. B. DOWSETT

Public Health Laboratory Service, Centre for Applied Microbiology and Research, Porton Down, Salisbury, Wiltshire, United Kingdom

Guinea pigs, rhesus monkeys (*Macaca mulatta*), and common marmosets (*Callithrix jacchus*) were infected with inhaled lung retention doses of *Legionella pneumophila* varying from 10^2 to 10^6 organisms (2). The small-particle (<5 μm) aerosols were generated from aqueous suspensions of buffered charcoal-yeast extract with α-ketoglutarate (BCYEα)-grown bacteria (4) by a Collison spray nebulizer in a mobile Henderson apparatus (3). The strains of *L. pneumophila* used were 74/81 and Corby (serogroup 1) and 166/81 (serogroup 3).

Susceptible animals of all three species developed fever and bronchopneumonia affecting all lung lobes from day 2. In guinea pigs and marmosets this was a lobular bronchopneumonia which became confluent and produced lobar consolidation and blood-stained pleural effusions at 3 to 4 days (1). Lesions in rhesus monkeys were less extensive, consisting of a few consolidated areas. High-dose infections produced high mortality in guinea pigs. The 50% lethal dose (LD_{50}) was $10^{4.5}$ for the 74/81 strain, $10^{4.0}$ for the 166/81 strain, and $10^{2.5}$ for the Corby strain.

Histopathologically, the changes were of an acute fibrinopurulent bronchopneumonia which involved principally the pulmonary acinus (1). Lesions arose first in respiratory bronchioles, alveolar ducts, and associated alveoli and spread rapidly into surrounding tissue. There were no lesions in the upper respiratory tract or bronchi, and only a small proportion of terminal bronchioles contained exudate and showed focal epithelial necrosis. Pneumonic areas were heavily infiltrated by polymorphs and macrophages in roughly equal proportions, and many of the cells underwent lysis. There was widespread exudation into alveoli of edema fluid and fibrin. Intraalveolar hemorrhage was variable in guinea pigs, but was a consistent feature in marmosets. Hilar and mediastinal lymph nodes were highly reactive and congested, with macrophage and polymorph accumulation in medullary sinuses. Lesions were not found in the brain, liver, kidneys, or other organs.

Electron microscopy of the early lesions showed numerous *L. pneumophila* organisms within phagolysosomes of the infiltrating phagocytes. In addition to the cellular infiltration, the most important lesion was damage to alveolar capillary endothelium. This was first apparent at 48 h. It consisted of cytoplasmic swelling, increased vesiculation, and focal necrosis and was responsible for the edema and fibrin leakage. Necrosis of alveolar epithelial cells was focal and infrequent. Alveoli and interalveolar septa contained masses of fibrin, granular edema, and cell debris. Many macrophages and polymorphs containing *L. pneumophila* had large cytoplasmic aggregates of β-glycogen. Bacteria survived and multiplied within phagocytes and after cell degeneration were released into alveoli in large numbers. They were not observed to penetrate pulmonary epithelial or endothelial cells at any stage, but were occasionally seen free in the lumen of capillaries.

The lungs of animals surviving after 6 days showed evidence of organization and repair, with sequestering of bacteria-laden and fibrin-containing macrophages in damaged areas. Fibroblast activation and proliferation were prominent, and special histological stains and ultrastructural studies indicated that after 1 to 3 months there was considerable deposition of collagen and elastic fibers. Small numbers of lymphocytes were also present in the interstitium and around some small bronchioles and blood vessels.

Bacteriological examination of guinea pigs showed that after a sublethal inhaled retention dose of $10^{3.5}$ bacteria, total viable lung counts were $10^{3.5}$ on day 1, $10^{2.9}$ on day 3, and negative by day 23. After 1 LD_{50} ($10^{4.5}$ retained), total lung counts in survivors were $10^{7.9}$ on days 1 to 3, $10^{4.8}$ on day 7, and negative by day 15. Doses of 10 LD_{50} ($10^{5.4}$) led to rapid multiplication of bacteria in the lungs to $10^{8.2}$ on day 1 and $10^{8.5}$ on day 3. At this stage the lungs were overwhelmed and the animals died.

Repeated infection of convalescent animals. To study the effects of repeated low-dose infection, which may occur in humans, guinea pigs which had recovered from low-dose ($10^{3.5}$) aerosol infection with *L. pneumophila* were challenged at monthly intervals with up to three aerosol infections ($10^{3.5}$) of the homologous strain. Pyrexia resulted from the first but not from subsequent rechallenge, and high levels of serum antibody developed. Lung lesions increased in size with each challenge, changing from the initial polymorph, macrophage, and fibrin exudation to lymphoreticular and fibrous proliferation. Several months after the initial infection, lesions still persisted as large accumulations of lymphocytes, macrophages, fibroblasts, collagen, and elastic fibers with distortion of adjacent alveoli. Small numbers of eosinophils and mast cells were also present. Bacteria were not detected by electron microscopy, but macrophages contained fibrin and residual bodies. Hyperplasia of type II alveolar epithelial cells did not occur. There were numerous mitotic figures in the septal and perivascular lymphoid tissue in the period after each challenge, which may be indicative of development of cell-mediated immunity within the lung.

Guinea pigs which had been exposed to three aerosol infections were not protected against a $10\ LD_{50}$ challenge. They died earlier than nonimmune controls, though with lower lung bacterial counts. The lungs contained the same acute fibrinopurulent bronchopneumonia superimposed on the existing chronic lesions.

Antibiotic therapy. The effects of antibiotic therapy on the course of aerosol *L. pneumophila* infection in guinea pigs were assessed by three parameters: survival, persistence and levels of bacteria in the lungs, and histopathological lesions (D. H. Gibson, R. B. Fitzgeorge, and A. Baskerville, J. Infect., in press). A comparison was made between the efficacy of erythromycin, gentamicin, and rifampin (Table 1). Dosage was based on the regimens required to produce optimum serum and lung tissue levels of the drugs, correlated with their minimum inhibitory and minimum bactericidal concentrations against *L. pneumophila* determined in vitro. Therapy was commenced 24 h after aerosol infection and was given intramuscularly every 4 h (erythromycin) 3 h (gentamicin), or 6 h (rifampin) for 48 h. Doses were either similar to those used in humans (5), on a weight comparison basis, or up to four times higher.

All three antibiotics prevented deaths after 1 LD_{50} infection (Table 1). The lungs of erythromycin-treated animals contained *L. pneumophila* for the same period as control animals (7 days), though at a lower level ($10^{3.8}$ compared with $10^{4.8}$). In contrast, gentamicin and rifampin

TABLE 1. Effect of antibiotic treatment on guinea pigs infected 1 day previously with *L. pneumophila*

Antibiotic (dose)	1 LD_{50} challenge		10 LD_{50} challenge	
	Survivors/ total	Last day bacteria detected in lungs	Survivors/ total	Mean time of death (h)
Control	3/6	7	0/6	48.3
Erythromycin (15 mg)	6/6	7	0/6	55.3
Gentamicin (4 mg)	6/6	4	0/6	51.8
Rifampin (5 mg)	6/6	2	3/6	74.7

cleared the lungs of infection after days 4 and 2, respectively. Histopathological lesions in the erythromycin-treated groups were only slightly less extensive than those in control animals, and the initial acute fibrinopurulent bronchopneumonia progressed to organization, repair, and residual fibrosis over a similar time course. Gentamicin-treated animals also had pulmonary lesions, but they were not so widespread as in those given erythromycin. Rifampin was also the most effective antibiotic as assessed by histopathological lesions, changes being confined to small foci of alveolar edema and cellular infiltration, which rapidly subsided to leave only a few insignificant fibrous scars. Rifampin was the only antibiotic which prevented death after $10\ LD_{50}$ infection.

LITERATURE CITED

1. **Baskerville, A., R. B. Fitzgeorge, M. Broster, and P. Hambleton.** 1983. Histopathology of experimental Legionnaires' disease in guinea pigs, rhesus monkeys and marmosets. J. Pathol. **139:**349–362.
2. **Baskerville, A., R. B. Fitzgeorge, M. Broster, P. Hambleton, and P. J. Dennis.** 1981. Experimental transmission of Legionnaires' disease by exposure to aerosols of *Legionella pneumophila.* Lancet **ii:**1389–1390.
3. **Druett, H. A.** 1969. A mobile form of the Henderson apparatus. J. Hyg. **67:**437–448.
4. **Edelstein, P. H.** 1981. Improved semiselective medium for isolation of *Legionella pneumophila* from contaminated clinical and environmental specimens. J. Clin. Microbiol. **14:**298–303.
5. **Kirby, B. D., K. M. Snyder, R. D. Meyer, and S. M. Finegold.** 1980. Legionnaires' disease: report of sixty-five nosocomially acquired cases and review of the literature. Medicine **59:**188–205.

Legionnaires Disease: a Hamster Model

S. W. RICHARDS, P. K. PETERSON, D. E. NIEWOHNER, AND J. N. HOIDAL

Department of Medicine, University of Minnesota, Minneapolis, Minnesota 55455

Legionnaires disease came to public attention after the first recognized outbreak of the disease in Philadelphia, Pa., in 1976. The causative organism is a fastidious gram-negative bacillus, *Legionella pneumophila*. Serogroup 1 *L. pneumophila* is the cause of most human disease, but other serogroups are also pathogenic (1, 3).

The disease in humans is characterized by a respiratory illness with typical features of a bacterial pneumonia. The histopathological changes seen in human Legionnaires disease are those of an acute fibrinopurulent pneumonia (2). Initially an acute inflammatory response of neutrophils (PMN) and alveolar macrophages (AM) is noted. The organism can be found within these inflammatory cells and is capable of intracellular multiplication (4, 5).

Pulmonary host defense mechanisms active in Legionnaires disease are poorly understood. Animal models of Legionnaires disease are needed to allow controlled conditions in which to study the pathogenesis of this disease and host defense mechanisms active against *L. pneumophila* infection.

The present study describes Legionnaires disease pneumonia produced in hamsters by intratracheal inoculation of *L. pneumophila*. This model closely parallels the pathogenesis seen in human Legionnaires disease, allows consistent quantification of inoculum size and 50% lethal dose, and permits description of important host defense mechanisms in an animal not known to harbor endogenous pulmonary infections.

Animals were infected with a serogroup 1 strain of *L. pneumophila* furnished by the Centers for Disease Control, Atlanta, Ga. *L. pneumophila* was grown on buffered charcoal-yeast extract agar for 3 days at 37°C in 5% CO_2 in air to yield fresh bacteria in the log phase of growth for each experiment. Colonies were washed off the plates with sterile phosphate-buffered saline and resuspended in phosphate-buffered saline to final concentrations of 10^5 to 10^9 CFU/ml.

Female golden Syrian hamsters (Harlan Industries, Indianapolis, Ind.), weighing 90 to 110 g, were housed in a dedicated facility with an average temperature of 22°C. They were allowed free access to hamster chow and distilled water without antibiotics. Animals were inoculated with *L. pneumophila* by intratracheal injection. Cultures of homogenized lung taken immediately after inoculation and at 24 h demonstrated viable *L. pneumophila* and no bacterial contamination.

The inflammatory response to *L. pneumophila* pulmonary infection was monitored by whole-lung lavage of excised hamster lungs. On day 0, just before *L. pneumophila* was inoculated intratracheally, 2×10^6 AM and no PMN were present in the lung lavage. A subsequent biphasic cellular response of AM and PMN to pulmonary infection with *L. pneumophila* was noted, with both AM and PMN reaching a peak level of 10^8 by day 4.

Histologically a fibrinopurulent pneumonia developed which was morphologically similar to that seen in human Legionnaires disease. The 50% lethal dose was consistently $10^{8.4 \pm 0.1}$ organisms.

An antibody response to *L. pneumophila* pulmonary infection was also demonstrated. Both immunoglobulin G and immunoglobulin M antibodies rose initially, reaching significant levels by 2 weeks after *L. pneumophila* inoculation and diminishing to approximately preinoculation levels by 6 weeks.

To evaluate the effect of local antibody availability on mortality, two groups of hamsters were inoculated intratracheally either with *L. pneumophila* preopsonized with specific anti-*L. pneumophila* antibody or with unopsonized *L. pneumophila*. There was no statistical difference in the 50% lethal dose between the two groups.

In summary, intratracheal inoculation of *L. pneumophila* in hamsters reproducibly caused a fibrinopurulent pneumonia that closely resembled Legionnaires disease in humans. Both cellular and humoral immune mechanisms were stimulated. An exuberant biphasic inflammatory cellular response by PMN and AM was elicited. Further, immunoglobulin M and immunoglobulin G antibodies were produced. Despite the demonstrated antibody response of the animals to *L. pneumophila*, preopsonized *L. pneumophila* was as virulent as unopsonized *L. pneumophila* when inoculated intratracheally. This model should prove to be useful in providing a better understanding of the pathogenesis and management of Legionnaires disease.

LITERATURE CITED

1. **Blackmon, J. A., F. W. Chandler, W. B. Cherry, A. C. England, J. C. Feeley, M. D. Hicklin, R. M. McKinney, and H. W. Wilkinson.** 1981. Legionellosis. Am. J. Pathol. **103:**429–465.
2. **Carrington, C. B.** 1979. Pathology of Legionnaires' disease. Ann. Intern. Med. **90:**496–499.
3. **Davis, G. S., W. C. Winn, and H. N. Beaty.** 1981. Legion-

naires' disease: infections caused by *Legionella pneumophila* and *Legionella*-like organisms. Clin. Chest Med. 2:145–166.

4. **Horwitz, M. A., and S. C. Silverstein.** 1980. Legionnaires' disease bacterium (*Legionella pneumophila*) multiplies in-

tracellularly in human monocytes. J. Clin. Invest. 66:441–450.

5. **Horwitz, M. A., and S. C. Silverstein.** 1981. Interaction of the Legionnaires' disease bacterium (*Legionella pneumophila*) with human phagocytes. J. Exp. Med. 153:386–397.

Legionella pneumophila Pneumonia in Athymic Nude Mice

D. J. DRUTZ, P. DeMARSH, P. EDELSTEIN, J. RICHARD, W. OWENS, R. ROLFE, AND S. FINEGOLD

Infectious Diseases Section, Audie L. Murphy Memorial Veterans' Hospital, Department of Medicine, University of Texas Health Science Center, San Antonio, Texas 78284, and Infectious Diseases Section, Wadsworth Veterans Administration Medical Center, Los Angeles, California 90073

Legionella pneumophila survives and multiplies in mononuclear phagocytes (5, 7) and shows increased pathogenicity for patients with impaired immunity (9). Furthermore, the microorganism is killed inefficiently by polymorphonuclear leukocytes and resists killing by antibody and complement (6). These features suggest that *L. pneumophila* may function as a facultative or obligate intracellular microorganism and may be under the control of cell-mediated immune mechanisms.

We have conducted studies of *L. pneumophila* pathogenicity in congenitally athymic nude mice to determine whether cell-mediated immunity plays an essential role in host defense against *L. pneumophila*.

The microorganism used in all studies was *L. pneumophila* 130b (serogroup 1), isolated in 1978 from a transtracheal aspirate specimen. Cultures were maintained on buffered charcoal-yeast extract agar supplemented with α-ketoglutarate (BCYEα [3]) and transferred to fresh medium at 3-day intervals. For some experiments, bacteria were passaged in guinea pigs at 3- to 4-week intervals to enhance virulence (2). For use in mouse experiments, the bacteria were grown on BCYEα for 3 days at 36°C. Colonies were harvested by loop, transferred to phosphate-buffered saline, enumerated by Petroff-Hausser chamber, and adjusted to the desired inoculum size. The inoculum was verified by colony counts.

BALB/c mice, bred in the Infectious Disease Laboratory at Audie L. Murphy Memorial Veterans' Hospital, were maintained under specific pathogen-free or germfree conditions. Two genotypes were employed: (i) nu/nu (nude) mice, homozygous for the nude gene and characterized by athymia, hairlessness, profoundly impaired cell-mediated immunity, and reduced T helper-cell function, and (ii) nu/+ mice, phenotypically normal and essentially immune intact, from the same litters (8). Mice, 6 to 10 weeks old and matched by age, weight, and sex, were inoculated with *L. pneumophila* by the intrana-

sal (i.n.), intraperitoneal (i.p.), or peroral route. Subsequent disease status was evaluated by the occurrence of clinical illness, respiratory distress, and weight loss. In animals that died or were sacrificed, tissue impression smears were stained for *L. pneumophila* by direct fluorescent antibody, using polyvalent antiserum. Histological sections were stained by Dieterle and Gimenez stains, tissues were cultured on BCYEα and standard microbial media, and serological response was measured by indirect fluorescent-antibody titer in pooled heart blood (1, 4).

L. pneumophila organisms passaged exclusively on BCYEα became filamentous and pleomorphic and lost their flagella. Although they maintained virulence for guinea pigs, such organisms were never infective for nu/nu or nu/+ mice by the i.n. route, even at inocula in excess of 10^9 bacteria. When given i.p., they produced sporadic mouse deaths, and pathogenicity was increased only slightly by the concomitant i.p. injection of mucin plus hemoglobin (8). Decomplementation by nine doses of cobra venom had no effect, and infection could not be produced by the peroral route in either specific pathogen-free or germfree, nu/nu or nu/+ mice. Despite the absence of clinically apparent infection, both nu/nu and nu/+ mice inoculated i.p. developed equivalent indirect fluorescent serum antibody responses (1:64 to 1:256 in pooled sera from i.p.-infected mice versus <1:16 in i.n.-infected mice). These data suggest that filamentous organisms administered i.n. may have never reached a locus where infection could be induced.

When *L. pneumophila* was passaged in guinea pigs and then subcultured on BCYEα, the organisms were short discrete rods with terminal flagella. Such organisms were regularly infective by the i.n. route (serological response), but clinical illness was apparent only when the inoculum size was at least 10^7 bacteria. At an inoculum size of 10^8 *L. pneumophila*, nu/nu mice became extremely ill and lost approximately 25% of their body weight. Surviving animals tended to remain ill until sacrificed 2 to 3 weeks later. In contrast, nu/+ mice rarely looked ill for more than 2 to 3 days and lost only 12.5% of their body weight, which (in contast to nu/nu mice) they quickly regained. Deaths occurred unpredictably in the 10^8 inoculum range, seldom exceeding 50% in nu/nu and 25% in nu/+ mice within a given experiment. Histological studies revealed extensive pneumonic consolidation in nu/nu, but only patchy consolidation in nu/+, mice; numbers of *L. pneumophila* identifiable by staining were not strikingly different. Extrapulmonary dissemination was not seen. When the inoculum size was raised to 10^9 organisms, 95% of nu/nu mice died within 2 weeks (50% by day 3) versus 45% of nu/+ ($P < 0.04$). Nu/nu mice implanted with a normal syngeneic thymus gland before *L. pneumophila* challenge developed an illness intermediate in severity between those of nu/nu and nu/+ mice.

Unlike the situation with exclusively BCYEα-passaged bacteria, guinea pig-passaged *L. pneumophila* stimulated an indirect fluorescent-antibody response in the i.n. challenge model, and the response was identical in nu/nu and nu/+ mice (1:256 in both groups, with a similar rate of titer rise).

These studies indicate that athymic nude mice are significantly more susceptible to *L. pneumophila* than their normal, syngeneic, thymus-bearing littermates. Immunity is at least partially restored by thymus transplantation. Preliminary studies suggest that passive immunization with antiserum from nu/+ or nu/nu mice that have recovered from legionellosis may be partially protective in nu/nu mice. In addition, nu/nu mice immunized i.p. with *L. pneumophila* show evidence of partial protection against subsequent i.n. challenge. Thus, thymus-mediated immunity plays a central role in host defense against *L.*

pneumophila. In addition, T-cell-independent antibody appears to assist in the immune response. Whether the principal role of the thymus is to influence macrophages or antibody production remains to be determined.

LITERATURE CITED

1. **Broome, C. V., W. B. Cherry, W. C. Winn, Jr., and B. R. MacPherson.** 1979. Rapid diagnosis of Legionnaires' disease by direct immunofluorescent staining. Ann. Intern. Med. **90**:1–4.
2. **Chandler, F. W., J. E. McDade, M. D. Hicklin, J. A. Blackmon, B. M. Thomason, and E. P. Ewing, Jr.** 1979. Pathologic findings in guinea pigs inoculated intraperitoneally with the Legionnaires' disease bacterium. Ann. Intern. Med. **90**:671–675.
3. **Edelstein, P. H.** 1981. Improved semiselective medium for isolation of *Legionella pneumophila* from contaminated clinical and environmental specimens. J. Clin. Microbiol. **14**:298–303.
4. **Edelstein, P. H., R. D. Meyer, and S. M. Finegold.** 1980. Laboratory diagnosis of Legionnaires' disease. Am. Rev. Respir. Dis. **121**:317–327.
5. **Horwitz, M. A., and S. C. Silverstein.** 1980. Legionnaires' disease bacterium (*Legionella pneumophila*) multiplies intracellularly in human monocytes. J. Clin. Invest. **66**:441–450.
6. **Horwitz, M. A., and S. C. Silverstein.** 1981. Interaction of the Legionnaires' disease bacterium (*Legionella pneumophila*) with human phagocytes. I. *L. pneumophila* resists killing by polymorphonuclear leukocytes, antibody and complement. J. Exp. Med. **153**:386–399.
7. **Kishimoto, R. A., M. D. Kastello, J. D. White, F. G. Shirey, V. G. McGann, E. W. Larson, and K. W. Hedlund.** 1979. In vitro interaction between normal cynomolgus monkey alveolar macrophages and Legionnaires disease bacteria. Infect. Immun. **25**:761–763.
8. **Lamborn, P. S., J. Cauthen, and D. J. Drutz.** 1982. Mechanism of resistance of nude mice to *Neisseria gonorrhoeae*, p. 67–76. *In* N. D. Reed (ed.), Proceedings of the 3rd International Workshop on Nude Mice. Gustav Fischer, New York.
9. **Saravolatz, L. D., K. H. Burch, E. Fisher, T. Madhaven, D. Kiani, T. Neblett, and E. L. Quinn.** 1979. The compromised host and Legionnaires' disease. Ann. Intern. Med. **90**:533–537.

SUMMARY

Pathology and Pathophysiology

A. BASKERVILLE

Public Health Laboratory Service, Centre for Applied Microbiology and Research, Porton Down, Salisbury, Wiltshire, England

The aim of this contribution is to bring together the pathological aspects of Legionnaires disease (LD) in humans and the pathological information now accumulating from the use of experimental animal models. Probably the most useful way in which to summarize present knowledge and that given at the Symposium is to deal with the various features of the pathophysiology of LD comparatively, that is, to compare the changes which occur in humans and experimental animals.

In general there is good correlation between human LD and the disease produced by aerosol infection of guinea pigs (Table 1), and hence this animal provides a suitable model. The common marmoset (*Callithrix jacchus*) is also a good subject, though with low mortality, but rhesus monkeys, rats, and hamsters are rather more resistant and develop a milder illness and less extensive pulmonary lesions (5, 12, 13).

Route of infection. Inhalation of aerosols containing *Legionella pneumophila* is thought to be the mode of infection in humans (21), as the Symposium papers have confirmed, and is the best method of producing pneumonic LD in animals. Small particles (<5 μm) seem to be necessary, perhaps because this size of particle ensures penetration to respiratory bronchioles and alveoli. Larger particles which remain in the upper respiratory tract may not initiate infection, or they may produce a different clinical disease. Experimental work is needed on the effects of *L. pneumophila* in particles of 5 to 50 μm. Aerosol infection of animals with *L. pneumophila*-infected amoebae is also required to establish the role of amoebae in the natural history of LD.

The possibility of oral infection with *L. pneumophila* is occasionally mentioned. The organism does occur in potable water (11), and there has been discussion at this meeting of a number of patients with intestinal lesions and peritonitis from which *L. pneumophila* was isolated. The experiments of Katz and co-workers (this volume), using oral dosing of guinea pigs with *L. pneumophila*, pose some intriguing questions. After ingestion of infected drinking water or gastric intubation the animals developed fever, macrophages and polymorphs (PMN) in alveolar walls, and *L. pneumophila* that could be isolated from the lungs and spleen. The illness was mild and self-limiting and clinically rather reminiscent of Pontiac fever in humans. A possible pathogenetic pathway here, particularly in the intubated group, might be penetration of the gastrointestinal tract and entry into the lymphatic system, followed by dissemination of bacteria to the lungs and spleen. The organisms are then probably cleared quite rapidly.

Fever. There is fever for several days in both humans and animals. The incubation period is shorter in animals than the human mean of 4 days, but this occurs in subjects given relatively high doses and it is capable of being influenced considerably by the inoculum size.

Mortality. Mortality varies from 10 to 20% in untreated human cases (14) in many published series. In guinea pigs, mortality is dose dependent and can be manipulated from 0 to 100% according to the infecting dose. This is obviously extremely useful for studying different ends of the disease spectrum.

Bacteremia. Bacteremia is intermittent and of unknown duration in both humans and animals. As a result, distribution of *L. pneumophila* to all organs and tissues is potentially possible.

Macroscopic lung lesions. Macroscopic lung lesions are similar in humans and aerosol-infected animals and are of a lobular bronchopneumonia which becomes confluent and frequently lobar in extent, with a variable pleural effusion. In over two-thirds of human cases there is multilobar involvement, with no preferential distribution (27). In aerosol-infected animals all lobes show consolidation to a similar degree. The initial distribution of foci, however, is much more even throughout all lobes than appears to occur in humans. This may reflect a certain artificiality of exposure to a uniform-particle-size aerosol. We have noticed this even distribution of scattered small foci after small-particle aerosol infection of different species with a number of agents, including *Francisella tularensis, Salmonella cholerae-suis*, and several viruses, such as influenza and Aujeszky's disease virus (2, 6).

TABLE 1. Comparison of natural and experimental LD

Humans (natural disease)	Guinea pigs (experimental disease)
Airborne infection	Small-particle aerosol infection
Pyrexia	Pyrexia
Dyspnoea	Dyspnoea
Fibrinopurulent broncho-pneumonia	Fibrinopurulent broncho-pneumonia
Mortality (untreated), 10 to 20%	Mortality, dose dependent (0 to 100%)
Bacteria in lungs	Bacteria in lungs
Bacteremia	Bacteremia
Antibody in survivors	Antibody in survivors
Evidence of cell-mediated immunity	Evidence of cell-mediated immunity

The quantitative analysis carried out by Winn and colleagues (this volume) showed that the percent area of the lungs affected in guinea pigs is directly related to the infecting dose of *L. pneumophila*. The greater the extent of the lesions, the poorer the animal's chances of survival. This is probably because much of the effect of the consolidation is to cause hypoxemia by interfering with the ventilation, diffusion, and perfusion functions of the lung, in addition to any toxemic effects there may be.

Histopathology. In humans and animals the lesions are very similar, though greater variation has been described in humans, probably because of intercurrent disease conditions, different ages of patients, and immunosuppression. As noted in the original work by the Centers for Disease Control, legionellosis is an acute fibrinopurulent pneumonia affecting the acinus (7), with no upper respiratory tract lesions and relatively little damage to the epithelium of bronchioles. In contrast, small-particle aerosol infection with other bacteria such as *F. tularensis* and salmonellae does cause nasal and upper respiratory tract lesions (6), so this particular histopathology does seem to be a property of *L. pneumophila* and not merely a feature of the size of the infecting particles.

The role of adherence or other attachment in the pathogenesis of LD is intriguing. In our studies so far, *L. pneumophila* appears to have little or no affinity for ciliated columnar epithelium. This is perhaps in accordance with its being an opportunistic, rather than an obligatory, pathogen, though some strains do possess fimbriae which could be used for attachment in the upper tract. Presumably they are not adapted for the particular receptor sites on the airway epithelial cell surfaces.

The alveolar cellular exudate in all species consists of PMN and macrophages in varying proportions, PMN often predominating. It was important that Winn and other workers (this volume) reminded us of the considerable PMN activity in LD, because otherwise research is liable to concentrate too narrowly on the macrophage. The PMN must play an important role in the disease process. In this respect the work of Richards and colleagues, in which they used anti-PMN serum to deplete hamsters of PMN and then exposed them to tobacco smoke to reduce the infectious and lethal doses of *L. pneumophila*, is highly relevant (this volume).

The most significant lesion in the lungs in human and animal LD is exudation into alveoli of edema fluid and fibrin (4, 8, 26). There is some focal alveolar epithelial necrosis, though not on the scale of some other bacterial or viral pneumonias. This seems to be more extensive in humans than in experimental animals and is perhaps not entirely caused by the bacteria, but is also partly due to the effects of supportive measures such as oxygen therapy.

Electron microscopy. At the ultrastructural level all of the above points are apparent, and again there is good correlation between the findings in humans and experimental animals (3, 9, 18, 25). Phagocytosis is by PMN and pulmonary macrophages, but many of the *L. pneumophila* survive, kill the cell, and are released in large numbers. There seems to be no direct penetration of lung parenchymal cells in any species.

The most spectacular and significant ultrastructural finding is leakage of edema fluid and fibrin from damaged capillaries. This has been described in human lung biopsy specimens (18) and is a consistent feature of the experimental disease in several species (3). The exudation interferes with gaseous exchange in the alveoli and is responsible for the hypoxia.

The observation by Horwitz (this volume) of accumulation of ribosomes around *L. pneumophila*-containing phagosomal membranes in macrophages in vitro may be of great physiological importance. We have also seen this in vivo in LD pneumonia in a number of species (3), but our initial interpretation was that it was caused by rapid expansion of the phagosome due to multiplying bacteria, the advancing phagosomal membrane simply pushing against ribosomes in its path. However, Horwitz showed that even small phagosomes soon after ingestion of *L. pneumophila* also acquired a layer of ribosomes. This illustrates well the importance of integrating in vitro and in vivo studies.

A topic related to this is the abnormally large aggregates of β-glycogen particles in the cytoplasm of legionella-containing macrophages and PMN in vivo (3). This must be indicative of changes taking place in the metabolism of infected phagocytes due to the presence in them of *L. pneumophila*. These changes lead to accumulation of glycogen, but whether this is due to an increased cellular requirement for energy, for which glycolysis is the main source, or to some specific property of the physiology of *L. pneumophila* is as yet unknown. Further work on the enzyme systems and metabolism of *L. pneumophila* is crucial to progress in this area.

Pathogenic mechanisms. Apart from the obvious effects, in the form of cellular infiltration, consolidation, and hypoxia, the mechanisms by which *L. pneumophila* produces cell and tissue damage in the lung and systemic illness are not understood. A number of toxins have now been isolated from *L. pneumophila*, including a hemolysin, cytotoxins, endotoxic activity (1, 16, 17), and a number of extracellular enzymes such as lipase, proteases, phosphatases, DNases, and a substance with chymotrypsin-like activity. Whether these substances are actually responsible for tissue damage is not proven, and to date little in vivo work has been attempted. The presence of *L. pneumophila* in the alveoli attracts both PMN and alveolar macrophages in large numbers, and phagocytosis occurs readily in the nonimmune host. However, there is evidence from the work of Friedman (this volume) that some *L. pneumophila* toxins inhibit many biochemical functions of PMN and may thus affect their capacity to deal with the bacteria. Horwitz (this volume) has also demonstrated that *L. pneumophila* prevents lysosome-phagosome fusion in macrophages and hence avoids exposure of the bacteria to degrading enzymes.

At present it is not known whether damage to tissues in the lung is caused solely by free, viable *L. pneumophila* organisms and their toxins after release from phagocytes, or whether dead bacteria provide another component in the response, possibly endotoxin. Again, some or all of the damage may be caused by PMN products or chemical mediators. Large numbers of degenerating PMN are certainly present in alveoli in the acute stages of LD in all species, and it is possible that proteases or other substances are released from them and play a part in the capillary damage. The fact that *L. pneumophila* gains access to blood vessels to produce a bacteremia is probably important in the development of fever and the systemic effects of LD.

Healing of lung lesions. Organization, repair, and fibrosis are the sequels to extensive, severe lung lesions in LD of humans and experimental animals. Fibrosis occurs even after erythromycin therapy in guinea pigs and humans, as shown by long-term radiographical and pathological follow-up (20). Physiological studies in humans have demonstrated reduced diffusion capacity for years (24). The animal models are thus available for work on the assessment and treatment of chronic LD lesions.

Clinical biochemistry. Clinical biochemistry has received relatively little attention during the Symposium, but it is nevertheless of considerable interest to clinicians responsible for the management of LD patients. In acute LD of humans and animals there are some changes in biochemical parameters (10, 19, 22), though many are nonspecific and probably only indicative of pyrexia and gram-negative septicemia. At the present time they do not appear to offer any very obvious clues to disease mechanisms, as was originally hoped. The most consistent change in humans is hyponatremia, and this also occurs in guinea pigs infected intraperitoneally (i.p.) with *L. pneumophila*. In aerosol infections, changes in all the parameters tend to be less marked and to occur slightly later (Hambleton, personal communication). In guinea pigs, as in humans, there is leukocytosis, followed by late leukopenia, an increase in the phenylalanine/tyrosine ratio (which develops in many infections with pyrexia), and an increase in serum Cu and enzymes such as sorbitol dehydrogenase, aspartate aminotransferase, alanine aminotransferase, lysozyme, and, transiently, creatine kinase (19). Synthesis of acute-phase proteins (α- and β-globulins) is also seen in guinea pigs, but there is a fall in Zn and Fe levels. Though elevated blood urea has been reported in humans, this has not been found in infected guinea pigs, nor has histological evidence of renal or hepatic damage.

Extrapulmonary manifestations. Although the extrapulmonary aspects of LD are secondary in importance to the pneumonia, extrapulmonary signs are intriguing and certainly important in individual patients. This is the area on which the animal models have hitherto shed least light. The conditions most commonly reported are diarrhea, renal and hepatic failure, and neurological disturbance. Some manifestations may be due to intercurrent conditions, and it has been suggested that others are systemic toxic effects or due to shock, rather than the result of direct tissue invasion by *L. pneumophila*.

None of these features is seen in the experimental animal models, perhaps in part because the animals are young, healthy, and free from lesions at the time of infection. We know, however, that bacteremia does exist in animals and legionellae can be detected by immunofluorescence in small numbers in extrapulmonary sites. Significant histological lesions have only occasionally been reported after aerosol infection of animals in extrapulmonary tissues (usually the spleen), and many of those in humans are either inconsistent with or not conclusively attributable to *Legionella* infection.

The discussion about creatine kinase increases in patients was interesting. We see these increases transiently after both aerosol and i.p. infection of guinea pigs, but the enzyme falls to subnormal levels after a few days (19). We have been unable so far to relate the increase to histologically proven rhabdomyolysis or myositis.

In summary, the clinical biochemical changes are in general minor and nonspecific and probably related to pyrexia, gram-negative septicemia, or shock. Currently, therefore, experimental LD offers no clues as to the reasons for the neurological disturbances reported, particularly the signs of cerebellar dysfunction, or for the diarrhea often described, although this could be due to endotoxemia.

Antibiotic therapy. Treatment of LD with antimicrobial agents is generally effective in humans and guinea pigs infected by aerosol or i.p. (15; D. H. Gibson, R. B. Fitzgeorge, and A. Baskerville, J. Infect., in press). The choice of antibiotic, the stage at which therapy is commenced, the dose given, and its duration are important factors which all influence the outcome of the infection. There is good radiographical and pathological evidence in humans that therapy with inappropriate agents, or even with erythromycin in some cases, still results in long-term pulmonary lesions. In our experience with experimental aerosol infections in guinea pigs, even high doses of erythromycin, which persists well in the lungs, do not prevent development of widespread lesions and fibrosis; rifampin, however, which penetrates well into macrophages to reach the organisms, confines lung lesions to a few insignificant foci (Gibson et al., in press). On the basis of present knowledge, therefore, a combination of both erythromycin and rifampin probably offers the best therapy, though studies on the development of *L. pneumophila* resistance to rifampin in vitro and in vivo are urgently needed to assess the risk of compromising this drug's efficacy against other organisms. In vivo experiments are also required with newly formulated antibiotics which have activity against *Legionella* spp.

Immunity. There is considerable overlap between the pathophysiology and immunology of LD, particularly because of the central role of the pulmonary macrophage in the pathogenesis of the infection and recovery from the disease. Since the immunological aspects of LD have been dealt with separately, only those features concerned with reinfection and repeated challenge will be considered here.

Serum antibodies to *L. pneumophila* develop in convalescent patients and in experimental animals. Probably cell-mediated immunity also occurs, as shown by positive delayed-type hypersensitivity skin reactions in some patients and animals recorded by several workers at this Symposium and previously (23, 28).

Initial reports of induction of immunity in experimental animals are not encouraging. The work of Eisenstein and colleagues presented here shows that i.p. immunization of guinea pigs with killed *L. pneumophila* gives little protection against i.p. challenge with 10 50% lethal doses (LD_{50}) of live organisms and none against 100 LD_{50} aerosol challenge. Guinea pigs which survived live sublethal i.p. infection were immune to challenge with 100 LD_{50} i.p. We have obtained similar results, but found it impossible to induce protection against aerosol challenge by live i.p. immunization (5a). We also found that guinea pigs immunized either intramuscularly with serotype-specific antigen or by repeated sublethal aerosol challenge and possessing antibody are not protected against even a 10 LD_{50} challenge. In fact, in all cases the animals die some 24 h earlier than nonimmune controls (5a). Histology shows that they have precisely the same acute fibrinopurulent pneumonia with no increase in eosinophils and no bronchiolospasm. The guinea pigs do not die of an anaphylactic reaction, but this histological pattern does not rule out the presence of an underlying type III or Arthus reaction. What we appear to have is a type of enhancement, reminiscent of that associated with some viral infections.

Repeated sublethal aerosol infection of guinea pigs is of interest pathologically because it induces lung lesions of increasing size. The nature of the lesions in this case changes from exudation of PMN, macrophages, edema, and fibrin to lymphoreticular and fibrous proliferation which persists at least for several months, if not permanently (5a). It is obviously difficult to know just how prevalent repeated infection of humans is. Epidemiological and clinical evidence at this Symposium indicates that it does occur. The sequels to reinfection seen in the guinea pig (that is, acute enhancement or chronic lesions, depending on the challenge dose) should therefore be borne in mind by clinicians.

Against this background, the immediate prospects for a safe and effective vaccine against LD do not appear promising. Much more work is required on the antigenic components of *L. pneumophila*, on the interaction between macrophages and the bacteria as facultative intracellular pathogens, and on the combined role of these features in inducing protective immunity, whether humoral, cellular, or a combination of the two. Analysis of the events occurring in immune animals on rechallenge, particularly the role of complement and chemical mediators, would be essential before human vaccination could be undertaken.

Pontiac fever. It is difficult to develop an experimental animal model for Pontiac fever because all the symptoms except pyrexia are subjective (headache, malaise, myalgia, and so on) and the lesions in humans are unknown, but presumably very minor and transient. It has been suggested for some time that the differences between LD and Pontiac fever may be due to differences in virulence of the strains of *L. pneumophila* involved. However, the work presented here by Huebner and co-workers shows that when guinea pigs were infected by aerosols of the Philadelphia and Pontiac isolates, both multiplied to the same degree in the lungs and spleen and the 50% infective dose and LD$_{50}$ were the same. This, of course, argues against LD and Pontiac fever being caused by organisms of different virulence. Is it possible that larger inhaled particles are involved in Pontiac fever and that the resulting different deposition pattern or inoculum size causes a different pathogenesis and clinical disease? Or is it that one organism is able to produce greater amounts of a toxin than the other and that this causes the Pontiac fever symptoms? The reported development of either LD or Pontiac fever by different individuals exposed to the same source of *L. pneumophila* in the same incident makes it difficult to escape the conclusion that some host factor is also involved. This could be the possession of some degree of immunity, perhaps cell-mediated or antibody, either circulating or in respiratory secretions.

LITERATURE CITED

1. **Baine, W. B., J. K. Rasheed, D. C. Mackel, C. A. Bopp, J. G. Wells, and A. F. Kaufmann.** 1979. Exotoxin activity associated with the Legionnaires disease bacterium. J. Clin. Microbiol. **9:**453–456.
2. **Baskerville, A.** 1971. The histopathology of pneumonia produced by aerosol infection of pigs with a strain of Aujeszky's disease virus. Res. Vet. Sci. **12:**590–592.
3. **Baskerville, A., A. B. Dowsett, R. B. Fitzgeorge, P. Hambleton, and M. Broster.** 1983. Ultrastructure of pulmonary alveoli and macrophages in experimental Legionnaires' disease. J. Pathol. **140:**77–90.
4. **Baskerville, A., R. B. Fitzgeorge, M. Broster, and P. Hambleton.** 1983. Histopathology of experimental Legionnaires' disease in guinea pigs, rhesus monkeys and marmosets. J. Pathol. **139:**349–362.
5. **Baskerville, A., R. B. Fitzgeorge, M. Broster, P. Hambleton, and P. J. Dennis.** 1981. Experimental transmission of Legionnaires' disease by aerosol infection with *Legionella pneumophila*. Lancet **ii:**1389–1390.
5a. **Baskerville, A., R. B. Fitzgeorge, J. W. Conlan, L. A. E. Ashworth, D. H. Gibson, and C. P. Morgan.** 1983. Studies on protective immunity to aerosol challenge with *Legionella pneumophila*. Zentralbl. Bakteriol. Parasitenkd. Infektionskr. Hyg. Abt. 1 Orig. Reihe A **255:**150–155.
6. **Baskerville, A., and P. Hambleton.** 1976. Pathogenesis and pathology of respiratory tularaemia in the rabbit. Br. J. Exp. Pathol. **57:**339–347.

7. **Blackmon, J. A., F. W. Chandler, W. B. Cherry, A. C. England, J. C. Feeley, M. D. Hicklin, R. M. McKinney, and H. W. Wilkinson.** 1981. Legionellosis. Am. J. Pathol. **103:**429–465.
8. **Blackmon, J. A., M. D. Hicklin, and F. W. Chandler.** 1978. Legionnaires' disease: pathological and historical aspects of a "new" disease. Arch. Pathol. Lab. Med. **102:**337–343.
9. **Chandler, F. W., J. A. Blackmon, M. D. Hicklin, R. M. Cole, and C. S. Callaway.** 1979. Ultrastructure of the agent of Legionnaires' disease in the human lung. Am. J. Clin. Pathol. **71:**43–50.
10. **Cordes, L. G., and D. W. Fraser.** 1980. Legionellosis, Legionnaires' disease: Pontiac fever. Med. Clin. North Am. **64:**395–416.
11. **Cordes, L. G., A. M. Wiesenthal, G. W. Gorman, J. P. Phair, H. M. Sommers, A. Brown, V. L. Yu, M. H. Magnussen, R. D. Myer, J. S. Wolf, K. N. Shands, and D. W. Fraser.** 1981. Isolation of *Legionella pneumophila* from hospital shower heads. Ann. Intern. Med. **94:**195–197.
12. **Davis, G. S., W. C. Winn, D. W. Gump, J. E. Craighead, and H. N. Beaty.** 1982. Legionnaires' pneumonia after aerosol exposure in guinea pigs and rats. Am. Rev. Respir. Dis. **126:**1050–1057.
13. **Fitzgeorge, R. B., A. Baskerville, M. Broster, P. Hambleton, and P. J. Dennis.** 1983. Aerosol infection of animals with strains of *Legionella pneumophila* of different virulence: comparison with intraperitoneal and intranasal routes of infection. J. Hyg. **90:**81–89.
14. **Fraser, D. W., T. F. Tsai, W. Orenstein, W. E. Parkin, H. J. Beecham, R. G. Sharrar, J. Harris, G. F. Mallison, S. M. Martin, J. E. McDade, C. C. Shepard, and P. S. Brachman.** 1977. Legionnaires' disease. Description of an epidemic of pneumonia. N. Engl. J. Med. **297:**1189–1197.
15. **Fraser, D. W., I. K. Wachsmuth, C. Bopp, J. C. Feeley, and T. F. Tsai.** 1978. Antibiotic treatment of guinea pigs infected with agent of Legionnaires' disease. Lancet **i:**175–177.
16. **Friedman, R. L., B. H. Iglewski, and R. D. Miller.** 1980. Identification of a cytotoxin produced by *Legionella pneumophila*. Infect. Immun. **29:**271–274.
17. **Fumarola, D., R. Monno, and I. Munno.** 1980. Toxic activities of *Legionella pneumophila*: current status of investigations. Microbiologica (Bologna) **3:**115–121.
18. **Glavin, F. L., W. C. Winn, and J. E. Craighead.** 1979. Ultrastructure of lung in Legionnaires' disease. Observations on three biopsies done during the Vermont epidemic. Ann. Intern. Med. **90:**555–559.
19. **Hambleton, P., A. Baskerville, R. B. Fitzgeorge, and N. E. Bailey.** 1982. Pathological and biochemical features of *Legionella pneumophila* infection in guinea pigs. J. Med. Microbiol. **15:**317–326.
20. **Kariman, K., J. D. Shelburne, W. Gough, M. J. Zacheck, and J. A. Blackmon.** 1979. Pathologic findings and long-term sequelae in Legionnaires' disease. Chest **75:**736–739.
21. **Lattimer, G. L., and R. A. Ormsbee.** 1981. Legionnaires' disease, p. 27–28. Marcel Dekker, New York.
22. **Lattimer, G. L., and L. V. Rhodes.** 1978. Legionnaires' disease. Clinical findings and one year follow-up. J. Am. Med. Assoc. **240:**1169–1171.
23. **Plouffe, J. F., and I. M. Baird.** 1981. Lymphocyte transformation to *Legionella pneumophila*. J. Clin. Lab. Immunol. **5:**149–152.
24. **Reingold, A. L., and J. D. Band.** 1982. Legionellosis, p. 217–239. *In* C. S. F. Easmon (ed.), Medical microbiology, vol. 1. Academic Press, Inc., New York.
25. **Rodgers, F. G., A. D. Macrae, and M. J. Lewis.** 1978. Electron microscopy of the organism of Legionnaires' disease. Nature (London) **272:**825–826.
26. **Winn, W. C., F. L. Glavin, D. P. Perl, J. L. Keller, T. L.**

Andres, T. M. Brown, C. M. Coffin, J. E. Sensecqua, L. N. Roman, and J. E. Craighead. 1978. The pathology of Legionnaires' disease, fourteen fatal cases from the 1977 outbreak in Vermont. Arch. Pathol. Lab. Med. **102**:344–350.

27. **Winn, W. C., and R. L. Meyerowitz.** 1981. The pathology of the Legionella pneumonias. A review of 74 cases and the literature. Hum. Pathol. **12**:401–422.

28. **Wong, K. H., P. R. B. McMaster, J. C. Feeley, R. J. Arko, W. O. Schalla, and F. W. Chandler.** 1980. Detection of hypersensitivity to *Legionella pneumophila* in guinea pigs by skin test. Curr. Microbiol. **4**:105–110.

IMMUNOLOGY

STATE OF THE ART LECTURE

Immunity to *Legionella pneumophila*

HERMAN FRIEDMAN, THOMAS KLEIN, AND RAYMOND WIDEN

Department of Medical Microbiology and Immunology, University of South Florida College of Medicine, Tampa, Florida 33612

Legionella pneumophila is an opportunistic facultative intracellular pathogen associated with human pulmonary infection. It is now well known that antigenic diversity exists among the legionellae. A variety of serogroups have been identified and characterized.

There is a growing body of information concerning the antigens of the legionellae. However, much less is known concerning the host immune response to these antigens and the relationship, if any, between humoral and cellular immune responses in resistance and susceptibility to *L. pneumophila*. During the last few years a number of studies have been performed in experimental animals to examine the mechanism of immunity either mediated by serum antibody or involving cellular immune responses (1, 4–6, 16, 17, 21, 23; M. A. Horwitz, Clin. Res. **30:**369A, 1982). Information from studies with patients indicates not only that serum antibody develops rapidly to these organisms, but also that cell-mediated immunity may be an important human response to these bacteria. In this review, a brief discussion of humoral and cellular immunity to *Legionella* spp., in both humans and experimental animals, will be presented.

Immune responses to *Legionella* spp. antigens. Early studies concerning the immune response to *Legionella* spp. showed that serum antibodies could be readily detected. Both direct and indirect immunofluorescent assays were utilized to identify those individuals with antibody to the bacteria and to identify the organisms per se (2, 5, 27). Serum from immunized experimental animals or from individuals who have recovered from legionellosis reacts in a variety of ways with the killed bacteria. The direct fluorescent-antibody test has been used mainly to demonstrate the presence of the bacteria in tissues, exudates, cultures, or environmental samples. The indirect fluorescent-antibody test has become the standard procedure for immunoserological diagnosis of patients who have or have had legionellosis. Titers of up to 1:128 or 1:256 have been readily detected, not only in patients who have evidence of Legionnaires disease but also in normal control subjects. Thus there appears to be a widespread distribution of "background" antibody response to these organisms in the general population.

For serodiagnostic purposes titers greater than 1:128 or 1:256 are necessary, or at least a fourfold or greater rise in titer for individuals with initial lower levels of serum antibody (17). A number of other serological tests have been developed in recent years which have been useful for demonstrating antibody to various serogroups or species of legionellae. Among these are the microagglutination technique, the microenzyme-linked immunoabsorbent assay, crossover immunoelectrophoresis, and immunoferritin labeling techniques (3, 9, 12, 18, 24, 25). Various studies involving these techniques have been of value in demonstrating serum antibody responses to *Legionella* spp., but have not provided much information concerning the role of such antibodies in protection.

Several studies have been carried out concerning delayed-type hypersensitivity to *Legionella* antigens in individuals recovering from legionellosis. Although a skin test for delayed cutaneous hypersensitivity, as developed for animal models (see below), would be advantageous, this has not yet been developed. Such skin tests would permit rapid screening of populations for epidemiological evidence of prior exposure or permit determination of the extent of clinical disease. Evidence of existence of cellular immunity by skin tests has not yet been reported for patients, but several procedures have been utilized with peripheral blood leukocytes to assess cellular immunity. For example, Plouffe and Baird showed that peripheral blood leukocytes from 17 patients recovering from Legionnaires disease evinced increased [³H]thymidine incorporation upon in vitro stimulation with *L. pneumophila* sonic extract as compared to blood leukocytes from individuals with no clinical history of Legionnaires disease (20). The mean stimulation indices showed significant differences between the two populations, but it is of interest that even control individuals had relatively high stimulation indices.

In follow-up studies, other "normal" controls showed significant incorporation of [³H]thymidine after stimulation with *L. pneumophila* antigen in vitro, suggesting either prior exposure to cross-reacting antigens or subclinical disease. In additional studies, it was found that the level of responsiveness of peripheral blood leukocytes from young, normal individuals was lower than that from older individuals (21, 22). Peak responses of such background blastogenesis occurred with individuals in their 40s, and the lowest responses occurred with cord blood from newborn infants. The interpretation of these results with regard to possible mechanisms involved is not clear, especially since the nature of the responding cell was not determined. Peripheral blood leukocyte preparations contain both B- and T-cell populations, although in humans about 80% are T cells. The investigators believe that the responding cells may be T cells. Blast cell transformation of T cells stimulated by specific antigen has been interpreted as evidence for cell-mediated immunity in other infectious diseases. Therefore, these studies have been tentatively interpreted as providing support for T-cell sensitization during *L. pneumophila* infection.

In a related series of studies, Horwitz and Silverstein showed that human monocytes from peripheral blood inhibited the growth of *L. pneumophila* when derived from patients who had recovered from disease or when activated with supernatants from concanavalin A-stimulated normal lymphoid cells (13–16). Such supernatants contained activating factors for macrophages. Human polymorphonuclear leukocytes readily ingested *Legionella* cells, especially after opsonization by antibody and complement. The ingested organisms were then killed by the polymorphonuclear leukocytes. No in vitro bactericidal activity was found by Horwitz and Silverstein for fresh, normal human serum. Recently, Horwitz also reported that peripheral blood mononuclear cells from patients who had recovered from legionellosis, when incubated with extracts from the same bacteria, release a soluble factor into the culture medium which activates other normal human monocytes to kill *Legionella* spp. in vitro (Horwitz, Clin. Res. **30**:369A, 1982). Thus it is likely that soluble factors derived from humoral peripheral blood monocytes may enhance cell-mediated immunity type response.

Immunity in experimental animals. A number of studies have shown that experimental animals injected with *Legionella* vaccine, or even with sublethal numbers of viable bacteria, develop anti-*Legionella* antibodies. For example, guinea pigs (considered a highly susceptible experimental animal for legionellosis) develop antibodies when injected with relatively small numbers of viable legionellae. Larger numbers of bacteria induce symptoms similar to legionellosis in hu-

mans, and the animals succumb to fatal pneumonia and septicemia. The pathobiology and immune response of guinea pigs infected with legionellae by the aerosol route or by intraperitoneal injection have been studied in detail (28, 29; W. C. Winn, Jr., G. S. Davies, D. W. Gump, et al., Lab. Invest. **44**:84, 1981). A majority of guinea pigs exposed to relatively small numbers of *L. pneumophila* cells (10^5 bacteria per animal) develop fatal pneumonia, with nearly 100% mortality (29). Serum antibody rapidly develops in those guinea pigs that survive the infection. Likewise, guinea pigs injected with a vaccine prepared from killed *L. pneumophila* in complete Freund adjuvant also develop serum antibody detectable either by indirect immunofluorescent procedures or by microagglutination assays. Antibody develops to both the cell wall antigens and flagella of legionellae in guinea pigs vaccinated with killed bacteria. Furthermore, it was recently found by Rolstad and Berdal that rats vaccinated with *L. pneumophila* also develop antibody and that passive administration of such antibody into normal rats protects the animals from challenge infection with viable organisms (23). Passive transfer of serum from immunized sheep was also found to be protective. Rats are generally much less susceptible than guinea pigs, and many more organisms are needed to kill them.

Skin reactions in guinea pigs immunized or infected with *L. pneumophila* have been demonstrated. In the initial studies by Wong et al., guinea pigs developed marked cutaneous hypersensitivity after sensitization with *L. pneumophila* antigen in complete Freund adjuvant (30). Positive skin reactivity occurred within 2 to 4 weeks after immunization, as shown by reactivity after intradermal challenge with purified cross-reactive antigen derived from the surface components of the bacteria. Such skin reactions were generally specific for *L. pneumophila*, but some cross-reactivity was evident. Furthermore, some nonspecific reactions occurred, because induration and erythema were evident, although to a lesser degree, even in nonimmunized guinea pigs challenged with this antigen. More recently, Wong and Feeley found that delayed cutaneous hypersensitivity could be induced by serogroup antigen mixed with the cross-reacting antigens isolated from *L. pneumophila* (31). When the cross-reacting antigen was used to immunize the guinea pigs, hypersensitivity occurred after challenge of the guinea pigs with a mixture of both cross-reacting and serogroup-specific antigen. The reaction in the guinea pigs given this mixture of antigens was essentially similar to that observed when guinea pigs were sensitized with killed bacteria in complete Freund adjuvant.

Recent studies by Widen et al. (26) showed that guinea pigs sensitized with killed *L. pneumophila* vaccine in complete Freund adjuvant developed marked skin hypersensitivity within 3 to 6 weeks, as evidenced by challenge skin reactions to crude sonic extracts as well as to purified extracts from *L. pneumophila*. The reaction was specific, and sensitivity persisted for at least 6 to 8 months (Table 1). Sensitization of guinea pig with killed vaccine, followed by a booster immunization several weeks later, resulted in even greater skin hypersensitivity. These skin reactions were also specific. Normal animals did not show reactivity by the skin test to sonic extracts or purified antigen. Histological examination of the reactions indicated that a mononuclear infiltration occurred, consistent with delayed hypersensitivity. However, there was some evidence of a polymorphonuclear cell infiltrate, suggesting a mixed dermal response possibly due to cell-mediated and humoral immune mechanisms. Guinea pigs sublethally infected with small doses of viable legionellae also developed skin hypersensitivity. Within 4 to 6 weeks after infection, marked skin reactions occurred upon challenge injection with *L. pneumophila* sonic extract, but this was not evident in normal control animals (Table 2). The largest skin reactions occurred in guinea pigs given a relatively small number (10^4 or 10^6 virulent *L. pneumophila* cells) as compared to less reaction in animals infected with a less virulent strain passaged in vitro.

In vitro correlates of cellular immunity are also apparent in guinea pigs either sensitized with *Legionella* antigen or sublethally infected with viable bacteria. The migration inhibitory factor (MIF) test is based on the inhibition of migration of macrophages in vitro and readily demonstrates development of sensitivity of guinea pigs to killed or viable bacteria. Injection of *Legionella* bacteria in complete Freund adjuvant into guinea pigs resulted in development of hypersensitivity, as shown by challenging spleen cells in vitro in a direct or indirect MIF assay (Table 3). For the direct assay, spleen cells from sensitized or normal control guinea pigs were mixed in medium containing 0.2% agarose and

TABLE 2. Skin test response to *L. pneumophila* sonic extract in guinea pigs infected with viable legionellae and tested with homologous sonic extract

Infection[a] (dose)	Skin test response (mm of response [range])[b]	
	Erythema	Induration
None (control)	5 (3–8)	2 (0–3)
Avirulent		
10^6	10 (9–12)	1 (0–3)
10^8	13 (11–15)	8 (6–12)
Virulent		
10^4	14 (7–22)	9 (5–13)
10^6	19 (15–22)	9 (5–13)

[a] Guinea pigs were injected intraperitoneally with the indicated dose of avirulent or virulent legionellae 45 days earlier.

[b] Mean response of four to six guinea pigs per group, injected intradermally with 10 μg of sonic extract 24 h previously.

TABLE 1. Sensitization of guinea pigs to *L. pneumophila* vaccine, assessed by skin test[a]

Time (days) after sensitization	Skin reaction[b] (no. positive/no. tested)			
	Saline control	*L. pneumophila* vaccine in:		
		Saline	IFA	CFA
7	0/3	0/3	0/3	0/3
15–25	0/3	0/3	0/3	1/3
30–40	0/3	0/3	0/3	3/3
45–60	0/3	0/3	0/3	3/3
75–120	0/3	0/3	0/3	3/3
150–200	0/3	0/3	0/3	3/3

[a] Groups of guinea pigs were sensitized by subcutaneous injection in the nucchal region with 3×10^9 Formalin-killed legionellae per kg in the indicated vehicle and then skin tested by intradermal injection with 10 μg of homologous *L. pneumophila* sonic extract. IFA, Incomplete Freund adjuvant; CFA, complete Freund adjuvant.

[b] Number of positive guinea pigs (induration and erythema of 10 mm or greater at 48 h postchallenge)/total number tested.

TABLE 3. Migration inhibition activity of spleen cells from *L. pneumophila*-sensitized guinea pigs as assessed by direct and indirect assays

Time (days) after sensitization[a]	Migration inhibition reaction (avg % inhibition) by assay:			
	Direct[b]		Indirect[c]	
	Vaccine	Sonic extract	Vaccine	Sonic extract
None (control)	<10	10	10	<10
7	<10	10	10	<10
10–20	<10	10	10	<10
25–35	<10	28.5	48.6	53.2
45–60	<10	30.5	40.5	35.5
75–100	<10	42.5	58.3	40.5
120–150	<10	32.5	40.2	30.5

[a] Groups of guinea pigs were sensitized by subcutaneous injection of 10^9 legionellae in complete Freund adjuvant at the indicated time before in vitro MIF assay with 10^6 killed legionellae or 5 μg of sonic extract.

[b] Guinea pig spleen cells after in vitro culture with indicated antigen.

[c] Indicator mouse peritoneal cells, treated with cell-free culture supernatants from guinea pig spleen cells stimulated in vitro with vaccine or sonic extract for 48 h.

placed in microwells in microtiter plates. To the medium in the wells was added either whole killed legionellae or sonic extract. Spleen cells from either immune or normal animals failed to show evidence of migration inhibition in response to the whole bacterial antigen in vitro, but there were marked reactions with the sonic extract as antigen. A dose of 10 to 100 μg of sonic extract per well resulted in marked inhibition of migration of spleen cells from guinea pigs which had been sensitized with *Legionella* vaccine in complete Freund adjuvant 4 weeks earlier. Little if any reactivity occurred when the guinea pigs were injected with *Legionella* vaccine in incomplete Freund adjuvant or saline. However, as with the skin test reactions, positive reactions were evident when guinea pigs were primed with killed bacteria in complete Freund adjuvant and given a second injection several weeks or months later with the *Legionella* antigen in incomplete adjuvant or saline. If, however, the primary injection consisted of legionellae in incomplete adjuvant or saline, no significant inhibition or migration occurred. It is of interest that only the sonic extract or the soluble extract of the bacteria could be used to demonstrate such direct migration inhibition.

Unlike the direct MIF assay, the indirect assay readily demonstrated reactivity for spleen cells from sensitized guinea pigs, even if whole bacteria were used as the test antigen (26). For this procedure, spleen cell suspensions from sensitized or normal untreated guinea pigs were incubated in vitro in small volumes with either bacteria or sonic extract. The cultures were centrifuged 48 h later, and cell-free supernatants were obtained. Samples of these supernatants were then tested for the presence of MIF by determining their ability to inhibit the migration of indicator peritoneal cells from normal mice in microdroplet agar cultures. The presence of MIF was readily demonstrable by inhibition of migration of the indicator murine peritoneal cells. In this procedure, spleen cell cultures from normal guinea pigs incubated with up to 100 μg of sonic extract or 10^8 bacteria failed to produce MIF in the supernatants. In contrast, when spleen cells were obtained after 3 to 4 weeks or longer from guinea pigs sensitized with *Legionella* organisms in complete Freund adjuvant, MIF reactivity was readily detected by the indirect assay. Similarly, when guinea pigs were sensitized with living legionellae, especially the virulent strain, MIF reactivity was apparent within 3 to 4 weeks after sublethal infection (7).

The lymphocyte blastogenic test has been utilized as a marker of cellular sensitivity to a wide variety of antigens, including those derived from bacteria. Extracts of gram-negative organisms especially are considered good mitogens for lymphoid cells from a wide variety of mammalian species. Thus the blastogenic test was also used as an indicator of sensitivity to *Legionella* antigens for both guinea pigs and mice. In the guinea pig system, spleen cells from animals injected with killed *L. pneumophila* in complete Freund adjuvant or sensitized by sublethal infection with *L. pneumophila* responded in a specific manner to *L. pneumophila* antigen in vitro (26a). Incubation of spleen cells obtained 3 to 4 weeks or longer after sensitization of guinea pigs with *L. pneumophila* sonic extract or whole-cell vaccine resulted in a two- to fourfold or greater stimulation index as compared to incubation of the same spleen cells for 3 to 6 days in vitro without antigen (Table 4). No such stimulation occurred with spleen cells from normal guinea pigs or guinea pigs sensitized with other antigens. Specificity was also readily demonstrable by incubation of spleen cells from guinea pigs sensitized with *Escherichia coli* lipopolysaccharide; no response occurred.

It is of interest that there was no background blastogenic response by spleen cells from normal guinea pigs with the *L. pneumophila* antigens. In contrast, spleen cells from normal nonsensitized mice incubated either with whole *Legionella* vaccine or sonic extract showed a marked degree of blastogenesis. In general, a stimulation index of 5 to 10 or more occurred with spleen cells from normal mice exposed to the *Legionella* antigen in vitro. Infection of mice with viable legionellae resulted in increased blastogenic responses, indicating that these bacteria sensitized the mice to heightened blastogenic responses similar to the increased responses noted by Plouffe and Baird with peripheral blood leukocytes from patients who had recovered from *L. pneumophila* infection (21). It is also of interest that the peak blastogenic response of spleen cells from normal mice stimulated in vitro with *Legionella* antigen was on day 3. There was very little residual stimulation, i.e., thymidine uptake, in the cultures by day 5. On the other hand, when spleen cells were derived from mice sensitized to *L. pneumophila* by prior infection, a heightened blastogenic response was observed not only on day 3 after culture initiation, but also on day 5. Thus, either there was a continuing proliferation of cells in vitro after exposure to legionellae, or an additional population of lymphoid cells in vitro was responding to the antigens in culture.

In studies of normal, noninfected mice the responding cells appear to be B lymphocytes, since prior treatment of the spleen cells with anti-theta serum plus complement to deplete T cells failed to reduce the responsiveness of the cultures to the *Legionella* antigen. Fractionation of the mouse spleen cells over nylon-wool col-

TABLE 4. Blastogenic responsiveness of spleen cells from *L. pneumophila*-sensitized guinea pigs, incubated in vitro with *Legionella* antigen[a]

Time (days) after infection	Blastogenic response[b]		
	cpm	SI	P
None (control)	683 ± 75	0.9	
7	820 ± 110	1.2	NS
10–20	1,025 ± 116	1.5	NS
25–40	1,912 ± 207	2.8	<0.01
45–60	2,459 ± 190	3.6	<0.01
70–100	2,116 ± 240	3.1	<0.01
120–150	2,051 ± 176	3.0	<0.01

[a] Guinea pigs were sensitized by subcutaneous injection with killed legionellae in complete Freund adjuvant on the day indicated before in vitro assay for blast cell transformation by *L. pneumophila* sonic extract.

[b] Average lymphocyte blastogenic reaction of 10^6 spleen cells from each indicated group of guinea pigs, stimulated in vitro with 10 µg of *L. pneumophila* sonic extract. SI, Stimulation index. NS, Not significant.

umns resulted in a cell population rich in adherent lymphocytes, mainly B cells, which responded well to the *Legionella* antigen in vitro. On the other hand, the nonadherent cell population, rich in T cells, failed to respond significantly to the *Legionella* antigen in vitro. These results pointed to B lymphocytes as the responding cells for the lymphocyte blastogenic test in vitro in studies of cells from normal mice. The cell type responsible for the increases seen in the previously infected mice remains to be defined.

Discussion and conclusions. Legionellosis is considered a devastating disease in susceptible individuals, especially those who are immunocompromised. However, it is important to note that because of the ubiquitous nature of these organisms and their presence in environmental niches associated with humans, most if not all individuals are either continuously or at least sporadically exposed to these bacteria. For example, the presence of legionellae in cooling water towers and air-conditioning systems, in freshwater ponds, especially during warm seasons, and even in circulating water supplies in homes and hospitals makes it likely that most individuals are often exposed to these bacteria. Nevertheless, except for development of serum antibody and positive in vitro lymphocyte blast cell transformations to *Legionella* antigens, there is little evidence that most individuals even develop a subclinical infection. Thus it may be presumed that under usual circumstances an individual may either inhale or ingest legionellae from environmental sources and will satisfactorily handle these organisms like many other similar nonpathogenic or saprophytic bacteria.

Because of the explosive outbreak of legionel-

losis among a relatively small number of conventioneers at the American Legion convention in Philadelphia, and the knowledge that nosocomial infections occur in hospitals or institutions where legionellae are present in the water supply or air-conditioning system, it seems likely that a failure of a host's defense mechanism to interact appropriately with the bacteria may contribute to the disease state. In this regard, studies concerning immune responsiveness, as well as mechanisms of nonspecific resistance, have relevance to the host-parasite relationship in legionellosis. It now seems quite evident that not only does serum antibody to these bacteria readily develop in experimental animals and humans, but also such antibody, at least in vitro, can interact with specific antigenic components of the microorganisms.

Since antibody is readily detectable in the serum of patients without evidence of legionellosis, there is some question as to whether such antibody is protective. The recent studies by Horwitz and Silverstein have shown that opsonization of legionellae by specific antibody leads to a more rapid phagocytosis of these bacteria by peripheral blood monocytes, but that such opsonized bacteria still have the capacity to replicate and kill the cell. Only monocytes from

TABLE 5. Comparative responses of guinea pigs to *Legionella* antigen as determined by in vivo and in vitro assays

Guinea pig treatment[a]	MIF reaction (% inhibition ± SD)[b]	Blastogenic reaction (SI)[c]	Skin test reaction (mm)[d]
None (controls)	<5	<1.0	<5.0
Legionellae injected			
7 days	7 ± 4	1.2 ± 0.3	<5.0
25–30 days	36 ± 18	1.8 ± 0.9	8.3 ± 1.2
40–60 days	52 ± 12	3.2 ± 1.1	10.4 ± 2.5
70–100 days	38 ± 16	3.1 ± 0.9	11.5 ± 2.9
Infected			
10–20 days	4 ± 2	1.0	5.0
30–60 days	39 ± 18	2.6 ± 0.7	9.6 ± 2.8

[a] Guinea pigs were treated as indicated with either 3 × 10^9 killed legionellae in complete Freund adjuvant injected subcutaneously, or 2 × 10^5 virulent bacteria injected intraperitoneally on the day indicated.

[b] Average percent migration inhibition ± SD for mouse peritoneal cells cultured in triplicate with supernatant from indicated guinea pig spleen cell cultures treated in vitro with *Legionella* sp. sonic extract, as compared to controls without antigen.

[c] Average stimulation index (SI) ± SD for three to six cultures of spleen cells from indicated guinea pigs after 6 days of incubation with *Legionella* sp. sonic extract, as compared to culture without antigen.

[d] Skin reaction (millimeters of induration) after skin testing with *Legionella* sp. sonic extract.

patients who have recovered from infection appear to have the capacity to inhibit the in vitro growth of *Legionella* spp. (15, 16; Horwitz, Clin. Res. **30**:369A, 1982).

Normal monocytes activated in vitro with concanavalin A-induced supernatant factors or treated with supernatants from *L. pneumophila*-stimulated mononuclear cells from legionellosis patients evince increased resistance to the bacteria (16). These and similar studies suggest that cell-mediated immunity may play an important role in the control of legionellosis. However, it should be pointed out that recent work has suggested that a toxin may also be related to the pathogenicity of *L. pneumophila* (8, 10, 11). Studies from the laboratories of B. H. Iglewski in Portland, Oreg., and K. W. Hedlund in Frederick, Md., suggest that a toxin present in these bacteria may have a role in promoting the pathogenicity of *L. pneumophila*, at least in immunocompromised individuals. The toxin identified by Iglewski's group in culture supernatants of *L. pneumophila* is a small-molecular-weight peptide and has the ability to interfere with the oxidative burst of phagocytic cells in vitro (8). Thus it is possible that once the *Legionella* organisms gain a foothold in macrophages, at least nonactivated ones, the toxin present in or on these bacteria interferes with the metabolic activity of the phagocyte which, under normal circumstances, should be capable of inhibiting the legionellae (19). Indeed, it has been shown in in vitro studies that legionellae are extremely susceptible to superoxide and lysosomal factors commonly present in macrophages. In this regard, it would be of interest to determine whether phagolysosome fusion is inhibited by legionellae in vitro or whether these organisms, during actual infection, prevent the normal functioning of phagocytes.

Studies in our laboratory, as well as in others, have shown that *Legionella* antigens are markedly immunostimulatory (6). The studies of Wong et al. indicated that specific antigenic determinants purified from *L. pneumophila* are capable of inducing enhanced serogroup-specific or common antigen-specific immune responses (31). Furthermore, killed legionellae, as well as their lipopolysaccharide-rich components, have been shown to be markedly immunostimulatory for both guinea pigs and mice (6). It is possible that enhanced antibody response to *L. pneumophila*, induced by surface immunostimulatory components, induces an inappropriate immune response in the host, such as a "blocking" antibody which may be non-complement fixing but still reactive with *L. pneumophila* surface antigens. Such enhanced non-bactericidal antibody could not only be protective for the legionellae per se, but also provide an increased opportunity for the bacteria to be phagocytized. In turn, such phagocytized bacteria could interfere with one of the most important defenses of the host, i.e., nonspecific killing by phagocytes.

There is little information available as to whether *Legionella* spp. components have an influence on specific or even nonspecific cell-mediated immune responses. However, it is evident from studies in a number of laboratories that delayed hypersensitivity reactions to these organisms, as assessed by cutaneous tests in sensitized guinea pigs, are readily induced. Furthermore, in vitro studies have shown that lymphocyte blastogenic and MIF responses to *Legionella* spp. may be induced, and these results correlate with in vivo skin reactions (Table 5). Whether such cellular immunity is protective or related to resistance to these bacteria is not yet known. It will be of value to determine whether legionellae per se or their components, including the toxin, interfere with cell-mediated immunity, as apparently occurs with the metabolic activity of monocytes.

It should be noted that if a toxin is important in the pathogenicity of *Legionella* spp., then antibody may play an important role in neutralizing the toxin and protecting individuals against disease. However, if intracellular parasitism is a more important component of the infectious process, then cell-mediated immunity would be expected to play a more important role in resistance. It is just as possible, however, that both types of immunity are important in the general resistance of individuals against legionellosis, and only when there is a systemic or possibly a local deficiency in one or both forms of immunity do these bacteria have the opportunity to establish a progressive infection.

It is important that to date most studies concerning cellular immunity to *Legionella* spp. have dealt with responses of peripheral blood leukocytes or, in experimental animal situations, with splenic or peritoneal cell populations. Since overt legionellosis is usually a pneumonia, it would be just as important to assess the role of alveolar macrophages and other alveolar lymphoid cells in resistance to or protection against legionellosis. In this regard there have been a number of recent studies concerning phagocytosis by alveolar macrophages. Continued studies of this type will be important to determine the role of alveolar macrophages versus macrophages from other sites or tissues in resistance to *Legionella* spp.

It is also important that experimental studies to date have mainly been performed with guinea pigs and in terms of humoral and cellular immunity to legionellae. Guinea pigs are considered highly susceptible to these organisms and indeed provide the main animal model for active infection with environmental or clinical specimens. However, it is evident that the marked susceptibility of these animals to legionellae contrasts sharply with the apparent relative resistance of humans to these organisms. If this is indeed the case, then experimental animal models with more innate resistance to Legionella spp. might be of value to assess the role of various factors in resistance. Winn et al. have recently shown that rats are highly resistant to pneumonia and other infections caused by L. pneumophila and also readily develop immune responses to these bacteria. Similarly, studies in our and other laboratories have shown that mice are also extremely resistant to infection by Legionella spp. Nevertheless, when mice are moderately immunosuppressed with drugs such as cyclophosphamide or cortisone, increasing numbers of animals become susceptible to infection. It is not yet clear whether a typical pneumonia may be induced in such animals. However, both immunosuppressed and normal mice vaccinated with L. pneumophila show increased resistance to challenge with this bacterium and develop both antibodies and sensitized T lymphocytes which directly interact with the bacteria and antigenic derivatives in vitro. Thus it seems likely that further analysis of immune responses in resistant species of animals, such as rats and mice, may provide further information concerning not only the role of humoral versus cellular immunity in resistance to legionellosis, but also the mechanisms involved.

LITERATURE CITED

1. **Arko, R. J., K. H. Wong, and J. Feeley.** 1979. Immunologic factors affecting the in-vivo and in-vitro survival of the Legionnaires' disease bacterium. Ann. Intern. Med. **90**:680–683.

2. **Cherry, W. B., and R. M. McKinney.** 1978. Detection in clinical specimens by direct immunofluorescence, p. 129–145. *In* G. L. Jones and G. A. Hebert (ed.), "Legionnaires": the disease, the bacterium and methodology. Centers for Disease Control, Atlanta, Ga.

3. **Collins, M. T., S.-N. Cho, N. Høiby, F. Espersen, L. Baek, and J. S. Reif.** 1983. Crossed immunoelectrophoretic analysis of *Legionella pneumophila* serogroup 1 antigens. Infect. Immun. **39**:1428–1440.

4. **Edson, D. C., H. E. Stiefel, B. B. Wentworth, and D. L. Wilson.** 1979. Prevalence of antibodies to Legionnaires' disease: a seroepidemiologic survey of Michigan residents using the hemagglutination test. Ann. Intern. Med. **90**:691–693.

5. **Friedman, A. P., and S. M. Katz.** 1981. The prevalence of serum antibodies to *Legionella pneumophila* in patients with chronic pulmonary disease. Am. Rev. Respir. Dis. **123**:238–239.

6. **Friedman, H., R. Widen, and T. Klein.** 1983. Cellular immunity to *Legionella* antigen—laboratory assays. Clin. Immun. Newsl. **4**:92–95.

7. **Friedman, H., R. Widen, I. Lee, and T. Klein.** 1983. Cellular immunity to *Legionella pneumophila* in guinea pigs assessed by direct and indirect migration inhibition reactions in vitro. Infect. Immun. **41**:1132–1137.

8. **Friedman, R. L., B. H. Iglewski, and R. D. Miller.** 1980.

Identification of a cytotoxin produced by *Legionella pneumophila*. Infect. Immun. **29**:271–274.

9. **Harrison, T. G., and A. G. Taylor.** 1982. A rapid microagglutination test for the diagnosis of *Legionella pneumophila* (serogroup 1) infection. J. Clin. Pathol. **35**:1028–1031.

10. **Hedlund, K. W.** 1981. Legionella toxin. Pharmacol. Ther. **15**:123–130.

11. **Hedlund, K. W., and R. Larson.** 1982. The identification of a *Legionella pneumophila* toxin with in vivo lethality, p. 279–284. *In* J. Robbins, J. Hill, and J. Sadoff (ed.), Bacterial vaccines. Thieme-Stratton Inc., New York.

12. **Holliday, M. G.** 1980. The diagnosis of Legionnaires' disease by counterimmunoelectrophoresis. J. Clin. Pathol. **33**:1174–1181.

13. **Horwitz, M. A., and S. C. Silverstein.** 1980. The Legionnaires' disease bacterium (*Legionella pneumophila*) multiplies intracellularly in human monocytes. J. Clin. Invest. **66**:441–450.

14. **Horwitz, M. A., and S. C. Silverstein.** 1981. Interaction of the Legionnaires' disease bacterium (*Legionella pneumophila*) with human phagocytes. I. *L. pneumophila* resists killing by polymorphonuclear leukocytes, antibody, and complement. J. Exp. Med. **153**:386–397.

15. **Horwitz, M. A., and S. C. Silverstein.** 1981. Interaction of the Legionnaires' disease bacterium (*Legionella pneumophila*) with human phagocytes. II. Antibody promotes binding of *L. pneumophila* to monocytes but does not inhibit intracellular multiplication. J. Exp. Med. **153**:398–406.

16. **Horwitz, M. A., and S. C. Silverstein.** 1981. Activated human monocytes inhibit the intracellular multiplication of Legionnaires' disease bacteria. J. Exp. Med. **154**:1618–1635.

17. **Klein, G. C., W. L. Jones, and J. C. Feeley.** 1979. Upper limit of normal titer for detection of antibodies to *Legionella pneumophila* by the microagglutination test. J. Clin. Microbiol. **10**:754–755.

18. **Kohler, R. B., S. E. Zimmerman, E. Wilson, S. D. Allen, P. H. Edelstein, L. J. Wheat, and A. White.** 1981. Rapid radioimmunoassay diagnosis of Legionnaires' disease. Ann. Intern. Med. **94**:601–605.

19. **Lockley, R. M., R. F. Jacobs, C. B. Wilson, W. M. Weaver, and S. J. Kebanoff.** 1982. Susceptibility of *Legionella pneumophila* to oxygen-dependent microbicidal systems. J. Immunol. **129**:2192–2197.

20. **Plouffe, J. F., and I. M. Baird.** 1981. Lymphocyte transformation to *Legionella pneumophila*. J. Clin. Lab. Immunol. **5**:149–152.

21. **Plouffe, J. F., and I. M. Baird.** 1982. Lymphocyte blastogenic responses to *L. pneumophila* in acute legionellosis. J. Clin. Lab. Immunol. **7**:43–44.

22. **Plouffe, J. F., and I. M. Baird.** 1982. Cord blood lymphocyte transformation responses to *Legionella pneumophila*. J. Clin. Lab. Immunol. **9**:119–120.

23. **Rolstadt, B., and P. B. Berdal.** 1981. Immune defenses against *Legionella pneumophila* in rats. Infect. Immun. **32**:805–812.

24. **Soriano, F., L. Aguilar, and J. L. Gomez-Garces.** 1982. Simple immunodiffusion test for detecting antibodies against *Legionella pneumophila* serotype 1. J. Clin. Microbiol. **15**:330–331.

25. **Tilton, R. C.** 1979. Legionnaires' disease antigen detected by enzyme-linked immunosorbent assay. Ann. Intern. Med. **90**:697–698.

26. **Widen, R., T. Klein, I. Lee, and H. Friedman.** 1983. Agar microdroplet assay for delayed hypersensitivity to *Legionella pneumophila* serogroup 1. J. Clin. Microbiol. **17**:819–823.

26a. **Widen, R., I. Lee, T. Klein, and H. Friedman.** 1983. Blastogenic responsiveness of spleen cells from guinea pigs sensitized to *Legionella pneumophila* antigens. Proc. Soc. Exp. Biol. Med. **173**:547–552.

27. **Wilkinson, H. W., D. D. Cruce, and C. V. Broome.** 1981. Validation of *Legionella pneumophila* in direct immuno-

fluorescence assay with epidemic sera. J. Clin. Microbiol. **13:**139–146.

28. **Winn, W. C., Jr.** 1983. Cellular inflammatory defenses against *Legionella pneumophila*. Clin. Immun. Newsl. **4:**83–87.

29. **Winn, W. C., Jr., G. S. Davis, D. W. Gump, J. E. Craighead, and H. N. Beaty.** 1982. Legionnaires' pneumonia after intratracheal inoculation of guinea pigs and rats. Lab. Invest. **47:**568–578.

30. **Wong, K. H., P. R. B. McMaster, J. C. Feeley, R. J. Arko, W. O. Schalla, and F. W. Chandler.** 1980. Detection of hypersensitivity to *Legionella pneumophila* in guinea pigs by skin test. Curr. Microbiol. **4:**105–110.

31. **Wong, K. H., W. O. Schalla, M. C. Wong, P. R. B. McMaster, J. C. Feeley, and R. J. Arko.** 1982. Biologic activities of antigens from *Legionella pneumophila*, p. 434–443. *In* J. Robbins, J. Hill, and J. Sadoff (ed.), Bacterial vaccines. Thieme-Stratton, Inc., New York.

Antigens of *Legionella* spp.

WILLIAM JOHNSON

Department of Microbiology, University of Iowa, Iowa City, Iowa 52242

The antigenic diversity of *Legionella pneumophila* became apparent when McKinney et al. (11) isolated a strain of *L. pneumophila* from a postmortem lung specimen of a patient with fatal pneumonia at the Veterans Administration Hospital in Togus, Maine. This isolate did not stain with the direct fluorescent-antibody reagent used to identify previous isolates of *L. pneumophila*. The Togus isolate thus became the second serogroup of *L. pneumophila*. Since then, additional serogroups have been identified, and *L. pneumophila* now comprises eight distinct serogroups which are identified by direct fluorescent-antibody reagents. Several additional species of *Legionella*—*L. bozemanii*, *L. dumoffii*, *L. gormanii*, *L. micdadei*, *L. longbeachae* (two serogroups), *L. jordanis*, and *L. oakridgensis*—have now been isolated. These organisms share many of the physiological characteristics of *L. pneumophila* but differ in their DNA homology and major cell surface antigens.

Serogroup-specific antigens of *L. pneumophila* have been isolated by a number of investigators (6, 8, 22). The high-molecular-weight serotypic antigen isolated by Wong et al. (22) contained all of the major branched-chain fatty acids characteristic of *L. pneumophila* and consisted of lipid-protein-carbohydrate complexes containing 7% protein and 2% carbohydrate. Flesher et al. (6) used EDTA to extract a serogroup-specific antigen from *L. pneumophila* serogroup 1. The antigen has a molecular weight of $\sim 4 \times 10^4$ and consists of <10% carbohydrate, 15% protein, 1.1% phosphate, and a lipid of unknown composition. Nuclear magnetic resonance studies showed that the antigen contains saturated hydrocarbons, and gas-liquid chromatography data suggest that the antigen also contains 2% fatty acids. Johnson et al. (9) have reported the isolation of serogroup-specific antigens from *L. pneumophila* serogroups 1, 2, 3, and 4. The high-molecular-weight antigen was composed of 35% carbohydrate, 2.6% protein, 1.8% phospholipid, and trace amounts of 2-keto-3-deoxyoctonate as measured by colorimetric analysis. Each of the serogroup-specific antigens from serogroups 1, 2, 3, and 4 reacted with antisera to the homologous antigen but not with antisera to heterologous antigen (Fig. 1). These studies have recently been extended to the isolation of the species-specific antigens of *L. bozemanii* and *L. gormanii*. The species-specific antigen from each of these organisms reacts with antisera to the homologous antigen but not with antisera to the heterologous antigen (Fig. 2). Although there is as yet no agreement on the exact chemical structure of the serogroup-specific and species-specific antigens, the results of Johnson et al. (9) suggest that the antigenic activity of the serogroup-specific antigens of *L. pneumophila* resides in the polysaccharide portion of the antigen. These results were confirmed by Flesher et al. (6), who showed that periodate oxidation destroyed the serologically active site of the serogroup 1-specific antigen.

The high-molecular-weight serogroup-specific antigens of *L. pneumophila* appear to be important in the host's antibody response to infection with *Legionella* spp. since preincubation of human convalescent-phase sera with the serogroup-specific antigen markedly reduces both the indirect fluorescent-antibody and microagglutination titers (8). This is not surprising in light of the fact that the serogroup-specific antigens are located on the cell surface (5) and are the major antigens detected by the direct fluorescent-antibody reagents. Immunoperoxidase-labeled antiserum to *L. pneumophila* serogroup 1 reacts with the cell surface of serogroup 1 cells (Fig. 3A), but immunoperoxidase-labeled antibody to *L. pneumophila* serogroup 2 does not react with the cell surface of serogroup 1 organisms (Fig. 3B).

Many features of the serogroup-specific antigen, including cell surface location, high molecular weight, and chemical composition, suggest similarities between the serogroup antigens of *L. pneumophila* and the lipopolysaccharide endotoxin which is classically associated with other gram-negative bacteria. Wong et al. (21) showed that *L. pneumophila* is positive in the *Limulus* assay for endotoxin, and they detected the presence of 2-keto-3-deoxyoctonate in *L. pneumophila* by colorimetric assay. However, no hydroxy fatty acids, which are commonly found in the lipid A portion of the classical endotoxin

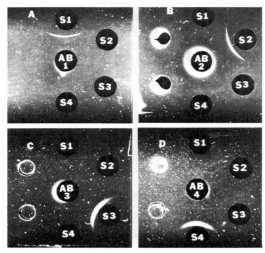

FIG. 1. Immunodiffusion reactions of serogroup-specific antigens with antisera to live cells of *L. pneumophila*. Antigens from serogroup 1 (S1), serogroup 2 (S2), serogroup 3 (S3), and serogroup 4 (S4) are in the outside wells. Rabbit antisera to live strains of (A) serogroup 1 (AB1), (B) serogroup 2 (AB2), (C) serogroup 3 (AB3), and (D) serogroup 4 (AB4) are in the center wells. (From reference 9.)

molecule, were detected in the *L. pneumophila* preparation. Also, the *L. pneumophila* preparations gave very weak pyrogenicity responses in rabbits, and polymyxin B, which reduces the toxicity of endotoxin by binding to lipid A, failed to reduce the toxicity of *L. pneumophila* preparations for mice. Although these results suggest that *L. pneumophila* may contain endotoxin activity, no attempt was made to determine whether this weak endotoxin-like activity was associated with the serogroup-specific antigen. Johnson et al. (8), using colorimetric analysis, detected the presence of 2-keto-3-deoxyoctonate in the serogroup-specific antigen isolated from serogroup 1 *L. pneumophila*. This antigen preparation also induced a localized Shwartzman reaction and a biphasic fever response in rabbits. However, the serogroup-specific antigen isolated by Flesher et al. (6) was devoid of both 2-keto-3-deoxyoctonate and heptose, both of which are associated with classical endotoxins isolated from gram-negative bacteria. The lipid portion of this antigen preparation contained low amounts of glucosamine and fatty acids, both of which are present in relatively high concentrations in lipid A. Schramek et al. (16) have reported the isolation of endotoxin-like material by phenol extraction of serogroup 1 *L. pneumophila*. Mouse toxicity studies confirmed the previous findings of Wong et al. (21) that the endotoxin isolated from *L. pneumophila* has a

relatively low toxicity for mice. Serological studies showed that the endotoxin of *L. pneumophila* contains a lipid which cross-reacts with antisera to lipid A isolated from other gram-negative organisms. Although these results suggest antigenic similarities between lipid A and the lipid isolated from *L. pneumophila*, no attempts were made either to ascertain the chemical structure of the lipid isolated from *L. pneumophila* or to determine whether the serotypic antigen was present on the endotoxin molecule. Collins et al. (2, 3), using crossed immunoelectrophoresis, have identified a serogroup-specific antigen with the characteristic electrophoretic mobility of endotoxin. The antigen was selectively precipitated with serogroup-specific antiserum and was highly reactive in the *Limulus* amoebocyte assay for endotoxin. No studies on the chemical structure or biological activity of this antigen have yet been reported.

It thus appears that *L. pneumophila* produces a compound which is similar but not identical to classical endotoxin in its chemical structure and biological activity. The confirmation of the presence of 2-keto-3-deoxyoctonate in *L. pneumophila* extracts by gas chromatography-mass spectroscopy will be a critical step in evaluating the similarities between these extracts and the classical lipopolysaccharide of other gram-negative bacteria. Since the complete chemical structure of the serogroup-specific antigens has not yet been elucidated, it is still uncertain whether the serotype-specific antigen is associated with the endotoxin activity of *L. pneumophila* or whether these are two distinct entities. Further studies on the chemical structure and biological activity of the serogroup-specific antigens are necessary.

The role of the serogroup-specific antigens of *L. pneumophila* in inducing immunity was first

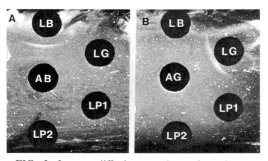

FIG. 2. Immunodiffusion reactions of species-specific antigens isolated from *L. bozemanii* and *L. gormanii*. Antigens from *L. bozemanii* (LB), *L. gormanii* (LG), *L. pneumophila* serogroup 1 (LP1), and *L. pneumophila* serogroup 2 (LP2) are in the outside wells. Rabbit antisera to (A) *L. bozemanii* (AB) and (B) *L. gormanii* (AG) are in the center wells.

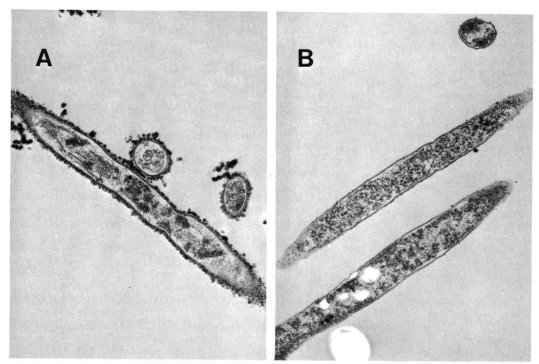

FIG. 3. Localization of the serogroup-specific antigen of *L. pneumophila* and demonstration of the serogroup specificity of antisera to the serogroup antigen by immunoperoxidase labeling. (A) Reaction of serogroup 1 antiserum with serogroup 1 cells; the immunoperoxidase label is associated with the cell surface. (B) Reaction of serogroup 2 antiserum with serogroup 1 cells; the serogroup 2 antiserum does not react with the serogroup 1 cells. (From reference 5.)

TABLE 1. Protection induced by F-1 fractions and heat-killed organisms of *L. pneumophila* serogroups 1, 2, 3, and 4 against challenge with serogroup 1 *L. pneumophila*[a]

Immunizing material	% Survival	Reciprocal microagglutination titer to:	
		Homologous antigen	Serogroup 1 F-1 antigen
F-1 antigen[b]			
Serogroup 1	100	512	512
Serogroup 2	25	256	<8
Serogroup 3	25	256	<8
Serogroup 4	0	256	<8
Control	0	0	ND[c]
Heat-killed cells[d]			
Serogroup 1	100	ND	ND
Serogroup 2	0	1,024	<8
Serogroup 3	0	ND	ND
Serogroup 4	0	ND	ND
Control	0	0	ND

[a] From reference 5.

[b] Guinea pigs were immunized with 1 mg of F-1 fraction in complete Freund adjuvant.

[c] ND, Not determined.

[d] Guinea pigs were immunized with approximately 10^8 heat-killed cells and challenged with 100 50% lethal doses.

investigated by Wong et al. (22), who showed that immunization with the serogroup-specific antigen isolated from serogroup 1 *L. pneumophila* protected guinea pigs against an intraperitoneal challenge with the homologous serotype. An immunoglobulin G fraction obtained from goats immunized with the serogroup 1 antigen was able to prolong the survival of guinea pigs challenged with the homologous serotype. Elliott et al. (5) extended these studies to show that guinea pigs immunized with the serogroup-specific antigen isolated from serogroup 1 *L. pneumophila* were protected against challenge with the homologous serotype (Table 1). However, guinea pigs immunized with the serogroup-specific antigen isolated from *L. pneumophila* serogroups 2, 3, and 4 were not protected against challenge with serogroup 1 *L. pneumophila*. Thus, the immunity developed after immunization with the serogroup-specific antigen protects only against challenge with the homologous serotype. The mechanism by which the serogroup-specific antigen induces immunity is not currently known. Although Arko et al. (1) first suggested that antibodies to serogroup 1 *L. pneumophila* enhanced complement-mediated bactericidal activity, Caparon and Johnson (M. G. Caparon and W. Johnson, Abstr. Annu. Meet. Am. Soc. Microbiol., 1983, E97, p. 92) have recently shown that immune human and guinea pig sera fail to demonstrate significant complement-mediated bactericidal activity against virulent and avirulent strains of *L. pneumophila*. Johnson et al. (9) were the first to show that antibody to the serogroup-specific antigen is required for phagocytosis of *L. pneumophila* (Fig. 4). These results have been confirmed by Horwitz and Silverstein (7) and Vildé et al. (17). The opsonization of *L. pneumophila* by macrophages does not appear to be complement dependent since fresh serum does not result in opsonization of the organisms in the absence of specific antiserum and does not enhance ingestion in the presence of specific antiserum. Further information on humoral and cell-mediated immunity in *L. pneumophila* infections is presented elsewhere in this volume.

Wilkinson et al. (18, 19) have shown that patients with legionellosis can respond with antibodies directed against antigens common to all serogroups of *L. pneumophila* and can respond to antigens common to both *L. pneumophila* and many other gram-negative bacteria. The presence of *L. pneumophila* antigens which cross-react with other gram-negative bacteria in direct fluorescent-antibody staining reactions was observed by Orrison et al. (12), who showed that the direct fluorescent-antibody reagent against serogroup 1 *L. pneumophila* cross-reacted with *Pseudomonas fluorescens, Pseudomonas alcaligenes*, and a member of the *Flavobacterium-Xanthomonas* group. The recent studies of Joly

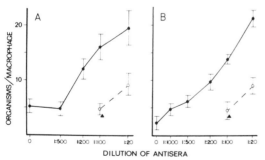

FIG. 4. Phagocytosis of ³H-labeled *L. pneumophila* serogroup 1 (A) and serogroup 2 (B) by macrophages in vitro in the presence of homologous and heterologous antisera. For both serogroups, antiserum to the homologous F-1 fraction (●) enhanced uptake of organisms, whereas antiserum to the heterologous F-1 fraction enhanced uptake of organisms only at low dilutions (○). Absorption of homologous antiserum with homologous F-1 fraction (▲) eliminated opsonic activity. Means ± standard error of results from four or more experiments. (From reference 9.)

and Kenny (10) and Collins et al. (2, 3), using cross immunoelectrophoresis, have confirmed the presence of antigens which are common to all six serogroups of *L. pneumophila*, in addition to antigens which are shared by *Legionella* spp. and other gram-negative bacteria and antigens which are *Legionellaceae* family specific. Each of the six serogroups of *L. pneumophila* was shown to contain a serogroup-specific antigen which formed an ill-defined broad precipitin line with a low peak height. At least 86% of the 82 antigens detected in *L. pneumophila* serogroup 1 were common to all six serogroups. One of these antigens may be the flagellar antigen which Elliott and Johnson (4) have shown to be present on *L. pneumophila* serogroups 1, 2, and 3. In the electrophoretic system of Collins et al., other species of *Legionella—L. bozemanii, L. dumoffii, L. gormanii*, and *L. micdadei*—cross-reacted with only 40 to 50% of the reference antigens of *L. pneumophila* serogroup 1 (3). One of the antigens identified in *L. pneumophila* cross-reacted with the common antigen of *Pseudomonas aeruginosa*, and another antigen was shown to cross-react with all gram-negative bacteria tested except *Bacteroides fragilis*. These antigens may be responsible for the cross-reactions observed by Orrison et al. (12) in direct fluorescent-antibody testing for *Legionella* spp., and the increase in the specificity of indirect fluorescent-antibody testing after absorption of sera with boiled extracts of *Escherichia coli*, as observed by Wilkinson et al. (18), may be due to absorption of antibodies directed against these cross-reaction antigens present in many gram-negative bacteria. Two antigens identified by Collins et al. (3) were shown to be common to all members of the family *Legionellaceae* and did not react with the other gram-positive and gram-negative organisms tested. The use of antigens which are specific for the family *Legionellaceae* would be of considerable value in antibody screening procedures to detect legionella infections if it were shown that there is a consistent antibody response to these antigens in individuals with legionellosis. Unfortunately, the antigens specific for the family *Legionellaceae* have not yet been isolated, so it is at present difficult to determine how useful these antigens may be for antibody screening tests.

Several of the antigens of *Legionella* spp. may also be of value as skin testing reagents. The serogroup-specific antigen isolated from *L. pneumophila* serogroup 1 has been shown to induce delayed-type skin reactions in sensitized guinea pigs (22). Unfortunately, the serogroup-specific antigen induced reactions in normal control animals, so no additional studies have been done to determine the sensitivity and specificity of the skin reactions to the serogroup-specific antigens isolated from other serogroups of *L. pneumophila* and other species of *Legionella*. A crude preparation of common antigens isolated from *L. pneumophila* serogroups 1, 2, 3, and 4 has been shown to elicit a combined immediate and delayed-type reaction in sensitized guinea pigs (20). The cross-reacting antigens elicited skin reactions in guinea pigs sensitized to any of the four serogroups of *L. pneumophila* tested, suggesting that the common antigens may be useful to detect hypersensitivity to all serogroups of *L. pneumophila*. Positive skin test reactions could be elicited up to 4 months after sensitization. Unfortunately, the specificity of these antigen preparations for detecting infection caused by *L. pneumophila* is not known. Also, there have been no studies to determine whether these antigen preparations will elicit delayed-type hypersensitivity reactions in humans infected with other *Legionella* spp. The identification of family- and species-specific antigens of *Legionella* spp. may now make it possible to use purified antigens to ascertain whether humans develop delayed hypersensitivity during legionella infection. The development of a skin test antigen specific for *Legionella* spp. infections may be important not only for the early diagnosis of legionellosis, but also to determine the prevalence of legionellosis in the population.

In a study on lymphocyte transformation to *L. pneumophila*, Plouffe and Baird (13) showed that peripheral blood lymphocytes from patients who had recovered from legionellosis underwent significant stimulation when incubated with a crude cell sonic extract of *L. pneumophila*. The blastogenic response was specific for *L. pneumophila* since sonic extracts from several other gram-negative bacteria failed to induce significant stimulation when incubated with the patients' lymphocytes. However, it was also found that a large percentage of the normal control patients have a response to *L. pneumophila* antigens which is considerably higher than background levels but significantly lower than that observed in patients with *L. pneumophila* infections. Although there is some nonspecific stimulation of cord blood lymphocytes by *L. pneumophila* antigens (15), it does not account for the high levels observed in normal control patients. It is likely that the high background levels in the control patients are due to the presence of antigens in the crude *L. pneumophila* sonic extract which are common to *L. pneumophila* and other gram-negative bacteria. Recent studies suggest that the blastogenic response to peripheral blood lymphocytes may occur earlier than measurable antibody production and that blastogenic assay may be useful in the early diagnosis of legionellosis (14).

Parts of these studies were supported by Public Health Service grants AI-15807 and AI-17159 from the National Institute of Allergy and Infectious Diseases.

LITERATURE CITED

1. Arko, R. J., K. H. Wong, and J. C. Feeley. 1979. Immunologic factors affecting the in-vitro and in-vivo survival of Legionnaires' disease bacterium. Ann. Intern. Med. 90:680–683.
2. Collins, M. T., S.-N. Cho, N. Høiby, F. Espersen, L. Baek, and J. S. Reif. 1983. Crossed immunoelectrophoretic analysis of *Legionella pneumophila* serogroup 1 antigens. Infect. Immun. 39:1428–1440.
3. Collins, M. T., F. Espersen, N. Høiby, S. Cho, A. Friis-Moller, and J. S. Reif. 1983. Cross-reactions between *Legionella pneumophila* (serogroup 1) and twenty-eight other bacterial species, including other members of the family *Legionellaceae*. Infect. Immun. 39:1441–1456.
4. Elliott, J. A., and W. Johnson. 1981. Immunological and biochemical relationships among flagella isolated from *Legionella pneumophila* serogroups 1, 2, and 3. Infect. Immun. 33:602–610.
5. Elliott, J. A., W. Johnson, and C. M. Helms. 1981. Ultrastructural localization and protective activity of a high-molecular-weight antigen isolated from *Legionella pneumophila*. Infect. Immun. 31:822–824.
6. Flesher, A. R., H. J. Jennings, C. Lugowski, and D. L. Kasper. 1982. Isolation of a serogroup 1-specific antigen from *Legionella pneumophila*. J. Infect. Dis. 145:224–233.
7. Horwitz, M. A., and S. C. Silverstein. 1980. Interaction of the Legionnaires' disease bacterium (*Legionella pneumophila*) with human phagocytes. II. Antibody promotes binding of *L. pneumophila* to monocytes but does not inhibit intracellular multiplication. J. Exp. Med. 153:398–402.
8. Johnson, W., J. A. Elliott, C. M. Helms, and E. D. Renner. 1979. A high molecular weight antigen in Legionnaire's disease bacterium: isolation and partial characterization. Ann. Intern. Med. 90:638–641.
9. Johnson, W., E. Pesanti, and J. Elliott. 1979. Serospecificity and opsonic activity of antisera to *Legionella pneumophila*. Infect. Immun. 26:698–704.
10. Joly, J. R., and G. E. Kenny. 1982. Antigenic analysis of *Legionella pneumophila* and *Tatlockia micdadei* (*Legionella micdadei*) by two-dimensional (crossed) immunoelectrophoresis. Infect. Immun. 35:721–729.
11. McKinney, R. M., B. M. Thomason, P. A. Harris, L. Thacker, K. R. Lewallen, H. W. Wilkinson, G. A. Hebert, and C. W. Moss. 1979. Recognition of a new serogroup of Legionnaires' disease bacterium. J. Clin. Microbiol. 9:103–107.
12. Orrison, L. H., W. F. Bibb, W. B. Cherry, and L. Thacker. 1983. Determination of antigenic relationships among legionellae and non-legionellae by direct fluorescent-antibody and immunodiffusion tests. J. Clin. Microbiol. 17:332–337.
13. Plouffe, J. P., and I. M. Baird. 1981. Lymphocyte transformation to *Legionella pneumophila*. J. Clin. Lab. Immunol. 5:149–152.
14. Plouffe, J. P., and I. M. Baird. 1982. Lymphocyte blastogenic responses to *L. pneumophila* in acute legionellosis. J. Clin. Lab. Immunol. 7:43–44.
15. Plouffe, J. F., and I. M. Baird. 1982. Cord blood lymphocyte transformation responses to *Legionella pneumophila*. J. Clin. Lab. Immunol. 9:119–120.
16. Schramek, S., J. Kazar, and S. Bazouska. 1982. Lipid A in *Legionella pneumophila*. Zentralbl. Bakteriol. Parasitenkd. Infektionskr. Hyg. Abt. 1 Orig. 252:401–404.
17. Vildé, T. L., E. Dournon, and T. Tram. 1981. Ingestion de *Legionella pneumophila* par le macrophage humains dérivés des monocytes: role des anticorps opsonisants et du complément. Ann. Immunol. 133:207–211.
18. Wilkinson, H. W., C. E. Farshy, B. J. Fikes, D. D. Cruce, and L. P. Yealy. 1979. Measure of immunoglobulin G-, M-,

and A-specific titers against *Legionella pneumophila* and inhibition of titers against nonspecific, gram-negative bacterial proteins in the direct immunofluorescent test for legionellosis. J. Clin. Microbiol. **10**:685–689.

19. **Wilkinson, H. W., B. J. Fikes, and D. D. Cruce.** 1979. Indirect immunofluorescence test for serodiagnosis of Legionnaires disease: evidence for serogroup diversity of Legionnaires disease bacterial antigens and for multiple specificity of human antibodies. J. Clin. Microbiol. **9**:379–383.

20. **Wong, K. H., P. R. B. McMaster, J. C. Feeley, R. J. Arko, W. O. Schalla, and F. W. Chandler.** 1980. Detection of hypersensitivity to *Legionella pneumophila* in guinea pigs by skin test. Curr. Microbiol. **4**:105–110.

21. **Wong, K. H., C. W. Moss, D. H. Hochstein, R. J. Arko, and W. O. Schalla.** 1979. "Endotoxicity" of the Legionnaires' disease bacterium. Ann. Intern. Med. **90**:624–627.

22. **Wong, K. H., W. O. Schalla, R. J. Arko, J. C. Bullard, and J. C. Feeley.** 1979. Immunochemical, serologic, and immunologic properties of major antigens isolated from the Legionnaires' disease bacterium. Ann. Intern. Med. **90**:634–638.

Interactions Between *Legionella pneumophila* and Human Mononuclear Phagocytes

MARCUS A. HORWITZ

Laboratory of Cellular Physiology and Immunology, The Rockefeller University, New York, New York 10021

Legionella pneumophila multiplies intracellularly in human mononuclear phagocytes (4; T. W. Nash, D. M. Libby, and M. A. Horwitz, Program Abstr. Intersci. Conf. Antimicrob. Agents Chemother. 22nd, Miami Beach, Fla., abstr. no. 93, 1982). This capability is central to the pathogenesis of Legionnaires disease and confronts the human immune defense system with a special challenge. This article summarizes in vitro studies on the interactions between *L. pneumophila* and human mononuclear and polymorphonuclear phagocytes. The article discusses the intracellular replication of *L. pneumophila* in human mononuclear phagocytes, the influence of antibody and complement on *L. pneumophila*-phagocyte interactions, the influence of phagocyte activation on *L. pneumophila*-mononuclear phagocyte interactions, the influence of antibiotics on intracellular *L. pneumophila*, phagocytosis of *L. pneumophila*, the formation of a novel phagosome in mononuclear phagocytes by *L. pneumophila*, and the capacity of *L. pneumophila* to inhibit phagosome-lysosome fusion.

Intracellular multiplication of *L. pneumophila.* *L. pneumophila* multiplies intracellularly in human peripheral blood monocytes (4) (Fig. 1) and in human alveolar macrophages obtained by bronchoalveolar lavage (Nash et al., 22nd ICAAC, abstr. no. 93). *L. pneumophila* does not multiply in the presence of human polymorphonuclear leukocytes, human lymphocytes, or tissue culture medium whether or not the medium is conditioned by human monocytes (4). Inside monocytes, the bacterium multiplies with a doubling time of about 2 h at midlogarithmic phase (4).

L. pneumophila is thus an intracellular parasite. Since the bacterium multiplies extracellularly on complex medium, it is a facultative intracellular parasite. *L. pneumophila* therefore belongs to a special group of bacterial pathogens that can evade host defenses by parasitizing mononuclear phagocytes. This group includes *Mycobacterium tuberculosis*, *Mycobacterium leprae*, *Listeria monocytogenes*, *Francisella tularensis*, *Brucella* spp., and *Salmonella* spp.

Role of humoral immunity. Patients with Legionnaires disease respond to the infection by producing antibody against *L. pneumophila*. In vitro studies have examined three potential functions of such antibody in promoting host defense. The studies have examined (i) whether antibody promotes killing of *L. pneumophila* by complement, (ii) whether antibody promotes phagocytosis and killing of *L. pneumophila* by human phagocytes, and (iii) whether antibody inhibits the multiplication of *L. pneumophila* in monocytes.

Regarding the first potential function of antibody in host defense, virulent egg yolk-grown *L. pneumophila*, Philadelphia 1 strain, is completely resistant to the bactericidal effects of human serum even in the presence of high-titer anti-*L. pneumophila* human or rabbit antibody (5).

Regarding the second potential function of antibody in host defense, antibody in the presence of complement markedly promotes phagocytosis of *L. pneumophila* by human polymorphonuclear leukocytes and human monocytes (5, 6). In the presence of antibody and complement, virtually all the bacteria in an inoculum are ingested and enclosed within membrane-bound vacuoles (5). In contrast, in the presence of antibody alone or complement alone, phagocytosis is very inefficient (5, 6).

Whereas phagocytosis of *L. pneumophila* coated with antibody and complement is very efficient, killing of these bacteria by human phagocytes is not. Human polymorphonuclear leukocytes and monocytes kill only about 0.5 log of an inoculum. In contrast, at the same time and under the same conditions, polymorphonuclear leukocytes very effectively kill an encapsulated serum-resistant strain of *Escherichia coli*, reducing the number of viable bacteria by 2.5 log (5) (Fig. 2). When *E. coli* and *L. pneumophila* are mixed together in the same test tube, polymorphonuclear leukocytes kill the *E. coli* as effectively as when the *E. coli* are present alone, indicating that the killing capacity of polymorphonuclear leukocytes is not reduced in cultures containing *L. pneumophila*.

Thus anti-*L. pneumophila* antibody in the

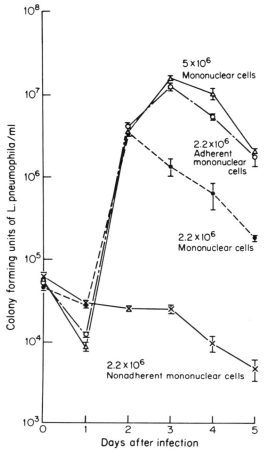

FIG. 1. *L. pneumophila* multiplies in the presence of monocytes (adherent cells) but not lymphocytes (nonadherent cells). *L. pneumophila* (10^5 CFU) was added to petri dishes containing 2.2 ml of RPMI, 15% serum, and either 2.2×10^6 or 5×10^6 mononuclear cells, 2.2×10^6 adherent mononuclear cells selected from 5×10^6 mononuclear cells, or 2.2×10^6 nonadherent mononuclear cells selected from the same mononuclear cell population. The cultures were incubated at 37°C in 5% CO_2–95% air on a gyratory shaker (100 rpm) for 3 h and under stationary conditions thereafter. CFU of *L. pneumophila* in each petri dish were determined daily. Each point represents the average for three replicate petri dishes ± standard error. (Reproduced from reference 4.)

presence of complement markedly promotes phagocytosis of *L. pneumophila* but promotes only a modest degree of killing by human phagocytes.

Regarding the third potential function of antibody in host defense, antibody does not significantly influence the rate of intracellular multiplication of *L. pneumophila* in human monocytes (Fig. 3). In cultures containing antibody and complement, a small proportion (0.25 to 0.5 log) of the bacterial inoculum is killed initially. *L. pneumophila* surviving the initial encounter with monocytes multiply at as rapid a rate as bacteria that enter monocytes in the absence of antibody (6).

Thus, in vitro studies have examined three major ways by which antibody might function to promote host defense. These studies have revealed (i) that specific antibody fails to promote killing of *L. pneumophila* by complement, (ii) that antibody and complement together fail to promote effective killing of *L. pneumophila* by polymorphonuclear leukocytes and monocytes, and (iii) that antibody and complement fail to inhibit the rate of multiplication of *L. pneumophila* in monocytes. These findings suggest that humoral immunity may not be an effective host defense against *L. pneumophila* and that a vaccine that results only in antibody production against *L. pneumophila* may not be efficacious (5, 6).

Role of cell-mediated immunity. Cell-mediated immunity plays an important role in host defense against many intracellular parasites (3). In vitro studies have examined the potential role of cell-mediated immunity in Legionnaires disease by investigating (i) whether activated human mononuclear phagocytes have the capacity to inhibit *L. pneumophila* multiplication and (ii) whether patients with Legionnaires disease develop cell-mediated immunity to *L. pneumophila*.

Regarding the microbistatic capacity of activated mononuclear phagocytes, human monocytes activated by incubation with human lymphocytes in the presence of the mitogen concanavalin A strongly inhibit the intracellular multiplication of *L. pneumophila* (7). Under these conditions, both lymphocytes and concanavalin A are required to activate the monocytes (7).

Human monocytes and human alveolar macrophages activated by incubation with cell-free filtered supernatants of concanavalin A-stimulated mononuclear cell cultures (cytokines) inhibit *L. pneumophila* multiplication (7; Nash et al., 22nd ICAAC, abstr. no. 93). The degree of inhibition is proportional to the concentration of supernatant added (7; Nash et al., 22nd ICAAC, abstr. no. 93) (Fig. 4). With monocytes, the degree of inhibition is proportional to the length of time the monocytes are preincubated with supernatant (48 > 24 > 12 h), and monocytes treated with supernatant daily after infection are more inhibitory than monocytes treated before infection only.

Monocytes activated with antigen-induced mononuclear cell cytokines also inhibit *L. pneumophila* multiplication (see below).

Activated monocytes inhibit *L. pneumophila*

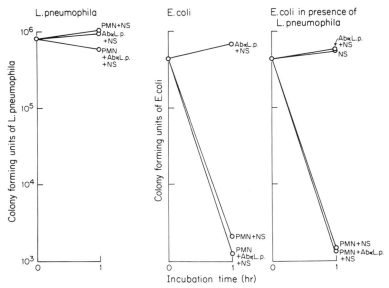

FIG. 2. *L. pneumophila* resists killing by polymorphonuclear leukocytes (PMN) under conditions in which polymorphonuclear leukocytes effectively kill a serum-resistant encapsulated strain of *E. coli*. Polymorphonuclear leukocytes (2.5×10^6) and 5×10^5 CFU each of *L. pneumophila*, *E. coli*, or both were incubated at 37°C for 1 h on a gyratory shaker in 0.9 ml of medium (final volume) containing 10% fresh normal human serum (NS) alone or 10% normal serum plus 8.5 agglutinating units of dialyzed rabbit anti-*L. pneumophila* antiserum (AbαL.p.) per ml. CFU of *E. coli* and *L. pneumophila* were determined initially and at the end of the incubation. (Reproduced from reference 5.)

multiplication in two ways. First, they phagocytize approximately 50% fewer bacteria than nonactivated monocytes, thereby restricting access of the bacteria to the intracellular milieu they require for multiplication (7). Second, they markedly slow the multiplication rate of those bacteria that are internalized; the doubling time is prolonged by greater than threefold (7).

Activated monocytes strongly inhibit *L. pneumophila* multiplication, but they do not kill *L. pneumophila* any better than nonactivated monocytes (7). As with nonactivated monocytes, activated monocytes kill *L. pneumophila* only in the presence of both specific antibody and complement, and even then they kill only a small proportion (<0.25 log) of an inoculum. In fact, activated monocytes consistently kill fewer bacteria than nonactivated monocytes, perhaps because they phagocytize less avidly (7).

Thus, inhibition of *L. pneumophila* multiplication is accomplished by activating the mononuclear phagocytes and not by coating the bacteria with specific antibody and complement. This indicates that cell-mediated immunity could play a major role in host defense against *L. pneumophila*.

Do patients with Legionnaires disease in fact develop cell-mediated immunity to *L. pneumophila*? In vitro studies indicate that they do. Mononuclear cells from patients recovered from Legionnaires disease respond to *L. pneumophila* antigens with both proliferation (lymphoproliferation), as measured by their capacity to incorporate [³H]thymidine, and the production of monocyte-activating cytokines (2). Such cytokines have the capacity to activate normal monocytes such that they inhibit *L. pneumophila* intracellular multiplication, and the degree of inhibition is dose dependent (2). The response of patient mononuclear cells to *L. pneumophila* antigens as measured by both assays (lymphoproliferation and production of monocyte-activating cytokines) is specific in the sense that mononuclear cells from patients recovered from Legionnaires disease respond more strongly to *L. pneumophila* antigens than to *M. leprae* antigens, whereas mononuclear cells from patients with tuberculoid leprosy respond more strongly to *M. leprae* antigens than to *L. pneumophila* antigens (2). Furthermore, mononuclear cells from patients recovered from Legionnaires disease respond much more strongly to *L. pneumophila* antigens in both assays than do mononuclear cells of age- and sex-matched persons with no history or serological evidence of Legionnaires disease (2).

Thus, cell-mediated immunity develops in patients with Legionnaires disease. The capability of activated mononuclear phagocytes to inhibit *L. pneumophila* intracellular multiplication indi-

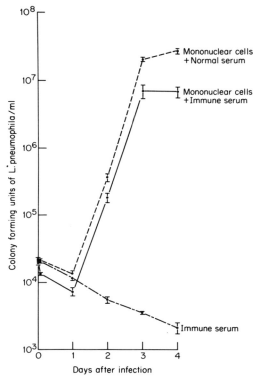

FIG. 3. Specific antibody fails to inhibit the multiplication of *L. pneumophila* in monocytes. *L. pneumophila* (2.5×10^4 CFU) was incubated at 37°C for 10 min in 33% fresh normal human serum or 33% fresh immune human serum. The bacteria were then incubated at 37°C in 5% CO_2–95% air with 5×10^6 mononuclear cells (or RPMI 1640 as control) in medium that contained a final concentration of 10% of the same type of serum to which the bacteria were initially exposed. The cultures were shaken for 1 h and incubated under stationary conditions thereafter for 4 days. CFU were determined at 0, 1, 24, 48, 72, and 96 h after the monocytes were infected. Each point represents the average for five replicate tubes ± standard error. (Reproduced from reference 6.)

cates that cell-mediated immunity plays a major role in host defense in Legionnaires disease.

Influence of antibiotics on intracellular *L. pneumophila*. Erythromycin and rifampin are the drugs of choice for the treatment of Legionnaires disease. Clinical experience and in vivo studies indicate that these drugs are efficacious in the treatment of Legionnaires disease. In vitro studies of *L. pneumophila* multiplying extracellularly on artificial medium show that such bacteria are highly susceptible to these antibiotics (8–10); erythromycin and rifampin inhibit the extracellular multiplication of *L. pneumophila* and kill these bacteria at relatively low concentrations of antibiotic (8–10). However, since *L.*

pneumophila is an intracellular pathogen, the influence of antibiotics on *L. pneumophila* multiplying intracellularly is of interest.

Erythromycin and rifampin inhibit the intracellular multiplication of *L. pneumophila* in human monocytes, at concentrations comparable to those which inhibit extracellularly multiplying bacteria (8). Multiplication of *L. pneumophila* in the logarithmic phase of growth in monocytes is inhibited within 1 h of the addition of these antibiotics (8).

Whereas intracellular *L. pneumophila* are inhibited from multiplying by concentrations of

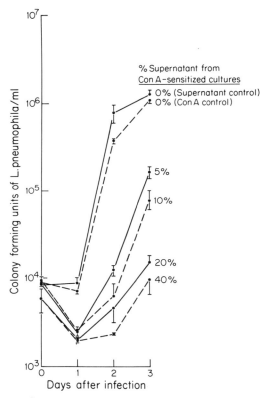

FIG. 4. Monocytes incubated with the supernatant of concanavalin A (ConA)-sensitized mononuclear cell cultures inhibit *L. pneumophila* multiplication. Monocytes in monolayer culture were incubated at 37°C in 5% CO_2–95% air in 2 ml of RPMI medium containing 15% fresh normal human serum and 0 to 40% cell-free supernatant from concanavalin A-sensitized mononuclear cell cultures. Control monolayers were incubated with supernatant from a mononuclear cell culture from which concanavalin A was omitted (Supernatant control) or with 15 μg of concanavalin A per ml (ConA control). After 24 h, *L. pneumophila* (2×10^4 CFU) was added to the cultures, and CFU in each culture were determined daily. Each point represents the average for three replicate petri dishes ± standard error. (Reproduced from reference 7.)

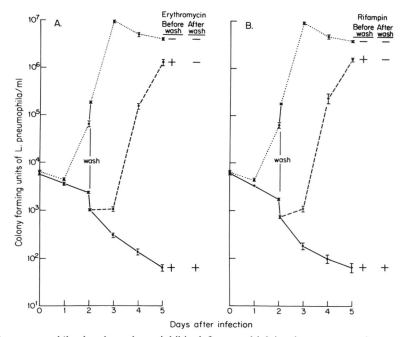

FIG. 5. *L. pneumophila* that have been inhibited from multiplying in monocytes by erythromycin and rifampin multiply after the antibiotics are removed. Mononuclear cells (5×10^6) in 2 ml of RPMI 1640 medium containing 15% fresh human serum were added to plastic tubes, infected with *L. pneumophila*, and incubated for 48 h without antibiotic or with 1.25 μg of erythromycin per ml (A) or 0.01 μg of rifampin per ml (B). After 48 h, the cultures were washed to remove antibiotics. Cultures that initially contained antibiotics were split into two groups; one group was incubated without antibiotic, and the other group was incubated with the same antibiotic as before washing. CFU of *L. pneumophila* in the medium were determined daily. Each point represents the mean for three replicate cultures ± standard error.

erythromycin and rifampin comparable to those that inhibit extracellular *L. pneumophila*, these bacteria exhibit markedly different susceptibilities to killing by these antibiotics. *L. pneumophila* organisms multiplying extracellularly are killed by concentrations of erythromycin and rifampin that are near the minimal inhibitory concentration; the minimal bactericidal concentration is 1 μg/ml for erythromycin and 0.009 μg/ml for rifampin (8). In contrast, *L. pneumophila* organisms multiplying intracellularly are resistant to killing by inhibitory concentrations of these antibiotics or by much higher concentrations equal to or greater than peak serum levels in humans, levels 12 times the minimal bactericidal concentration for erythromycin and 10^4 times the minimal bactericidal concentration for rifampin (8). By electron microscopy, intracellular bacteria that are inhibited from multiplying by antibiotics appear intact within membrane-bound vacuoles in the monocytes (8).

The inhibition of *L. pneumophila* multiplication in monocytes by erythromycin and rifampin is reversible (Fig. 5). When these antibiotics are removed from infected cultures after 2 days, *L.*

pneumophila resumes multiplying (8).

These findings indicate that patients with Legionnaires disease under treatment with erythromycin and rifampin require host defenses to eliminate *L. pneumophila*. Viewed in this way, the role of antibiotics in Legionnaires disease may be to buy time for the patient, i.e., inhibit *L. pneumophila* multiplication long enough for the patient to develop an effective immune defense against *L. pneumophila*. These findings also suggest that patients with inadequate host defenses may suffer relapse after cessation of therapy.

Phagocytosis and formation of a novel phagosome. *L. pneumophila* enters phagocytes by a highly unusual process. Long phagocyte pseudopods coil around the bacterium as the bacterium is internalized (unpublished data). After phagocytosis, *L. pneumophila* forms a novel phagosome in mononuclear phagocytes that involves an unusual sequence of cytoplasmic events that take place over 4 to 8 h (3a; M. A. Horwitz, 22nd ICAAC, abstr. no. 94).

Immediately after entry, a vacuolar membrane surrounds the bacterium; cellular organelles do not appear about the vacuolar membrane. By 15 min after entry, however, the majority of vacuoles are surrounded by smooth vesicles apparently fusing with or budding off from the vacuolar membrane. After 1 h, the majority of vacuoles are surrounded by mitochondria closely apposed to the vacuolar membrane. By 4 h, fewer smooth vesicles and mitochondria surround the vacuole, and now ribosomes and rough vesicles surround the majority of vacuoles. By 8 h, all vacuoles are studded with ribosomes that are separated from the vacuolar membrane by a gap of approximately 100 Å (10 nm) (Fig. 6A). The bacteria multiply in these ribosome-studded vacuoles, with a doubling time of about 2 h, until hundreds of organisms fill the vacuole (Fig. 6B); the monocyte becomes packed full with bacteria and ruptures.

Erythromycin, at concentrations that completely inhibit the intracellular multiplication of L. pneumophila, has no effect on vacuole formation (3a; Horwitz, 22nd ICAAC, abstr. no. 94).

In alveolar macrophages, as in monocytes, L. pneumophila multiplies in vitro in vacuoles studded with ribosomes (T. W. Nash, D. M. Libby, and M. A. Horwitz, unpublished data). This feature of the L. pneumophila vacuole is also seen in infected alveolar macrophages in lung biopsy specimens obtained from patients with Legionnaires disease (1).

Formalin-killed L. pneumophila organisms also reside in membrane-bound vacuoles after entry into monocytes. In contrast to the situation with live bacteria, vacuoles containing Formalin-killed L. pneumophila are not surrounded by cytoplasmic organelles at any time after entry (3a; Horwitz, 22nd ICAAC, abstr. no. 94). Formalin-killed L. pneumophila organisms are rapidly digested, and by 4 h few remain intact (3a; Horwitz, 22nd ICAAC, abstr. no. 94).

The L. pneumophila phagosome shares some features with certain other intracellular pathogens, notably Toxoplasma gondii and Chlamydia spp. (3a). These pathogens also share with L. pneumophila the capacity to inhibit phagosome-lysosome fusion (see below). This suggests that these organisms may share a common mechanism for phagosome formation and inhibition of phagosome-lysosome fusion.

Inhibition of phagosome-lysosome fusion. Some intracellular parasites, such as T. gondii, M. tuberculosis, and Chlamydia psittaci, inhibit fusion of phagosomes with lysosomes. Others,

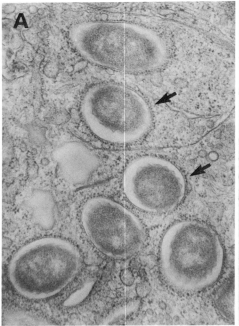 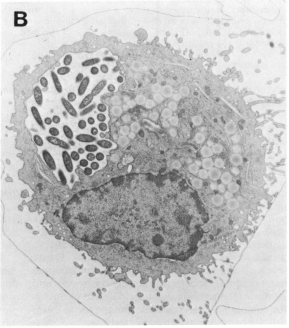

FIG. 6. Electron micrographs of human monocytes infected with L. pneumophila. (A) Six membrane-bound vacuoles, each containing a single L. pneumophila bacterium, are located in this portion of the monocyte cytoplasm. The vacuoles are studded with monocyte ribosomes (arrows) (×27,000). (B) Later in the course of infection, this monocyte contains a single large membrane-bound vacuole enclosing many L. pneumophila cells. Several bacteria appear to be in the process of dividing by binary fission. Although not apparent at this magnification, this vacuole is seen studded with ribosomes at higher magnification (×4,500).

such as *Leishmania* spp., do not inhibit fusion and are apparently able to survive and multiply within the normally inhospitable milieu of the phagolysosome. The interactions between the *L. pneumophila* phagosome and monocyte lysosomes have been investigated in vitro by prelabeling the lysosomes with thorium dioxide, an electron-opaque colloidal marker, and by acid phosphatase cytochemistry (Horwitz, 22nd ICAAC, abstr. no. 94; M. A. Horwitz, J. Exp. Med., in press).

Studies with thorium dioxide have revealed that phagosomes containing live *L. pneumophila* do not fuse with secondary lysosomes at 1 h after entry into monocytes, or at 4 or 8 h after entry, by which time the ribosome-studded vacuole has formed (Horwitz, 22nd ICAAC, abstr. no. 94; Horwitz, in press). In contrast, the majority of vacuoles containing Formalin-killed *L. pneumophila* organisms do fuse with secondary lysosomes. In the same experiments, vacuoles containing live *E. coli* and live *Streptococcus pneumoniae* also fuse with secondary lysosomes (Horwitz, in press).

Erythromycin, a potent inhibitor of bacterial protein synthesis, has no influence on fusion at concentrations which completely inhibit intracellular multiplication of *L. pneumophila*. However, coating the bacteria with antibody and complement or activating the monocytes promotes a modest degree of fusion of live *L. pneumophila* (Horwitz, in press).

Acid phosphatase cytochemistry has revealed that live *L. pneumophila* bacteria also do not fuse with primary lysosomes. In contrast to phagosomes containing live bacteria, the majority of phagosomes containing Formalin-killed *L. pneumophila* do fuse with lysosomes by acid phosphatase cytochemistry (Horwitz, in press).

Thus *L. pneumophila* inhibits phagosome-lysosome fusion. This capacity may be important to its survival in mononuclear phagocytes.

Conclusions. Humans are probably an incidental host for *L. pneumophila*, which likely acquired its capacity to multiply intracellularly in simpler beings such as amoebae or other protozoa. Like *L. pneumophila*, these unicellular organisms are ubiquitous in aquatic environments. *L. pneumophila* may utilize similar mechanisms to survive intracellularly in such disparate cells as amoebae and mononuclear phagocytes. These mechanisms remain to be elucidated, but the capacity of *L. pneumophila* to inhibit phagosome-lysosome fusion may be important.

Cell-mediated immunity appears to play a major role in host defense against *L. pneumophila*. Patients with Legionnaires disease develop cell-mediated immunity to *L. pneumophila*: their mononuclear cells respond to *L. pneumophila*

antigens with proliferation and with the generation of monocyte-activating cytokines. Mononuclear phagocytes activated by such cytokines inhibit the intracellular multiplication of *L. pneumophila*. Humoral immunity appears to play a modest role in host defense against *L. pneumophila*. Antibody does not alter the resistance of virulent *L. pneumophila* to complement lysis, nor does it influence the rate of intracellular multiplication of *L. pneumophila* in monocytes. Antibody and complement markedly promote phagocytosis of *L. pneumophila* by monocytes and polymorphonuclear leukocytes, but these phagocytes are able to kill only a limited proportion of an inoculum of antibody- and complement-coated bacteria.

Erythromycin and rifampin inhibit the intracellular multiplication of *L. pneumophila*, but these antibiotics do not kill intracellular bacteria even at very high concentrations. The inhibition of intracellular multiplication is reversible. When the antibiotics are removed, the bacteria resume multiplying in monocytes. This suggests that the role of antibiotics in Legionnaires disease may be to suppress multiplication of *L. pneumophila* long enough for the host to develop an effective immune defense against the bacterium.

L. pneumophila is phagocytized by an unusual process and then forms a novel phagosome in mononuclear phagocytes that is studded with ribosomes. Formation of this phagosome entails a remarkable sequence of cytoplasmic events that involves smooth vesicles, mitochondria, and ribosomes. How *L. pneumophila* orchestrates this sequence of cytoplasmic events and the role of the ribosome-studded vacuole in intracellular survival remain to be determined. Phagosomes of certain other intracellular pathogens have features in common with the *L. pneumophila* phagosome, and these pathogens also share with *L. pneumophila* the capacity to inhibit phagosome-lysosome fusion. This suggests that a common mechanism may underlie the capacity of these organisms to form a specialized phagosome and their capacity to inhibit phagosome-lysosome fusion.

I am supported by the John A. Hartford Foundation, The American Cancer Society, and National Institutes of Health grants AI 17254 and CA 30198.

LITERATURE CITED

1. **Glavin, F. L., W. C. Winn, Jr., and J. E. Craighead.** 1979. Ultrastructure of lung in Legionnaires' disease: observations of three biopsies done during the Vermont epidemic. Ann. Intern. Med. **90**:555–559.
2. **Horwitz, M. A.** 1982. Cell-mediated immunity in Legionnaires' disease. J. Clin. Invest. **71**:1686–1697.
3. **Horwitz, M. A.** 1982. Phagocytosis of microorganisms. Rev. Infect. Dis. **4**:104–123.
3a. **Horwitz, M. A.** 1983. Formation of a novel phagosome by

the Legionnaires' disease bacterium (*Legionella pneumophila*) in human monocytes. J. Exp. Med. **158**:1319–1331.

4. **Horwitz, M. A., and S. C. Silverstein.** 1980. The Legionnaires' disease bacterium (*Legionella pneumophila*) multiplies intracellularly in human monocytes. J. Clin. Invest. **66**:441–450.

5. **Horwitz, M. A., and S. C. Silverstein.** 1980. Interaction of the Legionnaires' disease bacterium (*Legionella pneumophila*) with human phagocytes. I. *L. pneumophila* resists killing by polymorphonuclear leukocytes, antibody, and complement. J. Exp. Med. **153**:386–397.

6. **Horwitz, M. A., and S. C. Silverstein.** 1980. Interaction of the Legionnaires' disease bacterium (*Legionella pneumophila*) with human phagocytes. II. Antibody promotes binding of *L. pneumophila* to monocytes but does not inhibit intracellular multiplication. J. Exp. Med. **153**:398–406.

7. **Horwitz, M. A., and S. C. Silverstein.** 1981. Activated human monocytes inhibit the intracellular multiplication of Legionnaires' disease bacteria. J. Exp. Med. **154**:1618–1635.

8. **Horwitz, M. A., and S. C. Silverstein.** 1983. The intracellular multiplication of Legionnaires' disease bacteria (*Legionella pneumophila*) in human monocytes is reversibly inhibited by erythromycin and rifampin. J. Clin. Invest. **71**:15–26.

9. **Thornsberry, C., C. N. Baker, and L. A. Kirven.** 1978. In vitro activity of antimicrobial agents on Legionnaires disease bacterium. Antimicrob. Agents Chemother. **13**:78–80.

10. **Thornsberry, C., and L. A. Kirven.** 1978. β-Lactamase of the Legionnaires' bacterium. Curr. Microbiol. **1**:51–54.

Growth of *Legionella pneumophila* Within Guinea Pig and Rat Macrophages

JOHN ELLIOTT AND WASHINGTON WINN, JR.

Department of Pathology, University of Vermont College of Medicine, Burlington, Vermont 05405

Legionella pneumophila is a gram-negative, facultative intracellular pathogen. Previous in vitro investigations of the interaction between *L. pneumophila* and host cells have revealed that these bacteria can grow within a variety of cells including cynomolgus monkey alveolar macrophages (7), guinea pig peritoneal macrophages (8), the MRC-5 line of human embryonic lung fibroblasts (10), normal but not activated human blood monocytes (4, 5), and normal but not activated human alveolar macrophages (T. W. Nash, D. M. Libby, and M. A. Horwitz, Program Abstr. Intersci. Conf. Antimicrob. Agents Chemother. 22nd, Miami Beach, Fla., abstr. no. 93, 1982). Quantitative bacterial plate counts revealed that *L. pneumophila* multiplied by 2 to 3 logs within these cells. Destruction of the monolayers occurred when bacterial growth was maximum. As useful as these in vitro cell systems may be for the investigation of certain aspects of the interaction of *L. pneumophila* with its host cell, each of these systems has limitations that preclude its use for extensive comparisons of in vivo and in vitro results. For example, the limited availability of human alveolar macrophages restricts extensive use. Also, any in vivo data pertaining to the intrapulmonary interaction of human alveolar macrophages with *L. pneumophila* can primarily be obtained from clinical reports and necropsies. No experimental modification of this system is possible. The best alternative, therefore, is to develop animal models that closely resemble human infections with *L. pneumophila* and to use cells from these animals for in vitro work. Well-characterized animal models for *Legionella* pneumonia have been described by Davis and co-workers (1, 2, 9), using guinea pigs and rats. These animal models appear to represent the susceptible (guinea pigs) and resistant (rat) human populations. This study is a report on the in vitro correlate of these animal models.

Normal resident macrophages were lavaged from the lungs and peritoneal cavities of Lewis rats (150 to 200 g) and Hartley strain guinea pigs (250 to 300 g) and suspended to approximately 10^6 cells per ml in RPMI 1640 containing 10% fetal bovine serum and cefazolin (5.8 µg/ml). A 2-ml volume of this suspension was added to 35-mm plastic petri dishes, incubated for 2 h at 37°C in 5% CO_2 and air, rinsed, and incubated for 18 to 24 h in RPMI 1640 with 10% fetal bovine serum without antibiotics. Previous results had shown that 5.8 µg of cefazolin per ml had no effect on *L. pneumophila* viability but was usually able to eliminate any bacterial contamination of the macrophage cultures. After overnight incubation there were approximately 2×10^5 to 8×10^5 macrophages per well which were 99 to 100% viable by nigrosin exclusion. The monolayers were then infected with 0.1 to 2.5 *L. pneumophila* organisms per macrophage, and total bacterial counts where determined at 1, 24, and 48 h. Results were obtained from monolayers infected with virulent Philadelphia 2 and Burlington, avirulent Philadelphia 2 and Burlington, and Pontiac strains of *L. pneumophila* serogroup 1.

Representative results with alveolar macrophages are shown in Fig. 1 and 2. With few exceptions, virulent *L. pneumophila* grew within guinea pig alveolar and peritoneal macrophages. The in vitro intracellular doubling times were approximately 2.5 h and 5 to 9 h, respectively. The virulent legionellae also grew in rat alveolar macrophages (doubling time, approximately 3.5 to 4 h), but not within rat peritoneal

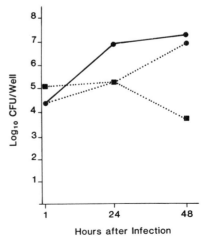

FIG. 1. Growth of *L. pneumophila* within guinea pig alveolar macrophages. Strains: (——) Burlington; (·····) Philadelphia 2; (●) virulent; (■) avirulent.

167

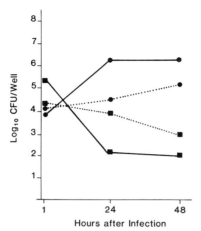

FIG. 2. Growth of *L. pneumophila* within rat alveolar macrophages. See legend to Fig. 1 for explanation of symbols.

macrophages (data not shown). The avirulent legionellae either did not grow or were killed by all macrophage populations studied. Growth of the Pontiac strain within guinea pig macrophages was similar to that observed for virulent Philadelphia 2 and Burlington strains (doubling time, approximately 3.6 h; data not shown) and is consistent with the reports that these animals develop pneumonia when exposed to aerosols of the Pontiac strain (3, 6).

The in vitro doubling time of virulent Burlington strain within guinea pig alveolar macrophages corresponds to the reported doubling time of these bacteria in guinea pig lungs (1). The in vitro doubling time of virulent Burlington strain within rat alveolar macrophages is slightly slower than the in vivo doubling time (3.5 h versus 2.2 h) and probably is due to differences between in vitro and in vivo environments.

The results of the in vitro growth of *L. pneumophila* appear to correlate well with the reported in vivo growth rates within guinea pig and rat lungs and also, most importantly, are comparable to the reported growth rates of these bacteria

within human blood monocytes and alveolar macrophages (4; Nash et al., 22nd ICAAC, abstr. no. 93). The guinea pig and rat systems, both in vivo and in vitro, should therefore provide complementary models which will be important for further investigations into the pathogenesis of *L. pneumophila* pneumonia.

LITERATURE CITED

1. Davis, G. S., W. C. Winn, Jr., D. W. Gump, J. E. Craighead, and H. N. Beaty. 1982. Legionnaires' pneumonia after aerosol exposure in guinea pigs and rats. Am. Rev. Respir. Dis. 126:1050–1057.
2. Davis, G. S., W. C. Winn, Jr., D. W. Gump, J. E. Craighead, and H. N. Beaty. 1983. Legionnaires' pneumonia in guinea pigs and rats produced by aerosol exposure. Chest 83S:15S–16S.
3. Glick, T. H., M. B. Gregg, B. Berman, G. Mallison, W. W. Rhodes, Jr., and I. Kassanoff. 1978. Pontiac fever—an epidemic of unknown etiology in a health department. I. Clinical and epidemiologic aspects. Am. J. Epidemiol. 107:149–160.
4. Horwitz, M. A., and S. C. Silverstein. 1980. Legionnaires' disease bacterium (*Legionella pneumophila*) multiplies intracellularly in human monocytes. J. Clin. Invest. 66:441–450.
5. Horwitz, M. A., and S. C. Silverstein. 1981. Activated human monocytes inhibit the intracellular multiplication of Legionnaires' disease bacteria. J. Exp. Med. 154:1618–1635.
6. Kaufman, A. F., J. E. McDade, C. M. Patton, J. V. Bennet, P. Skaliy, J. C. Feeley, D. C. Anderson, M. E. Potter, V. F. Newhouse, M. B. Gregg, and P. S. Brachman. 1981. Pontiac fever: isolation of the etiologic agent (*Legionella pneumophila*) and demonstration of its mode of transmission. Am. J. Epidemiol. 114:337–347.
7. Kishimoto, R. A., M. D. Kastello, J. D. White, F. G. Shirey, V. G. McGann, E. W. Larson, and K. W. Hedlund. 1979. In vitro interaction between normal cynomolgus monkey alveolar macrophages and Legionnaires disease bacteria. Infect. Immun. 25:761–763.
8. Kishimoto, R. A., J. D. White, F. G. Shirey, V. G. McGann, R. F. Berendt, E. W. Larson, and K. W. Hedlund. 1981. In vitro response of guinea pig peritoneal macrophages to *Legionella pneumophila*. Infect. Immun. 31:1209–1213.
9. Winn, W. C., G. S. Davis, D. W. Gump, J. E. Craighead, and H. N. Beaty. 1982. Legionnaires' pneumonia after intratracheal inoculation of guinea pigs and rats. Lab. Invest. 47:568–578.
10. Wong, M. C., E. P. Ewing, Jr., C. S. Callaway, and W. L. Peacock, Jr. 1980. Intracellular multiplication of *Legionella pneumophila* in cultured human embryonic lung fibroblasts. Infect. Immun. 28:1014–1018.

Cell-Mediated and Humoral Responses in Guinea Pigs Immunized with Antigen Preparation Isolated from *Legionella pneumophila*

MICHAEL G. CAPARON AND WILLIAM JOHNSON

Department of Microbiology, University of Iowa, Iowa City, Iowa 52242

Although several investigators have reported the induction of immunity to infection with *Legionella pneumophila* with serogroup-specific

antigens and heat-killed whole cells (1, 6), the mechanisms by which these antigen preparations induce immunity are not completely understood. The objective of this study was to determine the development of humoral and cell-mediated immunity in guinea pigs immunized with purified serogroup-specific antigens and heat-killed whole cells.

A virulent strain of *L. pneumophila* serogroup 1 obtained from the Centers for Disease Control, Atlanta, Ga., was maintained on charcoal-yeast extract agar. The 50% lethal dose of this strain has previously been shown to be 4.1×10^6 CFU. The serogroup-specific F-1 antigen was prepared as previously described (2), and whole cells harvested from 48-h growth on charcoal-yeast extract agar were killed by heat (101°C, 60 min). F-1 antigen and heat-killed whole cells were used to prepare the following immunogens: whole cells emulsified in Freund complete adjuvant (WC-A), whole cells with no adjuvant (WC), F-1 antigen emulsified in complete Freund adjuvant (F-1-A), and F-1 antigen without adjuvant (F-1). Groups of guinea pigs were immunized intramuscularly with 0.5 ml of each antigen preparation. Each group received a booster immunization 3 weeks after the primary immunization.

The potential for the development of cell-mediated immunity in immunized guinea pigs was assayed by a lymphocyte transformation assay of *L. pneumophila* sonic extracts, using a modification of the method of Plouffe and Baird (4). Single-cell suspensions of 10^5 spleen cells per 0.2 ml of culture obtained from immunized animals were cultured with varying concentrations of *L. pneumophila* sonic extract for 6 days in RPMI 1640 supplemented with 10% fetal calf serum and Click's additives. After 6 days of incubation, cultures were pulsed with 0.5 μCi of [^{3}H]thymidine for 4 h and harvested. Results are expressed as net counts per minute of [^{3}H]thymidine uptake (counts per minute of stimulated cultures minus counts per minute of unstimulated cultures) and as stimulation indices (counts per minute of stimulated cultures/counts per minute of unstimulated cultures).

The development of humoral immunity in immunized guinea pigs was assayed by microagglutination (MA) assay and by enzyme-linked immunosorbent assay (ELISA) for immunoglobulin G (IgG) and IgM titers. The ELISA consisted of coating purified F-1 antigen onto wells of polystyrene microtiter plates (Immulon II), adding appropriate dilutions of serum, and developing the reaction with IgM and IgG isotype-specific reagents. Results are expressed in terms of ELISA units, calculated by determining the equivalent dilution of the unknown serum corresponding to a standard serum arbitrarily assigned a given number of ELISA units.

Results obtained from the determination of MA titers over the immunization schedule indicated that MA titers increased rapidly after the primary immunization, reaching a peak by the end of week 1. Animals receiving WC-A and WC responded with significantly higher titers than animals receiving F-1-A and F-1. After the initial rapid increase, MA titers declined until a boost was administered after week 3. In animals receiving WC-A and WC, MA titers increased over the next 2 weeks to levels as high as or higher than their levels after week 1, whereas MA titers in animals receiving F-1-A and F-1 remained constant at their preboost levels.

In general, the development of the IgM response, as determined by ELISA, paralleled the development of MA titers with a rapid initial increase followed by a decline until the boost was administered at week 3. Titers then increased over the next 2 weeks to levels as high as or higher than the levels after week 1. As with MA titers, animals receiving WC-A and WC responded with significantly higher IgM titers than animals receiving F-1-A and F-1.

In contrast to the development of IgM titers, IgG titers, as determined by ELISA, increased slowly over the first 3 weeks. After the booster immunization at 3 weeks, IgG titers increased rapidly, reaching levels much higher than preboost levels by week 1 after the boost. As with IgM titers, animals receiving WC-A and WC responded with significantly higher IgG titers than animals receiving F-1-A and F-1.

Initial results obtained from the lymphocyte transformation assay showed that maximal blastogenic response was obtained when cells were stimulated with *L. pneumophila* sonic extract at 10 μg of protein per ml of culture. The results of the blastogenic assay indicated that, although animals immunized with WC-A and WC both responded with high levels of IgM and IgG, WC-A produced a significantly higher blastogenic response than WC (stimulation index of 151 versus 4.06). Similar results were obtained with F-1-A and F-1, with the response of animals immunized with F-1-A being significantly higher than that of animals immunized with F-1 (stimulation index of 13.9 versus 1.16).

The ability of the antigen preparations to provide protection was assayed by challenging immunized guinea pigs intraperitoneally or intratracheally with 100 50% lethal doses of virulent *L. pneumophila*.

All antigen preparations provided protection against the intraperitoneal route of challenge; animals developed no symptoms of infections, as determined by weight loss and fever (Table 1). Similar results were obtained when animals receiving WC-A and F-1-A were challenged intratracheally. Animals immunized with WC-A and F-1-A survived intratracheal challenge and

TABLE 1. Protective activity of *L. pneumophila* antigen preparations

Route of challenge	Immunogen	Titers			Protective activity	
		MA	IgM	IgG	Symptoms[a]	Survival[b] (days)
Intraperitoneal	None	<4	170	3.5	+	− (3.5)
	WC-A	≥8,192	6,581	1,546	−	+
	WC	2,048	6,664	1,435	−	+
	F-1-A	64	2,037	217	−	+
	F-1	64	650	77	−	+
Intratracheal	None	<4	170	3.5	+	− (2.5)
	WC-A	≥8,192	6,814	4,684	−	+
	WC	≥8,192	5,485	861	+	− (7)
	F-1-A	64	1,263	133	−	+
	F-1	1,024	2,363	114	+	+

[a] +, Animals showed symptoms of infection, including weight loss and fever.

[b] +, Animals survived the period of observation. Numbers in parentheses indicate mean time of survival in days.

showed no symptoms of developing an infection. However, when animals immunized with antigen preparations without adjuvant were challenged intratracheally, different results were obtained. Animals receiving F-1 and WC both developed symptoms of infection. Animals immunized with F-1 survived intratracheal challenge, whereas animals receiving WC did not survive intratracheal challenge even though these animals developed high circulating titers of IgM and IgG.

Next, a series of passive transfer experiments using inbred Strain 2 guinea pigs were designed to determine the mechanism by which the antigen preparations provide protection in immunized animals. Donor animals were immunized with WC-A since this route of immunization produced high MA, IgG, and IgM titers, and high blastogenic response and provided protection against all routes of challenge. The intratracheal route of challenge was chosen because this route simulates the natural route of infection. Animals receiving 1.4×10^8 spleen cells from immunized donors failed to survive intratracheal challenge with 100 50% lethal doses of virulent *L. pneumophila*, whereas 100% of animals receiving an intramuscular injection of serum from immune donors survived intratracheal challenge (Table 2).

Several factors may account for the failure of cells from immunized donors to transfer resistance to unimmunized recipients. The recirculation pathways of the transferred spleen cells may not have included the lung, so that they were unable to localize at sites of infection after intratracheal challenge. Another factor may be that a population of suppressor cells localizes in the spleens of immunized animals. These cells when transferred may suppress the ability of immune cells to transfer protection to normal animals. However, the results of the blastogenic

assay with this route of immunization suggest that suppressor cells are not responsible for the failure of spleen cells to protect. The failure of cells to transfer resistance to *L. pneumophila* may not be surprising since other investigators (1, 6), as well as data from our protection studies, show that immunity to infection with *L. pneumophila* may be induced with heat-killed cells as well as isolated antigen preparations. Induction of a protective immune response against facultative intracellular pathogens usually requires the use of live vaccine strains (3). The results of our studies clearly show that antigen preparations isolated from *L. pneumophila* can induce a blastogenic response. This is in agreement with several investigators who have shown that cell-mediated immune responses develop during infection and immunization with *L. pneumophila*. Wong et al. (5) have shown that delayed-type hypersensitivity reactions develop in immunized guinea pigs, and Plouffe and Baird (4) have shown that lymphocytes from patients with a history of legionellosis undergo blastogenesis in response to extracts

TABLE 2. Passive transfer of immunity to *L. pneumophila* in inbred guinea pigs by serum or cells from donors immunized with killed cells in adjuvant

Animals receiving:	Titers in recipients			% Survival
	MA	IgM	IgG	
Cells[a]	<4	394	8.8	0
Serum[b]	128	633	180	100
Immunized control	1,024	6,412	2,410	100
Unimmunized control	16	241	9.8	0

[a] Normal animals received 1.4×10^8 spleen cells from immunized donors 24 h before challenge.

[b] Normal animals received an intramuscular injection of 3 ml of serum from immunized donors 24 h before challenge.

of *L. pneumophila*. However, the fact that serum from immunized animals can transfer resistance against intratracheal challenge to unimmunized recipients suggests that humoral mechanisms of immunity may play a critical role in protecting the host against infection with *L. pneumophila*.

This investigation was supported by Public Health Service grant AI-15807 from the National Institute of Allergy and Infectious Diseases and by a grant from the American Legion of Iowa Foundation.

LITERATURE CITED

1. **Elliott, J. A., W. Johnson, and C. M. Helms.** 1981. Ultrastructural localization and protective activity of a high molecular weight antigen isolated from *Legionella pneu-*

mophila. Infect. Immun. 31:822–824.
2. **Johnson, W., J. A. Elliott, C. M. Helms, and E. D. Renner.** 1979. A high molecular weight antigen in Legionnaires' disease bacterium: isolation and partial characterization. Ann. Intern. Med. 90:638–641.
3. **Kearns, R. J., and D. J. Hinrichs.** 1977. Kinetics and maintenance of acquired resistance in mice to *Listeria monocytogenes*. Infect. Immun. 16:923–927.
4. **Plouffe, J. F., and I. M. Baird.** 1981. Lymphocyte transformation to *Legionella pneumophila*. J. Clin. Lab. Immunol. 5:149–152.
5. **Wong, K. H., P. R. B. McMaster, J. C. Feeley, R. J. Arko, W. O. Schalla, and F. W. Chandler.** 1980. Detection of hypersensitivity to *Legionella pneumophila* in guinea pigs by skin test. Curr. Microbiol. 4:105–110.
6. **Wong, K. H., W. O. Schalla, R. S. Arko, J. C. Bullard, and J. C. Feeley.** 1979. Immunochemical, serologic and immunologic properties of major antigens isolated from the Legionnaires' disease bacterium. Ann. Intern. Med. 90:634–638.

Multiplication of *Legionella micdadei* and *Legionella pneumophila* in Cultured Alveolar Macrophages from Susceptible and Immune Guinea Pigs

MICHAEL LEVI, A. WILLIAM PASCULLE, JOHN N. DOWLING, AND NEIL AMPEL

Department of Microbiology and Infectious Disease, Graduate School of Public Health, and Departments of Pathology and Medicine, School of Medicine, University of Pittsburgh, Pittsburgh, Pennsylvania 15261

Legionella micdadei has been shown to be similar in many respects to *Legionella pneumophila*. However, *L. micdadei* differs from *L. pneumophila* in that the former appears to cause pneumonia only in persons who are immunocompromised or otherwise debilitated (3, 4). This suggests that *L. micdadei* may lack some elements of virulence which *L. pneumophila* possesses. Since both organisms have been shown to be facultative intracellular bacteria, the differences between these two organisms may be reflected in the outcome of the interaction between the bacteria and host phagocytic cells.

We have begun to study the interaction between *L. micdadei* and phagocytic cells of normal and immune hosts in an effort to better understand the mechanisms of pathogenesis of and immunity to *L. micdadei* infection. To facilitate such studies, it was first necessary to develop an immune animal model. We previously developed an experimental model of *L. micdadei* pneumonia, using intratracheally infected guinea pigs (5), which has now been modified to enable us to produce large numbers of specifically immune animals in a relatively short period of time.

Previous studies in our laboratory (A. W. Pasculle and J. N. Dowling, Program Abstr. Intersci. Conf. Antimicrob. Agents Chemother. 22nd, Miami Beach, Fla., abstr. no. 95, 1982)

revealed that mortality among intratracheally infected guinea pigs can be reduced by treatment with erythromycin or the combination of sulfamethoxazole and trimethoprim for 1 week after infection. Since abortive infection with other intracellular pathogens has been shown to result in the induction of immunity (1, 6), we measured the effects of prior infection and antimicrobial therapy with either erythromycin or sulfamethoxazole-trimethoprim on the development of immunity to reinfection with *L. micdadei*. The abortive infection was produced by infecting guinea pigs intratracheally with 10^7 CFU of *L. micdadei*. Antimicrobial therapy with erythromycin or sulfamethoxazole-trimethoprim was begun 24 h later and was continued for 7 days. At various times after recovery from this primary infection, the animals were rechallenged intratracheally with 10^9 CFU of *L. micdadei*.

In four separate experiments involving 38 previously infected and treated animals and 27 normal littermates, animals which had recovered from *L. micdadei* infection were found to be immune to rechallenge with the same organism. Using a challenge inoculum which contained almost 2 logs more organisms than that used for the primary infection, we observed a mortality rate of 89% (24/27) among the normal animals as compared to 8% (3/38) in the previously infected animals. The recovered animals were immune as early as 3 weeks and as late as 5 weeks after their initial infection. In addition, challenge of one group of immune animals with *L. pneumophila* by the intratracheal route revealed that the animals' immunity appeared to be specific for *L. micdadei*.

Alveolar macrophages were collected by bronchial lavage and allowed to adhere to the surface of glass Leighton tubes to form monolayers containing 10^6 cells per tube. The monolayers were infected with either *L. micdadei* or *L. pneumophila* at a multiplicity of infection of 5 bacteria per macrophage. After a 2-h infection period, the monolayers were washed, and fresh tissue culture medium was added. This time point was designated as time 0. The number of bacteria present in the monolayers after various periods of additional incubation was determined after sonication of the cells to release the intracellular bacteria.

L. micdadei and *L. pneumophila* multiplied equally well in the alveolar macrophages of normal guinea pigs. The levels of initial phagocytosis of both organisms were similar, with uptake of 10^3 to 10^4 organisms per monolayer at the end of the period of initial adsorption of the bacteria. The bacteria multiplied exclusively within the cultured cells and not in cell-free culture medium. In cultures observed for up to 48 h there was an increase of 2.5 to 3 logs in the number of organisms for both species (Fig. 1).

The intracellular location of the bacteria was further confirmed both by direct fluorescent-antibody staining of the monolayers and by electron microscopy of the cultured cells. Electron microscopy revealed that *L. pneumophila* was contained in membrane-bound vacuoles the outer surfaces of which were studded with electron-dense particles resembling ribosomes, as

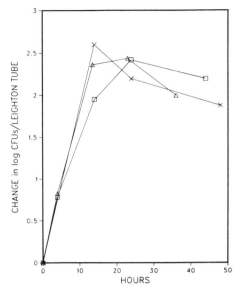

FIG. 2. Multiplication of *L. micdadei* in alveolar macrophages collected from immune guinea pigs (△) 2 weeks after primary infection, (×) 2 weeks after challenge, or (□) 5 weeks after challenge.

has been previously described (2; M. A. Horwitz, 22nd ICAAC, abstr. no. 94, 1982). *L. micdadei* was also found to be present within vacuoles, but in contrast to those containing *L. pneumophila*, ribosomal bodies did not appear to surround the vacuoles in the *L. micdadei*-infected cells. The significance of this observation, particularly in light of the similar rates of growth of the two organisms, remains to be determined.

Alveolar macrophages from immune guinea pigs, cultured in the absence of specific antibody and complement, did not appear to limit the growth of *L. micdadei*. The rate of multiplication of the bacteria in cells from immune animals was practically identical to that seen in those from susceptible animals. Cells collected from animals 2 weeks after the primary infection, when it could be demonstrated that the animals were solidly immune, did not limit the growth of the organism (Fig. 2). Additionally, cells collected 2 and 4 weeks after a second intratracheal challenge of the animals also did not limit the multiplication of the organism.

L. micdadei, like *L. pneumophila*, appears to be capable of multiplication within the alveolar macrophages of guinea pigs. Although a direct extrapolation between these in vitro data and the in vivo condition may not be possible, it appears that these findings support previous observations (4, 7) of large numbers of bacteria within alveolar macrophages in the lung tissue of patients with Pittsburgh pneumonia and suggest

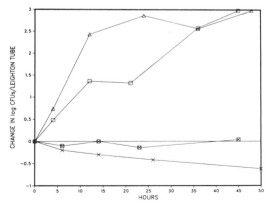

FIG. 1. Multiplication of *L. micdadei* and *L. pneumophila* in normal guinea pig alveolar macrophages. *L. micdadei* (△) growth and (×) culture medium control; *L. pneumophila* (□) growth and (⊠) culture medium control.

that these cells may be at least one site of multiplication of the bacteria within the susceptible host.

Recovery from pulmonary infection with *L. micdadei* confers solid immunity to reinfection with this organism but not with *L. pneumophila*. This is the first demonstration of immunity to *Legionella* pneumonia after recovery from a primary pulmonary infection with a strain of documented animal virulence. Pulmonary alveolar macrophages from immune guinea pigs do not by themselves limit the growth of *L. micdadei* in vitro and thus are not solely responsible for the immunity exhibited by the animals. The present experiments have not determined whether activation of these cells follows the development of immunity to reinfection, but they suggest that if such activation occurs it is not sufficient to limit the intracellular growth of *L. micdadei*. The present studies were carried out in the absence of specific antibody and complement, either of which may have additional stimulating effects on the ability of cells to limit the growth of this organism. Additionally, the role of other cells, such as monocytes and lymphocytes, in limiting the intracellular multiplication of this organism must be investigated.

LITERATURE CITED

1. **Germanier, R., and E. Furher.** 1975. Isolation and characterization of *Gal E* mutant Ty 21a of *Salmonella typhi*: a candidate strain for a live, oral typhoid vaccine. J. Infect. Dis. **131**:553–558.
2. **Katz, S. M., and S. Hashemi.** 1982. Electron microscopic examination of the inflammatory response to *Legionella pneumophila* in guinea pigs. Lab. Invest. **46**:24–32.
3. **Muder, R. R., V. L. Yu, and J. J. Zuravleff.** 1983. Pneumonia due to the Pittsburgh pneumonia agent: new clinical perspective with a review of the literature. Medicine **59**:188–204.
4. **Myerowitz, R. L., A. W. Pasculle, J. N. Dowling, G. J. Pazin, M. Puerzer, C. R. Rinaldo, Jr., R. B. Yee, and T. R. Hakala.** 1979. Opportunistic lung infection due to Pittsburgh pneumonia agent. N. Engl. J. Med. **301**:953–968.
5. **Pasculle, A. W.** 1981. Experimental studies of Pittsburgh pneumonia agent, p. 169–172. *In* D. Schlessinger (ed.), Microbiology—1981. American Society for Microbiology, Washington, D.C.
6. **Thorne, G. M., and S. L. Gorbach.** 1977. Shigella vaccines, shigella pathogenesis—Dr. Jekyll and Mr. Hyde. J. Infect. Dis. **136**:601–603.
7. **Winn, W. C., Jr., and R. L. Myerowitz.** 1981. The pathology of the *Legionella* pneumonias. Human Pathol. **12**:401–422.

Effect of Serum and Agar Passage on *Legionella micdadei* Interaction with Human Neutrophils

DAVID L. WEINBAUM, ROBERT R. BENNER, A. WILLIAM PASCULLE, J. N. DOWLING, AND GERALD R. DONOWITZ

University of Pittsburgh, Pittsburgh, Pennsylvania 15261, and University of Virginia, Charlottesville, Virginia 22908

Most investigations of host defense to the legionellae have concentrated on host response to *Legionella pneumophila*. Whereas *L. pneumophila* and *Legionella micdadei* resemble each other in many respects, *L. micdadei* appears to be primarily pathogenic for the immunosuppressed host (4). A comparative study of these two agents may elucidate basic differences in defense mechanisms to account for differing host susceptibility. In the present study we examined both the opsonic requirements for phagocytosis and the effect of repeated agar passage of *L. micdadei* on interactions with human neutrophils (PMN).

L. micdadei was isolated from human lung tissue, inoculated into embryonated hen eggs, and stored at −70°C (5). It was passed on buffered charcoal-yeast extract agar a specified number of times before use. *L. pneumophila* serogroup 1, Philadelphia strain, has been passaged 10 times on buffered charcoal-yeast extract agar. Human PMN and serum were obtained by standard separation techniques (1) from normal donors with no detectable antibody to the bacterium being studied. Serum was either used directly or stored at −70°C until used. Complement was inactivated by heating the serum at 56°C for 30 min immediately before use.

Phagocytosis and killing were determined by use of a differential centrifugation assay (2). PMN and bacteria at a ratio of 2 to 5 CFU per cell were added to 1 ml of Hanks balanced salt solution and 10% serum and incubated for 120 min. At timed intervals samples were removed, the PMN were lysed, and bacterial CFU were determined. Changes in these samples represent bacterial killing. At the same intervals a second sample was removed and centrifuged at $150 \times g$ for 5 min to separate cell-associated from non-cell-associated bacteria. The supernatant was decanted, and bacterial CFU were determined. Changes in CFU of the supernatant over time were used to determine phagocytosis. Additionally, phagocytic indices and electron micrographs were performed from the same samples.

Bacterial opsonization was determined by an

immunofluorescence assay (3). After incubation in 25% fresh or heat-inactivated serum, rhodamine-labeled anti-human C3 complement was added. After drying on a slide, the bacteria were stained with fluorescein-labeled anti-*L. micdadei* serum. Counts were made by fluorescence microscopy. Bacteria were identified under the fluorescein filter, and opsonization by C3 was then determined by observation with the rhodamine filter.

Results of phagocytosis and killing by human PMN of *L. micdadei* strains with different agar passage histories are shown in Fig. 1 and 2. Phagocytosis of multipassaged *L. micdadei* (97 ± 1.0%) was equivalent to that of *Staphylococcus aureus* (98 ± 1%). Twice-passaged (87 ± 1%) and once-passaged (75 ± 7%) bacteria were phagocytized significantly less than *S. aureus*. Killing of the multipassaged strain was equivalent to that of *S. aureus*, whereas killing of the twice-passaged and once-passaged organisms was negligible. Phagocytosis and not merely attachment was verified by electron microscopy.

Since the sera used in these experiments contained no detectable antibody to *L. micdadei*, we investigated the role of complement as an opsonin. *L. micdadei* activated the complement pathway and was opsonized by C3 in fresh serum without antibody. *L. pneumophila* was not opsonized by C3 in specific antibody-free serum. To clarify the role of complement in phagocytosis, we compared heat-inactivated serum with fresh serum for its opsonic properties.

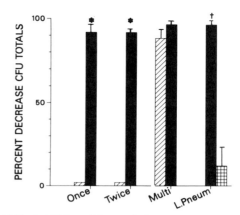

FIG. 2. Killing of *L. micdadei* (▨) or *L. pneumophila* (▦) by PMN in fresh serum without specific antibody, compared with killing of *S. aureus* (■). Numbers represent mean ± standard error of the mean. The agar passage history of *L. micdadei* is under each bar. *, $P < 0.001$; †, $P < 0.02$.

Heat-inactivated serum significantly decreased phagocytosis of twice-passaged *L. micdadei*, from 85 to 7%. In addition, the phagocytic index fell from 77 to 4% when heat-inactivated serum was used.

In summary, PMN do not appear to play a major role in host defense against *L. micdadei*. These bacteria are readily ingested but are not killed by PMN, with complement being the major opsonin. Agar passage destroys the ability of the bacteria to resist killing. Both the importance and mechanisms of survival within the neutrophil need further clarification.

We thank Judy Fossati for her excellent secretarial assistance.

This work was supported in part by grant Y-52 from the Health Research and Services Foundation, Pittsburgh, Pa.

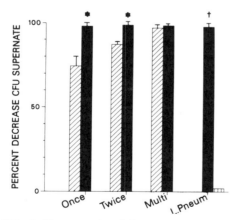

FIG. 1. Phagocytosis of *L. micdadei* (▨) or *L. pneumophila* (▦) by PMN in fresh serum without antibody, compared with phagocytosis of *S. aureus* (■). Numbers represent mean ± standard error of the mean. The agar passage history of *L. micdadei* is under each bar. *, $P < 0.05$; †, $P < 0.001$.

LITERATURE CITED

1. **Boyum, A.** 1968. Isolation of mononuclear cells and granulocytes from blood. Scand. J. Clin. Lab. Invest. **97**):77–89.
2. **Densen, P., and G. L. Mandell.** 1978. Gonococcal interactions with polymorphonuclear neutrophils. J. Clin. Invest. **62**:1161–1171.
3. **Horwitz, M. A., and S. C. Silverstein.** 1981. Interaction of the Legionnaires' disease bacterium (*Legionella pneumophila*) with human phagocytes. J. Exp. Med. **153**:386–397.
4. **Myerowitz, R. L., A. W. Pasculle, J. N. Dowling, G. J. Pazin, M. Puerzer, R. B. Yee, C. R. Rinaldo, and T. R. Hakala.** 1979. Opportunistic lung infection due to "Pittsburgh Pneumonia Agent." N. Engl. J. Med. **301**:953–958.
5. **Pasculle, A. W., J. C. Feeley, R. J. Gibson, L. G. Cordes, R. L. Myerowitz, C. M. Patton, G. W. Gorman, C. L. Carmack, J. W. Ezzell, and J. N. Dowling.** 1980. Pittsburgh pneumonia agent: direct isolation from human lung tissue. J. Infect. Dis. **141**:727–732.

Inhibitory Effects of *Legionella pneumophila* on Cell Spreading and Yeast Phagocytosis in Peritoneal Macrophage Cultures

JEANNE BECKER, ROBERT J. GRASSO, AND HERMAN FRIEDMAN

Department of Medical Microbiology and Immunology, University of South Florida College of Medicine, Tampa, Florida 33612

Macrophages are an important cell type which play a prominent role in the host's resistance against bacterial infections. Microorganisms, as well as their products, are capable of producing decreased host resistance by interfering with macrophage functions. For example, *Legionella pneumophila* is an intracellular pathogen that grows within monocytes and macrophages, apparently disrupting their metabolism (3, 4). Little is known, however, regarding the effects of these bacteria on normal macrophage activities, such as phagocytosis. Alteration of this important macrophage function may severely compromise host resistance, causing increased susceptibility to infection. This study was therefore initiated to determine whether killed *L. pneumophila* would modulate the ingestion of yeast particles and macrophage cell spreading in an in vitro model system. Our laboratory has shown previously that glucocorticoid steroids at pharmacological concentrations inhibit both functions (2). Thus, we also explored whether these bacteria would modulate the glucocorticoid-mediated suppression of these two macrophage activities.

Resident peritoneal leukocytes were lavaged from 18- to 20-g female BDF_1 mice, and cover slip cultures of adhered macrophages were established (designated day 0). The cultures were then divided into four groups. One group served as control and received culture medium. The second group was treated with medium containing 10^8 Formalin-killed *L. pneumophila* (serogroup 1) organisms per ml. The third was exposed to medium containing 1 µM dexamethasone. The remaining group received medium containing 10^8 Formalin-killed *L. pneumophila* organisms per ml plus 1 µM dexamethasone. The four groups were incubated for up to 5 days at 37°C in an atmosphere of 95% air–5% CO_2. On selected days, 15-min yeast phagocytosis assays were performed with heat-killed *Saccharomyces cerevisiae* particles as follows. The cover slips were prepared for light microscopy. The percentages of phagocytes in the populations were measured; a phagocyte was defined as a macrophage that had ingested at least one yeast particle. Phagocytic indices were determined by counting the number of intracellular yeasts and establishing the percentages of phagocytes that ingested one through eight or greater than eight

particles. Cell spreading measurements were performed on the same cover slips prepared for the phagocytosis assays. A spread macrophage was defined as one whose cell body or any of its processes extended beyond 20 µm.

When resident peritoneal macrophages are cultured in vitro, they exhibit increased cell spreading over a period of several days (1). We thus examined whether *L. pneumophila* would affect macrophage spreading over a 5-day incubation period. The results of these experiments demonstrated that on day 0, <5% of the control macrophages were spread. By day 1, this percentage increased to approximately 45%, whereas only ca. 25% of the cells in the population treated with bacteria were spread. Thereafter, the percentages of spread cells in these treated cultures continued to increase to almost 100%, similar to the controls on day 5. The percentages of spread macrophages in the cultures treated with only the steroid or the steroid together with the bacteria remained <40% over the 5-day test period.

To determine whether the presence of *L. pneumophila* in macrophage cultures modulates the ingestion of yeast particles, both the percentages of phagocytes and phagocytic indices were measured. The percentages of phagocytes in control cultures increased from ca. 50% to almost 100% between day 0 and day 5 (Fig. 1, A). In contrast, the percentages of phagocytes in cultures treated with *L. pneumophila* decreased to <30% after 1 day (Fig. 1, B). However, this effect was transient since the percentages in these treated cultures increased thereafter and were similar to those of controls on day 5. In cultures exposed to the steroid, the percentages of phagocytes remained at ca. 20% between days 1 and 5, regardless of whether the bacteria were also supplied to the cultures (Fig. 1, C).

In control cultures, phagocytic index measurements revealed that the percentages of the macrophage population that ingest greater than eight yeast particles increased from 35% on day 1 to ca. 90% on day 5 (Fig. 2). In contrast, most phagocytes ingested one to three particles on day 1 in cultures treated with only *L. pneumophila*; <5% ingested over eight particles. However, the percentages of these phagocytes that ingested greater than eight yeasts increased to ca. 20% by day 2 and to nearly 80% by day 5.

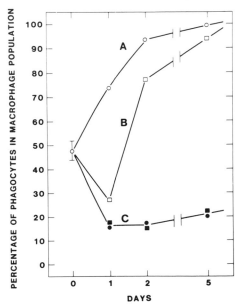

FIG. 1. Transient suppression of yeast phagocytosis in macrophage cultures treated with *L. pneumophila*. The symbols represent mean values in control cultures (○) and in cultures treated with 10^8 *L. pneumophila* organisms per ml (□), 1 μM dexamethasone (●), or both the bacteria and the steroid at these concentrations (■). *N* values were as follows: day 0 (*N* = 10), days 1 and 2 (*N* = 4), and day 5 (*N* = 4 to 6). Only standard errors of the mean ± >4% are shown.

These increases were accompanied by corresponding decreases in the percentages that ingested one to three yeast particles. In cultures that received dexamethasone alone or the glucocorticoid together with the bacteria, the majority of phagocytes ingested only one to three particles over the 5-day period.

Figures 1 and 2 demonstrate that, relative to controls, the phagocytic capacity of macrophages exposed to *L. pneumophila* was decreased between day 0 and day 1. We therefore examined whether this decrease occurred in a relatively short period of time after the addition of the bacteria to the cultures. The percentages of phagocytes and phagocytic indices were measured 1 and 2 h after the establishment of control and *L. pneumophila*-treated cultures on day 0. The results indicated that ca. 50% of the macrophages in both control and treated cultures were phagocytic at both time points. In addition, the majority of these phagocytes ingested one to three yeast particles, and <5% ingested greater than eight particles.

These data indicate that the presence of *L. pneumophila* in macrophage cultures for 1 day retards the rate of cell spreading, decreases the phagocytic population, and limits the uptake of larger numbers of yeast particles by the phagocytic subpopulation. Furthermore, these effects did not occur immediately but did so between 2 and 24 h of incubation. Both spreading and yeast phagocytosis were inhibited to the same extent in cultures exposed to the steroid plus the bacteria. This suggests that *L. pneumophila* does not interfere with the inhibitory response produced by the glucocorticoid. However, phagocytic indices reveal that on days 1 and 2, higher percentages of phagocytes in cultures which received both dexamethasone and bacteria ingest one and two yeast particles as compared to the percentages in cultures exposed to either the bacteria or the steroid. This effect was no longer apparent by day 5. Thus, *L. pneumophila* may exert a transient potentiating effect on the steroid-induced suppression of phagocytosis.

Our results demonstrate that the macrophage functions of yeast phagocytosis and cell spreading are both altered by the presence of *L. pneumophila*. Should these effects occur in vivo, host resistance could possibly be impaired by

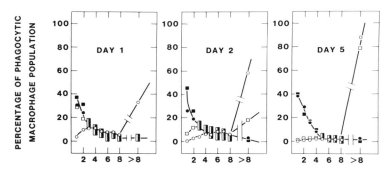

FIG. 2. Phagocytic indices in control and treated cultures. The *N* values and the symbols are the same as indicated in the legend to Fig. 1 except for the composite symbol (◨), which represents all treatments and was designed for clarity.

the modulation of normal macrophage processes by this ubiquitous pathogen.

This study was supported in part by Public Health Service grant AI 16618 from the National Institute of Allergy and Infectious Diseases.

LITERATURE CITED

1. **Cohn, Z. A., and B. Benson.** 1965. The differentiation of mononuclear phagocytes: morphology, cytochemistry and biochemistry. J. Exp. Med. **121:**153–160.

2. **Grasso, R. J., T. W. Klein, and W. Benjamin.** 1981. Inhibition of yeast phagocytosis and cell spreading by glucocorticoids in cultures of resident murine peritoneal macrophages. J. Immunopharmacol. **3:**171–192.

3. **Horowitz, M. A., and S. C. Silverstein.** 1980. Legionnaire's disease bacterium (*Legionella pneumophila*) multiplies intracellularly in human monocytes. J. Clin. Invest. **66:**441–450.

4. **Kishimoto, R. A., J. D. White, F. G. Shirey, V. G. McGann, R. F. Berendt, E. W. Larson, and K. W. Hedlund.** 1981. In vitro response of guinea pig peritoneal macrophages to *Legionella pneumophila*. Infect. Immun. **31:**1209–1213.

Interactions Between *Legionella pneumophila* and Human Monocyte-Derived Macrophages

JEAN-LOUIS VILDÉ, ERIC DOURNON, ANNE BURÉ, MARIE-CAROLINE MEYOHAS, AND PREMAVATHY RAJAGOPALAN

Service of Infectious Diseases, Central Laboratory of Microbiology, and INSERM U13, University Paris 7, Hospital Claude Bernard, Paris, France

Legionella pneumophila is a facultative intracellular (IC) bacterium multiplying within a variety of cells (1), including human monocytes (2). We have compared the ingestion and IC fate of a virulent and a nonvirulent strain of *L. pneumophila* serogroup 1 in human macrophages. These monocyte-derived macrophages were obtained by in vitro culture for 5 days of venous blood (4) from a healthy donor seronegative for *L. pneumophila*. These macrophages were inoculated on day 5 with the virulent or the nonvirulent strain in a ratio of about 100 bacteria per macrophage. The virulent strain, isolated from the lung of a patient who died from Legionnaires disease, has been frozen (−80°C) in portions after only two passages on buffered charcoal-yeast extract agar. The nonvirulent strain was derived from the virulent one after 30 passages on buffered charcoal-yeast extract agar over a 6-month period and was nonpathogenic for guinea pigs. Either fresh normal human serum or immune serum (IS) from a patient who recovered from Legionnaires disease (titer by immunofluorescence, 1:2,048) was added at a final dilution of 10%. The CFU of viable extracellular and IC bacteria were determined.

Ingestion was studied at 60 min. IS increased the uptake of virulent *L. pneumophila* from 6.5% of the inoculum with normal serum to 13 ± 7% with IS. The uptake with normal serum of bacteria preopsonized by heated (56°C for 30 min) IS was 16.5%. With the nonvirulent strain, rates of opsonization were higher with either normal serum or IS.

Enumeration of heat-killed bacteria by direct immunofluorescence also showed the enhancement of ingestion by IS (Fig. 1). Heating (56°C for 30 min) IS clearly decreased ingestion of the bacteria. Selective inactivation of the alternate pathway (50°C for 15 min) or of the classical pathway (10^{-2} M EGTA) showed that the latter plays the major role. Thus, specific antibodies and the classical pathway of complement are both required for optimal ingestion of virulent *L. pneumophila*. Influence of specific antibodies could not be demonstrated for ingestion of the nonvirulent strain.

IC multiplication was measured by comparing the total number of bacteria at different time intervals to the number of IC bacteria, taking 60 min as the base line. At 2 h the total number of bacteria was slightly lower than the base line for both the virulent and nonvirulent strains. At 12 and 24 h, however, the virulent bacteria had multiplied by 7- to 100-fold compared with the base-line count (Table 1); the nonvirulent strain multiplied also, but clearly to a lesser extent. The monolayer of macrophages was destroyed at 24 h by the virulent strain. No multiplication

TABLE 1. IC multiplication of *L. pneumophila*

Time (h)	Multiplication[a] of:	
	Virulent *L. pneumophila*	Nonvirulent *L. pneumophila*
2	×0.53 (*n* = 4)	×0.49 (*n* = 6)
12	×7 (*n* = 2)	ND[b]
24	×10 to ×100 (*n* = 8)	×3 (*n* = 4)
48	ND	×3 (*n* = 1)

[a] Results are expressed as the increase in the total number of bacteria (IC and extracellular) compared with the base line (the number of IC bacteria at 60 min).

[b] ND, Not done.

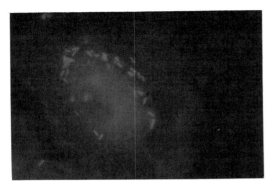

FIG. 1. Macrophage full of IC virulent *L. pneumophila* stained by specific immunofluorescence. Note the localization of most bacteria around the nucleus.

was observed in the control tubes without macrophages.

These results confirm a previous study (3) showing the enhancement of binding of *L. pneumophila* to mononuclear phagocytes by antibodies and complement and show that the IC multiplication of a nonvirulent strain within human mononuclear phagocytes is slower and of lower magnitude than that of a virulent strain, as has been shown for other facultative IC bacteria such as salmonellae (4).

LITERATURE CITED

1. **Daisy, J. A., C. E. Benson, J. McKitrick, and H. M. Friedman.** 1981. Intracellular replication of *Legionella pneumophila*. J. Infect. Dis. **143:**460–464.
2. **Horwitz, M. A., and S. C. Silverstein.** 1980. Legionnaires' disease bacterium (*Legionella pneumophila*) multiplies intracellularly in human monocytes. J. Clin. Invest. **66:**441–450.
3. **Horwitz, M. A., and S. C. Silverstein.** 1981. Interaction of the Legionnaires' disease bacterium (*Legionella pneumophila*) with human phagocytes. II. Antibody promotes binding of *L. pneumophila* to monocytes but does not inhibit intracellular multiplication. J. Exp. Med. **153:**398–406.
4. **Vildé, J. L., P. Lagrange, F. Vildé, and M. C. Blayo.** 1977. Some functional and metabolic properties of human monocyte-derived macrophages, p. 347–357. *In* F. Rossi, P. L. Patriarca, and D. Romeo (ed.), Movement, metabolism and bactericidal mechanisms of phagocytes. Piccin Medical Books, Padua, Italy.

Development of Cellular Immunity in Guinea Pigs Exposed to *Legionella pneumophila*

THOMAS KLEIN, RAYMOND WIDEN, KATHY CABRIAN, CAROLINE SEARLS, AND HERMAN FRIEDMAN

Department of Medical Microbiology and Immunology, University of South Florida College of Medicine, Tampa, Florida 33612

Little is known concerning the nature and mechanisms of the immune response in resistance to and recovery from infection by *Legionella pneumophila*. Patients recovering from legionellosis display a good antibody response to *L. pneumophila* antigens, and in animal models specific antibodies are opsonic for phagocytosis and protective against lethal infection. The development of specific cellular immunity in patients recovering from legionellosis has been suggested by studies demonstrating increased lymphocyte proliferation to *L. pneumophila* antigen in these individuals. Furthermore, in animal studies *L. pneumophila* antigens have been reported to induce a delayed-type hypersensitivity skin response in sensitized guinea pigs. In addition, immunologically activated macrophages have been shown to inhibit the intracellular multiplication of *L. pneumophila*. In the present work, we studied guinea pigs that were either immunized with killed *L. pneumophila* or infected with viable *L. pneumophila* and examined for the simultaneous development of three manifestations of cell-mediated immunity, including delayed-type hypersensitivity skin response, lymphocyte proliferation, and inhibition of macrophage migration.

Groups of guinea pigs were immunized subcutaneously with 3×10^7, 3×10^8, or 3×10^9 *L. pneumophila* organisms in Freund complete adjuvant and challenged intradermally either 2 or 4 weeks later with a clarified sonicate of *L. pneumophila*. Animals immunized with the two lower doses of *L. pneumophila* (3×10^7 and 3×10^8) failed to produce significant skin test responses. However, the group injected with 3×10^9 organisms displayed a significant 48-h skin test response when challenged 4 weeks after immunization. Boosting these animals with 1.5×10^9 organisms in Freund incomplete adjuvant followed by skin testing 2 weeks later resulted in an enhancement of the skin test response, while the response in nonboosted animals declined somewhat over the 2-week period. Control animals injected with complete adjuvant alone or *L. pneumophila* alone were uniformly skin test negative on all days tested. The skin lesions were histologically consistent with a delayed-

TABLE 1. Lymphocyte proliferation in response to *Legionella* antigens of spleen cells obtained from animals after primary and booster immunization

Stimulation antigen	Amt	Stimulation index[a]		
		Control[b]	Primary immunization	Primary plus booster immunization
Killed organisms	10^5	1.0 ± 0.3	1.1 ± 0.4	3.6 ± 0.7
	10^7	1.4 ± 0.1	2.0 ± 0.1	3.3 ± 0.7
Sonic extract	10 μg	1.0 ± 0.1	2.3 ± 0.4	2.3 ± 0.6
	20 μg	0.6 ± 0.1	2.1 ± 0.3	2.5 ± 0.7

[a] Stimulation index = counts per minute in antigen-treated cultures/counts per minute in cultures with medium only. Results are expressed as mean ± standard deviation; there were four animals per group.
[b] Control guinea pigs were injected with Freund complete adjuvant only.

type hypersensitivity reaction. These results indicate that guinea pigs may be primed for an enhanced skin test response to *L. pneumophila* antigens, suggesting the induction of immunological memory.

Along with skin testing, cell-mediated immunity to microbial antigens can be determined by in vitro methods such as lymphocyte proliferation and production of migration-inhibitory activity in response to microbial antigens. Accordingly, lymphocyte proliferation assays were performed, utilizing splenic lymphocytes from either primed guinea pigs or primed and boosted guinea pigs. The stimulating antigens used in culture were either whole killed *L. pneumophila* organisms or *L. pneumophila* sonic extract preparations. Splenocytes obtained from guinea pigs immunized with *L. pneumophila* displayed stimulation indices two- to threefold greater than control values (Table 1). The greatest proliferative response occurred in the boosted group after stimulation with killed *L. pneumophila*. In other experiments, the production of migration-inhibitory activity by splenocytes from immunized guinea pigs was determined. Splenocytes obtained from animals injected with *L. pneumophila* in adjuvant produced detectable amounts of inhibitory activity when incubated with *L. pneumophila* antigen (Table 2). Such activity was not detected in supernatants from control splenocytes.

In addition to immunization protocols involving killed *Legionella* organisms and adjuvant, we also performed experiments to determine the development of cell-mediated immunity after infection with living *L. pneumophila*. Initial experiments were performed with a relatively low-virulence, high-passage strain of *L. pneumophila* (serogroup 1) obtained from the Centers for Disease Control, Atlanta, Ga. The guinea pigs tolerated well an intraperitoneal injection of this organism of up to 10^8 CFU. Higher doses induced rapid death (i.e., 24 h) of the animals. Other studies were done with a high-virulence, low-passage clinical isolate obtained from a case

of legionellosis at Tampa General Hospital, Tampa, Fla. The 50% lethal dose of this strain of *L. pneumophila* (serogroup 1) was 10^6 CFU after intraperitoneal infection.

Studies were performed with groups of guinea pigs which were infected intraperitoneally with either 10^6 or 10^8 avirulent or 10^4 or 10^6 virulent *L. pneumophila* organisms. Animals were skin tested with killed *L. pneumophila* and sonic extract 45 days after infection and were subsequently sacrificed to obtain splenocytes for proliferation and migration assays. Challenge with either killed *L. pneumophila* or sonic extract elicited good skin test responses. Erythema and induration in response to sonic extract reached maximum proportions by 24 h after challenge and then declined slightly by 48 h in most groups. Challenge with killed *L. pneumophila*, on the other hand, resulted in a good skin test response at 24 h which either stayed the same or increased slightly by 48 h. On a dosage basis, infection with virulent organisms was more effective for sensitization, in that 10^4 virulent *L. pneumophila* resulted in a response greater than

TABLE 2. Production of migration-inhibitory activity by splenocytes from immunized guinea pigs in response to *L. pneumophila* antigen

Treatment	Guinea pig no.	Migration index[a]
Control	1	1.27
	2	1.70
	3	1.33
Immunized	1	0.68
	2	0.77
	3	0.32
	4	0.69
	5	1.09
	6	0.49

[a] The migration index is the ratio of the mean migration of six replicate droplets in the presence of culture supernatant to the mean migration in the presence of culture supernatant "spiked" with sonic extract after cultivation. A migration index of <0.80 was taken as significant migration inhibition.

that with 10^6 avirulent *L. pneumophila* and equivalent to that with 10^8 avirulent *L. pneumophila*. Also of interest was the finding that the skin test response to 10^4 virulent organisms was equivalent to that obtained with 10^6 virulent organisms (the 50% lethal dose). After skin testing, the guinea pigs were sacrificed and the spleen cells were harvested for lymphocyte proliferation assays. As was the case with skin testing, the best proliferative responses were obtained after infection with either 10^8 avirulent or 10^4 virulent *L. pneumophila* organisms.

These results suggest that in addition to induction of delayed-type hypersensitivity and macrophage activation, immunization and infection of guinea pigs with *L. pneumophila* can result in other manifestations of cell-mediated immunity such as specific lymphocyte proliferation and production of migration-inhibitory activity. Furthermore, it is suggested that infection with sublethal doses of virulent *L. pneumophila* is more effective in inducing cell-mediated immunity than is infection with either avirulent organisms or 50% lethal doses of virulent organisms.

Legionnaires Disease in Patients with Unexplained Pneumonia: Cellular Immunity in Patients with Legionnaires Disease, Listeriosis, Melioidosis, and Typhoid Fever

D. TANPHAICHITRA, S. SRIMUANG, AND A. BUSSAYANONDR

Infectious Disease and Host Defence Unit, Ramathibodi Medical School-Mahidol University Medical Center, Bangkok 10400, Thailand

Legionellosis is an acute bacterial infection of humans that is caused by *Legionella pneumophila* (Legionnaires disease [LD] and Pontiac fever strains), *Legionella micdadei* (*L. pittsburghensis*, TATLOCK, HEBA, Pittsburgh pneumonia agent), *Legionella bozemanii* (WIGA, ALLO 1 and 2, MI-15), *Legionella dumoffii* (ALLO 4, NY-13), and *Legionella gormanii* (ALLO 3, LS-13). LD is a newly described acute respiratory infection caused by an unusual aerobic gram-negative bacillus, namely, *L. pneumophila* (1, 2). Sera from 10 patients with unexplained pneumonia were tested for LD. Three patients had positive serology. T cells and subset were enumerated in patients with LD, melioidosis, listeriosis, and typhoid fever.

Sera from 10 patients with unexplained pneumonia were examined by using ether-treated antigen (Centers for Disease Control [CDC], Atlanta, Ga.) and formolized yolk sac antigen (Central Public Health Laboratories [CPHL], London, England). These patients were admitted to four hospitals in Bangkok, where blood for serological testing was obtained. Sixty-five control sera from patients with different kinds of infectious diseases were also tested for LD. The specific diagnosis of LD was confirmed by either a fourfold or greater serological titer rise to a reciprocal titer of at least 128 (CDC) or a titer of at least 64 (CPHL). In both cases titers were determined by indirect immunofluorescent-antibody testing (1, 3).

Complete leukocyte counts, including differential counts on 200 cells, were performed. Levels of membrane immunoglobulin were determined by direct immunofluorescence of viable cells incubated at 4°C for 30 min with fluorescein-conjugated goat anti-human immunoglobulin. The proportion of cells expressing the T-cell antigens (T-helper and T-suppressor) and pan-T cells was enumerated with the use of Lyt. monclonal antiserum (New England Nuclear Corp.) (4).

Sera of 10 patients with unexplained pneumonia were tested for the presence of antibody against the LD agent. Seven of the patients received prednisolone (20 to 60 mg/day). Only three patients had positive serology (determined by both CDC and CPHL). This series comprised two male patients and one female patient, ranging in age from 24 to 83 years. All three patients were febrile and had scanty sputum and roentgenographic signs of pneumonia ranging from a mild, patchy infiltrate to extensive pneumonitis. The two male patients died of acute respiratory failure. All cases were sporadic. One patient had probably been infected abroad, and two were on steroid therapy. Two were diagnosed retrospectively, and in only one was LD suspected during the illness. An analysis of the smoking histories of all three patients showed that at least two were current smokers smoking 20 or more cigarettes per day. Two were nonsmokers when they acquired the disease. An examination of the medical histories showed that two had preexisting disease and one had no relevant history. Of the two with a clinically significant previous medical history, one had cancer of the lung and the other had chronic obstructive pulmonary disease and tuberculosis. These two had been treated with prednisolone for cancer of the lung and with prednisolone plus antituberculous drugs, respectively. Data collected on the clinical presentation of the three cases are shown in Table 1. The common early symptoms were fever, cough (usually dry in those persons with or without underlying respiratory disease), malaise, and myalgia; two became confused during hospitalization. The time from onset of symptoms to hospital admission was 5 days in one case. Gastrointestinal symptoms were a feature in one case, and typhoid fever was also suspected in one patient; two complained of diarrhea before therapy was started with antibiotics. Other symptoms were, in order of frequency, anorexia, headache, and abdominal pain. All patients in this study were found to have pneumonia, mostly unilateral.

Four of the remaining seven cases of pneumonia were later proven to be caused by tuberculosis (one case), aspergillosis (two cases), and mycoplasma (one case). Three cases were unexplained.

Of 65 control sera, five patients, having *Haemophilus* spp. endocarditis (one case), toxoplasmosis retinitis (two cases), leptospirosis (one case), and melioidosis (one case), were positive for LD at low titer (CDC, 1:128; CPHL, <1:16) by indirect immunofluorescent-antibody testing.

The data obtained for T and B cells by using monoclonal antibodies are shown in Table 2. These data indicate that immunity may be depressed in patients with LD.

LD should be considered in the differential diagnosis of patients with unexplained pneumonia, negative routine bacteriological culture, and failure to respond to conventional antimicrobial treatment. Diagnosis of LD is often delayed because of the variability in clinical presentation and difficulty involved in the isolation of the organism. Serological testing of the serum becomes positive late in the disease and has little benefit in clinical judgement. Because *L. pneumophila* is ingested by monocytes, in which it can multiply, and from the results of this study, cellular immunity might be depressed and probably is a primary factor in defense against LD and other intracellular infections, including listeriosis, melioidosis, and typhoid fever. In addition to the fact that antigens are shared among the six serogroups of *L. pneumophila* and the five species of *Legionella*, we found that antigens are shared among legionellae, other genera of gramnegative bacilli, and leptospirae.

We thank the Centers for Disease Control, Atlanta, Ga., and the Central Public Health Laboratory, London, England, for serology.

This study was supported in part by the Rockefeller Foundation and the U.S. Agency for International Development.

LITERATURE CITED

1. **Chandler, F. W., M. D. Hicklin, and J. A. Blackmon.** 1977. Demonstration of the agent of Legionnaires' dis-

TABLE 1. Clinical features and details of three LD cases

Case	Age (yr)	Sex	Outcome	Underlying disorder	Dry cough	Diarrhea	Fever (°C)	Chest Rales	Chest X ray	Serum titer CDC	Serum titer CPHL	Time of diagnosis
1	24	Female	Survived	None	Yes, with blood streak	Yes	39	Left lower lobe	Patchy, mild	1:64 to 1:512	1:32 to 1:512	During illness
2	64	Male	Died	Lung cancer[a]	Yes, with blood streak	No	39.5	Both lungs	Patchy	1:128 to 1:1,024	1:32 to 1:512	Retrospectively
3	83	Male	Died	COPD, TB[a,b]	Yes	Yes	39.9	Lower lobe	Extensive abscess	1:1,024 to 1:64	1:512 to 1:16	Retrospectively

[a] Treated with steroids.
[b] Chronic obstructive pulmonary disease and tuberculosis.

TABLE 2. Monoclonal T- and B-cell enumeration in patients with different infections

Condition	No. of cases	T cells (%)	T-helper cells (%)	T-suppressor cells (%)	T-helper/ T-suppressor	B cells (%)	Lymphocytes/ mm³
Normal	20	73.9 ± 8.1	35 to 51	13 to 33	1.06 to 3.9	29.6 ± 3.4	1,700 to 5,800
LD	3	42.1 ± 3.5	8.5	28	0.29	31.8 ± 5.7	684
Melioidosis	7	40.1 ± 4.9	13.2 ± 5.5	24.30 ± 7.1	0.79 ± 0.68	20.41 ± 6.6	645 ± 72
Typhoid fever	21	43.4 ± 6.5	14.57 ± 5.63	26.95 ± 6.71	0.73 ± 0.50	24.41 ± 3.9	862.6 ± 164.3
Listeriosis	2	48.0	—[a]	—	—	31.5	654

[a] —, Not done.

ease. N. Engl. J. Med. **297:**1218–1220.
2. **Fraser, D. W., T. R. Tsai, W. Orenstein, W. E. Parkin, H. J. Beecham, R. G. Shararr, J. Harris, G. F. Mallison, S. M. Martin, J. E. McDade, C. C. Shepard, and P. S. Brachman.** 1977. Legionnaires' disease. Description of an epidemic pneumonia. N. Engl. J. Med. **297:**1189–1197.
3. **Lattimer, G. L., C. McCrone, and J. Galgon.** 1978. Diagnosis of Legionnaires' disease from transtracheal aspirate by DFA staining and isolation of the bacterium. N. Engl. J. Med. **299:**1172–1173.
4. **Tanphaichitra, D., A. Limsuwarn, R. Lasserre, V. Tanphaichitra, Y. Panupornprapongs, and P. Tanphaichitra.** 1978. Cellular immunity in common intracellular infections in the tropics, p. 151–152. *In* W. Siegenthaler and R. Lüthy (ed.), Current chemotherapy, vol. 1. American Society for Microbiology, Washington, D.C.

Mitogenic Effect of *Legionella pneumophila* on Murine Splenocytes

RAYMOND WIDEN, THOMAS KLEIN, HERMAN FRIEDMAN, CAROLINE SEARLS, AND KATHY CABRIAN

Department of Medical Microbiology and Immunology, University of South Florida College of Medicine, Tampa, Florida 33612

Legionella pneumophila and related organisms are newly recognized agents of pneumonia in humans. These bacteria are truly unique, possessing G+C ratios different from those of other groups of bacteria and having large amounts of branched-chain fatty acids. Since its initial isolation, numerous studies into the biochemistry and physiology of *L. pneumophila* have been performed. Additionally, many studies on the humoral immune response to this organism have been reported, and more recently some studies on cellular immunity to *L. pneumophila* have appeared in the literature. However, there have been no detailed studies into the immunomodulatory potential, if any, of *L. pneumophila* on normal leukocytes. Gram-negative bacteria as a group, and particularly the members of the *Enterobacteriaceae*, have been shown to possess a variety of immunomodulatory properties. One of these properties is the ability to induce blastogenesis of normal lymphocytes, particularly B-cells (1). Therefore, we describe here a series of studies performed to determine whether *L. pneumophila* has any mitogenic effect on splenocytes from normal mice or mice injected previously with viable legionellae. Mice were chosen for the animal model because most of the studies of immunomodula-

tion with gram-negative bacteria or their products have been performed on mice and because the immune system of mice is the best characterized of the animal species used in this type of study.

BDF_1 hybrid mice were used in all of the studies. *L. pneumophila* serogroup 1, strain Philadelphia 1, originally was supplied by Roger McKinney (Centers for Disease Control). Two preparations of *L. pneumophila* were used in most of the experiments; one was a Formalin-killed, washed, whole cell preparation (WC), and the other was a sonicated preparation prepared from the WC material. In some studies, purified outer membrane preparations, supplied by William Johnson (University of Iowa), and purified lipopolysaccharide (LPS) from *L. pneumophila*, supplied by K. H. Wong (Centers for Disease Control), were used. Induction of blastogenesis was monitored by determining the amount of [³H]thymidine incorporated by splenocytes during the last 18 to 24 h of a 72-h incubation, using liquid scintillation counting techniques.

Initial studies with splenocytes from normal BDF_1 mice indicated that both WC and sonicated preparations were capable of inducing increases in mitogenesis in a dose-dependent fash-

TABLE 1. Induction of blastogenesis of normal mouse splenocytes by *L. pneumophila*

Cell prepn	SI[a] tested with:				
	Medium control	*L. pneumophila* sonic extract	*L. pneumophila* WC	LPS	Concanavalin A
Nylon wool separation					
Untreated	—	12.2	5.5	26.3	117.7
Nylon wool adherent	—	12.5	8.2	55.0	85.0
Nylon wool nonadherent	—	2.6	1.6	7.9	327.2
Nude mouse splenocytes	—	3.3	13.3	12.4	2.3
Anti-Thy 1.2 plus complement					
Untreated	—	4.5	11.9	55.4	71.9
Complement treated	—	4.2	8.1	36.7	84.3
Anti-Thy 1.2 plus complement treated	—	2.8	5.4	40.6	3.8

[a] SI = counts per minute in test/counts per minute in control. Doses are as follows: sonic extract, 100 μg/ml; WC, 10^8/ml; LPS (*Escherichia coli*), 10 μg/ml; concanavalin A, 5 μg/ml.

ion, with stimulation indices (SIs) (counts per minute in test wells/counts per minute in control wells) of about 10 for 10^8 WC and 100 μg of sonic extract. The nature of the cells responding to the *Legionella* preparations was studied in another series of experiments. Studies using spleen cell preparations fractionated on nylon wool columns revealed that the nylon wool-adherent (B-lymphocyte-enriched) population responded well to the *Legionella* preparations; nylon wool-nonadherent cells (T-cell enriched) responded only minimally to legionellae (Table 1). Elimination of T-cells by treatment with anti-Thy 1.2 plus complement, which virtually eliminated responsiveness to the T-cell mitogen concanavalin A, had no effect on the blastogenic response to *L. pneumophila* (Table 1). Splenocytes from athymic nude mice exhibited positive blastogenic responses to the *Legionella* preparations, further indicating that T-cells were not the responding cells (Table 1). The mitogenic activity of the sonicated preparation was stable to heating at 56, 80, and 100°C, a characteristic similar

to that of LPS from other gram-negative bacteria. In this regard, purified LPS from *L. pneumophila* was mitogenic for normal mouse splenocytes, with an SI of approximately 20 with 10 μg/ml. Additionally, outer membrane components of *L. pneumophila* were potent mitogens for murine spleen cells, with an SI of approximately 15 with 10 μg/ml.

Other studies were performed to determine whether spleen cells from mice recovering from *L. pneumophila* infection possessed enhanced responses to legionellae in vitro. Splenocytes from mice injected 1 to 2 weeks earlier with 10^8 viable *L. pneumophila* demonstrated approximately two- to threefold greater responsiveness to *Legionella* antigens in vitro relative to cells from control mice (Table 2). However, by about 4 weeks postinfection, the blastogenic responses of the splenocytes were not significantly different from those of control mice (Table 2).

The results of this study demonstrate that components of *L. pneumophila* are mitogenic for normal mouse splenocytes. Although the

TABLE 2. Blastogenic responses of splenocytes from mice infected with *L. pneumophila*

Treatment[a]		SI[b] (mean ± SE) on postinfection day:		
In vivo	In vitro	7	15	25
None	None	—	—	—
	WC (10^6)	0.9 ± 0.1	1.7 ± 0.3	2.8 ± 0.8
	WC (10^8)	2.9 ± 0.4	10.7 ± 1.5	12.2 ± 4.0
	Sonic extract (10 μg)	1.2 ± 0.2	2.3 ± 0.1	2.9 ± 0.9
	Sonic extract (100 μg)	3.0 ± 0.5	7.8 ± 0.1	8.1 ± 1.9
L. pneumophila (10^8)	None	—	—	—
	WC (10^6)	1.3 ± 0.2	3.7 ± 0.2[c]	3.3 ± 1.0
	WC (10^8)	7.2 ± 0.4[c]	16.2 ± 1.4[c]	11.6 ± 1.1
	Sonic extract (10 μg)	2.7 ± 0.3[c]	5.0 ± 0.4[c]	2.2 ± 0.5
	Sonic extract (100 μg)	5.4 ± 0.4[c]	13.7 ± 1.3[c]	5.1 ± 0.6

[a] Mice were either untreated or injected with 10^8 viable *L. pneumophila* on the indicated days, and their splenocytes were treated for blastogenesis in response to the indicated materials in vitro.

[b] SI = counts per minute in infected mice/counts per minute in control mice.

[c] Significantly different from controls ($P < 0.05$ by Student's *t* test).

exact mechanism of interactions between legionellae and the splenocytes in this effect are not known, the responding cells are most likely B-lymphocytes, since responses were normal in cells from athymic nude mice and nylon wool-adherent populations and were not affected by treatment with anti-Thy 1.2 plus complement to remove T-lymphocytes. Additionally, the mitogenic activity of the sonicated preparation was stable to heat (100°C, 10 min), a characteristic similar to that of LPS from other gram-negative bacteria, and purified LPS from *L. pneumophila* was mitogenic for normal mouse splenocytes. LPS is known to be a potent B-cell mitogen with little or no mitogenic effect on T-cells.

Plouffee and co-workers (2) have reported that human peripheral blood leukocytes respond with increased incorporation of [^{3}H]thymidine when incubated with *L. pneumophila* in vitro, with cells from patients recovering from *Legionella* infections displaying higher SI than cells from individuals with no known previous exposure. We noted similar occurrences in mice, i.e., considerable background blastogenic responses in normal mice with elevated responses after sublethal infection with *L. pneumophila*. In recent preliminary studies, splenocytes from mice incubated for 5 rather than 3 days in vitro displayed enhanced mitogenic responses when the cells were from mice injected intraperitoneally with live legionellae 35 days earlier; control mice had an SI of less than 2, suggesting that sensitization of the infected mice had occurred. The nature of the cells exhibiting the enhanced responses remains to be determined.

LITERATURE CITED

1. **Behling, U. H., and A. Nowotny.** 1981. Immunomodulation by LPS and its derivatives, p. 165–179. *In* H. Friedman, T. Klein, and A. Szentivanyi (ed.), Immunomodulation by bacteria and their products. Plenum Publishing Corp., New York.
2. **Plouffee, J. F., and J. M. Baird.** 1981. Lymphocyte transformation to *Legionella pneumophila*. J. Clin. Lab. Immunol. **5**:149–152.

Intracellular and Extracellular Replication of *Legionella pneumophila* in Cells In Vitro

FRANK G. RODGERS AND LOUISE J. OLDHAM

Department of Microbiology and PHLS Laboratory, University Hospital, Nottingham, England

As a facultative intracellular pathogen, *Legionella pneumophila* replicates in cells in vivo (2, 5) and in cells in cultures (6). The purpose of this study was to evaluate the attachment, intracellular growth, and release of *L. pneumophila* in cell cultures in vitro.

A high-titer suspension of low-passage *L. pneumophila* serogroup 1, strain Nottingham N7, was inoculated onto cell monolayers of human diploid lung fibroblasts (MRC-5 cells) and of the continuous transformed cell lines Vero (simian) and HEp-2 (human). Viable counts were performed in duplicate on harvested culture fluids daily for 7 days, and the CFU per milliliter was estimated by using 10-fold dilutions (Rodgers and Elliott, this volume) inoculated onto enriched blood agar (1). Intracellular replication of organisms was also determined on lysed culture cells previously washed to remove adhering bacteria. The rate of release of organisms from infected cells was determined by removing the inoculum after 24 h of absorption and washing the monolayers before incubation and subsequent assay of extracellular fluids. The cytopathic effect (CPE) of *L. pneumophila* on each cell line was examined by light microscopy, and the attachment, penetration, and intracellular development of organisms were investigated by thin-section and scanning electron microscopy (SEM) by methods previously outlined (2).

Viable counts showed that MRC-5 cells took up organisms rapidly within 24 h; this was followed by replication, with a peak of intracellular organisms on day 2. Thereafter, the number of intracellular bacteria declined, and, because of release, the extracellular numbers increased. During this time a CPE developed in the monolayers, with foci of granulation and pronounced cell lysis. This progressed until by 5 days 95% of the cells were destroyed. In contrast, in the HEp-2 cells a low rate of organism uptake occurred in the first 24 h, followed by slow intracellular replication. This was maximal at 6 or 7 days postinoculation, and ultimately the intracellular CFU per milliliter was highest in this cell line. At 5 days of incubation, a CPE, consisting of foci of granulation, became discernible. Unlike in the other two cell lines, release of *L. pneumophila* from HEp-2 into the extracellular fluids progressed along with continued intracellular replication. With MRC-5 and Vero cells, the extracellular CFU per milliliter increased due to cell lysis and a consequent reduction in intracellular bacteria. HEp-2 cells appeared tolerant to *L. pneumophila*, and a more chronic infection developed. Intermediate in intracellular replication and cell degeneration were Vero cells, with an organism uptake rate similar to that for MRC-5 cells. Intracellular replication peaked at 5 days postinoculation, with an increasing rate of organism release over the 7 days. A CPE of focal degeneration appeared by day 3 of the lytic infection and increased slowly from day 5 to 7. From the CPE studies, the three cell lines appeared to occupy a spectrum of resistance to the lytic action of *L. pneumophila*, with MRC-5 cells most susceptible and HEp-2 cells most resistant. Although viable counts showed that *L. pneumophila* replicated equally well in the three cell types, differences in cell lysis rates outlined variations in the bacterial growth kinetics. *L. pneumophila* did not replicate in the cell controls containing tissue culture media and no viable cells.

By SEM, many legionellae, no doubt aided by motility (4), were seen reversibly associated with all cell lines within 1 h of inoculation. Initial adherence was by nonspecific electrostatic forces, and although this binding resisted washing, involvement of cell receptors or bacterial adhesins such as pili or flagella was not visible by SEM. Engulfment was mediated by the production of cellular microvilli and filopodia which appeared to extend from the cell surfaces toward and in response to the bacterial presence

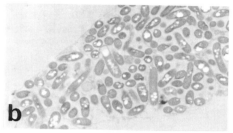
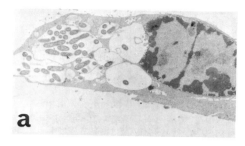

FIG. 2. Thin sections of cell monolayers inoculated with 3×10^7 CFU of *L. pneumophila* per ml. (a) HEp-2 cell at day 6 of infection. Note large vacuoles with dividing organisms surrounded by clearly defined electron-lucent zones. Magnification, $\times 2,600$. (b) MRC-5 cell at day 3 of infection, showing extensive intracellular replication. Organisms occupy the entire cytoplasmic contents. Magnification, $\times 5,200$.

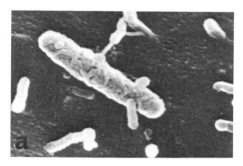

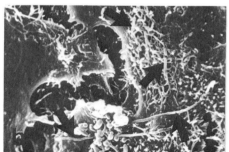

FIG. 1. SEM of cell monolayers inoculated with 3×10^7 CFU of *L. pneumophila* per ml. (a) MRC-5 cell at 2 h postinoculation. Note microvilli extending toward the organism. Magnification, $\times 24,000$. (b) Organisms (arrows) on HEp-2 cells. Engulfment is through extensive microvilli; note smooth areas of cells in absence of bacteria. Magnification, $\times 1,800$.

(Fig. 1a). Internalization of organisms was by these microvilli (Fig. 1b). Cell surfaces without bacteria were comparatively smooth. *L. pneumophila* appeared by SEM as previously described (2, 3), with rounded or tapered ends, often with terminal or lateral blebs and ruffled surfaces.

By thin-section electron microscopy of the three cell lines, organisms were seen attached to cell membranes which became thickened subjacent to the bacteria, and this was followed in the early stages by the appearance of organisms within close-fitting vacuoles surrounded by cellular ribosomes. Later, vacuoles were large and contained numerous replicating organisms surrounded by circumscribed clear zones (Fig. 2a). As intracellular replication progressed, toxic effects were noted on the cells and their organelles. Although nuclei were often disrupted, organisms were never detected within these structures. Lysis of cell contents followed extensive replication of the legionellae, and these often occupied the entire contents of the cytoplasm (Fig. 2b) before release into the extracellular fluids. No direct cell-to-cell dissemination was seen.

Morphological and viable count studies showed that *L. pneumophila* replicated within the three cell lines in vitro, but not in the extracellular fluids. If similar findings also occur in studies of clinical legionellosis, they may offer an explanation for the efficacy of erythromycin, an antibiotic which is actively taken up by eucaryotic cells but is of lower potency in vitro (Rodgers and Elliott, this volume). That the lung-derived cells were more susceptible to infection and lysis by *L. pneumophila* was not surprising. The rapid lytic infection in MRC-5 cells and the more chronic process in HEp-2 cells offer useful models of infection with this organism at the cellular level.

LITERATURE CITED

1. **Dennis, P. J., J. A. Taylor, and G. I. Barrow.** 1981. Phosphate buffered, low sodium chloride blood agar medium for *Legionella pneumophila*. Lancet ii:636.
2. **Rodgers, F. G.** 1979. Ultrastructure of *Legionella pneumophila*. J. Clin. Pathol. **32:**1195–1202.
3. **Rodgers, F. G., and M. R. Davey.** 1982. Ultrastructure of the cell envelope layers and surface details of *Legionella pneumophila*. J. Gen. Microbiol. **128:**1547–1557.
4. **Rodgers, F. G., P. W. Greaves, A. D. Macrae, and M. J. Lewis.** 1980. Electron microscopic evidence for flagella and pili on *Legionella pneumophila*. J. Clin. Pathol. **33:**1184–1188.
5. **Rodgers, F. G., A. D. Macrae, and M. J. Lewis.** 1978. Electron microscopy of the organism of Legionnaires' disease. Nature (London) **272:**825–826.
6. **Wong, M. C., E. P. Ewing, C. S. Callaway, and W. L. Peacock.** 1980. Intracellular multiplication of *Legionella pneumophila* in cultured human embryonic lung fibroblasts. Infect. Immun. **28:**1014–1018.

Age-Related Immunity to *Legionella pneumophila* in a Community Population

JOSEPH F. PLOUFFE AND IAN M. BAIRD

The Ohio State University, Columbus, Ohio 43210, and Riverside Methodist Hospital, Columbus, Ohio 43214

Legionella pneumophila appears to be ubiquitous in our environment (2, 3). Sporadic outbreaks of Legionnaires disease pneumonia have occurred in many locations throughout the world (4). The attack rates in these outbreaks range from <1 to 5% (1). Why certain people acquire the disease and others do not is not understood. This study was designed to look at both humoral and cellular immunity to *L. pneumophila* antigens in an age-stratified community population in Columbus, Ohio. Subjects known to be immunosuppressed were excluded. The antibody titers to *L. pneumophila* serogroups 1 through 6 were measured in 661 subjects by the indirect fluorescent-antibody technique (7) with reagents kindly supplied by the Bureau of Biologics, Centers for Disease Control. Titers of <16 were considered to be 1 for calculation of geometric means. The method of lymphocyte transformation has been reported previously (5, 6). Briefly, Ficoll-Hypaque-purified mononuclear cells are cultured for 5 days with sonicated antigens of *L. pneumophila* serogroups 1 through 6 (10 to 100 µg of protein/ml of culture), and blastogenesis is determined by [^{3}H]thymidine incorporation. The results from 633 subjects are reported as the stimulation index (SI), which is the counts per minute of triplicate stimulated cultures/counts per minute of triplicate unstimulated cultures.

Antibody titers were highest to *L. pneumophila* serogroup 1. The mean geometric titers rose from 3 in the cord blood specimens to 8 in decade 2, peaked at 19 in decade 3, fell to 11 in decade 4 and to 10 in decade 5, and stabilized at 5 to 6 in decades 6 through 9.

In the blastogenic assay with sonicated *L. pneumophila* antigens, there was a gradual increase in the mean SI ± 1 standard deviation (Table 1) from 2.73 ± 2.88 in neonates through 5.76 ± 7.28 in decade 5. It then leveled off through decade 8 and declined in decade 9. If the data are expressed as the percentage of the age groups with SI > 5 (Fig. 1), then a linear relationship ($y = 10.9 + 0.56x$) is found through decade 7 ($r = 0.97$), where 45% of the population have SI > 5 to *L. pneumophila* serogroup 1 antigen. The percentage of individuals in decades 8 and 9 with SI > 5 declined to 20 to 25%. Approximately 10% of those ≥20 years of age had SI > 10, and 1 to 5% had SI > 20.

The normal community population without a history of Legionnaires disease in an area where

TABLE 1. Lymphocyte blastogenesis with crude *L. pneumophila* serogroup 1 antigens

Age	No.	SI (mean ± 1 SD)
Neonates	213	2.73 ± 2.88
20–29	53	4.29 ± 5.24 [a]
30–39	62	4.40 ± 4.42[b]
40–49	85	5.76 ± 7.28[b]
50–59	83	5.58 ± 5.43[b]
60–69	69	5.33 ± 6.34[b]
70–79	43	6.63 ± 15.00[b]
≥80	27	4.06 ± 7.2[b]

[a] $P < 0.01$ compared with neonates.
[b] $P < 0.001$ compared with neonates.

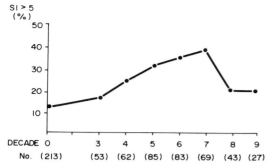

FIG. 1. Lymphocyte blastogenesis to *Legionella* antigens in a community population. The percentage of individuals in each age group with SI > 5 is plotted versus age. There is a linear relationship ($y = 10.9 + 0.56x$) through decade 7.

or whether a different clinical illness occurs is not known. We have recognized different subtypes of *L. pneumophila* serogroup 1 which appear to have differences in virulence (J. F. Plouffe, M. Para, B. Hackman, L. Webster, and W. Maher, this volume). Perhaps less virulent environmental strains of *L. pneumophila* infect subjects without causing severe disease and confer protection against the more virulent strains of *L. pneumophila*.

This work was supported by Public Health Service grant AI-17552 from the National Institutes of health.

We thank Judy Smiley, Norman Delzeith, and Barbara Hackman for their assistance and Sharon Venters for typing the manuscript.

L. pneumophila is prevalent in the environment appears to have both humoral and cellular sensitization to *Legionella* antigens. The mean antibody levels as measured by the indirect fluorescent-antibody technique peak during decade 3 and then decline in the older population. The lymphocyte studies show stepwise increases in the percentage of the population with blastogenic responses of >5 through decade 7 and then a decline in the older population.

These data suggest that normal subjects are exposed to *Legionella* antigens in their environment and develop an immune response without acquiring typical Legionnaires disease. Whether these subjects are asymptomatically infected with legionellae (or a closely related organism)

LITERATURE CITED

1. **Broome, C. V., and D. W. Fraser.** 1979. Epidemiologic aspects of legionellosis. Epidemiol. Rev. **1**:1–16.
2. **Fliermans, C. B., W. B. Cherry, L. H. Orrison, S. J. Smith, D. L. Tison, and D. H. Pope.** 1981. Ecological distribution of *Legionella pneumophila*. Appl. Environ. Microbiol. **41**:9–16.
3. **Fliermans, C. B., W. B. Cherry, L. H. Orrison, and L. Thacker.** 1979. Isolation of *Legionella pneumophila* from nonepidemic-related aquatic habitats. Appl. Environ. Microbiol. **37**:1239–1242.
4. **Myer, R. D.** 1983. Legionella infections: a review of five years of research. Rev. Infect. Dis. **5**:258–278.
5. **Plouffe, J. F.** 1979. Lymphocyte cultures with small numbers of cells. J. Immunol. Methods **29**:111–115.
6. **Plouffe, J. F., and I. M. Baird.** 1981. Lymphocyte transformation to *Legionella pneumophila*. J. Clin. Lab. Immunol. **5**:149–152.
7. **Wilkinson, H. W., B. J. Fikes, and D. D. Cruce.** 1979. Indirect immunofluorescence test for serodiagnosis of Legionnaires disease: evidence for serogroup diversity of Legionnaires disease bacterial antigens and for multiple specificity of human antibodies. J. Clin. Microbiol. **9**:379–383.

Immunostimulation by *Legionella pneumophila* Antigens

HERMAN FRIEDMAN, THOMAS KLEIN, AND RAYMOND WIDEN

Department of Microbiology and Immunology, University of South Florida College of Medicine, Tampa, Florida 33612

Legionella pneumophila is considered the etiological agent of Legionnaires disease in humans (4). Although much is now known about this organism, especially morphological and physiological characteristics as well as environmental and ecological niche, much less is known about host immune responses to this bacterium (1a). Serum antibody is readily induced in humans and experimental animals to this organism. Such antibody reacts with surface antigenic components (5).

Legionella organisms are now considered to be facultative, intracellular, opportunistic pathogens which readily replicate within macrophages (2). However, if macrophages are derived from

immune individuals or nonspecifically activated, they rapidly kill or inhibit the legionellae (2). During the host defense responses to these bacteria, serogroup-specific antigens on the surface of the microorganisms induce group-specific immunity (1). An acid-soluble protein toxin elaborated from the organisms and present in extracts may also induce cross-protection to different serogroups. The effects of antibody to these antigens in promoting phagocytosis and abrogating toxicity recently has been studied by Wong et al. (6).

Serogroup-specific antigenic complexes have been found to serve as a potent adjuvant in modulating host immune responses to legionellae as well as to other antigens. In the present study, the immunomodulatory properties of antigenic preparations derived from legionellae were investigated. Enhanced antibody formation to an unrelated antigen, i.e., sheep erythrocytes (SRBCs), was induced in mice injected with either *Legionella* whole cell vaccine or a soluble sonic extract thereof. An antigenic extract rich in lipopolysaccharide (LPS)-like components was also found to be immunostimulatory.

For these studies, BALB/c mice were injected with killed *Legionella* vaccine or a sonic extract derived from the organisms either simultaneously, before, or after immunization with SRBCs as antigens. The antibody response to the SRBCs was assessed by the localized antibody-plaque assay, using a microplaque technique (3). In brief, spleen cells from SRBC-immunized mice were mixed with 1% agarose and plated for plaque-forming cells (PFC) detected by the subsequent addition of complement. In vitro assays were also performed whereby spleen cells from normal mice were incubated with various amounts of *Legionella* preparations and immunized with SRBCs. After several days of incubation in vitro, the number of PFC were determined by the plaque assay.

Mice injected with graded numbers of killed *Legionella* organisms showed a variable degree of enhanced antibody responses to the SRBCs, depending upon the dose of antigen (Table 1). As little as 10^6 or 10^7 legionellae resulted in a two- to threefold enhancement of the antibody response. This concentration of the vaccine had a slight to moderate effect on the spleen weight of the animal. Higher doses of *Legionella* vaccine moderately suppressed the antibody response, but did not appear significant.

The sonic extract (1 to 100 µg) had a similar enhancing activity for the antibody response of mice to SRBCs (Table 1). A dose of 10 to 50 µg of sonic extract resulted in a two- to fourfold increase in the antibody response and also had a slight enhancing effect on the spleen cellularity. The enhancement induced by *Legionella* antigen was generally similar to that induced by purified LPS extract from *Escherichia coli*. In additional experiments it was found that an LPS-rich extract provided by K. H. Wong, Centers for Disease Control, Atlanta, Ga., had similar immunoenhancing activity for the mouse spleen cells. Heating the LPS at 100°C for 10 min had no effect on immunostimulatory activity, suggesting that the immunoenhancement was indeed due to the endotoxin per se.

Relatively similar effects were noted when the *Legionella* vaccine or sonic extract was added to spleen cell cultures immunized with SRBCs in vitro. A vaccine of 10^6 to 10^7 killed *Legionella* organisms resulted in a three- to fourfold en-

hancement of the antibody response on the peak day (day 5) of the culture (Table 2). Even greater enhancement occurred with 10 to 100 µg of *Legionella* sonic extract added to the cultures in vitro. Similar to the in vivo results, the *Legionella* LPS, as well as *E. coli* LPS, had immunoenhancing activity.

It is noteworthy that the immunoenhancement was evident not only on the peak day of the immune response, both in vivo and in vitro, but also at earlier and later times (data not shown). Thus, the cytokinetics of the antibody response to the SRBCs were relatively similar regardless of whether *Legionella* antigen was administered, but the total numbers of PFC developing on each day of assay were higher than observed in the untreated controls.

Maximum PFC enhancement occurred when the legionellae, as either vaccine, sonic extract, or LPS, were added on the day of culture initiation or 1 day thereafter, but not at a later time. Thus, immunoenhancement appeared to be related to stimulation of early events in antibody formation, culminating in enhancement of the total antibody response. Furthermore, *Legionella* antigen, given either in vivo or in vitro, also resulted in increased "background" antibody formation in unimmunized animals or cultures (data not shown). Such enhancement of the background response indicated that the *Legionella* antigen could serve as a polyclonal B cell stimulator, probably similar to the polyclonal B cell stimulator activity of many other

TABLE 1. Effect of *Legionella* antigen on the antibody response of mice immunized with SRBCs

Bacterial prepn injected[a]	Amt	Antibody response	
		PFC/10^6 spleen cells (mean ± SD)[b]	% of control
None (control)		706 ± 182	—
Killed legionellae	10^5	1,038 ± 95	147
	10^6	1,630 ± 250	231
	10^7	1,980 ± 136	275
	10^8	630 ± 190	89
	10^9	510 ± 60	72
Legionella sonic extract	1.0 µg	1,280 ± 162	181
	10.0 µg	1,680 ± 95	238
	50.0 µg	2,982 ± 150	423
Legionella LPS	100.0 µg	2,850 ± 193	418
E. coli LPS	100.0 µg	1,650 ± 178	233

[a] The indicated amount of bacterial antigen was injected intraperitoneally into groups of mice on the day of intraperitoneal immunization with 2×10^8 SRBCs.
[b] For three or four mice per group 5 days after immunization.

TABLE 2. Effect of *Legionella* antigen on in vitro antibody response of mouse spleen cells immunized in vitro with SRBCs

Bacterial prepn added[a]	Amt	Antibody response	
		PFC/10^6 spleen cells (mean ± SD)[b]	% of control
None (control)		548 ± 120	—
Killed legionellae	10^4	620 ± 180	113
	10^5	897 ± 95	164
	10^6	1,760 ± 285	321
	10^7	2,381 ± 190	435
	10^8	1,960 ± 240	285
Legionella sonic extract	1.0 μg	950 ± 130	174
	10.0 μg	1,830 ± 420	336
	50.0 μg	4,050 ± 360	739
	100.0 μg	6,500 ± 360	1,186
Legionella LPS	1.0 μg	2,060 ± 420	368
	10.0 μg	3,150 ± 460	552
E. coli LPS	10.0 μg	950 ± 160	173
	100.0 μg	1,280 ± 95	234

[a] The indicated amount of bacterial antigen was added to cultures of 5×10^6 spleen cells from normal mice immunized in vitro with 2×10^6 SRBCs.

[b] For three to five cultures per group 5 days after culture incubation.

gram-negative bacterial antigens.

To examine the possible mechanisms involved, coculture experiments were performed in which spleen cell suspensions were first incubated with the *Legionella* antigen and then fractionated into glass-adherent (macrophage-rich) and glass-nonadherent (lymphocyte-rich) cell preparations. Upon coculture of each of the cell fractions with the corresponding fraction from normal spleen cell populations which had not been first stimulated with the *Legionella* antigen, it was apparent that the macrophages appeared to be most involved in the enhanced response (data not shown). For example, adherent cell populations from spleen cells incubated for a period of 4 to 12 h with *Legionella* antigen, after separation and washing, enhanced the antibody response of nonadherent lymphoid cell populations from normal mice immunized in vitro with SRBCs. Enhancement was at least two- to fourfold. Supernatants from spleen cells or macrophage-rich cell populations incubated with *Legionella* antigen in vitro for 3 to 5 days also were immunoenhancing. Since macrophages mediated such enhancement, this suggests that antibody helper factors from these cells were being induced by the *Legionella* antigen.

The cells responding to the antibody helper factor(s) appeared to be B lymphocytes, since spleen cells deficient in B cells after treatment with anti-immunoglobulin serum plus complement, but not anti-theta serum plus complement, showed much less enhancement of antibody responses to SRBCs when treated either with *Legionella* antigen or with soluble supernatant factors from macrophage cultures treated with *Legionella* antigen in vitro. Thus, it appears likely that the *Legionella* antigen, like other gram-negative bacterial preparations, is a marked enhancer of antibody formation. This may also be related to the mitogenic properties of this *Legionella* antigen, as shown in previous studies. The enhancing factor(s) in the *Legionella* preparations appeared to be heat-stable endotoxin which, when added to spleen cell or macrophage cultures in vitro, induces the release of an antibody helper factor(s), presumably interleukin 1 or a similar lymphocyte-activating factor(s). Thus the factor(s) appears to have the ability to stimulate enhanced polyclonal B cell activity.

The relationship of immunostimulation induced by *Legionella* antigens and the pathobiology of the infection caused by legionellae is not clear. However, legionellae enhance antibody function not only to unrelated antigens such as SRBCs and bovine serum albumin (as shown by Wong and Feeley) but also to the organism itself. Antibody is readily detectable in both humans and experimental animals after exposure to viable legionellae or after vaccination with killed *Legionella* vaccine or extracts. Also, antibody to legionellae appears to enhance the phagocytosis of these organisms by host phagocytes. However, as shown by Horwitz and Silverstein (2), after phagocytosis of opsonized legionellae by peripheral blood monocytes from nonimmune individuals, the legionellae still replicate rapidly and may kill the monocytes.

It is plausible that an inappropriate immune response to legionellae, such as development of noninhibitory but opsonizing antibody, may subvert the host's immune defense to the organism. Cell-mediated immunity may be a more important mechanism of killing these bacteria in vivo, and thus stimulation of B lymphocytes to produce higher levels of antibody, rather than stimulation of killer or helper T cells, may be an advantage to the bacterium and disadvantage to the host. Although major gaps exist in the knowledge of the pathogenesis of *Legionella* species, a further understanding of factors responsible for the resistance or susceptibility of individuals is an important goal for understanding immunity to these bacteria.

LITERATURE CITED

1. **Elliot, J. A., W. Johnson, and C. M. Helms.** 1981. Ultrastructural localization and protective activity of high-molecular-weight antigen isolated from *Legionella pneumophila*. Infect. Immun. 31:822–824.

1a.Friedman, H., R. Widen, and T. Klein. 1983. Cellular immunity to *Legionella* antigen—laboratory assays. Clin. Immunol. Newsl. 4:92–95.

2. Horwitz, M. A., and S. C. Silverstein. 1980. Legionnaire's disease bacterium (*Legionella pneumophila*) multiplies intracellularly in human monocytes. J. Clin. Invest. 68:441–450.

3. Kamo, I., S.-H. Pan, and H. Friedman. 1976. A simplified procedure for in vitro immunization of dispersed spleen cells. J. Immunol. Methods 11:55–62.

4. Mayer, R. D., and S. M. Finegold. 1980. Legionnaire's disease. Annu. Rev. Med. 31:219–232.

5. Wilkinson, H. W., D. D. Cruce, and C. V. Broome. 1981. Validation of *Legionella pneumophila* indirect immunofluorescence assay with epidemic sera. J. Clin. Microbiol. 13:139–146.

6. Wong, K. H., W. O. Schalla, M. C. Wong, P. R. B. McMaster, J. C. Feeley, and R. J. Arko. 1982. Biologic activities of antigen from Legionella pneumophila, p. 434–443. In J. Robbins, J. Hill, and S. Sadoff (ed.), Bacterial vaccines. Thieme-Stratton Co., New York.

Longitudinal Immunological Studies in Patients with Legionnaires Disease

JOSEPH F. PLOUFFE AND IAN BAIRD

The Ohio State University, Columbus, Ohio 43210, and Riverside Methodist Hospital, Columbus, Ohio 43214

Legionnaires disease has been recognized to be endemic in Columbus, Ohio, since 1977 (2). Previous reports from this laboratory have shown that patients with Legionnaires disease develop sensitized lymphocytes capable of undergoing blastogenesis when cultured with *Legionella* antigens (4). We have also shown that this response occurs fairly quickly after the onset of clinical disease and before the production of antibody (5). We have demonstrated minimal nonspecific stimulation when cord blood lymphocytes are exposed to *Legionella* antigens (6). In this paper we report the results of longitudinal immunological studies in our population of patients after acute Legionnaires disease. We report that the lymphocyte response is mainly T cell mediated and present preliminary data on delayed-hypersensitivity skin tests with *Legionella* antigens.

Patient populations. Seventy-five patients with Legionnaires disease were studied. The diagnosis was made by positive cultures for *Legionella pneumophila*, direct fluorescent-antibody stain of pulmonary secretions, a fourfold rise in reciprocal titer to ≥128, or a convalescent reciprocal titer of ≥256 (1). All patients had disease caused by *L. pneumophila* serogroup 1 (SG-1). After informed consent, blood was obtained during the acute phase of disease (within 2 weeks of onset), at 1 month, and then at 6-month intervals thereafter. All samples were not obtained from each subject.

Antibody studies. Antibodies to *L. pneumophila* were measured by indirect fluorescence assay by the method of Wilkinson et al. (7), using reagents for *L. pneumophila* SG-1 through SG-6 kindly supplied by the Bureau of Biologics, Centers for Disease Control. The reciprocal geometric mean titers to *L. pneumophila* SG-1 antigens are presented in Table 1. Thirty patients had 87 samples obtained from 6 months to 5.5 years after disease and could be grouped into three categories of antibody responders. Seventeen patients had mean peak titers of 338 that by 18 months fell to ≤128. A second group ($n = 8$), with initial peak titers of 588, had titers of 512 at 1 to 1.5 years and then maintained stable titers of 256 up to 3 years after disease. A third group of high responders ($n = 5$) had mean titers at 1 month of 675 and then maintained titers of ≥1,024 for at least 2 years.

Lymphocyte studies. Mononuclear cells were separated from heparinized peripheral blood by density gradient centrifugation. In several experiments T cells were depleted from mononuclear cells by sheep erythrocyte rosetting, and B cells were depleted by immunoadsorption using a column with immunoglobulin G. Triplicate microcultures were established with 50,000 lymphocytes in 0.1-ml volumes with 80% RPMI 1640 with HEPES (N-2-hydroxyethylpiperazine-N'-2-ethanesulfonic acid) buffer and 20% normal pooled plasma in microtiter plates (3). Antigens were added in 10-μl volumes. The cultures were incubated for 5 days at 37°C in 5% CO_2 and air. [^{3}H]thymidine (0.5 μCi) was added for the final 6 h of incubation. The cultures were harvested, and the counts per minute of tritium were determined by liquid scintillation spectrometry.

Antigens. The following strains were obtained from the Centers for Disease Control: *L. pneumophila* SG-1 Philadelphia 1, *L. pneumophila* SG-2 Togus 1, *L. pneumophila* SG-3 Bloomington 2, *L. pneumophila* SG-4 Los Angeles 1, *L. pneumophila* SG-5 Dallas 1, and *L. pneumophila* SG-6 Chicago 2. The organisms were harvested from charcoal-yeast extract agar, Formalin killed, dialyzed, and sonicated. Preliminary experiments in known positive controls determined the concentrations which gave optimal stimulating activity (10 to 100 μg of protein/ml of culture).

In five Legionnaires disease patients a mononuclear cell population gave mean stimulation indices (SI) (counts per minute of stimulated cultures/counts per minute of background) to *L. pneumophila* SG-1 antigens of 26 ± 11 (mean ± standard deviation). T cell populations gave SI of 26 ± 21, and B cell populations gave 2 ± 2. Longitudinal studies with Knoxville antigen are presented in Table 2. The data are presented as mean SI and percentages of patients with SI of >5, >10, and >20. Fairly consistent activity remained up to 4 years postinfection. These data can be compared with data from community controls (Plouffe and Baird, this volume) with mean SI of approximately 5. Cross-reactivity in the lymphocyte transformation assay with other *L. pneumophila* serogroups was high (data not shown), indicating shared antigens among the *L. pneumophila* serogroups.

Skin tests. Preliminary delayed-hypersensitivity skin test experiments were performed. The antigen used was the initial peak from DEAE chromatography of sonicated *L. pneumophila* SG-1 (Knoxville), as described by Wong et al. (8, 9). Concentrations of between 10 and 100 μg of protein/0.1 ml were injected intradermally; this elicited erythema and induration in 5 subjects with previous Legionnaires disease and no induration in 10 controls. The higher concentration produced an exaggerated response in one subject, and the lower concentrations only gave a positive response (>10 mm of induration) when the subjects were retested 1 week later. Skin biopsy in one patient demonstrated a mononuclear cell inflammatory reaction consistent with a delayed-hypersensitivity response. Further studies are being performed to deter-

TABLE 2. Lymphocyte blastogenesis to *L. pneumophila* SG-1 antigens

Time after disease	No. of subjects	SI			
		Mean ± 1 SD	% > 5	% > 10	% > 20
≤2 wk	39	23 ± 48	51	36	31
1 mo	38	33 ± 37	92	68	50
6 mo	21	19 ± 21	76	57	38
1 yr	16	20 ± 19	88	63	38
1.5 yr	12	25 ± 20	100	92	42
2 yr	12	21 ± 21	83	67	42
2.5 yr	11	43 ± 37	82	82	64
3 yr	8	12 ± 8	75	50	25
3.5 yr	3	63 ± 101	67	33	33
4 yr	8	26 ± 40	63	50	25

mine the concentration which will give a positive response on initial exposure.

Our results indicate that both humoral and cellular immune responses develop to *L. pneumophila* antigens after clinical Legionnaires disease and persist for at least 4 years.

This work was supported by Public Health Service grant AI-17552 from the National Institutes of Health.

We thank Judy Smiley, Norman Delzeith, and Barbara Hackman for their assistance and Sharon Venters for typing the manuscript.

LITERATURE CITED

1. **Centers for Disease Control.** 1978. Legionnaires' disease: diagnosis and management. Ann. Intern. Med. **88:**363–365.
2. **Marks, J. S., T. F. Tsai, W. J. Martone, R. C. Baron, J. Kennicott, F. J. Holtzhauer, I. Baird, D. Fay, J. C. Feeley, G. F. Mallison, D. W. Fraser, and T. J. Halpin.** 1979. Nosocomial Legionnaires' disease in Columbus, Ohio. Ann. Intern. Med. **90:**565–569.
3. **Plouffe, J. F.** 1979. Lymphocyte cultures with small numbers of cells. J. Immunol. Methods **29:**111–115.
4. **Plouffe, J. F., and I. M. Baird.** 1981. Lymphocyte transformation to *Legionella pneumophila.* J. Clin. Lab. Immunol. **5:**149–152.
5. **Plouffe, J. F., and I. M. Baird.** 1982. Lymphocyte blastogenic response to *L. pneumophila* in acute legionellosis. J. Clin. Lab. Immunol. **7:**43–44.
6. **Plouffe, J. F., and I. M. Baird.** 1982. Cord blood lymphocyte transformation to *Legionella pneumophila.* J. Clin. Lab. Immunol. **9:**119–120.
7. **Wilkinson, H. W., B. J. Fikes, and D. D. Cruce.** 1979. Indirect immunofluorescence test for serodiagnosis of Legionnaires disease: evidence for serogroup diversity of Legionnaires disease bacterial antigens and multiple specificity of human antibodies. J. Clin. Microbiol. **9:**379–383.
8. **Wong, K. H., P. R. B. McMaster, J. C. Feeley, R. J. Arko, W. D. Schalla, and F. W. Chandler.** 1980. Detection of hypersensitivity to *Legionella pneumophila* in guinea pigs by skin test. Curr. Microbiol. **4:**105–110.
9. **Wong, K. H., W. D. Schalla, R. J. Arko, J. C. Bullard, and J. C. Feeley.** 1979. Immunochemical, serologic and immunologic properties of major antigens isolated for the Legionnaires' disease bacterium. Observations bearing on the feasibility of a vaccine. Ann. Intern. Med. **90:**634–638.

TABLE 1. Longitudinal antibody titers to *L. pneumophila* SG-1

Time after disease	No. of subjects	Reciprocal geometric mean titer
<2 wk	75	5
1 mo	67	511
6 mo	13	372
1 yr	14	116
1.5 yr	10	274
2 yr	11	128
2.5 yr	11	169
3 yr	8	210
≥3.5 yr	18	170

In Vitro Influence of Lipopolysaccharide from Two Species of Legionella on Human Blood Leukocyte Procoagulant Activity

DONATO FUMAROLA, CARLO MARCUCCIO, AND GIUSEPPE MIRAGLIOTTA

Institute of Medical Microbiology, University of Bari, 70124 Bari, Italy

Legionellosis is a multisystemic disease often associated with various hemostatic complications, including thrombocytopenia, deep venous thrombosis, and disseminated intravascular coagulation (2, 4, 5, 8, 10). Activation of the coagulation pathway contributes in additional ways to the damage. However, the reason(s) for abnormal clotting during legionellosis is not yet completely understood. In previous works (3, 7) we have demonstrated that normal human peripheral blood mononuclear cells generate a potent procoagulant activity (PCA), identified as tissue factor, when exposed in vitro to various members of the genus Legionella. The evidence that several Legionella strains have some of the in vivo and in vitro biological properties associated with endotoxin of typical gram-negative bacteria and the morphological similarity revealed by electron microscopy studies between the bacterium associated with Legionnaires disease and gram-negative bacteria might suggest that the presence of an endotoxin-like substance(s) in the outer membrane of the organism is the major reason for the production of mononuclear cell PCA. Indeed, in the above-mentioned works Legionella organisms behaved like the typical gram-negative bacterium Escherichia coli.

In the present paper we report the in vitro PCA production after exposure of blood samples to lipopolysaccharide (LPS) extracted by the procedure of Westphal et al. (11) from Legionella pneumophila (Togus strain) and Legionella bozemanii (WIGA strain) (kindly provided by O. Lüderitz, Max Planck Institute, Freiburg, Federal Republic of Germany). We carried out our investigation with both aqueous- and phenolic-phase LPS extracted from these Legionella strains.

The study was performed as follows. Citrated whole blood was mixed with each of the LPS preparations (final concentration, 20 µg/ml) or with a similar volume of sterile isotonic saline (as a control) and incubated at 37°C in plastic tubes. Directly after mixing and after 4 h of incubation, production of PCA was measured in whole blood-LPS mixtures and in mononuclear cells isolated from these mixtures by the Ficoll-Hypaque technique (1). This was performed by measurements of the one-stage recalcification time (9) for the whole blood-LPS mixtures and by determining the capacity of isolated cells to shorten the clotting time of human plasma, using

a mixture of 0.1 ml of plasma, 0.1 ml of cell suspension, and 0.1 ml of 0.025 M $CaCl_2$ (6), for the mononuclear cells. The results are summarized in Table 1. Both L. pneumophila- and L. bozemanii-extracted LPS strongly induced the generation of mononuclear cell PCA after 4 h of incubation with whole blood. Indeed, mononuclear cells isolated from these mixtures shortened the recalcification time of normal plasma. No significant difference was found between the two LPS preparations and between the phenolic and the aqueous phase of each LPS. These results closely parallel those obtained by measuring PCA generation in whole blood-LPS mixtures. No PCA generation was found in control samples incubated with sterile saline. LPS induced the classical factor VII-dependent PCA (tissue factor), since no PCA was produced when plasma from factor VII-deficient patients was used (data not shown).

The experiments described here indicate that LPS from two species of Legionella is able to activate human mononuclear cell PCA production and that there is no significant difference between them in this regard. The finding that no significant difference was found between aqueous- and phenolic-phase LPS extracts is not surprising, since their main difference is that the latter is more lipophilic than the former, whereas the biological effects of LPS (except the anticomplementary effect and O antigen specificity) are due to lipid A. To the best of our knowledge, the composition of lipid A is the same for aqueous- and phenolic-phase LPS from Legionella species. However, our findings may have

TABLE 1. PCA of human mononuclear cells (5,000/µl) in whole blood-LPS mixtures and after isolation

Test material[a]	Recalcification time (s)	
	Whole blood	Isolated cells
L. pneumophila LPS		
Water phase	61–90	70–135
Phenol phase	54–86	54–100
L. bozemanii LPS		
Water phase	58–82	47–98
Phenol phase	57–80	50–80
Saline	294–358	300–360

[a] Ten experiments were performed with each material.

implications contributing to understanding of the mechanism(s) underlying the thrombotic complications or disseminated intravascular coagulation sometimes associated with severe legionellosis.

This work was supported in part by a grant from Ministero della Pubblica Istruzione.

LITERATURE CITED

1. Böyum, O. 1968. Separation of leucocytes from blood and bone marrow with special reference to factors which influence and modify sedimentation properties of hematopoietic cells. Scand. J. Clin. Lab. Invest. 21(Suppl. 97):9–19.
2. Dowling, J. N., F. J. Kroboth, M. Karpf, R. B. Yee, and A. W. Pasculle. 1983. Pneumonia and multiple lung abscesses caused by dual infection with Legionella micdadei and Legionella pneumophila. Am. Rev. Respir. Dis. 127:121–125.
3. Fumarola, D., and G. Miragliotta. 1983. Legionella longbeachae and disseminated intravascular coagulation. Can. Med. Assoc. J.128:782.
4. Gasper, T. M., P. A. Farndon, and R. Davies. 1978. Thrombocytopenia associated with Legionnaires' disease. Br. Med. J. 2:1611–1612.
5. Lam, S., J. A. Smith, J. D. Burton, R. G. Evelyn, B. Harper, V. Huckell, and E. A. Jones. 1982. Legionella longbeachae pneumonia diagnosed by bronchial brushing. Can. Med. Assoc. J. 127:223–224.
6. Miragliotta, G., D. Fumarola, M. Colucci, and N. Semeraro. 1981. Platelet aggregation and stimulation of leucocyte procoagulant activity by rickettsial lipopolysaccharides in rabbits and in man. Experientia 37:47–48.
7. Miragliotta, G., N. Semeraro, L. Marcuccio, and D. Fumarola. 1982. Legionella pneumophila and related organisms induce the generation of procoagulant activity by peripheral mononuclear cells in vitro. Infection 10:215–218.
8. Oldenburger, D., J. P. Carson, W. J. Gundlach, F. I. Ghaly, and W. H. Wright. 1979. Legionnaires' disease. Association with Mycoplasma pneumoniae and disseminated intravascular coagulation. J. Am. Med. Assoc. 241:1269–1270.
9. Østerud, B., and E. Bjørklid. 1982. The production and availability of tissue thromboplastin in cellular populations of whole blood exposed to various concentrations of endotoxin. Scand. J. Haematol. 29:175–184.
10. Segnestam, K., B. Gästrin, R. Lundström, and A. Lövestad. 1980. Case of Legionnaires' disease with deep venous thrombosis. Infection 8:126–127.
11. Westphal, O., O. Lüderitz, and F. Bister. 1952. Über die Extraction von Backterien mit Phenol/Wasser. Z. Naturforsch. 7B:148–155.

Legionella Vaccine Studies in Guinea Pigs: Protection Against Intraperitoneal but Not Aerosol Infection

T. K. EISENSTEIN, R. TAMADA,† AND J. MEISSLER

Department of Microbiology and Immunology, Temple University School of Medicine, Philadelphia, Pennsylvania 19140

The literature supports the concept that legionellae are facultative intracellular pathogens. They have been shown to grow in vitro in phagocytes as well as in tissue-cultured lines (2, 7, 8, 14, 15). In general, control of growth of other obligate or facultative intracellular pathogens is believed to occur via macrophage activation mediated by lymphokines released from sensitized T cells that have contacted the antigens of the organism to which the T cells have been previously exposed. The activated macrophage then acquires enhanced microbicidal and tumoricidal capacity (1, 10). This effector pathway has been termed cell-mediated immunity. For some of these organisms, such as Mycobacterium tuberculosis and Listeria monocytogenes, antibodies appear to play no role in acquired resistance (11, 13). However, for other organisms, such as Salmonella spp. (4) and Toxoplasma gondii (9), antibody can be shown to afford significant protection. The role of antibody has an important bearing on vaccine strategy, as it is generally believed that killed vaccines

induce antibody, but that only live vaccines (or killed organisms in complete Freund adjuvant) can induce cell-mediated immunity. Therefore, if antibody can confer significant protection against legionellae, nonviable extracts or whole killed cells may be considered candidate vaccines. However, if the only mechanism of immunity is via cell-mediated immunity, then only viable, attenuated legionellae may be a satisfactory vaccine.

The vaccination studies reported in this paper were carried out with guinea pigs as an experimental model of Legionella infection.

To assess protection by various vaccines against pulmonary infection, an aerosol model of guinea pig infection was established by using a Tri-R airborne-infection apparatus (Tri-R Instruments, Inc., Rockville Center, N.Y.) of the type originally designed by Middlebrook (12). Animals were placed in a wire basket in the chamber of the instrument, and an aqueous suspension of Legionella organisms was put into the nebulizer. Aerosolization was carried out for 30 min, the chamber was flushed with fresh air for 10 min, and the surface of the fur was disinfected with UV light for 15 min. The density of Legionella

† Present address: Department of Surgery, Kyushu University, Fukuoka, Japan.

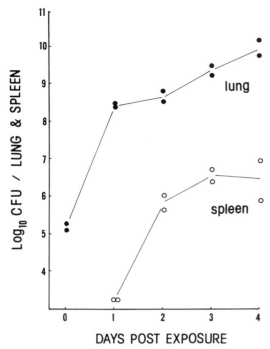

DAYS POST EXPOSURE

FIG. 1. Growth of *L. pneumophila* in aerosol-infected guinea pig lungs and spleens. Two animals were sacrificed at each time point.

organisms in the chamber was estimated by drawing 1,000 ml of air through an impinger filled with distilled water and determining the number of viable CFU in 0.1 ml of the water on BCYEα agar (charcoal-yeast extract agar plates buffered with 1% ACES buffer [*N*-(2-acetamido)-2-aminoethanesulfonic acid] and 0.1% α-ketoglutaric acid). By using the formula of Guyton (6) to calculate the respiratory volume of each animal, the theoretical number of organisms inhaled in 30 min was determined. The actual number of organisms in a lung was veri-

fied by sacrificing the animal immediately after aerosolization, homogenizing the lungs, and plating appropriate dilutions on BCYEα agar plates. From the number of CFU per plate the number of organisms per lung was calculated. There was excellent agreement between the number of CFU per lung calculated from the impinger counts with the Guyton formula and the actual number of organisms recovered after aerosolization on necropsy.

Using this system, an aerosol 50% lethal dose (LD_{50}) of 2×10^4 organisms per lung was established for 350-g strain 2 guinea pigs (Murphy Breeding Laboratories, Plainfield, Ind.). Animals succumbed in 3 to 6 days. Figure 1 shows the numbers of legionellae, Philadelphia 1 strain, recovered from the lungs and spleens of aerosol-infected guinea pigs. Two animals were sacrificed at each time point. Rapid growth of organisms was seen in the lungs during the first 24 h after aerosolization, with apparent later dissemination, as legionellae were first detected in the spleen 24 h after infection. Infected animals showed a peak elevation of mean rectal temperature of 1.4°C above controls, to 41°C on day 2 of the infection, and had a progressive weight loss starting 24 h after aerosolization. At 4 days, surviving infected animals had an average weight of 286 ± 1.0 g as compared with 344 ± 24 g for controls. Thus, a progressive infection in guinea pigs has been established by the aerosol route, in which legionellae multiply in the lung and disseminate systemically, producing fever and weight loss.

Using our aerosol model, we carried out vaccination experiments with whole heat-killed organisms (100°C, 1 h) and compared the survival of these animals with that of similarly immunized animals challenged intraperitoneally. The intraperitoneal LD_{50} was 1.1×10^6 cells. Groups of four guinea pigs were vaccinated with two doses of 10^9 heat-killed *Legionella pneumophila*, Philadelphia 1 strain, given 2 weeks apart. Each

TABLE 1. Effect of immunization with heat-killed cells on protection against intraperitoneal and aerosol infection

Guinea pigs[a]	Intraperitoneal infection[b]			Aerosol infection[b]		
	Challenge dose		Survivors/ total	Challenge dose		Survivors/ total
	No. of cells	No. of intraperitoneal LD_{50} doses		No. of cells	No. of aerosol LD_{50} doses	
Immunized	3.1×10^7	29	4/4	4.0×10^5	20	0/6
	1.5×10^7	14	4/4	—[c]	—	—
	6.2×10^6	6	4/4	1.3×10^5	6.5	1/6
Control	6.2×10^6	6	0/4	1.3×10^5	6.5	0/4

[a] Immunized guinea pigs were vaccinated with two doses of 10^9 heat-killed cells, given 14 days apart. Control animals were given injections of saline. Injections were intraperitoneal.

[b] Challenge was 16 days after the booster dose.

[c] —, Not done.

animal was bled by cardiac puncture before vaccination and before infection. Sixteen days after the second injection, vaccinated and control animals were challenged intraperitoneally or by aerosol.

Immunized animals were able to withstand 29 LD_{50}s (3.1×10^7 cells) given intraperitoneally, but there was no protection against 6.5 LD_{50}s (1.3×10^5 cells) given by aerosol (Table 1). Vaccinated pigs had high titers of antibody as detected by an enzyme-linked immunosorbent assay (3) to an outer membrane antigen (5) (courtesy Alan Flesher, Abbott Laboratories, Chicago, Ill.) (geometric mean titer = 4.505) and by indirect fluorescence to heat-killed cells (geometric mean titer = 3.840).

These results show that killed-cell vaccination induces high titers of circulating antibody to *Legionella* surface antigens and protects against intraperitoneal infection but that the same vaccination protocol fails to protect against aerosol infection. At present we do not know whether failure to protect against aerosol challenge is (i) due to a requirement for secretory antibody, (ii) due to a requirement for antibody to an antigen not displayed on heat-killed cells, or (iii) because humoral immunity is ineffective against aerosol infection and cellular immunity is required.

This research was supported by Public Health Service grant AI 16804 from the National Institute of Allergy and Infectious Diseases.

LITERATURE CITED

1. **Collins, F. M.** 1979. Cellular antimicrobial immunity. Crit. Rev. Microbiol. **7:**27–91.
2. **Daisy, J. A., C. E. Benson, J. McKitrick, and H. M. Friedman.** 1981. Intracellular replication of *Legionella pneumophila*. J. Infect. Dis. **143:**460–464.
3. **Eisenstein, T. K., B. J. De Cueninck, D. Resavy, G. D. Shockman, R. B. Carey, and R. M. Swenson.** 1983. Quantitative determination in human sera of vaccine-induced antibody to type-specific polysaccharides of group B streptococci using an enzyme-linked immunosorbent assay. J. Infect. Dis. **147:**847–856.
4. **Eisenstein, T. K., and B. M. Sultzer.** 1983. Immunity to *Salmonella* infection, p. 261–296. *In* T. K. Eisenstein, P. Actor, and H. Friedman (ed.), Host defenses to intracellular pathogens. Plenum Publishing Corp., New York.
5. **Flesher, A. R., and D. L. Kasper.** 1982. Development of an enzyme-linked immunosorbent assay using an isolated outer membrane-associated antigen from *Legionella pneumophila*, p. 444–449. *In* J. B. Robbins, J. C. Hill, and J. C. Sadoff (ed.), Bacterial vaccines. Thieme-Stratton, New York.
6. **Guyton, A. C.** 1947. Measurement of respiratory volumes. Am. J. Physiol. **150:**70–77.
7. **Horwitz, M. A., and S. C. Silverstein.** 1980. Legionnaires' disease bacterium (*Legionella pneumophila*) multiplies intracellularly in human monocytes. J. Clin. Invest. **66:**441–450.
8. **Kishimoto, R. A., M. D. Kastello, J. D. White, F. G. Shirey, V. G. McGann, E. W. Larson, and K. W. Hedlund.** 1979. In vitro interaction between normal cynomolgus monkey alveolar macrophages and Legionnaires disease bacteria. Infect. Immun. **25:**761–763.
9. **Krahenbuhl, J. L., L. Ruskin, and J. S. Remington.** 1972. The use of killed vaccine in immunization against an intracellular parasite: *Toxoplasma gondii*. J. Immunol. **108:**425–431.
10. **Mackaness, G. B.** 1971. Resistance to intracellular infection. J. Infect. Dis. **123:**439–445.
11. **Mackaness, G. B., R. V. Blanden, and F. M. Collins.** 1968. The immunology of antituberculous immunity. Am. Rev. Respir. Dis. **97:**337–344.
12. **Middlebrook, G.** 1952. An apparatus for airborne infection. Proc. Soc. Exp. Biol. Med. **80:**105–110.
13. **Miki, K., and G. B. Mackaness.** 1964. The passive transfer of acquired resistance to *Listeria monocytogenes*. J. Exp. Med. **120:**93–104.
14. **Rinaldo, C. R., Jr., A. W. Pasculle, R. L. Myerowitz, F. M. Gress, and J. N. Dowling.** 1981. Growth of the Pittsburgh pneumonia agent in animal cell cultures. Infect. Immun. **33:**939–943.
15. **Wong, M. C., E. P. Ewing, Jr., C. S. Callaway, and W. L. Peacock, Jr.** 1980. Intracellular multiplication of *Legionella pneumophila* in cultured human embryonic lung fibroblasts. Infect. Immun. **28:**1014–1018.

ROUND TABLE DISCUSSION

Host Defenses to Legionellosis

Moderator: HERMAN FRIEDMAN

Participants: M. A. Horwitz, B. P. Berdal, P. H. Edelstein, T. K. Eisenstein, J. C. Feeley, B. H. Iglewski, W. Johnson, A. W. Pasculle, J. K. Spitznagle, R. Van Furth, and W. C. Winn, Jr.

Legionellosis, in its most serious form, manifests itself as a severe respiratory disease that may culminate as a fatal pneumonia. It is widely accepted that most individuals are exposed to the Legionnaires disease bacterium in the environment, and yet very few individuals develop clinical symptoms. This session, dealing with host defenses to legionellae, was concerned with this paradox; i.e., why are some individuals extremely susceptible to Legionnaires disease while the majority is not? Most individuals are likely to be exposed to these bacteria without developing clinical symptoms, but they do respond by producing serum antibodies and, presumably, sensitized lymphoid cells. During the discussion on host defenses to legionellosis, the panel discussed various aspects of the nature and mechanisms by which individuals react immunologically when exposed to legionellae.

The legionellae are intracellular bacteria that gain access to phagocytic cells. Among the questions that arise is what mediates the uptake of these bacteria by phagocytic cells. At present the mechanism of phagocytic uptake of most, if not all, intracellular microbes is not well understood. Another question is how microorganisms, including legionellae, survive the bactericidal activity of mononuclear phagocytes. In the case of legionellae, there is now some evidence that a toxin may be important in protecting the bacterium from the antimicrobial activity of phagocytes. A related question is how an activated mononuclear phagocyte inhibits the multiplication of legionellae and other intracellular pathogens.

Information was presented concerning the sequence of cytoplasmic events which take place after the entry of various microorganisms, including bacteria, into monocytes. Phagosomes containing *Legionella pneumophila* share some of the characteristics of phagosomes containing other intracellular microorganisms, notably *Chlamydia* sp. and *Toxoplasma gondii*. Mitochondria appear around the phagosomes containing all three pathogens, and ribosomes are associated with *L. pneumophila*- and *T. gondii*-containing phagosomes in mononuclear phagocytes.

There was discussion of how cytotoxins produced by legionellae may provide a mechanism for the organism to survive intracellularly, especially when the monocytes produce oxygen metabolites to which the organisms are exquisitely sensitive. A low-molecular-weight toxin accumulating in the culture fluid of legionellae growing in vitro appears to be a peptide containing six amino acid residues, with three different amino acids in a ratio of 3:2:1. The toxin is heat stable and resists a wide range of acid and alkaline pH. The toxin kills a variety of tissue culture cells in a dose-dependent fashion and also kills chicken embryos. These doses of the toxin do not impair cell viability or phagocytic capacity of human mononuclear cells but do inhibit the respiratory burst that normally follows particle ingestion. Human monocytes are quite sensitive to the toxin, and the respiratory burst is also prevented. The enzymes responsible for superoxide production and hexose monophosphate shunt activity are not affected by the toxin. For example, toxin-treated monocytes or polymorphonuclear leukocytes evince normal respiratory burst activity when treated with soluble stimulators such as concanavalin A, pokeweed mitogen, etc. Thus, the toxin does not appear to have any effect on the system per se but rather on particle activation as occurs when leukocytes interact with bacteria. Thus, the final result of toxin interaction with leukocytes renders the cells incapable of killing bacteria that normally would be susceptible to oxygen products. It has also been shown that the toxin prevents chemoluminescence and membrane fluidity changes which normally occur after interaction of leukocytes with a microorganism. Also, there is some evidence that the heat-stable toxin is a potent chemoattractant for polymorphonuclear cells. Thus, it appears that the toxin produced by legionellae attracts leukocytes and interacts with these cells at the membrane level, altering their ability to produce oxygen products which would normally kill the legionellae.

It seems likely that an animal model would be useful in examining how legionellae can infect macrophages and yet not cause disease in most exposed individuals. The guinea pig model appears to be the most suitable one at present to analyze mechanisms of resistance or susceptibility to infection. Both guinea pigs and rats have been used recently for both infectious disease studies with legionellae and studies concerning humoral and cellular immune responses. There has also been some interest in examining the role of nonimmunological inflammatory mediators such as chemoattractants and the role of complement in resistance of guinea pigs to legionellae. It is apparent from in vitro studies that legionellae do not bind complement well. These organisms may also have chemoattractant activity because of either cell surface components or extracellular products such as a toxin. The pathophysiology of the inflammatory process in the pulmonary vasculature and interstitium has been examined in some detail in guinea pigs, as well as in rats. Many inflammatory cells appear in the lung after active infection with legionellae. These may release lysosomal enzymes that damage the lung tissue. Animal models provide useful clues about the pathophysiology of disease in humans.

Although some studies with guinea pigs have suggested that immunosuppression is an important risk factor in legionellosis, it is important to note that this animal model may not completely reflect the situation in humans. Guinea pigs are highly susceptible to legionellosis; humans appear to be much more resistant. A rat model system has been developed in which only about 10% of the animals develop pneumonia after either intraperitoneal or aerosol exposure. However, some inflammatory disease occurs in almost all rats exposed to *Legionella* organisms. Furthermore, both humoral and cellular immune responses can be detected in these animals. The rat model will permit detailed examination of predisposing factors to *Legionella* infection, including the role of immunosuppression and pre-existing lung disease. Such an animal model should also provide information as to the apparent dichotomy between the susceptibility of monocytes to infection by these organisms and the likelihood that these cells are the prime cellular defense against these bacteria. A large number of species and strains that have been identified must be studied experimentally in terms of host resistance. Furthermore, the vagaries of multiple passage of the organisms in vitro are well known to microbiologists. It is now generally accepted that virulence for guinea pigs may change drastically after culture of the bacteria in vitro in various media. The virulence of the organisms for macrophages in vitro also may change after in vitro passage. Nevertheless, it is acknowledged that the high-virulence legionellae are quite lethal for macrophages.

Awareness that legionellae, like many other bacteria, produce toxins in vitro as well as other extracellular material such as proteinases provides a starting point to examine the question of whether these substances are involved in the pathogenicity of the disease process. It is also important to understand how and why these materials are produced and whether they are released by living bacteria or are only a product of the cell lytic processes. The proteinases secreted by legionellae are quite active in vitro and may alter other products or enzymes secreted by the legionellae. Also, the proteinases may influence the host immune defense system, since now it is widely recognized that some proteinases from bacteria may have specificity for immunoglobulin molecules, even in vivo. Thus, host defenses to legionellae may involve antibodies directed to the toxin, to proteinases, and to other extracellular products, as well as to the organism itself. In this regard, the relative contribution of humoral versus cellular immunity to resistance to legionellae is just beginning to be examined. Several panel members discussed mechanisms of humoral versus cellular immunity to various other intracellular pathogens to determine what lessons can be drawn from these other host defense systems in regard to legionellosis.

There is much diversity of behavior of the host toward various intracellular pathogens. The ability of an intracellular pathogen to replicate in a macrophage is an important virulence factor. The activation of macrophages, either specifically or nonspecifically, to retard the multiplication of these organisms appears to be important for host defense responses. Legionellae apparently fall into the category of organisms that can multiply in macrophages unless T-cell mediator factors enhance the bacteriostatic or bactericidal efficacy of macrophages. Various observations in vivo and in vitro concerning the role of macrophages and T-cell immunity in defense against these organisms indicate that cell-mediated immunity may be important for defense. However, equal importance should be given to antibody in host resistance, especially since antibody is readily demonstrable in the serum of both humans and animals before and after infection. Except for the rat model, there have been no studies to indicate that serum antibody can be passively transferred to naive animals and protect them against infection by legionellae. Indeed, this is the classic experiment whereby cell-mediated immunity was first discovered; i.e., the inability to transfer anti-tuberculosis immunity to guinea pigs by serum was demonstrated. Thus, in both tuberculosis and legionellosis, serum antibody does not appear to be protective for susceptible guinea pigs, but cells may be. How this relates to the situation in humans is not known at present, and it is necessary to develop models in experimental animals relevant to humans.

Much new information about antigens of legionellae is being developed. Some of the cruder antigenic preparations obtained from these bacteria are quite useful for immunodiagnostic procedures. Although these bacteria obviously now constitute a large family of organisms which are considered intracellular opportunistic pathogens, common antigens exist which may be useful for diagnostic purposes. Antibodies to the common or cross-reacting antigens, as well as to type-specific antigens, have been utilized for a wide variety of immunoserological tests, including direct and indirect immunofluorescent procedures, microhemagglutination inhibition procedures, enzyme-linked immunosorbent assays, and cross-over electrophoresis procedures. However, there are many gaps that must still be filled. For example, one very important diagnostic test for legionellosis still lacking is a useful skin test reagent. It is possible that epidemiologically a skin test, similar to the skin test used for tuberculosis, would be useful. In vitro correlates of delayed hypersensitivity have been developed which can be considered an aid for epidemiological studies, based on cell immune responses to *Legionella* antigens. For example, the lymphocyte blastogenic test has been utilized to demonstrate the sensitivity of peripheral blood leukocytes to *Legionella* antigens, in both convalescent patients and normal individuals.

Many individuals, especially during the middle decades of life, develop detectable serum antibodies to legionellae and sensitized T cells. Thus, it appears unlikely that a vaccine will be necessary for all individuals. Nevertheless, a useful vaccine that can stimulate both antibody and cellular immunity might be important for individuals at risk, such as kidney transplant recipients or those who are immunosuppressed because of other conditions. In this regard, there has been a good deal of work concerning the immunochemistry of serogroup-specific antigens of legionellae. Analysis and description of such antigens for possible antigen detection methods, using specific antibody, certainly appear valid. Nevertheless, it is still important to isolate and purify serogroup-specific antigens and to determine the specificity and sensitivity of a test designed to detect these antigens in fluids of patients who may be infected. Efforts must still be made to identify those antigens from different groups of legionellae which may be related to virulence or pathogenicity. Environmental studies concerned with isolation of legionellae from hospitals, water supplies, etc., are quite tedious, and it is not possible to determine whether an organism found in an environmental sample is indeed virulent unless animal studies are performed. A useful immunoserological test to identify infectious versus avirulent or saprophytic legionellae would be of value.

Purification of various antigens for potential vaccines also is important. Studies are now being performed to dissect *Legionella* antigens into various subcellular components, including surface lipopolysaccharide-rich factors, flagellar antigen, pili antigens, etc. Interactions of these antigens with antibody from patients or from immunized experimental animals are being analyzed. It can be anticipated that additional studies will provide useful new information concerning products of legionellae that will be of value for more fundamental studies concerning the role of humoral and cellular immunity in host defenses to *Legionella* infection.

It is important also to examine in some detail the genetic aspect of resistance to legionellae and to determine whether host resistance may be related to genetic control of cellular or humoral immune responses to these bacteria or to nonspecific resistance, possibly based on natural killer cells or macrophage function. A number of important points still must be examined in detail, including the relationship among the various antigens of the different species of *Legionella* to host immune responsiveness. Also, the role of surface components of legionellae, including pili, flagella, and somatic antigens, in attachment, colonization, and infection of humans as a primary or secondary host must be examined.

The relationship of various strains of legionellae found in the environment to infection is still unknown. It is apparent that in some of the outbreaks attributed to *L. pneumophila*, only nonfatal, relatively minor disease occurred, whereas in the Philadelphia, Pa., outbreak among American Legion conventioneers, a rapidly fulminating fatal pneumonia occurred. Even in the latter case, only a small percentage of the Legionnaires exposed to the air-handling system developed clinical symptoms of disease. Thus, it is possible that life-style, temporary alterations in host defenses, etc., may play a role in susceptibility to legionellosis. A further understanding of resistance to *Legionella* infection, as well as of the mechanisms of acquired immunity, still requires a great deal of analysis and experimental work. Nevertheless, it was evident from the discussion during the panel on host defenses to legionellosis that a beginning has been made in understanding the complexity of the interaction between legionellae and the host defense system.

SUMMARY

Some Views on the Immune Responses in *Legionella* Infections

R. van FURTH

Department of Infectious Diseases, University Hospital, 2333 AA Leiden, The Netherlands

This summation of what is known about the immunological aspects of legionellosis will not be a summary of the presentations made during this symposium. Instead, I shall make some statements and formulate some questions that require resolution in the near future, and I shall also indulge in some speculations.

Let us start with the microorganism. We know that legionellae are ubiquitous in the environment; for instance, the bacteria may be present in potable water, and we may be exposed to them when we drink water or inhale aerosols produced by showers, faucets, and cooling towers of air-conditioning systems.

The minimum infective dose for apparently healthy individuals is unknown. The next question concerns the route of entry. Is it the nasal mucosa, the oropharyngeal mucosa, the lower respiratory tract, or the digestive tract? Most people accept the view that an aerosol of small droplets (<5 μm) is responsible for infection, but I shall discuss other possibilities as well. Not much is known about the adherence of legionellae to the mucosal membranes. Are there differences between virulent and avirulent strains in this respect, and are there differences between various *Legionella* species?

Next, we must take into consideration specific local defense mechanisms, such as antibodies. Mucosal secretions contain secretory immunoglobulin A (IgA), which prevents adherence, but also IgG and IgM. Do these immunoglobulins have antibody activity against legionellae? If so, what is known about the avidity of these antibodies and their functional activities? For example, does agglutination occur such that larger particles are formed, which can be removed by the mucociliary system of the respiratory tract? Are there other humoral factors, present in the mucosal secretions, that play a role in the defense?

The next phase to be considered is the early inflammatory response in the lungs after legionellae have been inhaled. It is highly probable that in this phase granulocytes play an important role. The process starts when substances released by the bacteria initiate events that lead to the migration of granulocytes from the capillaries into the alveoli together with the exudation of plasma proteins. Both are required for the defense reaction, the immunoglobulins and complement for opsonization and the granulocytes for removal of the bacteria by phagocytosis. In this early phase there are no specific antibodies in the circulation, but normal serum immunoglobulins are capable of opsonizing virulent legionellae. Recently, we found that normal human serum has less opsonic activity than serum with high titers of antibodies to virulent *Legionella pneumophila*. Complement enhances the ingestion of legionellae, and there is evidence that these microorganisms activate the complement pathway via the alternative route. However, at this meeting it has been shown that activation can also occur via the classical pathway. The question as to whether both activation pathways are involved must be answered.

The next step in the sequence of defense mechanisms is the intracellular killing of bacteria. I agree with Horwitz that granulocytes kill ingested legionellae. This is a very important defense mechanism, not only for the elimination of bacteria but also for the next phase, which involves more specific immune reactions. When we consider the early specific antibody response and the sensitization of T lymphocytes needed to initiate cell-mediated immunity, we must keep in mind that intact microorganisms are unable to induce either reaction. The bacteria must first be ingested, killed intracellularly, and degraded before small antigen fragments become available to stimulate B and T lymphocytes. The early presence of antigen in the urine of patients with legionellosis shows that this process of bacterial disintegration has already started in an early phase of the infection, but it is not known whether it occurs in the lung or elsewhere in the body. Let us assume for a moment that the lung is not involved. It has been mentioned several times during this meeting that bacteremia occurs often in the early phase of the disease, but this has not been documented very precisely. When bacteremia occurs, the bacteria are removed from the circulation by the Kupffer cells and the macrophages in the spleen and in the lung. This process also requires antibodies and complement. The macrophages in these organs are able to ingest, kill, and digest the microorganisms, but the rate at which these processes occur is not yet known. Degradation of bacteria at these sites will also make antigen available for the stimulation of lymphocytes. My hypothesis raises another remote possibility with respect to the initiation of pneumonia. It is conceivable that the route of entry of legionellae is the oropharynx and that after penetration of the oropharyngeal mucosa bacteremia occurs and the bacteria then reach the lungs via circulation, where the infection evolves. Local conditions in the lung, which do not occur in the liver and spleen, may then favor the outgrowth of legionellae. Although this route seems rather unlikely, there is reason to think that the possibility should be investigated in an animal model.

Alveolar macrophages also play an important role in the removal of legionellae from the lung. It is known that these cells derive from circulating monocytes. Recently, we showed that during the early phase of inflammation, a factor released at the affected site augments the production of monocytes in the bone marrow. This increased production of monocytes is required to ensure that enough cells are available to migrate to the site of infection and form a cellular exudate. We have characterized this factor, which is called factor increasing monocytopoiesis; it proved to be a protein with a molecular weight of 20,000. It is not the same molecule as colony-stimulating factor, it is cell line specific (i.e., it does not stimulate granulocytopoiesis or lymphocyte production [other regulatory factors have been described for granulocytes]), and it is synthesized and secreted by macrophages. This is an example of positive feedback loop regulation.

It has been shown that monocytes and macrophages ingest opsonized legionellae and that they do so more effectively when specific antibodies and complement are present. However, the ability of these phagocytes to kill ingested legionellae is very limited. This ability improves after stimulation of T lymphocytes by specific *Legionella* antigens and the formation of lymphokines. Horwitz has shown that activated macrophages kill ingested legionellae much more efficiently than do nonactivated cells (this volume). It is not known where sensitization of T lymphocytes occurs; it may be locally in the lung, in the local lymph nodes, or in the spleen. The results of studies on lymphocyte sensitization are confusing. Because we do not have reagents to define guinea pig and rat lymphocytes, the T-cell response in these animals has not been studied in detail. The situation is simpler for an infection in humans, because reagents are available. Studies may have been done already on the changes in the total number of T and B lymphocytes and the ratio between T-helper and T-suppressor lymphocytes during the course of legionellosis. Another point that deserves more attention is the precise time sequence with respect to the exudation of granulocytes and monocytes, the formation of specific IgM and IgG antibodies, and T-lymphocyte sensitization.

To obtain more insight into the occurrence of cell-mediated immunity, we also need a good antigen for skin testing. This is a rapid, inexpensive, and reliable technique to demonstrate cell-mediated immunity in vivo.

It is well known that patients who suffer from granulocytopenia (due, for example to acute leukemia, aplastic anemia, or treatment with a cytotoxic drug), patients with defective cell-mediated immunity (for example, in Hodgkin's disease, non-Hodgkin lymphoma, and other lymphoreticular malignancies), and patients being treated with glucocorticosteroids and azathioprine are more susceptible to pneumonia caused by legionellae. From the foregoing it is evident that in these patients the cellular defense mechanisms (granulocyte, macrophage, and lymphocyte responses) are impaired. In this connection it is of interest that patients with acquired immunodeficiency syndrome, who show a severe depression of cell-mediated immunity, have not been reported to have legionellosis.

Patients with severely decreased host resistance, including those with a renal or bone marrow transplant, should be nursed in a protective environment. However, since legionellae may be present in potable water, there is a question of whether special precautions are indicated.

The question arises whether prophylaxis with antimicrobial drugs given to reduce the numbers of potentially pathogenic microorganisms—with nonadsorbable antibiotics, as done in our hospital and called partial antibiotic decontamination, or with co-trimoxazole—can also prevent the occurrence of *Legionella* infections. In Leiden, for example, no leukemic patients who were potentially exposed to legionellae via tap water and received prophylaxis with partial antibiotic decontamination developed nosocomial *Legionella* pneumonia, whereas during the same period and in the same hospital environment 11 patients with leukemia or Hodgkin's disease who were not given partial antibiotic decontamination prophylaxis developed pneumonia caused by *L. pneumophila*.

Finally, there is the question of the process of immune modulation, i.e., immunization to prevent legionellosis. What kind of vaccine is needed, one that will stimulate the production of circulating antibodies, antibodies in the secretions, including secretory IgA, or antibodies against specific "toxins" produced by legionellae? We have heard that the "toxins" of legionellae affect the functions of the macrophages and perhaps have other deleterious effects. If these toxins were inactivated with specific antibodies, it might be possible to overcome legionellosis. Once a vaccine becomes available, we will have to consider the route of administration. For antibodies in secretions the parenteral route is not optimal. Administration by aerosolization could be dangerous if antibodies are already present in the circulation. Baskerville has shown that rechallenge of animals with legionellae leads to a severe inflammatory reaction in the lung (this volume). This might well be an Arthus reaction, i.e., a reaction between circulating antibodies and antigen, as is the case with respiratory syncytial virus infections in young children with circulating antibodies.

A final speculation concerns the way in which healthy persons acquire immunity. Presumably, we all are exposed to low numbers of legionellae, for example, when we drink water. Since legionellae appear to be resistant to acid pH and therefore will not be destroyed by gastric acidity, they might reach the gut and stimulate the lymphoid cells of the intestinal tract to produce antibodies locally. Antibody-producing cells from the intestinal tract are known to circulate, but they have a specific pathway. These cells reappear at mucosal sites and other places where secretory immunoglobulins are produced (for example, the mammary glands) and produce the same antibodies there as at the primary site in the intestines. With respect to anti-*Legionella* antibodies it is conceivable that antibody-producing cells of the gastrointestinal tract migrate to the oral mucosa, salivary glands, and respiratory tract, and there continue to produce these antibodies and establish a low grade of local immunity.

EPIDEMIOLOGY

STATE OF THE ART LECTURE

Current Issues in Epidemiology of Legionellosis, 1983

CLAIRE V. BROOME

Division of Bacterial Diseases, Center for Infectious Diseases, Centers for Disease Control, Atlanta, Georgia 30333

The 2nd International Symposium on *Legionella* is an appropriate setting for evaluating the progress that has been made since the first symposium four and one-half years ago. At that time, the descriptive epidemiology of legionellosis had been defined; subsequent investigations have confirmed the essential correctness of those assessments. However, there are still questions regarding the epidemiology of legionellosis. The first issue is the incidence of legionellosis and its importance as a cause of pneumonia. This includes the question of the relative frequency of all *Legionella* species that have been characterized to date, as well as the possible role that as-yet-undiscovered members of this genus may play as etiological agents in the substantial portion of all pneumonia for which an etiology cannot be determined. The second unresolved question is the relationship between the two clinical syndromes of legionellosis, the pneumonic and nonpneumonic forms, commonly referred to as Legionnaires disease and Pontiac fever after the first two described outbreaks. Finally, further information is needed to define the various modes of transmission of legionellosis, their relative importance, and optimal control measures.

Our knowledge of the descriptive epidemiology of legionellosis has expanded since the first symposium, but the basic conclusions have not changed. Investigations of outbreaks and sporadic cases have shown that attack rates are highest in males, with risk approximately 3 times greater for males, and in older age groups (1–3, 5, 6, 14, 17, 23, 29, 33, 34). Cases have been reported in infants and children, but the frequency of disease in these age groups seems to be lower (27). A number of studies have identified other risk factors for acquisition of Legionnaires disease, including cigarette smoking (2, 14, 31), heavy alcohol consumption (31), and preexisting disease resulting in immunocompromise (8, 17).

Secular trends in the occurrence of legionellosis have been difficult to determine due to differences in availability of diagnostic capabilities for the disease. Cases of legionellosis have been identified as having occurred as early as 1943 and 1947 (18, 25), and an outbreak has been

documented which occurred in 1957 (28). Since diagnostic methods were not available before 1977, it is impossible to assess whether there has been an increase in the incidence of legionellosis in recent years.

In outbreaks in which it has been possible to define an incubation period, the median has been 5 to 6 days, with a range of 2 to 13 days (1, 6, 14, 33, 34).

Definition of the geographic distribution of legionellosis has improved. Cases have now been identified on six of the seven continents. It is unclear, however, whether the frequency of disease is similar in different geographic locations. The passive surveillance system maintained by the Centers for Disease Control suggests that there is some variation in attack rate, as 10 of the 13 states with the highest attack rates of reported cases are east of the Mississippi River (8). Another approach to this question is to examine the proportion of pneumonia cases due to *Legionella* species in various studies of the etiology of pneumonia. Table 1 summarizes studies which have been undertaken since diagnostic methods for legionellosis have been available. The only study which gives a population-based estimate of incidence is the Seattle, Wash., study, in which it was found that 1% of pneumonia cases could be ascribed to *Legionella pneumophila*, yielding an attack rate of approximately 1.2 cases of pneumonia due to *L. pneumophila* per 10,000 persons per year. However, the population represented in this study may have been biased toward an oversampling of patients with pneumonia that could be treated on an outpatient basis (12). The other studies summarized in Table 1 are difficult to interpret as they were hospital-based studies with different criteria for inclusion of pneumonia cases and different methods of diagnosis. Furthermore, at least one of the series is from a hospital with a nosocomial legionellosis outbreak (38). However, in spite of these methodological problems, it is striking that between 1 and 13% of all pneumonias seen in hospitals in the United States, Canada, the United Kingdom, and Germany are caused by legionellae (21, 22, 24, 27, 38). Although more work needs to be done in defining the incidence of legionellosis in population-based studies, these hospital-based studies suggest that it is not an uncommon cause of pneumonia.

The relative importance of *Legionella* species other than *L. pneumophila* as causes of pneumonia is addressed by three of the studies listed in Table 1 (21, 22, 38). In the United States and Europe, *Legionella micdadei* was identified as a possible etiological agent in approximately 1% of pneumonias. Wilkinson et al. (37) have also addressed the issue of the relative frequency of disease due to various *Legionella* species by studying 444 pairs of sera submitted from patients with suspected legionellosis from 28 states in 1979 and 1980. They found that 3.2% of the pairs demonstrated a "species-specific" seroconversion to *L. pneumophila* in the indirect fluorescent-antibody assay to a reciprocal titer of ≥128. The second most common species identified was *L. micdadei*, with a frequency of 0.9%. Although it is difficult to identify with certainty an etiological agent on the basis of serological results alone, this study at least presents preliminary data on the relative importance of the different species. Reingold et al. (this volume) also address this issue by using direct fluorescent-antibody conjugates that may increase the specificity of the findings. However, further work needs to be done, preferably in population-based studies, employing culture and direct fluorescent-antibody techniques to establish definitively the relative frequency of disease due to *Legionella* species other than *L. pneumophila*. The importance of incorporating culture

methods in prospective studies of the etiology of pneumonia needs to be emphasized, as it is highly likely that additional unnamed species of *Legionella* or related organisms may be isolated. In studies of pneumonia, investigators frequently have been unable to determine an etiology for up to 30 to 40% of pneumonia cases documented by radiographic changes (9, 10, 32); it is reasonable to speculate that some of these cases might be due to *Legionella* species.

The basis for the two different clinical and epidemiological presentations of legionellosis is not yet understood. Since the first symposium, three additional outbreaks of Pontiac fever have been described in the literature, and two others were presented in this Symposium. The distinguishing characteristics of Pontiac fever, including a high attack rate among exposed persons (many without underlying disease), a mean incubation period of approximately 36 h, and a relatively benign, self-limited clinical course have been seen in all outbreaks (13, 15, 20; L. A. Herwaldt et al., this volume; E. E. Jones et al., submitted for publication; K. Spitalny et al., Program Abstr. Intersci. Conf. Antimicrob. Agents Chemother. 22nd, Miami Beach, Fla., abstr. no. 87, 1982). Outbreaks have resulted from airborne transmission. The source of the aerosols has included an air conditioning system, presumably contaminated with water from the evaporative condenser (15), a steam turbine condenser (13, 20), recreational whirlpools (Spitalny et al., 22nd ICAAC, abstr. no. 87; Jones et al., submitted for publication), and an industrial coolant system (Herwaldt et al., this volume).

TABLE 1. *Legionella* species as etiological agents of pneumonia

Location	Years of study	Species sought[a]	No. of cases studied	No. positive	% of pneumonias	Comment	Reference
Seattle, Wash.	1963–75	*L. pneumophila* 1	500	5	1	1.2 cases per 10,000 person-years	12
Pittsburgh, Pa.	1979–80	*L. pneumophila* 1 to 4 *L. micdadei*	142	18 2 (?)	13	Nosocomial hyper-endemic	38
Halifax, Nova Scotia, Canada	1979–80	*L. pneumophila* 1 to 4	27	7	26	Excluded "bacterial" pneumonias	24
Los Angeles, Calif.	1979–80	*L. pneumophila* 1 and 2	110	1	1	Pediatric only	27
Nottingham, U.K.	1980–81	*L. pneumophila* 1 to 6 *L. micdadei, L. boze-manii, L. gormanii, L. dumoffii*	127	16 3	13 2		22
Germany	1980–81	*L. pneumophila* 1 to 6 *L. micdadei*	110	8 1	7 1		21

[a] Numbers indicate serogroups.

Probably the most striking finding, however, is that Pontiac fever can be caused by different serogroups and species of *Legionella*, including *L. pneumophila* serogroup 6 (Spitalny et al., 22nd ICAAC, abstr. no. 87; Jones et al., submitted for publication) and a previously undescribed *Legionella* species (Herwaldt et al., this volume). To date, no characteristics of either the organisms or the epidemiological settings have been identified which can explain the different clinical and epidemiological findings. Speculation has focused on the possibility that Pontiac fever may be primarily a response to nonviable organisms or that there is a virulence factor specific to the organisms that determines which syndrome is produced.

Four and one-half years ago the first report of isolation of *L. pneumophila* from the environment was presented at the First International Symposium on Legionnaires Disease (26). Since that initial report, investigators have discovered that *L. pneumophila* can be found in numerous environmental niches, including natural freshwater, potable water, and cooling-tower water (1, 4, 6, 11, 26, 35, 36). The apparent ubiquity of the organism makes it important to establish a specific association with a postulated means of transmission, as the isolation of the organism from a particular environment does not necessarily imply that the positive site was the source of infection. Determining the circumstances under which an environmental source of legionellae may result in human disease is also important, because the methodology for eradicating legionellae from environmental sites is not fully developed. Currently available methods such as biocides, water temperature elevation, and hyperchlorination may not be effective and may have unexpected adverse effects.

The mode of transmission of the organism was investigated extensively in the initial outbreaks. Evidence from the initial Philadelphia, Pa., outbreak suggested that the disease was not transmitted by ingestion of food or water or spread person to person (14). These findings have been substantiated by subsequent investigations (1, 2, 5, 6). The finding in Philadelphia that transmission of infection was most likely to have occurred by the airborne route is consistent with findings in subsequent investigations (1, 3, 6, 13–15; Spitalny et al., 22nd ICAAC, abstr. no. 87; Jones et al., submitted for publication; Herwaldt et al., this volume).

The source of airborne transmission has not always been well documented; for example, in the Philadelphia outbreak, an unequivocal source could not be identified (14). As noted above, recreational whirlpools or an industrial coolant system functioned as the source of an aerosol which produced a Pontiac fever outbreak. The initial outbreak of Pontiac fever demonstrated the role of a contaminated heat rejection system in production of an aerosol (15). In this outbreak, water from the evaporative condenser had contaminated the air intake for the air conditioning system of the building, either through cross-connections between the ductwork or through the close proximity between the exhaust for the evaporative condenser and the air intake for the air conditioning system. The role of heat rejection devices in outbreak of pneumonic legionellosis has subsequently been demonstrated in three separate outbreaks. In Atlanta, Ga., in July 1978, an outbreak of pneumonia occurred among golfers at a country club. A case-control study showed that persons who had spent more time on the golf course were at a higher risk for disease. These persons had increased exposure to exhaust from an evaporative condenser; *L. pneumophila* was subsequently isolated from the evaporative condenser (3). In August and September 1978, an outbreak of nosocomial pneumonic legionellosis incurred in Memphis, Tenn. The outbreak began shortly after an auxiliary cooling tower and air conditioning system had been turned on due to flooding damage to the usual air conditioning system. Cases ceased to occur 9 days after the use of the auxiliary system was discontinued. Multiple separate ventilation systems were present in the hospital; air intakes for two areas were in close proximity to exhaust from the auxiliary cooling tower, whereas other air intakes were located on the other side of the 13-story building. A case-control study showed that case patients were more likely to have been hospitalized in the part of the hospital receiving air from the intakes near the cooling-tower exhaust. *L. pneumophila* serogroup 1 was isolated from the cooling tower (6). In Eau Claire, Wis., in 1979, an outbreak of pneumonic legionellosis occurred in visitors to a hotel. Investigation revealed that disease was significantly more likely to have occurred in visitors who had been in one particular meeting room; this meeting room was under negative pressure relative to the outside air, and air intake occurred through a chimney. The chimney was located on the roof in close proximity to a cooling tower contaminated with *L. pneumophila* serogroup 1. Air circulating to the other four meeting rooms did not come from the same area of the roof (1). These three outbreaks suggest that contaminated heat rejection devices can serve as a means of transmission for epidemic legionellosis. Their possible role in transmission of sporadic or endemic legionellosis is less clear. Contaminated potable water may also serve as a reservoir for legionellosis (see Bartlett, this volume). Investigators need to remain alert to the possibility of other means of transmission.

The evidence that contaminated cooling towers may play a role in the transmission of legionellosis raises immediate questions about the need for monitoring towers for the presence of the organism and the feasibility of eradicating the organism when it is found. Initial studies regarding means of eradicating the organism concentrated on laboratory studies of biocides currently employed in cooling towers. Several biocides, including a preparation that yielded 3.3 mg of free chlorine per ml, didecyldimethylammonium chloride (a quaternary ammonia compound), 2,2-dibromo-3-nitrilopropionomide, and a combination of N-alkyl dimethyl benzoyl ammonium chloride and bis(tri-n-butyltin) oxide were found to be effective in in vitro testing at concentrations which should be achievable in a cooling tower (16, 30). Two studies have examined the effectiveness of such compounds in cooling towers and found that use of these compounds in cooling towers has not always resulted in eradication of the organism (7, 19). The failure of in vitro susceptibility to predict actual biocide activity is not surprising, since cooling-tower water frequently may contain additional organic material and be exposed to light and aeration, all of which may affect the activity of the biocides. Because of the difficulty of eradicating the organism and the unknown toxicity of altering currently recommended cooling-tower treatment protocols, current recommendations call for routine maintenance of cooling towers and do not suggest surveillance culturing for legionellae. However, in an outbreak, cooling towers must be considered as a possible source for transmission of the organism. If an epidemiological investigation documents that a cooling tower may be implicated in transmission, cleaning and hyperchlorination appear to be effective in decreasing the risk of disease. More information is needed about methods for decontaminating systems which have been implicated in transmission of human disease.

LITERATURE CITED

1. Band, J. D., M. LaVenture, J. P. David, G. F. Mallison, P. Skaliy, P. S. Hayes, W. L. Schell, H. Weiss, D. J. Greenberg, and D. W. Fraser. 1981. Epidemic Legionnaires' disease: airborne transmission down a chimney. J. Am. Med. Assoc. 245:2404–2407.
2. Broome, C. V., S. A. J. Goings, S. B. Thacker, R. L. Vogt, H. N. Beaty, D. W. Fraser, and the Field Investigation Team. 1979. The Vermont epidemic of Legionnaires' disease. Ann. Intern. Med. 90:573–577.
3. Cordes, L. G., D. W. Fraser, P. Skaliy, C. A. Perlino, W. R. Elsea, G. F. Mallison, and P. S. Hayes. 1980. Legionnaires' disease outbreak at an Atlanta, Georgia, country club: evidence for spread from an evaporative condenser. Am. J. Epidemiol. 111:425–431.
4. Cordes, L. G., A. M. Wiesenthal, G. W. Gorman, J. P. Phair, H. M. Sommers, A. Brown, V. L. Yu, M. H. Mag-nussen, R. P. Meyer, J. S. Wolf, K. N. Shands, and D. W. Fraser. 1981. Isolation of Legionella pneumophila from hospital shower heads. Ann. Intern. Med. 94:194–199.
5. Dondero, T. J., Jr., H. W. Clegg II, T. F. Tsai, M. Weeks, E. Duncan, J. Strickler, C. Chapman, G. F. Mallison, B. Politi, M. E. Potter, and W. Schaffner. 1979. Legionnaires' disease in Kingsport, Tennessee. Ann. Intern. Med. 90:569–573.
6. Dondero, T. J., Jr., R. C. Rendtorff, G. F. Mallison, R. M. Weeks, J. S. Levy, E. W. Wong, and W. Schaffner. 1980. Outbreak of Legionnaires' disease associated with a contaminated air conditioning cooling tower. N. Engl. J. Med. 302:365–370.
7. England, A. C. III, D. W. Fraser, G. F. Mallison, D. C. Mackel, P. Skaliy, and G. C. Gorman. 1982. Failure of Legionella pneumophila sensitivities to predict culture results from disinfectant-treated air conditioning cooling towers. Appl. Environ. Microbiol. 43:240–244.
8. England, A. C., D. W. Fraser, B. D. Plikaytis, T. F. Tsai, G. Storch, and C. V. Broome. 1981. Sporadic legionellosis in the United States: the first thousand cases. Ann. Intern. Med. 94:164–170.
9. Fekety, F. R., Jr., J. Caldwell, D. Gump, J. E. Johnson, W. Maxson, J. Mulholland, and R. Thoburn. 1971. Bacteria, viruses and mycoplasmas in acute pneumonia in adults. Am. Rev. Respir. Dis. 104:499–507.
10. Fiala, M. 1969. A study of the combined role of viruses, mycoplasmas and bacteria in adult pneumonia. Am. J. Med. Sci. 257:44–51.
11. Fliermans, C. B., W. B. Cherry, L. H. Orrison, S. J. Smith, D. L. Tison, and D. H. Pope. 1981. Ecological distribution of Legionella pneumophila. Appl. Environ. Microbiol. 41:9–16.
12. Foy, H. M., C. V. Broome, P. S. Hayes, I. Allen, M. K. Cooney, and R. Tobe. 1979. Legionnaires' disease in a prepaid medical care group in Seattle 1963–1975. Lancet i:767–770.
13. Fraser, D. W., D. C. Deubner, D. L. Hill, and D. K. Gilliam. 1979. Non-pneumonic, short-incubation-period legionellosis (Pontiac fever) in men who cleaned a steam turbine condenser. Science 205:691–692.
14. Fraser, D. W., T. F. Tsai, W. Orenstein, W. E. Parkin, H. J. Beecham, R. E. Sharrar, J. Harris, G. F. Mallison, S. M. Martin, J. E. McDade, C. C. Shepard, P. S. Brachman, and the Field Investigation Team. 1977. Legionnaires' disease: description of an epidemic of pneumonia. N. Engl. J. Med. 297:1189–1197.
15. Glick, T. H., M. D. Gregg, B. Berman, G. Mallison, W. W. Rhodes, Jr., and I. Kassanoff. 1978. Pontiac fever: an epidemic of unknown etiology in a health department. I. Clinical and epidemiologic aspects. Am. J. Epidemiol. 107:149–160.
16. Grace, R. D., N. E. Dewar, W. G. Barnes, and G. R. Hodges. 1981. Susceptibility of Legionella pneumophila to three cooling tower microbicides. Appl. Environ. Microbiol. 41:233–236.
17. Haley, C. E., M. L. Cohen, J. Halter, and R. D. Meyer. 1979. Nosocomial Legionnaires' disease: a continuing common-source epidemic at Wadsworth Medical Center. Ann. Intern. Med. 90:583–586.
18. Hebert, G. A., C. W. Moss, L. K. McDougal, F. M. Bozeman, R. M. McKinney, and D. J. Brenner. 1980. The rickettsia-like organisms Tatlock (1943) and Heba (1959): bacteria phenotypically similar to but genetically distinct from Legionella pneumophila and the Wiga bacterium. Ann. Intern. Med. 92:45–52.
19. Kurtz, J. B., C. L. R. Bartlett, U. A. Newton, R. A. White, and N. L. Jones. 1982. Legionella pneumophila in cooling water systems: report of a survey of cooling towers in London and a pilot trial of selected biocides. J. Hyg. Camb. 88:369–381.
20. Lauderdale, J. F., and C. C. Johnson. 1983. An outbreak of acute fever among steam turbine condenser cleaners. Am. Ind. Hyg. Assoc. J. 44:156–160.
21. Lode, H., H. Schafer, and G. Ruckdeschel. 1982. Legion-

arskrankheit: Prospektive studie zur Haufigkeit, Klinik and Prognose. Dtsch. Med. Wschr. **107:**326–331.

22. **MacFarlane, J. T., R. G. Finch, M. J. Ward, and A. D. MacRae.** 1982. Hospital study of adult community-acquired pneumonia. Lancet **i:**255–258.

23. **Marks, J. S., T. F. Tsai, W. J. Martone, R. C. Baron, J. Kennicott, F. J. Holtzhauer, I. Baird, D. Fay, J. C. Feeley, G. F. Mallison, D. W. Fraser, and T. J. Halpin.** 1979. Nosocomial Legionnaires' disease in Columbus, Ohio. Ann. Intern. Med. **90:**565–569.

24. **Marrie, T. J., E. V. Haldane, M. A. Noble, R. S. Faulkner, R. S. Martin, and S. H. S. Lee.** 1981. Causes of atypical pneumonia: results of a 1 year prospective study. Can. Med. Assoc. J. **125:**1118–1123.

25. **McDade, J. E., D. J. Brenner, and F. M. Bozeman.** 1979. Legionnaires' disease bacterium isolated in 1947. Ann. Intern. Med. **90:**659–651.

26. **Morris, G. K., C. M. Patton, J. C. Feeley, S. E. Johnson, G. Gorman, W. T. Martin, P. Skaliy, G. F. Mallison, B. D. Politi, and D. C. Mackel.** 1979. Isolation of the Legionnaires' disease bacterium from environmental samples. Ann. Intern. Med. **90:**664–666.

27. **Orenstein, W. A., G. D. Overturf, M. J. Leedom, R. Alvarado, M. Geffner, A. Fryer, L. Chan, V. Haynes, T. Store, and B. Portnoy.** 1981. The frequency of legionella infection prospectively determined in children hospitalized with pneumonia. J. Pediatr. **99:**403–406.

28. **Osterholm, M. T., T. D. Y. Chin, D. O. Osborne, H. B. Dull, A. G. Dean, D. W. Fraser, P. S. Hayes, and W. N. Hall,** 1983. A 1957 outbreak of Legionnaires' disease associated with a meat-packing plant. Am. J. Epidemiol. **117:**60.

29. **Politi, B. D., D. W. Fraser, G. F. Mallison, J. V. Mohatt, G. K. Morris, C. M. Patton, J. C. Feeley, R. D. Telle, and J. V. Bennett.** 1979. A major focus of Legionnaires' disease in Bloomington, Indiana. Ann. Intern. Med. **90:**587–591.

30. **Skaliy, P., T. A. Thompson, G. W. Gorman, G. K. Morris, and H. V. McEachern.** 1980. Laboratory studies of disinfectants against *Legionella pneumophila.* Appl. Environ. Microbiol. **40:**697–700.

31. **Storch, G. A., W. B. Baine, D. W. Fraser, C. V. Broome, H. W. Clegg II, M. L. Cohen, S. A. J. Goings, B. D. Politi, W. A. Terranova, T. F. Tsai, B. D. Plikaytis, C. C. Shepard, and J. V. Bennett.** 1979. Sporadic community-acquired Legionnaires' disease in the United States: a case control study. Ann. Intern. Med. **90:**596–600.

32. **Sullivan, R. J., W. R. Dowdle, W. M. Marine, and J. C. Hierholzer.** 1972. Adult pneumonia in a general hospital. Arch. Intern. Med. **129:**935–942.

33. **Terranova, W., W. L. Cohen, and D. W. Fraser.** 1978. 1974 outbreak of Legionnaires' disease diagnosed in 1977: clinical and epidemiological features. Lancet **ii:**122–124.

34. **Thacker, S. B., J. V. Bennett, T. F. Tsai, D. W. Fraser, J. E. McDade, C. C. Shepard, K. H. Williams, Jr., W. H. Stuart, H. B. Dull, and T. C. Eickhoff.** 1978. An outbreak in 1965 of severe respiratory illness caused by the Legionnaires' disease bacterium. J. Infect. Dis. **138:**512–519.

35. **Tobin, J. O., J. Beare, M. S. Dunnill, S. Fisher-Hoch, M. French, R. G. Mitchell, P. J. Morris, and M. F. Myers.** 1980. Legionnaires' disease in a transplant unit: isolation of the causative agent from shower baths. Lancet **ii:**118–121.

36. **Wadowsky, R. M., R. B. Yee, L. Mezmar, E. J. Wing, and J. N. Dowling.** 1982. Hot water systems as sources of *Legionella pneumophila* in hospital and non-hospital plumbing fixtures. Appl. Environ. Microbiol. **43:**1104–1110.

37. **Wilkinson, H. W., A. L. Reingold, B. J. Brake, D. L. McGiboney, G. W. Gorman, and C. V. Broome.** 1983. Reactivity of serum from patients with suspected legionellosis against 29 antigens of *Legionellaceae* and *Legionella*-like organisms by indirect immunofluorescence assay. J. Infect. Dis. **147:**23–31.

38. **Yu, V. L., F. J. Kroboth, J. Shonnard, A. Brown, S. McDearman, and M. Magnussen.** 1982. Legionnaires' disease: new clinical perspective from a prospective pneumonia study. Am. J. Med. **73:**357–361.

STATE OF THE ART LECTURE

Potable Water as Reservoir and Means of Transmission

CHRISTOPHER L. R. BARTLETT

PHLS Communicable Disease Surveillance Centre, London NW9 5EQ, England

That hospital water systems other than cooling circuits could serve as sources of *Legionella* spp. was first put forward by Tobin and colleagues in 1980 (28). During the investigation of two cases of Legionnaires disease the previous year in a transplant unit in Oxford, England, they isolated similar strains of *Legionella pneumophila* from both of the patients and from water samples taken from the shower unit in the postoperative cubicle the patients had occupied. The isolates seemed to be related to serogroup 3 by immunofluorescence, but later work showed that they were all serogroup 6, a serogroup that is recovered very infrequently from clinical or environmental specimens in the United Kingdom. Thus, strong circumstantial evidence had been produced to indicate that shower units could serve as sources or vehicles of infection. Other papers substantiating these findings followed shortly afterwards.

In the United Kingdom the same year another outbreak of Legionnaires disease, this time associated with a hotel, provided further grounds to implicate piped water systems (27). Four cases of *L. pneumophila* serogroup 1 pneumonia and several pneumonias of unknown etiology occurred in 1 month among guests who had stayed in the hotel. *L. pneumophila* serogroup 1 was isolated from various parts of both the hot and cold water systems including storage tanks, taps, and shower units but from no other environmental samples collected in or around the hotel. Furthermore, the building did not have an air-conditioning system and there was no cooling tower in its vicinity; on this indirect evidence, therefore, it was concluded that the domestic water systems were the likely source, and control measures were introduced.

Two years later, another case of *L. pneumophila* serogroup 1 pneumonia occurred in a man who became ill 5 days after staying in one of the hotel's bedrooms in which it had not been possible, for technical reasons, to achieve a recommended hot water temperature of at least 50°C, although there was satisfactory chlorination of the cold water (3). This suggested that the hot water system may have been the source of infection, and indeed *L. pneumophila* serogroup 1 was found in high titer in water taken from the hot taps in the room. The circulating hot water in other parts of the hotel had been maintained at 55 to 60°C, and the organism was not isolated from these. This evidence, albeit circumstantial, does point to the possibility of the multiplication of *L. pneumophila* in the peripheral parts of hot water systems.

Cordes and colleagues in the United States studied three sites of nosocomial infection in late 1979 and early 1980 to evaluate the role of showers as possible disseminators of the bacterium (7). In three hospitals across the country they cultured from shower fittings *L. pneumophila* of the same serogroup as had infected patients. They recovered the organism more frequently from shower-head specimens taken from wards associated with nosocomial legionella pneumonias than from those taken from wards or hospitals in which such cases had not been recognized. The paper provided another pointer toward water systems as sources but in common with the other reports did not present data to show that the infection was acquired by showering or bathing. Such evidence is lacking also in many later reports of outbreaks, including one caused by *Legionella micdadei* (25; M. Best, V. L. Yu, J. Stout, A. Goetz, and R. Muder, Program Abstr. Intersci. Conf. Antimicrob. Agents Chemother. 22nd, Miami Beach, Fla., abstr. no. 88, 1982) in which water systems were considered to be the source of infection. Such outbreaks have occurred in various parts of the world, and all have been associated with either hotels or hospitals (Table 1). Internationally, engineers appear to use different terms to describe water distribution systems; Fig. 1 illustrates the terminology used in this paper.

Several studies have shown that legionellae are frequently present in domestic water systems, particularly hot water circuits and cooling water systems, in large institutions not known to be associated with Legionnaires disease, as well as in natural collections of water (6, 9, 13, 14, 19, 29, 30). Clearly, hot or cold domestic water systems in institutions are widespread potential sources of legionellae, but can we be confident yet that they do lead to infection? We know that congruent species or serogroups have been isolated from water systems implicated as sources

TABLE 1. Outbreaks and case clusters in which domestic water systems have been implicated as sources[a]

Years	Place	Site	No. of cases
1973–80	Benidorm, Spain	Hotel	42+
1977–80	Los Angeles, Calif.	Hospital	180+
1978–83	Leiden, Netherlands	Hospital	17
1979	Oxford, England	Hospital	2
1979	Corby, England	Hotel	4+
1979–80	Kingston, England	Hospital	13
1979–81	Pittsburgh, Pa.	Hospital	100+
1979–80	Ballarat, Australia	Hospital	2
1980–82	Lido di Savio, Italy	Hotel	23+
1980–83	Quarteira, Portugal	Hotel	12
1981	Iowa City, Iowa	Hospital	24
1981–83	Paris, France	Hospital	37+

[a] Shown in chronological order rather than date of publication of paper. Data from references 3, 8, 10, 20, 23, 27, 28; Best et al., 22nd ICAAC, abstr. no. 88, 1982; Helms et al., 22nd ICAAC, abstr. no. 89, 1982; Shands et al., Clin. Res. 29:260A, 1981; Dournon, personal communication; Meenhorst, personal communication.

of epidemic or endemic legionella infections, but the finding of a ubiquitous organism in one such location does not, of course, establish proof. What additional evidence is there, therefore, to support the hypothesis?

First, in one hospital a chance survey of selected parts of the water system, prior to a nosocomial outbreak, found *L. pneumophila* in all of the sites tested (26). The number of sites yielding the organisms decreased during the

month after the outbreak, although no control measures had been introduced.

Second, during the investigation of another nosocomial outbreak, *L. pneumophila* was isolated from tap water specimens and from respiratory therapy devices which had been filled with tap water (1). A comparison of cases with control patients in the hospital at the same time showed that exposure to a combination of aerosolized tap water from respiratory devices and systemic corticosteroid therapy was significantly associated with Legionnaires disease.

Third, in several institutions in which outbreaks of Legionnaires disease have occurred (Table 2) intervention studies have been undertaken with periods of extended surveillance for Legionnaires disease after the application of measures to control the growth of legionellae in hot or both hot and cold water systems. Legionnaires disease had been shown to be associated with all but one of these establishments over a period of at least 2 years and, in one instance, 7 years (3, 22) before the intervention. Various control strategies were tried: two centers used a combination of continuous chlorination and raised hot water temperatures (3, 10); another used shot chlorination and intermittent raised hot water temperatures (C. Helms, R. Massanari, R. Zeitler, S. Streed, M. Gilchrist, N. Hall, W. Hausler, W. Johnson, L. Wintermeyer, J. Sywassink, and W. Hierholzer, 22nd ICAAC, abstr. no. 89, 1982); one relied on continuous chlorination alone (K. N. Shands, J. L. Ho, G. W. Gorman, R. D. Meyer, P. H. Edelstein, S. M. Finegold, and D. W. Fraser, Clin. Res.

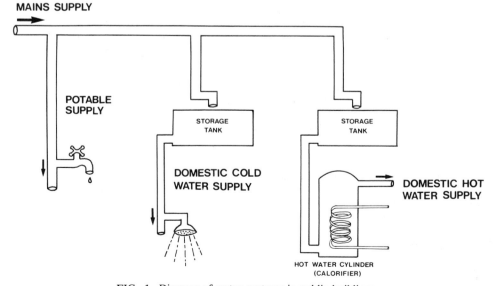

FIG. 1. Diagram of water systems in public buildings.

TABLE 2. Intervention studies in establishments associated with Legionnaires disease

Place	Site	Cases in	Control measures began	Principal methods	No. of cases	
					Before intervention	After intervention
Benidorm, Spain[a]	Hotel	1973–80	1980	Temp ↑ , chlorination	42+	0
Kingston, England[a]	Hospital	1979–80	1980	Temp ↑ , chlorination	11	2
Los Angeles, Calif.[b]	Hospital	1977–80	1980	Chlorination	175+	6
Pittsburgh, Pa.[c]	Hospital	1979–81	1981	Temp ↑ intermittently	100+[d]	12
Iowa City, Iowa[e]	Hospital	1981	1981	Chlorination, temp ↑ intermittently	24	0

[a] Data from references 3, 10, and 22.
[b] Shands et al., Clin. Res. **29:**260A, 1981.
[c] Best et al., 22nd ICAAC, abstr. no. 88, 1982.
[d] L. pneumophila and L. micdadei.
[e] Helms et al., 22nd ICAAC, abstr. no. 89, 1982.

29:260A, 1981); and the last used intermittent raised hot water temperatures (Best et al., 22nd ICAAC, abstr. no. 88, 1982). In three of the institutions postintervention surveillance has been maintained for more than 2 years (3, 10; Shands et al., Clin. Res. **29:**260A, 1981), and in the other two surveillance has been maintained for at least a year (Best et al., 22nd ICAAC, abstr. no. 88, 1982; Helms et al., 22nd ICAAC, abstr. no. 89, 1982). No associated legionella infections have been detected since the intervention in two of the establishments, and dramatic falls in incidence of infection were seen in the other three. Indeed, later modifications of control measures in two of these then virtually eliminated nosocomial legionella infection.

Fourth, in three of the incidents there is the additional circumstantial evidence of work on the water system preceding the outbreak. In Wadsworth Memorial Hospital, Los Angeles, Calif., there was a significant drop in the hospital water pressure shortly before a considerable increase in the number of nosocomial cases (Shands et al., Clin. Res. **29:**260A, 1981). There were complaints of "black water" for several weeks after the pressure drop. One month later, tap water samples in another wing of the hospital were collected before and after an experimental lowering of the water pressure. Samples collected after this pressure drop were darker in color and yielded 20-fold more organisms than those obtained before. In Kingston Hospital, Kingston upon Thames, England, after an outbreak of 11 cases of Legionnaires disease, the introduction of control measures appeared initially to be successful in preventing further cases, but a year later another nosocomial case occurred (11). This man had been admitted to the hospital on the day that a reserve calorifier (heating cylinder) was brought into operation in the hospital. Members of the staff had complained that for several weeks after this event the hot water supply in the wards showed a brownish discolor-

ation. A sample of hot water taken at that time was found to contain L. pneumophila serogroup 1. Examination of another standby calorifier in the hospital found that there was a thick brown liquid deposit in the base from which 54×10^7 organisms of L. pneumophila serogroup 1 were isolated per liter. It is likely that the person with Legionnaires disease, who took a shower and a bath on the day of admission, was exposed to large numbers of organisms which had arisen from the stagnant deposits in the newly connected calorifier. In a hotel-associated epidemic also there is a history of work on the water supply immediately before the outbreak (3). The first case occurred within a week of the reconnection of a supplementary water supply from a well which had been out of commission because of a pump failure.

Fifth, there have been recent reports of Legionnaires disease and Pontiac fever associated with the use of whirlpools which had been fed directly from potable supplies (R. S. Remis, E. E. Jones, K. A. Tait, H. B. McGee, B. B. Wentworth, G. W. Gorman, B. J. Brake, H. W. Wilkinson, A. W. Hightower, and C. V. Broome, Abstr. Annu. Meet. Am. Soc. Microbiol. 1983, New Orleans, La., C359, p. 371; K. Spitalny, R. Vogt, L. Witherwell, L. Orciari, L. Orrison, P. Etkind, and L. Novick, 22nd ICAAC, abstr. no. 87, 1982).

Although most of it is indirect, the weight of the evidence above indicates domestic water systems in institutions as sources of legionellosis. Not only has the agent been isolated from cases and implicated water systems, but these isolations have also been associated in time. On occasion there has been a plausible explanation for the start of the epidemic, and the application of control measures has prevented further cases. The one important major criterion which needs to be met to establish water systems as sources is the demonstration of an epidemiological association. No case control study to date has shown significant differences in exposure rates of showering or bathing, perhaps due to the difficulties involved in taking account of host susceptibility variation and the fact that high concentrations of the organism may be released from water outlets only very intermittently.

If domestic water is the source, does the patient become infected through drinking it, inhaling it, or some other means? There is certainly much to support the hypothesis that inhalation is the principal route. Airborne spread from cooling water systems is well documented for Pontiac fever and Legionnaires disease (2, 10, 16, 18). Then there are the recently reported associations between legionellosis and exposure to aerosols produced by respiratory therapy equipment and high-energy whirlpool systems. However, it is not clear whether taps or even showers can produce dense aerosols of particles small enough to be drawn into the lower respiratory tract. In this respect it is interesting that in one nosocomial outbreak seven of the cases are known not to have showered at any time in hospital although they had had conventional baths (10). Perhaps a fine-particle aerosol is not essential for the spread of legionellosis. Could it be, as suggested by Rowbotham, that the infective particles are the legionella-containing free-living amoebae commonly found in tap water (24)? Nevertheless, there is clear-cut experimental evidence that small-particle inhalation is a mode of transmission in animals (4; P. Meenhorst, personal communication), so it seems likely, although not yet proved, that in humans this is the mechanism for infection from showers or taps.

There is little to indicate that infection can be acquired through ingestion, and there are no reports to date of this mode of transmission in animals. In the investigation of the Philadelphia outbreak of Legionnaires disease in 1976 an association was found with drinking tap water in one hotel. However, this was not shown to be independent of confounding variables such as length of stay in the hotel, and 38% of cases denied drinking water there (15). No other case control study has shown an increased rate of water drinking in cases, and of course many outbreak investigations have implicated hot water (3, 12; Best et al., 22nd ICAAC, abstr. no. 88, 1982; Helms et al., 22nd ICAAC, abstr. no. 89, 1982), which is not likely to have been drunk. On the other hand, there is a report of the isolation of *L. pneumophila* from the gut and peritoneum (E. Dournon, personal communication), so perhaps we should not yet dismiss ingestion as an occasional route of infection. Nosocomial cases often occur in individuals not well enough to have a shower or bath, so it may be that ingestion of water containing the organism leads, particularly in the immunocompromised, to colonization of the respiratory tract. There is no evidence in humans or animals of other mucous membranes, such as the conjunctivae, serving as portals of entry. Finally, the parenteral route must be considered in view of a paper describing a patient with a dialysis shunt site infected by *L. pneumophila* (17). The author concluded justifiably that the infection was probably blood-borne from the respiratory tract. The lack of similar reports indicates that parenteral transmission must occur very rarely.

Another intriguing question yet to be answered is why, in view of the widespread distribution of legionellae in water systems, is endemic and epidemic infection in institutions not recognized more frequently? In the case of nosocomial disease, it is evident that many legionella infections are still being missed. In a recent prospective study in a community hospital in Pittsburgh, 14.3% of nosocomial pneumonias were found to be Legionnaires disease, although this infection had not previously been recognized there (21). Water systems in hotels, as well as hospitals, may be important sources of legionella infection, but associated case clusters may only be recognized if there is an efficient regional or national reporting scheme. One national surveillance scheme found that 24 of the 32 identified outbreaks from 1979 to 1982 were associated with hotels, the majority of which were located in popular tourist resorts in other countries (3).

Failure to look for the infection is certainly not the only explanation. Detailed investigations of some nosocomial outbreaks have shown that they may occur predominantly in just one block or wing of a hospital (5, 10, 28). This could be explained by the fact that such infections occur mainly in the wards housing patients who are particularly susceptible; in some incidents, however, similar groups of patients in other parts of the hospital have remained free of the disease (5, 10, 28).

There must be explanations, therefore, other than differences in susceptibility. It could be that many of the legionellae colonizing water systems are nonvirulent strains. Brown and colleagues (5) found that the majority of environmental isolates in one hospital contained a single 80-megadalton plasmid, although plasmidless strains were also isolated. During an outbreak of Legionnaires disease, the attack rate was higher on wards in which the shower units and tap samples revealed isolates without plasmids. All of the clinical isolates were plasmidless as well. Brown et al. suggested that plasmids present in some strains might modify virulence or environmental adaptability. This is a fascinating observation which warrants further study.

The infective dose, and thus also the concentration of the organism present in the water, may be critical for legionella infection. In this respect the design of water systems may be important. Some calorifier designs may encourage the growth of legionellae if they allow constant warm temperatures and poor flow in the sump. This part of the cylinder often contains a sludge of corrosion products rich in iron which may provide a nutritional source for legionellae (11). Other design characteristics such as rate of flow and pressure differentials may also be relevant, as may be the composition of the materials used in the construction. The way in which engineers maintain and operate systems may be important, particularly in their use of stand-by procedures for pump calorifiers and temperature settings. There may well be other factors yet to be determined such as the chemistry of the water or chemical methods used for its treatment, or perhaps lack of them in the light of several outbreaks of Legionnaires disease associated with the use of untreated borehole water (3, 23).

Inevitably, I have dealt in this review largely with the information that has resulted from studies of epidemics and endemic nosocomial infection. Whether the conclusions drawn from these are applicable to sporadic infection in the general community is a moot point. It would be surprising if domestic water systems in houses, apartment buildings, other establishments, and industrial sites were not sources of legionella infection, but their contribution in relation to other reservoirs, such as cooling water systems, is not known. Work on this aspect is needed if rational control programs are to be devised. The relative importance of the different sources may depend on climatic factors and thus is likely to vary from country to country.

Turning finally to the issue of prevention of legionellosis: is control practicable, and what are its indications? There is general agreement that control measures are indicated when a water system is identified as a source through epidemiological investigation. It is now really a matter of deciding on the best strategy, be it continuous or intermittent chlorination, temperature control, or some other method.

The application of control measures to the majority of water systems, however, cannot be justified at present, although good engineering practices in their maintenance and operation should be ensured. More needs to be known about the pathogenesis of legionellosis and the ecology of water systems before the question of primary prevention can be properly addressed. Perhaps environmental surveys will reveal that inexpensive changes in the design of plumbing systems, or the materials used in their construction, may be all that is required to discourage the multiplication of the organism. More realistically, any proposed control measure will have to be carefully evaluated, with the costs, disadvantages, and perhaps hazards being weighed against the burden of morbidity and mortality from domestic water systems, once these have been estimated. Many of the questions will not be answered immediately, but the small but definite possibility of primary prevention of legionellosis, and perhaps related infections as yet unidentified, is an exciting prospect.

LITERATURE CITED

1. **Arnow, P. M., T. Chou, D. Weil, E. N. Shapiro, and C. Kretzschmar.** 1982. Nosocomial Legionnaires' disease caused by aerosolized tap water from respiratory devices. J. Infect. Dis. **146:**460–467.
2. **Band, J. D., M. LaVenture, J. P. Davis, G. F. Mallison, P. Skaliy, P. S. Hayes, W. L. Schell, H. Weiss, D. J. Greenberg, and D. W. Fraser.** 1981. Epidemic Legionnaires' disease. Airborne transmission down a chimney. J. Am. Med. Assoc. **245:**2404–2407.
3. **Bartlett, C. L. R., and L. F. Bibby.** 1983. Epidemic legionellosis in England and Wales 1979–1982. Zentralbl. Bakteriol. Mikrobiol. Hyg. Abt. 1 Orig. Reihe A **255:**64–70.
4. **Baskerville, A., R. B. Fitzgeorge, M. Broster, P. Hambleton, and P. J. Dennis.** 1981. Experimental transmission of Legionnaires' disease by exposure to aerosols of *Legionella pneumophila.* Lancet ii:1389–90.
5. **Brown, A., R. M. Vickers, E. M. Elder, M. Lema, and G. M. Garrity.** 1982. Plasmid and surface antigen markers of endemic and epidemic *Legionella pneumophila* strains. J. Clin. Microbiol. **16:**230–235.
6. **Brown, A., V. L. Yu, M. H. Magnussen, R. M. Vickers, G. M. Garrity, and E. M. Elder.** 1982. Isolation of Pittsburgh pneumonia agent from a hospital shower. Appl. Environ. Microbiol. **43:**725–726.
7. **Cordes, L. G., A. M. Wiesenthal, G. W. Gorman, J. P. Phair, H. M. Sommers, A. Brown, V. L. Yu, M. H. Magnussen, R. D. Meyer, J. S. Wolf, K. N. Shands, and D. W. Fraser.** 1981. Isolation of *Legionella pneumophila* from hospital shower heads. Ann. Intern. Med. **94:**195–197.
8. **de Lalla, F., G. Rossini, G. Giannattasio, R. Giura, G. Nessi, and D. Santoro.** 1980. Legionnaires' disease in an Italian hotel. Lancet ii:1187.
9. **Dennis, P. J., R. B. Fitzgeorge, J. A. Taylor, C. L. R. Bartlett, and G. I. Barrow.** 1982. *Legionella pneumophila* in water plumbing systems. Lancet i:949–951.
10. **Dondero, T. J., R. C. Rendtorff, G. F. Mallison, R. M. Weeks, J. S. Levy, E. W. Wong, and W. Schaffner.** 1980. An outbreak of Legionnaires' disease associated with a contaminated air-conditioning cooling tower. N. Engl. J. Med. **302:**365–370.
11. **Fisher-Hoch, S. P., C. L. R. Bartlett, J. O'H. Tobin, M. B. Gillett, A. M. Nelson, J. E. Pritchard, M. G. Smith, R. A. Swann, J. M. Talbot, and J. A. Thomas.** 1981. Investigation and control of an outbreak of Legionnaires' disease in a District General Hospital. Lancet i:932–936.
12. **Fisher-Hoch, S. P., M. G. Smith, and J. S. Colbourne.** 1982. *Legionella pneumophila* in hospital hot water cylinders. Lancet i:1073.
13. **Fliermans, C. B., W. B. Cherry, L. H. Orrison, S. J. Smith, D. L. Tison, and D. H. Pope.** 1981. Ecological distribution of *Legionella pneumophila.* Appl. Environ. Microbiol. **41:**9–16.
14. **Fliermans, C. B., W. B. Cherry, L. H. Orrison, and L. Thacker.** 1979. Isolation of *Legionella pneumophila* from nonepidemic-related aquatic habitats. Appl. Environ. Microbiol. **37:**1239–1242.
15. **Fraser, D. W., T. R. Tsai, W. Orenstein, W. E. Parkin,**

H. J. Beecham, R. G. Sharrar, J. Harris, G. S. Mallison, S. M. Martin, J. E. McDade, C. C. Shepard, P. S. Brachman, and the Field Investigation Team. 1977. Legionnaires' disease: description of an epidemic of pneumonia. N. Engl. J. Med. 297:1189–1197.

16. Glick, T. H., M. B. Gregg, B. Berman, G. Mallison, W. W. Rhodes, and I. Kassanoff. 1978. Pontiac fever: an epidemic of unknown etiology in a health department. I. Clinical and epidemiologic aspects. Am. J. Epidemiol. 107:149–160.

17. Kalweit, W. H., W. C. Winn, T. A. Rocco, and J. C. Girod. 1982. Hemodialysis fistula infections caused by Legionella pneumophila. Ann. Intern. Med. 96:173–175.

18. Kaufmann, A. F., J. E. McDade, C. M. Patton, J. V. Bennett, P. Skaliy, J. C. Feeley, D. C. Anderson, M. E. Potter, V. F. Newhouse, M. B. Gregg, and P. S. Brachman. 1981. Pontiac fever: isolation of the etiologic agent (Legionella pneumophila) and demonstration of its mode of transmission. Am. J. Epidemiol. 114:337–347.

19. Kurtz, J. B., C. L. R. Bartlett, U. A. Newton, R. A. White, and N. L. Jones. 1982. Legionella pneumophila in cooling water systems. Report of a survey of cooling towers in London and a pilot trial of selected biocides. J. Hyg. 88:369–380.

20. Makela, T. M., S. J. Harders, P. Cavanagh, G. Rouch, and R. Ayre. 1981. Isolation of Legionella pneumophila (serogroup one) from shower-water in Ballarat. Med. J. Aust. 1:293–294.

21. Muder, R. R., V. L. Yu, J. K. McClure, F. J. Kroboth, S. D. Kominos, and R. M. Lumish. 1983. Nosocomial Legionnaires' disease uncovered in a prospective pneumonia study. J. Am. Med. Assoc. 249:3184–3188.

22. Reid, D., N. R. Grist, and R. Najera. 1978. Illness associated with "package tours": a combined Spanish-Scottish study. Bull. WHO 56:117–122.

23. Rosmini, F., D. Greco, M. Castellani Pastoris, and A. Zampieri. 1983. Istituto Superiore di Sanita, Roma. ISTISAN 83/7.

24. Rowbotham, T. J. 1980. Preliminary report on the pathogenicity of Legionella pneumophila for freshwater and soil amoebae. J. Clin. Pathol. 33:1179–1183.

25. Stout, J., V. L. Yu, R. M. Vickers, and J. Shonnard. 1982. Potable water supply as the hospital reservoir for Pittsburgh pneumonia agent. Lancet i:471–472.

26. Stout, J., V. L. Yu, R. M. Vickers, J. Zuravleff, M. Best, A. Brown, R. B. Yee, and R. Wadowsky. 1982. Ubiquitousness of Legionella pneumophila in the water supply of a hospital with endemic Legionnaires' disease. N. Engl. J. Med. 306:466–468.

27. Tobin, J. O'H., C. L. R. Bartlett, S. A. Waitkins, G. I. Barrow, A. D. Macrae, A. G. Taylor, R. J. Fallon, and F. R. N. Lynch. 1981. Legionnaires' disease: further evidence to implicate water storage and distribution systems as sources. Br. Med. J. 282:573.

28. Tobin, J. O'H., J. Beare, M. S. Dunnill, S. P. Fisher-Hoch, M. French, R. G. Mitchell, P. J. Morris, and M. F. Muers. 1980. Legionnaires' disease in a transplant unit: isolation of the causative agent from shower baths. Lancet ii:118–121.

29. Tobin, J. O'H., R. A. Swann, and C. L. R. Bartlett. 1981. Isolation of Legionella pneumophila from water systems: methods and preliminary results. Br. Med. J. 282:515–517.

30. Wadowsky, R. M., R. B. Yee, L. Mezmar, E. J. Wing, and J. N. Dowling. 1982. Hot water systems as sources of Legionella pneumophila in hospital and nonhospital plumbing fixtures. Appl. Environ. Microbiol. 43:1104–1110.

Nosocomial Legionnaires Disease: Difference in Attack Rates Associated with Two Strains of *Legionella pneumophila* Serogroup 1

JOSEPH PLOUFFE, MICHAEL PARA, BARBARA HACKMAN, LINDA WEBSTER, AND WILLIAM MAHER

The Ohio State University, Columbus, Ohio 43210

Nosocomial Legionnaires disease has been reported from many cities, including Columbus, Ohio (9). Environmental sampling in several hospitals has revealed *Legionella pneumophila* in the shower heads and hot-water systems (4, 10, 11). The organism has also been found in various aquatic environments where Legionnaires disease is not prevalent in the neighboring community (7). We have found a skewed distribution of cases of nosocomial Legionnaires disease between two hospital buildings in our medical center, with 19 cases (1.03/1,000 discharges) occurring in building UH and 1 case (0.08/1,000 discharges) occurring in building RH. The patients were exposed to the hospital from 2 to 10 days before developing an acute pneumonia. Sixteen patients had positive cultures for *L. pneumophila*, one only had positive direct fluorescent-antibody stain of pulmonary secretions, two had a fourfold reciprocal antibody titer rise to ≥512, and one had a single convalescent reciprocal titer of 2,048 (2).

Our acute-care medical and surgical patients are housed in either of two adjoining hospital buildings, UH (442 beds), built in 1951, or RH

(355 beds), built in 1980. The domestic water of both buildings is supplied by the same municipal main. Each building has separate plumbing and hot-water storage tanks. The patient populations in both buildings are similar except that RH houses the majority of hematology-oncology patients and renal transplant patients. The diagnosis of Legionnaires disease was considered frequently in RH patients, as 484 *Legionella* serology tests were ordered in a 9-month period in 1982 compared with 172 serology tests for UH patients.

Hot-water taps from patient areas in both buildings were cultured weekly for the presence of *L. pneumophila*. The initial 50 ml of hot water was collected, the inside of the faucet was swabbed, and the swab was rinsed in the water. The sample was centrifuged (1,000 × *g* for 30 min), and the sediment was acid treated (1) and plated onto buffered charcoal-yeast extract agar (5, 6). Typical *Legionella* colonies were confirmed by the direct fluorescent-antibody reagent (3) provided by the Centers for Disease Control. Table 1 lists the results of surveillance cultures for *L. pneumophila* serogroup 1 from

TABLE 1. Surveillance cultures for *L. pneumophila* serogroup 1 and nosocomial cases of Legionnaires disease

Yr	Mo	UH			RH		
		No. of surveillance sites	No. (%) positive	No. of nosocomial cases	No. of surveillance sites	No. (%) positive	No. of nosocomial cases
1982	January	7	3 (43)	1	30	12 (40)	0
	February	12	6 (50)	0	11	5 (45)	0
	March	11	4 (36)	0	8	4 (50)	0
	April	16	6 (38)	0	11	6 (55)	0
	May	14	6 (43)	0	10	7 (70)	0
	June	16	8 (50)	0	8	5 (63)	0
	July	32	22 (69)	5	11	3 (27)	0
	August	33	14 (42)	5	6	3 (50)	1
	September	75	24 (32)	0	8	5 (63)	0
	October	89	14 (16)	1	20	5 (25)	0
	November	88	21 (24)	0	46	11 (24)	0
	December	73	10 (14)	5	43	14 (33)	0
1983	January	99	19 (19)	1	48	0 (0)	0
	February	106	30 (28)	1	44	2 (5)	0
Total		671	187 (28)	19	304	82 (27)	1

TABLE 2. Subtypes of *L. pneumophila* serogroup 1

Subtype	Colony type	No. of plasmids	LP-I-81 agglutination	No. of environmental isolates (%)[a]		No. of clinical isolates (%)	
				UH	RH	UH	RH
UH-1	I	0	+	3,941 (94)	0 (0)	12 (80)	0 (0)
UH-2	II	1 (45 Mdal[b])	+	190 (5)	0 (0)	3 (20)	0 (0)
RH-1	III	2 (45 and 85 Mdal)	−	54 (1)	1,995 (100)	0 (0)	1 (100)

[a] Obtained between January 1982 and June 1983.
[b] Mdal, Megadaltons.

hot-water taps in both buildings. Both had similar percentages of positive samples (28% in UH, 27% in RH). From the 187 positive cultures in UH, 3,465 colonies of *L. pneumophila* serogroup 1 were counted. A total of 67% of the cultures had 1 to 10 colonies per plate, 19% had 11 to 50 colonies per plate, and 14% had >50 colonies per plate. In RH the 82 positive cultures had 1,772 colonies, with 68% having 1 to 10 CFU per plate, 21% having 11 to 50 CFU per plate, and 11% having >50 CFU per plate. Subtle differences were noted among colonies that stained with the direct fluorescent-antibody reagent for *L. pneumophila* serogroup 1. Typical colonies (type I) were 3- to 6-mm white, convex colonies at 48 to 72 h which had a purple hue on subculture. Type II was a 3- to 6-mm white, flat colony which also had a purple hue on subculture. Type III was a 2- to 3-mm translucent colony which had a blue tint on subculture. These different colony types were subjected to plasmid analysis by the alkaline sodium dodecyl sulfate method of Kado and Liu (8; W. E. Maher, J. F. Plouffe, and M. F. Para, this volume). The organisms were tested for microagglutination in the presence of LP-I-81 monoclonal antibody, which is specific for *L. pneumophila* serogroup 1 (M. F. Para, J. F. Plouffe, and W. E. Maher, this volume).

Table 2 lists the colony types, plasmid profile, and monoclonal antibody reactivity for the subtypes of *L. pneumophila* serogroup 1 isolated from the hot-water systems and from patients in UH and RH between January 1982 and June 1983. The colony morphology consistently predicted the plasmid content and microagglutination results. Most strains isolated from UH were UH-1, with no plasmids. Occasional UH-2 strains were isolated from a room in UH which had been occupied by patients whose clinical isolates were UH-2. Both strains agglutinate with the monoclonal antibody LP-I-81.

Although we have not been able to isolate *L. pneumophila* from the source water, our suspicion is that the organisms enter the hot-water systems and that unknown differences in the individual hot-water systems select for the different subtypes. To date, we have not found

differences in water temperature, pH, or free chlorine residual (9a) or in initial determinations of concentrations of the trace elements calcium, magnesium, aluminum, cadmium, chromium, copper, iron, lead, manganese, nickel, selenium, and zinc.

The difference in attack rates in the two buildings colonized by similar numbers of *L. pneumophila* serogroup 1 suggests a difference in virulence between the subtypes found in UH and RH. Fifteen of the sixteen clinical isolates contained the surface antigen recognized by LP-I-81. This antigen may be important in determining the relative virulence of different *L. pneumophila* serogroup 1 strains.

This study was supported in part by a grant from the Division of Water, Columbus, Ohio, and by Public Health Service grant AI-17552 from the National Institutes of Health.

We appreciate the assistance of Paul Edelstein with initial culturing methodology and the helpful suggestions of Victor Yu, and we thank Sharon Venters for typing the manuscript.

LITERATURE CITED

1. Bopp, C. A., J. W. Sumner, G. K. Morris, and J. G. Wells. 1981. Isolation of *Legionella* spp. from environmental water samples by low-pH treatment and use of a selective medium. J. Clin. Microbiol. 13:714–719.
2. Center for Disease Control. 1978. Legionnaires' disease: diagnosis and management. Ann. Intern. Med. 88:363–365.
3. Cherry, W. B., B. Pittman, P. P. Harris, G. A. Hébert, B. M. Thomason, L. Thacker, and R. E. Weaver. 1978. Detection of Legionnaires disease bacteria by direct immunofluorescence staining. J. Clin. Microbiol. 8:329–338.
4. Cordes, L. G., A. M. Wiesenthal, G. W. Gorman, J. P. Phair, H. M. Sommers, A. Brown, V. L. Yu, M. H. Magnussen, R. D. Myer, J. F. Wolf, K. N. Shands, and D. W. Fraser. 1981. Isolation of *Legionella pneumophila* from hospital shower heads. Ann. Intern. Med. 94:195–197.
5. Edelstein, P. H. 1981. Improved semiselective medium for isolation of *Legionella pneumophila* from contaminated clinical and environmental specimens. J. Clin. Microbiol. 14:298–303.
6. Feeley, J. C., R. J. Gibson, G. W. Gorman, N. C. Langford, J. K. Rasheed, D. C. Mackel, and W. B. Baine. 1979. Charcoal-yeast extract agar: primary isolation medium for *Legionella pneumophila*. J. Clin. Microbiol. 10:436–441.
7. Fliermans, C. B., W. B. Cherry, L. H. Orrison, and L. Thacker. 1979. Isolation of *Legionella pneumophila* from nonepidemic-related aquatic habitats. Appl. Environ. Microbiol. 37:1239–1242.

8. Kado, C. I., and S. T. Liu. 1981. Rapid procedure for detection of large and small plasmids. J. Bacteriol. 145:1365–1373.

9. Marks, J. S., T. F. Tsai, W. J. Martone, R. C. Baron, J. Kennicott, F. J. Holtzhauer, I. Baird, D. Fay, J. C. Feeley, G. F. Mallison, D. W. Fraser, and T. J. Halpin. 1979. Nosocomial Legionnaires' disease in Columbus, Ohio. Ann. Intern. Med. 90:565–569.

9a. Plouffe, J. F., L. R. Webster, and B. Hackman. 1983. Relationship between colonization of hospital buildings with Legionella pneumophila and hot water temperatures.

Appl. Environ. Microbiol. 46:769–770.

10. Stout, J., V. L. Yu, R. M. Vickers, J. Zuravleff, M. Best, A. Brown, R. Yee, and R. Wadowski. 1982. Ubiquitousness of Legionella pneumophila in the water supply of a hospital with endemic Legionnaires' disease. N. Engl. J. Med. 306:466–468.

11. Tobin, J. H., M. S. Dunnill, M. French, P. F. Morris, J. Beare, S. Fisher-Hoch, R. G. Mitchell, and M. F. Muers. 1980. Legionnaires' disease in a transplant unit: isolation of the causative agent from shower baths. Lancet ii:118–121.

Prospective Survey of Acquisition of Legionnaires Disease

RICHARD D. MEYER, GREGORY H. SHIMIZU, RONALD FULLER, JAMES SAYRE, JOHN CHAPMAN, AND PAUL H. EDELSTEIN

Research, Medical, and Nursing Services, Wadsworth Veterans Administration Medical Center, Los Angeles, California 90073; Department of Medicine, Cedars-Sinai Medical Center, Los Angeles, California 90048; and UCLA School of Medicine and Graduate School of Public Health, Los Angeles, California 90024

Nosocomial Legionnaires disease (LD) occurs in both sporadic and outbreak patterns. Several known nosocomial outbreaks have been recorded (2, 5); by late 1979 over 120 cases of nosocomial LD had occurred in our hospital, and the source was not yet known. Previous studies with Centers for Disease Control had shown that about 0.4% of patients admitted developed LD and that length of stay and compromised immunity were risk factors for acquisition (4). An active surveillance survey was undertaken in 1979 in an attempt to detect as many cases as possible and to assess risk factors.

Admitted patients were followed by infection control personnel over a 3-month period from November 1979 through January 1980. Acute and convalescent (≥2 weeks and, in some patients, 6 weeks) serum specimens were obtained. When clinically indicated, submission of other specimens was encouraged by infection control personnel. Immunosuppressed patients were included to be followed whenever possible. Indirect fluorescent-antibody (FA) testing was performed by screening with a polyvalent heat-killed antigen and then with serogroup 1, 2, 3, and 4 Legionella pneumophila monovalent antigens. A case of LD was defined as signs and symptoms of infection with radiographic evidence of pneumonia and one or more of the following: (i) $\geq$fourfold rise in the indirect FA titer to $\geq1:128$, (ii) positive direct FA test on respiratory tract secretions, or (iii) positive culture for L. pneumophila. Lower respiratory tract infection was defined as two or more of the following: purulent sputum, leukocytosis, or radiographic infiltrate developing in a patient >72 h after admission or within 3 weeks of discharge.

Patient records were available in approximately 88% of instances and were reviewed retrospectively for data on selected underlying diseases, treatment, demographic factors, and clinical features; infection control records were available for the rest of the instances. Data were coded for computer analysis. Statistical analysis was performed by the chi-square test with Yate's correction or by Student's t-test.

A total of 819 patients, approximately one-third of the 2,393 patients admitted, were followed; paired serum specimens were available for 494 of the 819 patients. Seroconversion was found in 29 members of this cohort but LD was confirmed by seroconversion in only 9 of these with radiographic infiltrates; of 325 patients without serological specimens, five other cases were diagnosed by other laboratory means (Table 1). Three additional cases were documented in the other 1,574 hospitalized patients not followed; thus, 17 of 2,393 admissions (0.7%) and 14 of 819 patients in the cohort (1.7%) were found to have LD. LD accounted for 15 of 144 (10.4%) of nosocomial lower respiratory tract infections (Table 1).

The distribution of demographic and potential risk factors for LD acquisition showed no significant differences between cases and others with regard to selected underlying diseases (malignancy, chronic renal disease, chronic obstructive pulmonary disease, or myocardial infarction), age, sex, smoking history, alcohol use, immunosuppressive drugs, corticosteroid administration, radiation therapy, a previous outpatient clinic visit, or prior admission ≥2 weeks. There was, however, a significant difference between cases of LD and non-seroconverting noncases with regard to adult-onset diabetes mellitus (4 of 14 versus 41 of 450; $\chi^2 = 3.86$, $P = 0.05$) and a trend toward a prior admission within the last 2 weeks (2 of 14 versus 10 prior admissions of 412; $\chi^2 = 3.17$, $P = 0.075$). The three groups did not differ with regard to length of hospitalization; means were 2.07, 2.15, and 1.71 months, respectively.

A comparison of clinical features of cases of LD and seroconverters without pneumonia identified several constitutional, respiratory, and neurological features as being possibly associated with LD (Table 2). Many of these are also symptoms of pneumonic LD (5).

These data confirm that diabetes mellitus may be a risk factor for nosocomial acquisition of LD (2) and are consistent with exposure to an endemic area where, subsequently, potable water was shown to have been contaminated (K. Shands, G. W. Gorman, R. D. Meyer, P. H. Edelstein, and S. M. Finegold, Clin. Res. 29:260A, 1981). Failure to confirm other nosocomial risk factors in LD previously reported by the Centers for Disease Control and by our center, such as end-stage renal disease, immunosuppressive therapy, cancer, and chronic obstructive pulmonary disease, (4, 6), is likely due to the relatively small number of LD cases considered here and to suspension of the renal transplantation program in this setting. Multiregression analysis might also link risk to other factors, including some reported in sporadic cases (3). The incidence of LD was significant but appeared to be lower than in another outbreak where different diagnostic criteria were used (1, 7).

Intensive surveillance with indirect FA testing appears to be sensitive but is limited by lack of specificity for diagnosis of LD pneumonia even in an endemic area; all diagnostic methods are needed to detect cases. Intensive surveillance is labor intensive and, if resources are limited, the most efficient use of them should be directed toward high-risk patients.

TABLE 2. Clinical features of seroconverters[a]

Feature	No. of cases (n = 8)	No. of seroconverter noncases (n = 7)[b]	χ^2	P
Fever	8	3	3.65	0.056
Malaise	5	0	4.05	0.044
Cough	6	0	5.90	0.015
Chest pain	4	1		NS[c]
Dyspnea	6	0	5.90	0.015
Myalgia	0	0		NS
Headache	1	0		NS
Diarrhea	0	2		NS
Nausea/vomiting	1	0		NS
Abdominal pain	5	1		NS
Neurological disorder	6	1	3.36	0.067
Bacteremia	1	0		NS

[a] With complete charts available for retrospective review. All charts and radiographs were reviewed at time of seroconversion.
[b] No radiographic evidence of pneumonia.
[c] NS, Not significant.

TABLE 1. Summary of cases of LD and respiratory tract infections

Mo	Discharges[a]	Respiratory tract infections	Cases of LD[b] In study	Cases of LD[b] Total
November 1979	788	34	5	5
December 1979	807	49	4	5
January 1980	798	61	3	5

[a] Does not include intermediate-care patients.
[b] In addition, two patients followed in January 1980 had onset of LD in February 1980.

LITERATURE CITED

1. Brown, A., V. L. Yu, E. M. Elder, M. H. Magnussen, and F. Kroboth. 1980. Nosocomial outbreak of Legionnaire's disease at the Pittsburgh Veterans Administration medical center. Trans. Assoc. Am. Physicians 93:52–59.
2. England, A. C. III, and D. W. Fraser. 1981. Sporadic and epidemic legionellosis in the United States. Epidemiologic features. Am. J. Med. 70:707–711.
3. England, A. C. III, D. W. Fraser, B. D. Plikaytis, T. F. Tsai, G. Storch, and C. V. Broome. 1981. Sporadic legionellosis in the United States: the first thousand cases. Ann. Intern. Med. 94:164–170.
4. Haley, C. E., M. L. Cohen, J. Halter, and R. D. Meyer. 1979. Nosocomial legionnaires' disease: a continuing common-source epidemic at Wadsworth medical center. Ann. Intern. Med. 90:583–586.
5. Meyer, R. D. 1983. Legionella infections: a review of five years of research. Rev. Infect. Dis. 5:258–278.
6. Serota, A. I., R. D. Meyer, S. E. Wilson, P. H. Edelstein, and S. M. Finegold. 1981. Legionnaires' disease in the postoperative patient. J. Surg. Res. 30:417–427.
7. Yu, V. L., F. J. Kroboth, J. Shonnard, A. Brown, S. McDearman, and M. Magnussen. 1982. Legionnaires' disease: a new clinical perspective from a prospective pneumonia study. Am. J. Med. 73:357–361.

Prospective One-Year Study of Legionnaires Disease in a German University Hospital

H. LODE, R. GROTHE, H. SCHÄFER, G. RUCKDESCHEL, AND H. E. MÜLLER

Medical Department of Klinikum Steglitz, Freie Universität Berlin, West Berlin; Institute of Clinical Microbiology, University of Munich, Munich; and Institute of General Clinical Microbiology, Braunschweig, West Germany

Legionnaires disease is playing an increasingly important role in the etiology of ambulant and nosocomial pneumonia in Europe. In a prospective study, the frequency, clinical picture, course, and prognosis of Legionnaires disease were recorded over 1 year in a university hospital with 1,300 beds. From April 1980 to April 1981, 110 unselected continuous patients were studied who had severe and moderately severe pneumonia; these patients were from the internal and operative intensive-care units as well as the pulmonological unit. The diagnosis of pneumonia was made on clinical and radiological grounds, and the majority of infections were acquired nosocomially. Additional evidence for the presence of *Legionella* species was obtained by means of indirect immunofluorescence for detection of antibodies against six serological groups of *Legionella pneumophila* and against *Legionella micdadei*. For this purpose, serum samples were taken from each patient during the first days of pneumonia, after 2 to 3 weeks, and again during convalescence. Single titers of or exceeding 1:256 or a fourfold titer increase to at least 1:128 was considered significant.

Twelve diagnoses of Legionnaires disease were recorded in 11 of the 110 patients investi-

gated. The patient group consisted of four women and seven men ranging in age between 24 and 74 years, with a mean age of 57 years. One patient contracted the disease twice (May 1980 and January 1981). All patients had severe underlying diseases (Table 1). Four infections (cases 1, 3, 11, and 12) were identified as having been acquired outside the hospital, whereas eight were classified as nosocomial. All patients required intensive medical monitoring and treatment over long periods of their illness. All patients had fever (between 38.8 and 40.4°C). Maximal leukocyte values fluctuated between 6.8×10^9 and 28.8×10^9/liter, the lowest value (case 2) being due to prior immunosuppressive treatment. Liver function values (serum glutamic pyruvic transaminase and alkaline phosphatase) were slightly to moderately increased in 10 and 7 cases, respectively. The principal and most serious complications were acute renal failure and respiratory decompensation. Four patients suffered acute renal failure requiring hemodialysis. Four other patients showed a transient renal insufficiency, with a marked course in two cases. Nine patients had to be under artificial respiration for different periods of time, some for several weeks. Eight of the

TABLE 1. Clinical characteristics of 11 patients with Legionnaires disease pneumonia[a]

Case	Age (yr)	Sex[b]	Date of admission	Underlying or previous disease	Pulmonary symptoms	Intestinal symptoms	Course
1	62	M	4/13/80	Chronic obstructive lung disease	+	+ (D/O)[c]	Died 5/15/80
2	61	F	4/7/80	Kidney transplantation	+	+ (D)	Died 6/5/80
3	67	M	5/23/80	Chronic obstructive lung disease	+	+ (O)	Discharged, later died
4	68	M	4/30/80	Diabetes mellitus, operated colon carcinoma	+	−	Died 6/12/80
5	74	F	6/11/80	Operated hepatic cysts	+	+ (D)	Discharged
6	24	M	6/12/80	Open craniocerebral trauma	+	−	Died 7/9/80
7	49	M	7/25/80	Hepatic cirrhosis, shunt operation	+	−	Discharged
8	41	F	10/10/80	Operated meningioma	+	+ (D)	Regression
9	45	M	12/1/80	Polytrauma	+	−	Died 1/16/81
10	51	M	1/11/81	Late reanimation after myocardial infarct	+	+ (D)	Died 4/8/81
11	68	M	2/17/81	Chronic obstructive lung disease, bronchial carcinoma	+	+ (O)	Discharged, later died
12	72	F	4/11/81	Chronic obstructive lung disease, myogenic cardiac insufficiency	+	+ (O)	Died 5/25/81

[a] Cases 3 and 11 were the same patient.
[b] M, Male; F, female.
[c] D, Diarrhea; O, obstipation.

TABLE 2. Course of immunofluorescence titer in 11 patients with Legionnaires disease[a]

Case	Titer		
	Acute	Midcourse (maximum)[b]	Convalescent
1	1:64	1:512 (1)	—[c]
2	Negative	1:2,048 (1, 3, 4)	—
3	1:64	1:256 (2)	1:32
4	1:64	1:512 (1)	—
5	1:128	1:256 (2)	Negative
6	1:32	1:512 (L. micdadei)	—
7	Negative	1:256 (4)	1:512
8	1:64	1:1,024 (1)	1:256
9	ND[d]	1:512 (L. micdadei)	—
10	1:64	1:256 (1)	1:64
11	Negative	1:1,024 (1)	1:256
12	1:64	1:256 (4)	—

[a] Cases 3 and 11 were the same patient.
[b] Numbers in parentheses are the L. pneumophila serogroup.
[c] —, Patient died before convalescence.
[d] ND, Not detected.

eleven patients died, four of them (cases 1, 2, 4, and 12) as a consequence of the septic effects involved in Legionnaires disease pneumonia. Two patients (cases 6 and 9) died of cerebral failure, one patient (case 10) died of complications associated with decerebration, and another patient (cases 3 and 11) died of a peripheral bronchial carcinoma 6 months after a second manifestation of Legionnaires disease. The courses of the immunofluorescence titers are presented in Table 2. It is evident that the threshold titer of 1:256 was reached in all patients.

Radiological findings in 12 cases showed an initial predominant involvement of both lower lobes, which were affected a total of 11 times. In one case, a multilobular diffuse infiltration picture was already observed at the onset of the disease. In the course of the legionellosis, however, multilobular diffuse spreading was frequently seen; right-sided pleuritis developed in one case, and bilateral pneumothorax developed in another.

Surprised at the relative frequency of Legionnaires disease (11%) in our patients, we carried out 208 environmental examinations in both intensive care units. Culture tubes (for rectal diagnostics) were used to examine diverse water systems, infusion solutions, drainage systems, and respirators. Three samples positive for L. pneumophila serogroup 5 (confirmed by the Centers for Disease Control) were obtained from two water taps with perlator sieves and from one water tap orifice without a sieve.

This first prospective study at a German university hospital on the frequency of Legionnaires disease shows that legionellae are responsible for pneumonia and should receive increased diagnostic and therapeutic attention in Germany.

LITERATURE CITED

1. Holzer, E., and G. Ruckdeschel. 1979. Legionärskrankheit in Deutschland. Infection 7:149–151.
2. Lode, H., H. Schäfer, and G. Ruckdeschel. 1982. Legionärskrankheit—Prospektive Studie zur Häufigkeit, Klinik und Prognose. Dtsch. Med. Wschr. 107:326–331.
3. Macfarlane, J. T., R. G. Finck, M. J. Ward, and A. D. Macrae. 1982. Hospital study of adult community-acquired pneumonia. Lancet ii:255–258.

Epidemiology of Two Outbreaks of Legionnaires Disease in Burlington, Vermont, 1980

DOUGLAS N. KLAUCKE, RICHARD L. VOGT, DENISE LaRUE, LINDEN E. WITHERELL, LILLIAN ORCIARI, KENNETH C. SPITALNY, RAYMOND PELLETIER, WILLIAM B. CHERRY, AND LLOYD F. NOVICK†

Vermont Department of Health, Burlington, Vermont 05401, and Field Services Division Epidemiology Program Office and Center for Infectious Diseases, Centers for Disease Control, Atlanta, Georgia 30333

In May 1980, the infectious disease division of a regional medical center in Burlington (1980 population, 37,712; area, 41 km²), Vt., noticed that a greater than usual number of hospitalized patients had atypical pneumonia. No similar outbreaks of pneumonia had been identified in any of the 15 other hospitals located throughout the state, despite active surveillance for cases of atypical pneumonia.

† Present address: Agency of Human Services, Waterbury, VT 05676.

Case findings at the medical center consisted of reviewing the 24-h hospital admission list to identify persons with a diagnosis of pneumonia, reviewing X-ray reports to identify patients with new pulmonary infiltrates, identifying patients who had sputum specimens with polymorphonuclear leukocytes but no bacterial organisms, and reviewing the hospital pharmacy log to identify all patients receiving erythromycin.

All persons with illness onset between 1 May and 5 June 1980 or between 1 July and 15 August 1980 were considered to have possible outbreak-associated cases. All cases were laboratory confirmed by one of the following diagnostic methods: (i) a positive culture for *Legionella pneumophila*, (ii) a ≥fourfold serum antibody rise to 1:128 or greater (6), or (iii) a positive direct fluorescent-antibody test on pleural fluid or lung tissue (1). All persons with suspected *L. pneumophila* were tested for serogroups 1 through 4.

A matched-pair case-control study was done to examine personal and environmental risk factors for patients in each of the outbreaks. The case-control study included all cases. Controls were matched for sex and age (±5 years) and randomly selected from persons who were admitted to the hospital on the same day, but persons admitted to the pediatric or obstetric wards were excluded.

All study participants were queried about their activities inside and outside the hospital for each of the 14 days before onset of symptoms in the persons with Legionnaires disease (LD). Specific questions about exposures to the medical center, respiratory therapy, X rays, surgery, physical therapy, and showers were asked.

The hospital and the area around the hospital were surveyed for possible source of *L. pneumophila*. Water samples were collected from several cooling towers in Burlington, the Burlington municipal water supply, and multiple sites in the hospital. At the hospital, environmental samples were collected from shower heads, sink faucets, room air conditioners, respiratory therapy and dialysis equipment, air compressors, and hot-water storage tanks. All environmental samples were screened for *L. pneumophila* serogroups 1 through 4 by guinea pig intraperitoneal inoculation (5, 7), which was the method of choice at the time of the outbreak.

Records from a nearby National Weather Service station were reviewed for April through September 1980 to identify conditions that might favor a particular source or sources.

Using 1980 census data (8), attack rates were calculated for different neighborhoods in Burlington and Chittenden County served by different water systems. The number of full-time employees at the University of Vermont, the medical center, the city of Burlington, and the immediate county (Chittenden County) was determined. LD attack rates were calculated for employees in these different areas.

A serological survey of asymptomatic cooling-tower maintenance workers was done to assess the prevalence of *Legionella* antibodies in workers exposed to a culture-positive tower versus those exposed to a culture-negative tower. Two cooling-tower maintenance workers who had legionellosis were not included in this study (4). The two groups of workers had similar age

distributions, but the workers exposed to the culture-positive tower reported more exposure to the hospital than workers exposed to the culture-negative tower. The blood samples were tested simultaneously for antibodies to *L. pneumophila* serogroups 1 through 4.

A total of 85 confirmed cases of *L. pneumophila* serogroup 1 legionellosis were identified in two outbreaks, 24 in the first outbreak and 61 in the second. Both outbreaks occurred in Burlington, and persons with LD had illness onset between 1 May and 15 August 1980 (Fig. 1). Eighty-three patients (98%) had characteristic LD pneumonia, and two had a nonpneumonic, milder illness. Sixteen (19%) of the persons considered to have cases died. All persons with cases had spent time in Burlington during their incubation period.

Patients with inpatient-acquired LD had no different inpatient exposures as compared with the control group. Six persons with other than inpatient-acquired LD had only a single day of exposure to the hospital before the onset of their illness. Based on this exposure, their incubation period ranged from 4 to 8 days, with a mean of 6 days.

The hospital had no cooling tower, but there were three cooling towers within 300 m of the hospital. *L. pneumophila* serogroup 1 was isolated from five of the environmental samples. Three of the positive samples were from one medical school cooling tower (tower A), the fourth was from another cooling tower (tower B) on the university campus, and the fifth was from one of the hot-water heater tanks in the hospital. LD attack rates were lower on wards served by the hot-water tank that was positive for *L. pneumophila* than on wards that were served by hot-water tanks found negative for legionellae.

Positive environmental samples of tower A coincided with the occurrence of cases of LD in both outbreaks (Fig. 1). Cases of legionellosis occurred after cessation of chlorination of tower A water.

In May and July 1980, the prevailing winds for the Burlington area were from the south. Therefore, the hospital was upwind of tower A but not of tower B. Cooling tower A was the only tower with an unobstructed path to the hospital; the other towers have buildings or trees in their path.

Of the 42 persons with other than inpatient-acquired cases, 27 had contact with an area within 200 m of tower A which also included the medical center hospital. Persons with other than inpatient-acquired cases were much more likely to have been inside the hospital during their incubation period than the study controls (chi-square = 14.04 by matched-pair analysis; $P < 0.0001$).

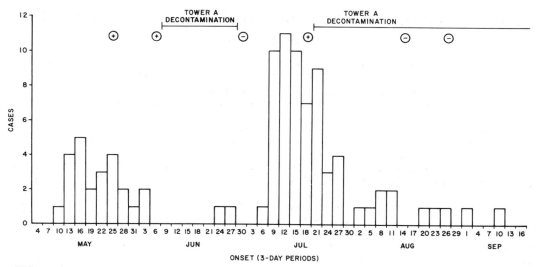

FIG. 1. Cases of LD associated with a cooling tower, by date of onset, Vermont, May to September 1980. ○, Culture.

There were 15 of 42 persons with other than inpatient-acquired cases who did not have contact with an area within 200 m of tower A (which included the medical center and the medical center hospital) during the 14 days before their illness. Thirteen lived in Burlington and two lived in the surrounding county. Persons living in surrounding neighborhoods to the north and northwest (downwind) of the medical center had higher estimated attack rates than those living in other areas of the city served by the same water supply. The relative risk for persons working <200 m from the outbreak-associated cooling tower as compared with other persons working full-time in Burlington is estimated to be 22.10 (P < 0.0001 by binomial distribution, two tailed).

The serological testing of the asymptomatic cooling-tower maintenance workers revealed that those who worked with tower A had significantly higher *Legionella* titers than those who worked with a tower outside the area. The geometric means of the antibody titers for the tower A workers and the downtown workers were 1:89 and 1:25, respectively; the Student *t*-test value on log-transformed data for this difference is 2.438 (P = 0.029, two tailed).

Through a variety of environmental and epidemiological analyses, tower A was linked with the development of legionellosis. This is the first report of the development of cases of legionellosis after cessation of chemical (chlorine) treatment of a cooling tower. The distance (at least 200 m) of airborne transmission of *L. pneumo-*

phila in this outbreak was much greater than previously reported (2, 3).

LITERATURE CITED

1. **Cherry, W. B., B. Pittman, P. P. Harris, G. A. Hébert, B. M. Thomason, L. Thacker, and R. E. Weaver.** 1978. Detection of Legionnaires disease bacteria by direct immunofluorescent staining. J. Clin. Microbiol. **8:**329–338.
2. **Cordes, L. G., D. W. Fraser, P. Skaliy, C. A. Perlino, W. R. Elsea, G. F. Mallison, and P. S. Hayes.** 1980. Legionnaires' disease outbreak at an Atlanta, Georgia, country club: evidence for spread from an evaporative condenser. Am. J. Epidemiol. **111:**425–431.
3. **Dondero, T. J., R. C. Rendtorff, G. F. Mallison, R. M. Weeks, J. S. Levy, E. W. Wong, and W. Schaffner.** 1980. An outbreak of Legionnaires' disease associated with contaminated air-conditioning cooling tower. N. Engl. J. Med. **302:**365–370.
4. **Girod, J. C., R. C. Reichman, W. C. Winn, D. N. Klaucke, R. L. Vogt, and R. Dolin.** 1982. Pneumonic and nonpneumonic forms of legionellosis as a result of a common-source exposure to *Legionella pneumophila*. Arch. Intern. Med. **142:**545–547.
5. **Jones, G. L., and G. A. Hébert (ed.).** 1979. "Legionnaires' '': the disease, the bacterium, and the methodology. Centers for Disease Control, Atlanta, Ga.
6. **McDade, J. E., C. C. Shepard, D. W. Fraser, T. F. Tsai, M. A. Redus, and W. R. Dowdle.** 1977. Legionnaires' disease: isolation of a bacterium and demonstration of its role in other respiratory disease. N. Engl. J. Med. **297:**1197–1203.
7. **Morris, G. K., C. M. Patton, J. C. Feeley, S. E. Johnson, G. Gorman, W. T. Martin, P. Skaliy, G. F. Mallison, B. D. Politi, and D. C. Mackel.** 1979. Isolation of Legionnaires' disease bacterium from environmental samples. Ann. Intern. Med. **90:**664–666.
8. **U.S. Bureau of the Census.** 1980. Census of population and housing block statistics, Burlington, Vt. Bureau of the Census, U.S. Department of Commerce, Washington, D.C.

Community Hospital Legionellosis Outbreak Linked to Hot-Water Showers

JOHN P. HANRAHAN, DALE L. MORSE, VIRGINIA B. SCHARF, JACK G. DEBBIE, GEORGE P. SCHMID, ROGER M. McKINNEY, AND MEHDI SHAYEGANI

New York State Health Department, Albany, New York 12237; Centers for Disease Control, Atlanta, Georgia 30333; and Alice Hyde Memorial Hospital, Malone, New York 12953

In September 1982 a cluster of legionellosis cases was noted among patients at an 89-bed community hospital in Malone, N.Y. The ensuing epidemiological investigation focused on identification, confirmation, and classification of cases; ascertaining and testing potential environmental sources of exposure; and initiating control measures. Additional cases were sought by reviewing records of previous pneumonia cases in patients, employees, and nursing home residents and by collecting acute and convalescent sera for *Legionella* antibody testing on all persons with recent pneumonias.

The investigation identified seven serologically diagnosed ($\geq$fourfold indirect fluorescent-antibody serological titer rises to *Legionella pneumophila* serogroup 1) cases of legionellosis that occurred between February and September 1982. Four other persons with nonbacterial pneumonias during the same period were classified as possible cases on the basis of single convalescent titers of 256. Four of the seven serologically diagnosed cases were diagnosed within a 6-week period in which there were 325 medical and surgical discharges, giving a rate of 1.2 cases per 100 discharges.

The seven persons with legionellosis ranged in age from 50 to 66 years, with a mean of 59 years. Five were male, two were female. Six were patients, one was an employee. The patient cases all appeared to have been nosocomially acquired, as respiratory symptoms began 8 to 15 days after admission, with a mean of 10.5 days. All of the patients were ambulatory and had few health problems before admission; after admission all developed pneumonia, but none died of legionellosis.

A comparison of the 6 hospital cases with 21 controls matched for sex, age within 5 years, and hospital admission within 3 days showed no statistical differences in reason for admission, ambulatory status, smoking, steroid use, respiratory therapy, surgery, or shower use. However, statistical differences were found between cases and controls for length of hospital stay (mean of 10.5 days for cases versus 4.6 days for controls, $P < 0.01$ by Mann-Whitney U test) and for distance of hospital room from the shower (mean of 1.5 room lengths for cases versus 4.1 lengths for controls, $P < 0.05$ by Mann-Whitney U test).

Potential environmental sources of *L. pneumophila* were considered. There was no nearby excavation or water-cooled cooling tower, nor was respiratory equipment used in the management of any of the cases. The water supply of the hospital was surface water obtained from a spring-fed brook; it was treated with ammonia and chlorine to 1 ppm, stored in open reservoirs, and then distributed with a negligible chlorine residual. Twenty-eight water specimens were collected from various hot- and cold-water taps at different hospital sites over several weeks. Seven of the specimens from five sites were positive for *L. pneumophila* serogroup 1; these included specimens from four showers and one hot water heater. Showers on the two patient floors were positive on multiple occasions. Multiple cold-water specimens were negative.

To test the hypothesis that aerosolized hot water was the source of the outbreak, the one human, seven environmental hot-water, and four control *L. pneumophila* serogroup 1 isolates were sent under double-blind conditions to the Centers for Disease Control for monoclonal antibody subtyping (1). All of the outbreak-related isolates, including the patient isolate, were of the same monoclonal antibody pattern and differed from the control specimens. Attempts to use air-monitoring equipment to isolate from aerosols at various distances from the showers were unsuccessful.

Control measures were initiated early and included temporarily discontinuing use of the showers, followed by drainage and flushing of the hot water tank, shock superchlorination to 10 ppm, hyperchlorination to 3 ppm for 10 days, and then maintenance at 2 ppm. With the exception of one occasion when the chlorine residual dropped below 1 ppm, subsequent cultures have been negative. No additional cases have occurred.

This outbreak further implicates hot water containing legionellae as a source of nosocomially acquired legionellosis. This conclusion is supported by the epidemiological investigations (proximity of cases to showers), the monoclonal antibody testing linking human to environmental isolates, and the lack of cases after control measures. The high rate of illness demonstrates the extent of nosocomial illness that can be caused by this organism.

LITERATURE CITED

1. McKinney, R. M., L. Thacker, D. E. Wells, M. C. Wang, W. J. Jones, and W. F. Bibb. 1983. Monoclonal antibodies to *Legionella pneumophila* serogroup 1: possible applica- tions in diagnostic tests and epidemiologic studies, p. 90–95. *In* F. J. Fehrenbach (ed.), Proceedings of Workshop Conference on Legionnaires' Disease. West German Public Health Reports, West Berlin, Germany.

Nosocomial Legionnaires Disease Associated with Exposure to Respiratory Therapy Equipment, Connecticut

ELLEN JONES, PATRICIA CHECKO, ANECIA DALTON, JOHN COPE, JAMES BARBAREE, GEORGE KLEIN, WILLIAM MARTIN, AND CLAIRE BROOME

Centers for Disease Control, Atlanta, Georgia 30333, and State Department of Health Services, Hartford, Connecticut 06010

Numerous investigators have suggested a role for potable water in outbreaks of nosocomial Legionnaires disease (LD) (1, 2, 3). An investigation of an outbreak of nosocomial LD has allowed us to extend previous observations concerning the role of contaminated potable water among immunosuppressed patients receiving respiratory therapy.

Between December 1980 and December 1982, 36 cases of nosocomial LD occurred in a 400-bed Connecticut hospital. However, only 18 of these patients had been hospitalized continuously during the 10 days before onset of illness, so only these were included in further analysis of the causes of this nosocomial outbreak. Of these 18 cases, 17 met standard criteria for the laboratory confirmation of LD; one was a presumptive case, with a standing titer of 1,024. All cases occurred over a 2-year period. Five of these cases resulted in death, giving a fatality rate of 26%. Hospital charts on each patient and three controls (matched for age, sex, and date of admission) were reviewed.

Univariate analysis of host factors and in-hospital exposures revealed that steroid therapy (odds ratio = 8.9, $P = 0.007$), antacids (odds ratio = 4.9, $P = 0.010$), nebulizer (odds ratio = 11.1, $P = 0.028$) and intermittent positive-pressure breathing (IPPB) therapy (odds ratio = 6.6, $P = 0.025$), and use of a room humidifier (odds ratio, 6.6, $P = 0.025$) were associated with illness. Patients with LD also had significantly more underlying diseases, such as adult-onset diabetes, cancer, renal failure, collagen vascular disease, and chronic obstructive pulmonary disease, as indicated by a score for underlying disease. Cases did not differ from controls in ward of hospitalization, exposure to a particular heating, ventilation, and air-conditioning unit, smoking history, or use of other respiratory devices such as aerosol masks, respirators, or nasal cannulae.

Although steroid therapy, antacid use, and exposure to nebulizer were significant risk factors on univariate analysis, they were highly correlated and it was difficult to evaluate them as independent variables. Further analysis of the relationship between steroids and antacids, nebulizers and steroids, and antacids and nebulizers also was limited by the small number of cases. However, examining the data in an unmatched fashion was useful in examining trends among variables.

Among those receiving steroids, antacid use increased the risk of illness (88 versus 17%) (Table 1). However, among those not being treated with steroids, antacid therapy did not appear to have an effect (20 versus 17%).

The role of antacids is unclear. Antacid use was a significant variable even after controlling for the index of underlying disease or steroid use. It is interesting to hypothesize that antacids may have played a role in facilitating infection with *Legionella pneumophila* and that the interaction of antacids and steroids may have been the important variable, rather than just steroids as is commonly believed. One mechanism of infection may be ingestion of the organism, followed by aspiration into the lungs and subsequent pneumonia. A neutralized gastric pH may facilitate this process.

Steroid and antacid therapy as well as the presence of underlying diseases represent the significant host susceptibility factors in this outbreak; possible sources of exposure to *L. pneumophila* identified in the univariate model included nebulizers and room humidifiers. Again, examining the data in an unmatched fashion by a two-by-two table in an attempt to sort out the interactions between steroids and nebulizer use

TABLE 1. Stratification of antacid use by steroid therapy

Antacid use	No. receiving steroids (% receiving antacids)		No. not receiving steroids (% receiving antacids)	
	Cases	Controls	Cases	Controls
+	7 (88)	1 (17)	2 (20)	8 (17)
−	1	5	8	40

TABLE 2. Odds ratio univariate and multivariate analysis

Analysis	Odds ratio	Confidence interval		Log likelihood
		Lower limit	Upper limit	
Univariate				
Score	1.370	1.084	1.731	−20.51
Nebulizer	11.074	1.298	94.490	−21.23
Steroids	8.916	1.833	43.381	−20.23
Antacids	4.881	1.449	16.440	−21.38
Steroids × antacids	21.000	2.585	170.628	−17.976
Nebulizer × antacids	—	4.633	—	−16.636
Steroids × nebulizer	13.14	1.513	114.114	−21.006
Nebulizer × antacids × antacids	—	3.656	—	−18.02
Multivariate				
Nebulizer × steroids × antacids	—	—	—	—
Score	1.322	1.030	1.698	−15.18[a]
Nebulizer × antacids	—	—	—	−14.263
Score	1.300	1.009	1.675	—
Steroids × antacids	13.653	1.614	115.466	−16.022
Score	1.272	0.988	1.639	—

[a] Test for improvement, multivariate versus univariate analysis: $\chi^2 = 5.674$, $P = 0.02$.

showed that patients with LD, whether or not they were receiving steroids, were twice as likely to have used a nebulizer as were controls (62 versus 33% and 20 versus 10%, respectively). If we look at the interaction between antacids and nebulizers, it becomes evident that among those treated with antacids, a much larger percentage of cases were exposed to nebulizers than were controls. Among those not treated with antacids, exposure to a nebulizer had very little effect. A multivariate analysis (Table 2) includes all these interactions. This model shows a significantly better fit than all the other multivariate models examined.

The next consideration was how the nebulizer or room humidifier might have become contaminated with L. pneumophila. The answer was most obvious for the room humidifiers, which are routinely filled with potable water. Multiple samples obtained from the potable water system, including tap water in patient rooms and showers, and from the hot water heating system grew L. pneumophila serogroups 1, 4, and 5. No samples obtained from the cooling tower adjacent to the hospital, the oxygen or condensed-air lines, or the heating, ventilation, and air-conditioning units yielded L. pneumophila.

The way in which nebulizers became contaminated was less obvious. However, it was noteworthy early in the investigation that although nebulizers, IPPB equipment, and room humidifiers could be implicated, aerosol masks, nasal cannulae, and mechanical ventilators could not. Furthermore, the aerosol masks, nasal cannulae, and mechanical ventilators all included a humidification system, while the implicated respiratory therapy equipment did not. The answer seemed to be that the reservoirs used with aerosol masks, nasal cannulae, and mechanical ventilators were filled with sterile rather than potable water. Although they do not have humidification reservoirs, nebulizers and IPPB equipment are rinsed between treatments with potable water.

An alternative hypothesis relates to the compressed-air system. Nebulizers and IPPB can be operated with compressed air, whereas nasal cannulae, aerosol masks, and mechanical ventilators uniformly utilize oxygen. We do know that the compressed-air lines became contaminated in March 1982, when a pump diaphragm broke, allowing potable water into the lines. However, this event followed the onset of illness in 16 of the 18 cases.

In conclusion, our results concur with previous studies which have demonstrated the importance of underlying disease in the acquisition of LD. Our results also raise the possibility that the role of steroid therapy in promoting LD may be correlated with concurrent antacid therapy, and they confirm the suggestion that aerosolization of potable water by respiratory therapy equipment is associated with illness.

LITERATURE CITED

1. **Arnow, P. M., T. Chou, D. Weil, E. N. Shapiro, and C. Kretzschmar.** 1982. Nosocomial Legionnaires' disease caused by aerosolized tap water from respiratory devices. J. Infect. Dis. **146:**460–467.
2. **Fisher-Hoch, S. P., C. L. R. Bartlett, J. O'H. Tobin, M. B.**

Gilbert, A. M. Nelson, J. E. Pritchard, M. G. Smith, R. A. Swann, J. M. Talbot, and J. A. Thomas. 1981. Investigation and control of an outbreak of Legionnaires' disease in a district general hospital. Lancet i:932–936.

3. Haley, C. E., M. L. Cohen, J. Halter, and R. D. Meyer. 1979. Nosocomial Legionnaires' disease: a continuing common-source epidemic at Wadsworth Medical Center, Ann. Intern. Med. 90:583–586.

An Ongoing Outbreak of *Legionella micdadei*

JENNIFER E. RUDIN, EDWARD J. WING, AND ROBERT B. YEE

Montefiore Hospital and University of Pittsburgh School of Medicine and Graduate School of Public Health, Pittsburgh, Pennsylvania 15213

Legionella micdadei is a cause of severe pneumonia in immunosuppressed patients and has a high mortality in patients with nosocomially acquired disease. We have seen 26 cases of hospital-acquired *L. micdadei* infection over the past 4 years (1a, 2; J. E. Rudin, T. Evans, and E. J. Wing, Am. J. Med., in press). By contrast, infection with *Legionella pneumophila* has been noted in only two patients during this same period. Since 1981, environmental surveillance of the water system and respiratory assist equipment has been conducted in an effort to define a possible point source for these infections.

A selective medium of buffered charcoal-yeast extract agar with glycine, vancomycin, and polymyxin B added was used to culture the *Legionella* species (1). All cultures were incubated at 37°C and examined daily for 7 days.

Legionella species were identified by colony morphology and by direct immunofluorescence with reagents provided by the Centers for Disease Control. These reagents included fluorescein-conjugated antisera against *L. pneumophila* serogroups 1 to 6 and against *L. micdadei*.

The diagnosis of *L. micdadei* infection was made in patients with pneumonia in whom *L. micdadei* was identified in respiratory secretions or lung tissue (2). In 22 of 26 cases, the organism was recovered by culture.

Each of the 26 patients who developed *L. micdadei* pneumonia had been hospitalized during the 2 weeks before acquisition of the disease. A total of 25 of 26 patients had immunosuppressive underlying diseases, and 22 patients were being treated with corticosteroids. Clinically, each patient had a disease characterized by high

TABLE 1. Selected environmental cultures before and after heating and flushing

Location	Site[a]	No. of *L. pneumophila* colonies on the following date:			
		7/81	9/22/81	9/25/81[b]	10/27/81
Room 1	Faucet spout	ND[c]	>1,000	0	0
	CW valve seat/stem	ND	0	0	0
	HW valve seat/stem	ND	300	0	0
	Shower head	ND	0	0	0
	Shower pipe	ND	3	0	142
Room 2	Faucet spout	ND	300	0	0
	CW valve seat/stem	ND	0	0	0
	HW valve seat/stem	ND	200	0	0
	Shower head	ND	0	0	184
	Shower pipe	ND	0	0	225
Room 3	Faucet spout	ND	200	0	465
	CW valve seat/stem	ND	0	0	0
	HW valve seat/stem	ND	23	0	0
	Shower head	ND	4	1	0
	Shower pipe	ND	72	0	0
HW tank[d]	Left	1,110	ND	50	6
	Right	390	ND	20	60

[a] CW, Cold water; HW, hot water.
[b] Date on which heating and flushing was done.
[c] ND, Not done.
[d] Temperature, 110°F.

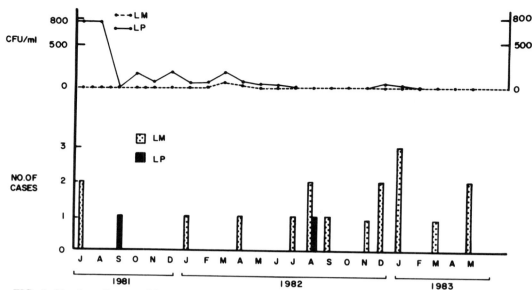

FIG. 1. Number of cases of *Legionella* pneumonia from 1981 to 1983, with corresponding concentrations of *L. micdadei* (LM) and *L. pneumophila* (LP) in the hot-water tanks.

fever, pleuritic chest pain, and pleural-based alveolar infiltrates. Although erythromycin therapy was effective in the majority of patients, the mortality rate was 30%. During this same time, only two cases of *L. pneumophila* infection were noted.

Epidemiological studies were undertaken in an attempt to locate a source for the infection. Ongoing evaluation has revealed no association of our cases with location in the hospital, contact with physicians, or prior use of respiratory equipment.

An investigation of the plumbing system and its distribution was begun in the summer of 1981. The potable water system consists of two water mains fed by a central Pittsburgh reservoir. Each main serves one of the two towers of the hospital where all inpatient rooms are located. Seven hot-water tanks are served by these mains.

Environmental culturing was initially undertaken in June 1981 and revealed relatively high numbers of *Legionella* species in the hot-water tanks and outlet valves and particularly in the sediment in the bottom of the tanks. The initial counts of legionellae as measured in various parts of the water distribution system are shown in Table 1. The vast majority of the isolates were *L. pneumophila*. *L. micdadei*, although ultimately isolated from the plumbing fixtures, was present in much lower concentrations.

In September 1981 the temperature of the hot-water tanks was raised from 110 to 130°F (ca. 43 to 54°C), and the entire system was flushed for approximately 30 min. The colony counts were markedly reduced by this maneuver (Table 1).

Thereafter the tanks were kept at 130°F, and the sediment was drained weekly. In addition, the entire system was flushed bimonthly. Since February 1983, the flushing has been carried out at 150°F.

Figure 1 shows the number of cases of *L. micdadei* pneumonia in our hospital and the average concentration of legionellae in hot-water tanks during the time of our environmental surveillance. There was no good correlation between the average number of organisms and the occurrence of infection. In fact, one may have expected larger numbers of cases of *L. pneumophila* infection. Because of these data, possible sources of water contamination other than the plumbing system were surveyed in 1983. Whirlpools, faucets, air conditioning vents, and ice machines in selected patient care areas all remained negative for legionellae.

Our experience is unusual in that cases of *L. micdadei* infection have increased during a time when the concentration of legionellae in the potable water has decreased. In addition, we have observed far greater numbers of cases of *L. micdadei* infection than of *L. pneumophila* infection, despite the presence of greater numbers of *L. pneumophila* in the potable water supply. In conclusion, our findings suggest that nosocomial infection with *L. micdadei* may not parallel water contamination and that further investigation is needed to determine the reservoir and mode of transmission of this organism.

In summary, there has been an ongoing outbreak of *L. micdadei* in our hospital since 1980, while infection with *L. pneumophila* has been minimal. Environmental surveillance of plumbing equipment initially showed significant concentrations of *L. pneumophila*. These concentrations have been reduced as a result of heating with sequential flushing. The presence of *L. micdadei* in our water has been intermittent and in very low numbers. The other possible factors in the transmission of *L. micdadei*, such as respiratory equipment, location in the hospital, physician contact, stagnant water systems, or air conditioning and heating vents, do not appear to be related to our cases.

LITERATURE CITED

1. **Wadowsky, R. M., R. B. Yee, L. Mezmar, E. J. Wing, and J. N. Dowling.** 1982. Hot water systems as sources of *Legionella pneumophila* in hospital and nonhospital plumbing fixtures. Appl. Environ. Microbiol. 43:1104–1110.
1a. **Wing, E. J., F. J. Schaefer, and A. W. Pasculle.** 1981. Successful treatment of *Legionella micdadei* (Pittsburgh Pneumonia Agent) pneumonia with erythromycin. Am. J. Med. 71:836–840.
2. **Wing, E. J., F. J. Schafer, and A. W. Pasculle.** 1982. The use of tracheal and pulmonary aspiration to diagnose *Legionella micdadei* pneumonia with erythromycin. Chest 82:705–707.

Seroepidemiological Study of *Legionella* Infections in Renal Transplant Recipients

JOHN N. DOWLING, A. WILLIAM PASCULLE, FRANK N. FROLA, M. KATHRYN ZAPHYR AND ROBERT B. YEE

Departments of Medicine and Pathology, School of Medicine, and Department of Microbiology, Graduate School of Public Health, University of Pittsburgh, Pittsburgh, Pennsylvania 15216

In the original description of disease caused by *Legionella micdadei*, we reported on seven cases of pneumonia seen in Pittsburgh, Pa. (3). Five of the patients described in that report were renal allograft recipients who acquired their disease while in Presbyterian-University Hospital. This outbreak of nosocomial cases occurred from October 1977 to November 1978 and then apparently ended. To determine whether further, unrecognized cases of *L. micdadei* or other *Legionella* pneumonia occurred subsequently, serum specimens were systematically collected from renal transplant recipients at Presbyterian-University Hospital from October 1978 through January 1981. During the 28-month study a total of 103 patients underwent renal transplantation. We attempted to obtain a serum sample from each transplant recipient before transplantation and approximately once each month thereafter, either for a 6-month period or until the allograft failed and immunosuppressive therapy was discontinued. No sera were obtained from two recipients who underwent early, acute rejections and lost their grafts within a few days of transplantation. Thus, at least one serum sample was available from 101 of the 103 transplant recipients. A single serum was also obtained from 67 kidney donors. Since 23 of these supplied one kidney to each of two recipients, the 67 represented the renal donors to 90 of the 101 recipients. These banked sera were examined for antibody to *Legionella pneumophila* serogroups 1 to 6 and *L. micdadei* by indirect immunofluorescence testing, and the results were correlated with the clinical courses of the recipients.

The first serum sample obtained from 22 (21.8%) of the 101 transplant recipients con-

tained antibody to one of the seven antigens tested at a titer of ≥1:64. Sixteen recipients had an elevated pretransplant titer to *L. pneumophila* serogroup 6; two, to *L. micdadei*; two, to *L. pneumophila* serogroup 5; one, to *L. pneumophila* serogroup 1; and one, to *L. pneumophila* serogroup 3. When the clinical records of these 22 recipients were reviewed, few if any of the elevated pretransplant titers could be accounted for by episodes of pneumonia. Only 5 of the 22 had a history of pneumonia, at any time in the past, that possibly could have been legionellosis. None had pneumonia in the year preceding transplantation or during the immediate post-transplant period. To assess whether the prevalence of elevated titers among recipients before transplantation was increased, kidney donors were utilized as a comparison group. Kidney donors have some contact with hospitals and might be more similar to recipients in that respect than the normal population. Of the 67 donors, 9 (13.4%) had a titer of at least 1:64 against one of the antigens, a prevalence no different from that of recipients. The distribution of antibody in the pretransplant sera of the recipients was also similar to that of donor sera. Six donors had a titer against *L. pneumophila* serogroup 6, and three had a titer against *L. micdadei*.

Serum specimens were obtained for 1 month or less after transplantation from 12 of 101 recipients. In 10 instances further sera were not obtained because the recipient underwent early allograft rejection; in two cases the recipient was lost to follow-up. Considering those recipients who were followed for more than 1 month after transplantation, 21 of 89 (23.6%) underwent ser-

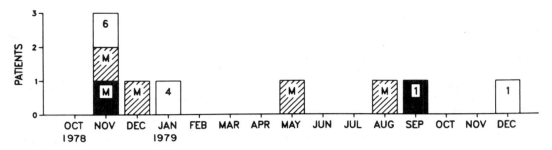

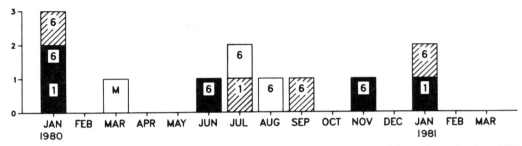

FIG. 1. Frequency of *Legionella* seroconversions among renal transplant recipients from October 1978 through January 1981, by month. Each bar represents a recipient who underwent seroconversion either related to a proven *Legionella* pneumonia (■), associated with a febrile pneumonia (▧), or in the absence of pneumonia (☐). The letter or numeral within each bar indicates the antigen to which seroconversion occurred, as follows: M, *L. micdadei*; 1, 4, or 6, *L. pneumophila* serogroup 1, 4, or 6, respectively.

oconversion, defined as a greater-than-fourfold increase in titer to at least 1:128, to one of the legionellae. When the clinical courses of the recipients who underwent seroconversion were reviewed, they could be classified into one of three groups: (i) proven *Legionella* pneumonia, (ii) seroconversion associated with febrile pneumonitis, and (iii) asymptomatic seroconversion. Of the 21 seroconverters, 7 were diagnosed as having *Legionella* pneumonia at the time of the seroconversion. In six instances the diagnosis was established by the isolation of legionellae from a lung biopsy specimen, and in one instance the diagnosis was by demonstration of the organism in the lung at autopsy by direct fluorescent-antibody testing. Three of the proven cases were caused by *L. pneumophila* serogroup 1, three were due to *L. pneumophila* serogroup 6, and one was due to *L. micdadei*. Six of these seven proven cases were nosocomial; that is, the patient had been in the hospital for at least 5 days before the onset of pneumonia.

Of the remaining 14 seroconverters, 8 had fever and pneumonitis which at the time were ascribed to other causes, but which in retrospect were due to *Legionella*. Four recipients with febrile pneumonia seroconverted to *L. micdadei*, three seroconverted to *L. pneumophila* serogroup 6, and one seroconverted to *L. pneu-*

mophila serogroup 1. All instances of febrile pneumonitis in association with seroconversion were nosocomial.

The final six patients who underwent seroconversion after transplantation had no contemporaneous clinical evidence of an infectious pneumonia. Three of the seroconversions were to *L. pneumophila* serogroup 6, and one each was to *L. micdadei*, *L. pneumophila* serogroup 1, and *L. pneumophila* serogroup 4. Since the timing of the event which resulted in seroconversion for these asymptomatic patients was unknown, it was not strictly possible to determine whether the event was nosocomial. However, in all but a single instance it appeared that the seroconversion was related to a period of hospitalization.

The temporal occurrence of seroconversions among recipients over the 28-month study period is shown in Fig. 1. The patient included in this study who had proven *L. micdadei* pneumonia in November 1978 was the last recognized case among renal transplant recipients in the original series (3). Four additional renal transplant recipients who developed *L. micdadei* pneumonia with onset of disease between November 1978 and August 1979 were discovered in the present study. An additional patient underwent seroconversion to *L. micdadei* in March 1980. After that time, all seroconversions, whether associated with pneumonia or not, were due to *L. pneumophila* serogroup 6 or serogroup 1.

This seroepidemiological investigation revealed that 24% of 89 renal transplant recipients seroconverted to *L. micdadei* or *L. pneumophila* during the 6-month period after transplantation and the initiation of immunosuppressive therapy. In 71% of these instances, the seroconversion occurred in relation to an episode of clinical pneumonia, and virtually all of these were nosocomial. Approximately one-half of these pneumonias were caused by legionellae. That the etiology of the remainder was not identified at the time was due largely to the relatively primitive techniques available for the diagnosis of *Legionella* pneumonia during much of the period studied. The overall incidence of *Legionella* pneumonia after renal transplantation in the present study (15 of 89, 17%) is similar to the 14% rate found at the University of Vermont (2). However, at our hospital the majority of cases were due to *L. micdadei* or *L. pneumophila* serogroup 6. Of great interest are the six in-stances in which recipients seroconverted in the absence of any pneumonia. Seroconversion to *L. pneumophila* among hospitalized patients who did not develop clinical pneumonia has been noted previously (1). Presumably, such seroconversions represent *Legionella* infection without disease, although other explanations, including serological cross-reactions to other organisms, cannot be excluded. Regardless of the interpretation of these asymptomatic sero-conversions, their occurrence engenders a sense of caution in utilizing seroconversion alone to establish a diagnosis of legionellosis.

LITERATURE CITED

1. **Haley, C. E., M. L. Cohen, J. Halter, and R. D. Meyer.** 1979. Nosocomial Legionnaires' disease: a continuing common-source epidemic at Wadsworth Medical Center. Ann. Intern. Med. **90**:583–586.
2. **Marshall, W., R. S. Foster, Jr., and W. Winn.** 1981. Legionnaires' disease in renal transplant patients. Am. J. Surg. **141**:423–429.
3. **Myerowitz, R. L., A. W. Pasculle, J. N. Dowling, G. J. Pazin, M. Puerzer, R. B. Yee, C. R. Rinaldo, Jr., and T. R. Hakala.** 1979. Opportunistic lung infection due to "Pittsburgh pneumonia agent." N. Engl. J. Med. **301**:953–958.

Legionella Infections in Renal Transplant Recipients

LOUIS D. SARAVOLATZ, DONALD POHLOD, KAY HELZER, BERTTINA WENTWORTH, AND NATHAN LEVIN

Henry Ford Hospital, Detroit, Michigan 48202, and Michigan Department of Public Health, Lansing, Michigan 48909

Legionella infections have been recognized as a cause of pneumonia among immunosuppressed patients and especially among renal transplant recipients; however, the exact role of *Legionella* species in infection of renal transplant recipients has been studied in only a limited way. In this study, we examined the epidemiology of *Legionella* infections over a 6-year period (1977 to 1982) by using a case-control study to determine patient risk factors and a prospective cohort study to evaluate colonization. Also, we performed environmental culturing to determine the effect of environmental contamination on the cases.

All renal transplant recipients from 1977 to 1982 were reviewed for pneumonia. Evaluation included the patient's location when the pneumonia was acquired, etiology, demographic features, clinical presentation, associated infection, renal graft function, and clinical outcome. A case-control study was performed between 16 patients with *Legionella* pneumonia and 16 patients with non-*Legionella* pneumonia. *Legionella* infection was diagnosed when the patient demonstrated at least a fourfold increase in *Legionella* antibody titer by the indirect fluorescent-antibody technique or when legionellae were isolated from lung tissue or lower respiratory secretions and there was a concurrent diagnosis of pneumonia (2–4). Serology tests included antibody directed against Formalin-prepared antigens of *L. pneumophila* serogroups 1 to 6, *L. micdadei*, *L. bozemanii*, *L. gormanii*, *L. longbeachae* serogroups 1 and 2, and *L. dumoffii*.

During the 6-year period, 213 patients received kidney transplants, and 40 (18.8%) subsequently developed pneumonia. A *Legionella* species was the etiological agent in 40% (16 of 40). There were 13 cases of *L. pneumophila* (serogroup 1, nine cases; serogroup 3, two cases; serogroup 5, two cases), two cases of *L. micdadei*, and one case of *L. bozemanii* infection. Over the 6-year period, the incidence of *Legionella* pneumonia in renal transplants declined from 9.68% in 1977 to 6.38% in 1982. This decline was not statistically significant ($P > 0.05$). In addition, all documented bacterial pneumonias in transplants declined from 25.81% in 1977 to 12.77% in 1982.

Seasonal variation in pneumonias among renal

transplant recipients was striking. *Legionella* infection never occurred in January to March during the 6 years, whereas non-*Legionella* pneumonias in renal transplant recipients occurred most frequently during these months (*P* = 0.002 by Fisher's exact test). *Legionella* and cytomegalovirus infections occurred more frequently during the first 2 months after transplantation, whereas fungal pathogens and *Pneumo-cystis carinii* occurred more often after the first 2 months.

The results of the case-control study are shown in Table 1. The risk factors associated with *Legionella* pneumonias were similar to those for non-*Legionella* pneumonias. Male cigarette smokers past the fourth decade of life were at greatest risk for developing pneumonia. Also, *Legionella* pneumonias occurred nosoco-

TABLE 1. Case-control study of *Legionella* and non-*Legionella* pneumonias

Parameter	No.[a] with:		P
	Legionella pneumonia (*n* = 16)	Non-*Legionella* pneumonia (*n* = 16)	
Age (yr)	40.44 ± 11.32[b]	38.94 ± 11.85[b]	NS[c]
Male/female ratio	10:6	9:7	NS
Cadaver kidney recipient	14	15	NS
Diabetes	3	1	NS
Smoker	10	7	NS
Previous pneumonia	0	3	NS
Previous transplants	1	2	NS
Nosocomial pneumonia	11	10	NS
Interval from last hospitalization:			
<12 days	2	2	NS
12 to 30 days	1	2	NS
30 to 180 days	2	1	NS
>1 year	0	1	NS
Symptoms			
Cough	9	15	NS
Nonproductive	7	5	NS
Productive	2	10	0.009
Gastrointestinal	9	8	NS
Diarrhea	1	2	NS
Other	8	6	NS
Neurological	3	4	NS
Laboratory parameters			
Hepatic dysfunction	5	4	
Hypophosphatemia (<2.5 meq/liter)	4	1	
Hyponatremia (<125 meq/liter)	0	1	NS
Leukocytosis (>10,000/mm³)	7	1	
Leukopenia (<4,000/mm³)	5	5	
Hematuria	4	4	
Culture results			
Multiple pathogens	12	8	NS
Bacteremia	3	3	NS
Graft response			
Early rejection	4	3	NS
Irreversible rejection	5	3	NS
Functioning graft	5	7	NS
Clinical response			
Days to respond to antibiotics	9.66 (1–33)[d]	4.55 (1–24)[d]	0.04
Mortality	6	5	NS

[a] Except as noted, results are expressed as the number of patients, of a total of 16 in each group.
[b] Expressed as mean ± standard deviation.
[c] NS, Not significant.
[d] Expressed as mean number of days, with range in parentheses.

mially as frequently as non-*Legionella* pneumonias. *Legionella* pneumonia patients had a productive cough much less frequently than non-*Legionella* pneumonia patients ($P = 0.009$ by Fisher's exact test) even though leukopenia occurred at the same frequency in both groups. Other symptoms, signs, and laboratory parameters were similar for *Legionella* and non-*Legionella* pneumonias. Mixed infections were common in both *Legionella* and non-*Legionella* pneumonias; cytomegalovirus and fungal agents were the most commonly associated pathogens. Although the graft response and mortality were similar, *Legionella* pneumonia patients took longer to respond to appropriate antimicrobial agents than non-*Legionella* patients ($P = 0.04$ by Student's t test).

A cohort of renal transplant recipients ($n = 70$) and hemodialysis patients ($n = 62$) followed prospectively for 2 years revealed colonization with a *Legionella*-like organism in 2.86% (2 of 70) of the renal transplant recipients and in 4.84% (3 of 62) of the hemodialysis patients. Respiratory secretions were evaluated by direct immunofluorescence. They were considered positive if more than one swab demonstrated ≥2 organisms which fluoresced with the conjugates of *L. pneumophila* serogroups 1 to 6, *L. micdadei, L. bozemanii, L. longbeachae* serogroups 1 and 2, or *L. dumoffii.* Colonization lasted 2 to 24 weeks and occurred only with *L. micdadei* and *L. dumoffii.* No patient developed pneumonia or an antibody rise concurrent with colonization of *Legionella*-like organisms. An attempt was made to culture legionellae from specimens positive by the direct fluorescent-antibody test, but this was unsuccessful. Although this raises the question of cross-reacting organisms colonizing patients, cross-reacting organisms rarely have been reported with the conjugates for legionellae (1).

Environmental water samples, including both potable and cooling-tower water, were obtained between 1979 and 1983, and culture techniques varied as new methods became available. The initial cooling-tower water was inoculated into guinea pigs and cultured on Mueller-Hinton agar supplemented with 2% IsoVitaleX and 1% hemoglobin and on buffered charcoal-yeast extract agar. Subsequent environmental cultures, including both potable and cooling-tower water samples, were cultured on buffered charcoal-yeast extract agar supplemented with antibiotics (cephalothin, 4 µg/ml; colistin, 16 µg/ml; vancomycin, 0.5 µg/ml; cycloheximide, 80 µg/ml, and α-ketoglutarate, 0.1%). Multiple water samples were obtained from sinks, showers, and bathtubs in all rooms where *Legionella* infection occurred, as well as in other rooms, selected at random, where *Legionella* infection did not occur.

Although *L. pneumophila* serogroup 5 was recovered in 1979 from cooling-tower water, only one of four *Legionella* cases that occurred during the time of environmental contamination was caused by the serogroup 5 pathogen. Cooling-tower contamination was controlled by chlorination. Potable water was subsequently identified as contaminated with *L. pneumophila* serogroup 3 and *L. dumoffii.* However, during the period of potable water contamination, no cases of *Legionella* infection caused by these pathogens occurred. Environmental water samples had <5 CFU/ml, except for one specimen which had *L. dumoffii* at 92 CFU/ml. No temporal association was found between environmental isolation and clinical cases of *Legionella* infection. Environmental contamination resolved spontaneously without treatment of potable water.

Legionella sp. was the most common etiological agent of pneumonias in renal transplant recipients in a 6-year study. The demographic features, laboratory parameters, and clinical presentation were similar except for the absence of productive cough in *Legionella* pneumonia. Colonization may occur in both renal transplant recipients and hemodialysis patients. *Legionella* infections represent a major pathogen for renal transplant recipients and should be included in the differential diagnosis and therapy of pneumonia in renal transplant recipients.

This study was supported by the Michigan Department of Public Health, Renal Disease Project.

LITERATURE CITED

1. **Cherry, W. B., and R. M. McKinney.** 1978. Detection in clinical specimens by direct immunofluorescence, p. 129–146. *In* G. L. Jones and G. A. Hébert (ed.), "Legionnaires'": the disease, the bacterium, and methodology. Centers for Disease Control, Atlanta, Ga.
2. **Edelstein, P. H.** 1983. Legionnaires' disease laboratory manual. National Technical Information Service, Springfield, Va.
3. **McDade, J. E., C. C. Shepard, D. W. Fraser, T. R. Tsai, M. A. Reders, and W. R. Dowdle.** 1977. Legionnaires' disease: isolation of a bacterium and demonstration of its role in respiratory disease. N. Engl. J. Med. 297:1197–1203.
4. **Saravolatz, L. D., G. Russell, and D. Cvitkovich.** 1981. Direct immunofluorescence in the diagnosis of Legionnaires' disease. Chest 79:566–570.

Legionnaires Disease in Renal Transplant Recipients

RODNEY ZEITLER, CHARLES HELMS, NANCY HALL, NANCY GOEKEN, GARY ANDERSON,
AND ROBERT CORRY

*Departments of Medicine, Pathology, and Surgery, University of Iowa College of Medicine, and University
Hygienic Laboratory, Iowa City, Iowa 52242*

Since the first description of Legionnaires disease (LD), several reports have appeared reporting the occurrence of *Legionella pneumophila* infection in renal transplant recipients (1, 5–7, 9, 10). We undertook a retrospective study to determine the impact of *L. pneumophila* pneumonia on a renal transplant program.

A renal transplantation service has been in operation at the University of Iowa since 1969. Between November 1969 and March 1981, 467 patients received renal allografts. A retrospective analysis of discharge diagnoses and clinical records revealed 119 episodes of pneumonia occurring in 98 patients with functioning renal transplants.

Since the initiation of the transplant program, sequential sera have been obtained from transplant patients at intervals after transplantation. Indirect fluorescent-antibody assays (11) were utilized to search for serological evidence of LD in 60 cases in which paired sera were available. Biopsy and autopsy materials obtained at the time of pneumonia were available in 45 cases and were examined for the presence of *L. pneumophila* antigens by the direct fluorescent-antibody technique (2). Indirect and direct fluorescent-antibody tests were available in 76% of the pneumonias.

Thirteen cases of LD were identified. LD was the most common cause of bacterial pneumonia, accounting for 14% of all pneumonias. The LD attack rate for the renal transplant population was 1.0 cases per 100 patient-years at risk. The attack rate varied by year. The highest rate of 20 cases per 100 patient-years occurred in 1971. No cases were detected in 1972, 1977, 1978, 1979, and 1980. The attack rate for the period 1971 through 1975 (4.5 cases per 100 patient-years) was significantly greater than that during the period 1976 through 1981 (0.2 cases per 100 patient-years) ($P < 0.001$ by the chi-square test). Similarly, the mean prednisone dosage at onset of pneumonia was greater in the period 1971 through 1975 than after 1975 ($P = 0.027$). This suggests an association between the LD attack rate and the degree of immunosuppression.

LD attack rates varied by recipient sex and donor type, but these rates were not significantly different (male, 1.4 cases per 100 patient-years; female, 0.6 cases per 100 patient-years; and cadaver, 1.4 cases per 100 patient-years; living relative, 0.5 cases per 100 patient-years).

The median age of LD cases was 32 years (range, 21 through 56 years). The ratio of males to females was 10:3, which was not significantly different from that of the overall transplant population. Cough was a presenting complaint in only nine cases, and sputum production was noted in only four patients. The median maximum temperature was 39.5°C (range, 38.4 to 40.5°C). Selected laboratory tests revealed median creatinine at onset of pneumonia of 6.0 mg/dl (range, 1.3 to 9.1 mg/dl) and median maximum leukocyte count of 6,600/mm^3 (range, 2,000 to 17,900/mm^3). By chest radiography, the lobe most frequently involved at diagnosis was the left upper lobe (five cases). This was followed in decreasing frequency by the right lower lobe (four cases), right middle lobe (three cases), left lower lobe (two cases), and right upper lobe (two cases). Three cases had multiple lobe involvement at presentation. The complications of respiratory failure and shock occurred in five cases. LD pneumonia was an important cause of mortality in this population. The case fatality rate of 46% for LD pneumonia was significantly greater than that for other pneumonias (19%) ($P = 0.036$).

Factors in pneumonia cases that showed a significant association with the diagnosis of LD are listed in Table 1. These risk factors for LD included nosocomial acquisition, occurrence in summer or fall, occurrence within 100 days of renal transplantation, pneumonia development in patients in acute rejection, prednisone dosage of greater than 30 mg at onset of pneumonia, and serum creatinine of greater than 4 mg/dl at onset of pneumonia. There was no association between LD and the type of transplant, history of multiple transplantation attempts, azathioprine dosage, or leukopenia (leukocyte count of <4,000/mm^3) ($P > 0.5$).

L. pneumophila appears to be ubiquitous in nature (4). The development of infection in humans seems to be dependent on exposure intensity, immunosuppression, or both (7).

A significant association between LD pneumonia and daily prednisone dosage, nosocomial acquisition, acute rejection, season of occurrence, serum creatinine, and occurrence within 100 days of transplant was found. An association of LD with immunosuppression has been noted previously (3, 7, 8). With the exception of season of occurrence, all of the identified risk factors may be covariates. Renal transplant patients tend to develop acute allograft rejection

TABLE 1. Risk factors for LD

| Risk factor | Ratio for patients with: | | P^a |
	LD	Other pneumonias	
Pneumonia type (nosocomial/community acquired)	9:1	0.7:1	0.004
Seasonal occurrence (winter-spring/summer-fall)	0.08:1	1.5:1	0.003
Pneumonia occurrence within 100 days of transplant (≤100/>100)	12:1	1:1	0.004
Acute rejection (yes/no)	2.3:1	0.3:1	0.003
Daily prednisone dosage (≤30 mg/>30 mg)	0.3:1	2.3:1	0.001
Serum creatinine (≤4.0 mg/dl/>4.0 mg/dl)	0.2:1	1.9:1	<0.001

^{a} By chi-square test.

with renal failure shortly after transplantation, often while hospitalized in the immediate postoperative period. Subsequently, patients developing rejection often require hospitalization for management. High doses of steroids are used in the therapy of allograft rejection to suppress the host immune response. We suspect that this immunosuppression is the key factor predisposing the host to the development of *L. pneumophila* pneumonia. In our population the attack rate for LD was high in the early period of the program but has fallen in recent years. The decreased LD incidence associated with the use of lower doses of prednisone in the last 5 years at our institution provides evidence supporting the immunosuppression hypothesis.

It has been popularly speculated that the lowering of hospital hot-water temperature recommended by the Joint Commission for the Accreditation of Hospitals may have led to a higher incidence of LD in hospitals with water systems containing legionellae. The hot-water temperature at the University of Iowa Hospitals was lowered from 140 to 120°F in January 1978. After lowering of the hot-water temperature, only one case of LD has occurred in the renal transplant population. It would appear that lowering the hot-water temperature was not associated with an increased incidence of LD in the renal transplant population at our institution.

LD was the most common cause of bacterial pneumonia in the renal transplant population at the University of Iowa. The attack rate was highest in the early years of the program, at a time when these patients were receiving higher doses of prednisone. LD was an important cause of mortality, with a case fatality rate of 46%. Factors associated with LD included nosocomial acquisition, occurrence in summer or fall, occurrence within 100 days of transplant, acute rejection, elevated serum creatinine, and high-dose prednisone therapy. The influence of steroid dosage on incidence of LD and on enhancing susceptibility to LD deserves further study.

LITERATURE CITED

1. Bock, B. V., P. H. Edelstein, K. M. Snyder, C. M. Hatayama, R. P. Lewis, B. D. Kirby, W. L. George, M. L. Owens, C. E. Haley, and R. D. Meyer. 1978. Legionnaires' disease in renal-transplant recipients. Lancet i:410–413.
2. Cherry, W. B., B. Pittman, P. P. Harris, G. A. Hebert, B. M. Thomason, C. Thacker, and R. E. Weaver. 1978. Detection of Legionnaires disease by direct immunofluorescent staining. J. Clin. Microbiol. 8:329–338.
3. England, A. C., III, D. W. Fraser, B. D. Plikaytis, T. F. Tsai, G. Storch, and C. V. Broome. 1981. Sporadic legionellosis in the United States: the first thousand cases. Ann. Intern. Med. 94:164–170.
4. Fliermans, C. B., W. B. Cherry, L. H. Orrison, S. J. Smith, D. L. Tyson, and D. H. Pope. 1981. Ecological distribution of *Legionella pneumophila*. Appl. Environ. Microbiol. 41:9–16.
5. Foster, R. S., Jr., W. C. Winn, Jr., W. Marshall, and D. W. Gump. 1979. Legionnaires' disease following renal transplantation. Transplant. Proc. 11:93–95.
6. Gump, D. W., R. O. Frank, W. C. Winn, R. S. Foster, C. V. Broome, and W. B. Cherry. 1979. Legionnaires' disease in patients with associated serious disease. Ann. Intern. Med. 90:538–542.
7. Haley, C. E., M. L. Cohen, J. Halter, and R. D. Meyer. 1979. Nosocomial Legionnaires' disease: a continuing common-source epidemic at Wadsworth Medical Center. Ann. Intern. Med. 90:583–586.
8. Helms, C. M., R. M. Massanari, R. Zeitler, S. Streed, M. Gilchrist, N. Hall, W. Hausler, L. Wintermeyer, and W. Hierholzer. 1983. Legionnaires' disease associated with a hospital water system: a cluster of 24 nosocomial cases. Ann. Intern. Med. 99:172–178.
9. Marshall, W., R. S. Foster, and W. Winn. 1981. Legionnaires' disease in renal transplant patients. Am. J. Surg. 141:423–429.
10. Taylor, R. J., F. N. Schwentker, and T. R. Hakala. 1981. Opportunistic lung infections in renal transplant patients: a comparison of Pittsburgh pneumonia agent and Legionnaires' disease. J. Urol. 125:289–292.
11. Wilkinson, H. W., D. D. Cruce, B. J. Fikes, L. T. Yealy, and C. E. Farshy. 1979. Indirect immunofluorescent test for Legionnaires' disease, p. 111–116. *In* G. L. Jones and G. A. Hébert (ed.), "Legionnaires' ": the disease, the bacterium, and methodology. Centers for Disease Control, Atlanta, Ga.

Epidemiology and Control of *Legionellaceae* in State Developmental Centers

THOMAS R. BEAM, JR., DANIEL MORETON, THOMAS A. RAAB, WILLIAM HEASLIP, MARIO MONTES, JOHN HANRAHAN, MICHELLE BEST, AND VICTOR YU

Buffalo VA Medical Center and SUNY at Buffalo, Buffalo, New York 14215; J. N. Adam Developmental Center, Perrysburg, New York 14129; New York State Department of Health, Albany, New York, 12237; and Pittsburgh VA Medical Center and University of Pittsburgh, Pittsburgh, Pennsylvania 15240

New York State Developmental Centers provide living accommodations and care for persons with severe mental or physical handicaps. Residents range in age from 10 to 92 years. Although Down's syndrome is one of the most common diagnostic classifications for these residents, the majority have no specific diagnoses. Mechanical problems in airway clearance predispose these individuals to repeated bouts of aspiration pneumonia.

Respiratory illness involving both the upper and lower respiratory tract is a significant cause of morbidity and mortality in these patients. Illness occurs both as sporadic cases and as epidemic outbreaks. Because a history usually cannot be obtained and because of poor compliance in producing sputum for analysis, a diagnosis of respiratory infection is usually based upon signs such as fever and cough, X-ray results, leukocyte count, and response to empiric antimicrobial therapy. Etiology is established in a minority of cases.

The purposes of this study were to (i) establish the incidence of infection caused by *Legionellaceae*, (ii) identify environmental sources of contamination, (iii) use available methods to decontaminate these sources, (iv) evaluate the impact of decontamination procedures on the incidence of infections caused by *Legionellaceae*.

Water at the J. N. Adam Developmental Center (center A) is obtained from a series of 12 wells. Usually, six or more wells are active at any one time. The water source is an aquifer. Well water is filtered and delivered to a large (1,000,000-gallon) or small (500,000-gallon) uncovered reservoir. It is then chlorinated and delivered via 12-in. (ca. 30-cm) or 6-in. (ca. 15-cm) pipes to the hospital. Hot water is generated by three large tanks for use throughout the facility.

In July 1981, 20 of 22 residents on a single ward developed a respiratory infection; 15 showed seroconversion for *Mycoplasma pneumoniae*. Testing for Legionnaires disease was not performed. In August 1981 a prospective evaluation began, and 22 of 24 residents on a different ward developed respiratory illness. Five had sputa positive for legionellae by the direct fluorescent-antibody (DFA) test, but the number of fluorescent bacilli was fewer than 25,

and fluorescence was nonspecific except in one case. There were no seroconversions, and cultures of sputum and numerous environmental sources were negative. One resident developed pneumonia and died despite treatment with erythromycin. Postmortem laboratory examination of spleen tissue gave positive DFA results for *Legionella micdadei*.

A third outbreak occurred in September 1981. Sixteen of nineteen residents in a third ward were ill. Five had DFA-positive sputa for *Legionella dumoffii*; three showed mixed fluorescence positivity. All cultures of sputum and environmental sources were negative, and there were no seroconversions.

In March and April of 1982 both epidemic and sporadic cases of respiratory tract illness occurred. On the basis of reports of isolation of *Legionella pneumophila* from new environmental sources (2), improved methods for isolation of *Legionella* species from water samples (1), and modification of the agar medium (3), we modified our procedures. Positive environmental cultures were obtained from hot- and cold-water sources for *L. pneumophila* group 6 and *L. micdadei*. Seven patients had DFA-positive sputa for either *L. pneumophila* group 6 or *L. micdadei*. One patient was culture positive, and one showed seroconversion.

Hot-water tanks were heated to 160°F (ca. 71°C) for 72 h, and the lines were flushed for a minimum of 15 min. All thermostatic controls had to be removed from the system and replaced after the heat treatment. Heat treatments were repeated in June, July, and August 1982 because of regrowth of *Legionella* species.

Chlorine treatment was attempted next. The normal chlorine levels in the system averaged 0.5 ppm. Chlorine was added to achieve a final concentration of 2 ppm; this caused sediment to leach from the iron pipes and discolor the water. A repeat attempt in September was beset with similar problems. However, the prevalence of respiratory illness declined, and subsequent environmental cultures were negative.

New cement liners were installed in the hot-water tanks during fall 1982. The first cultures from the newly lined tanks were obtained in January 1983 and grew *Legionella* species. In February and March 1983, a major outbreak of respiratory illness occurred among both residents and employees. Analysis of samples is incomplete. There were 22 DFA-positive sputum samples for *L. dumoffii*, 4 DFA-positive samples for *L. micdadei*, 2 for both *L. dumoffii* and *L. micdadei*, and 1 for *L. pneumophila* group 6. Environmental cultures were only positive for *L. pneumophila* group 6. At least five cases of *M. pneumoniae* infection were documented by fourfold rise in complement fixation antibody titer. These individuals had DFA-positive sputum for *L. dumoffii*, negative cultures, and no seroconversion for *Legionella* species.

The West Seneca Development Center (center B) houses 1,062 residents, compared with 250 in center A. An outbreak of respiratory illness occurred in September 1981. Eleven residents were evaluated serologically for *Legionella* infection; none seroconverted. In June 1982, an employee was diagnosed as infected with *Legionella longbeachae* by fourfold rise in titer. The water for this facility is obtained from Lake Erie and purified by the Erie County Water Authority. Thirty cultures obtained in July from various sites were negative for legionellae. In August, 2 of 53 cultures were positive. One showed *L. pneumophila* group 5 in a hot-water tank, and one showed *L. pneumophila* group 6 in a shower head. Although several areas were DFA positive for *L. longbeachae*, this organism was not grown. Because the facility has separate hot-water tanks for each of the more than 80 buildings, heat treatment was applied to culture-positive tanks. Chlorine was used in other areas

of the facility. All subsequent environmental cultures have been negative.

A retrospective and prospective analysis of respiratory tract infection was conducted. Sixteen employees and thirty residents had a clinical illness consistent with Legionnaires disease. Only one employee showed seroconversion. An additional 26 episodes were not evaluated because the clinical course was inconsistent with lower respiratory tract infection with *Legionella* species.

We conclude that *Legionellaceae* appear to be present in the water systems in both centers A and B. Although a number of individuals were DFA positive, there were very low rates of culture positivity and seroconversion, raising major questions about the meaning of the DFA results. It is possible that some DFA results were false-positive. Environmental decontamination measures did not result in long-term culture negativity.

LITERATURE CITED

1. **Bopp, C. A., J. W. Sumner, G. K. Morris, and J. G. Wells.** 1981. Isolation of *Legionella* spp. from environmental water samples by low-pH treatment and use of a selective medium. J. Clin. Microbiol. **13:**714–719.
2. **Cordes, L. G., A. M. Wilsenthal, G. W. Gorman, J. P. Phair, H. M. Sommers, H. Brown, V. L. Yu, M. H. Magnussen, R. P. Myer, J. S. Wolf, K. N. Shands, and D. W. Fraser.** 1981. Isolation of *Legionella pneumophila* from hospital shower heads. Ann. Intern. Med. **94:**195–197.
3. **Vickers, R. M., A. Brown, and G. M. Garrity.** 1981. Dye-containing buffered charcoal-yeast extract medium for differentiation of members of the family *Legionellaceae*. J. Clin. Microbiol. **13:**380–382.

Recurrent Legionnaires Disease from a Hotel Water System

C. L. R. BARTLETT, R. A. SWANN, J. CASAL, L. CANADA ROYO, AND A. G. TAYLOR

Public Health Laboratory Service Communicable Disease Surveillance Centre and Central Public Health Laboratory, Colindale, London, United Kingdom; John Radcliffe Hospital, Oxford, United Kingdom; and Centro Nacional de Microbiologia, Virologia e Immunologia Sanitarias and Ministerio de Sanidad y Seguridad Social, Madrid, Spain

Travel-associated cases of *Legionella* pneumonia are identified in England and Wales by means of a national surveillance scheme which was established by the Public Health Laboratory Service in 1977. Cases of *Legionella* pneumonia occurring among British tourists who had stayed at hotel A in Benidorm, Spain, were reported in each of the years 1977 to 1980, including a cluster of four cases in July 1979. A retrospective study of stored specimens found one case each in 1975 and 1976 associated with the hotel, and Reid et al. reported a large outbreak which had occurred at the same hotel in 1973 (2). A further outbreak was identified in the summer of 1980, the epidemiological investigation of which,

and the control methods used, is presented here.

Cases of pneumonia associated with Benidorm were identified through the national surveillance scheme and by case searching in collaboration with hospital and Public Health Laboratory Service microbiological laboratories and infectious-disease physicians. Pneumonia cases were considered to be Legionnaires disease (LD) if they showed a fourfold or greater rise in indirect fluorescent-antibody titer or a titer of 128 or more in convalescent specimens with Colindale *Legionella pneumophila* serogroup 1 formolized yolk sac antigen (3). Random controls were selected from the hotel register, and an equal number were selected who were

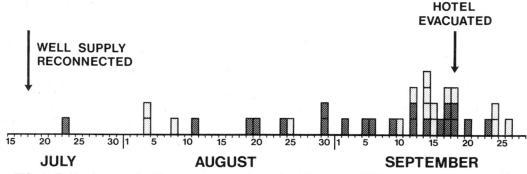

FIG. 1. Epidemic curve for 37 cases of pneumonia, July to September 1980. ▨, *Legionella* pneumonia; ▤, pneumonia of unknown etiology.

matched for age group, sex, and month of stay. Histories were collected by means of a self-completed questionnaire which was sent through the mail.

A total of 22 water samples were collected from the hotel under investigation and two other hotels in Benidorm. These consisted of 20-liter samples from showers, hot and cold taps, and storage tanks, and other points in the hotel water supply system, including the well. In addition, a 5-liter sample of water from a nearby cooling tower was collected. All samples were filtered and inoculated into pairs of guinea pigs for *Legionella* isolation and serology (4). Isolates were identified by whole-cell fatty acid analysis by gas-liquid chromatography (1) and sero-grouped by direct immunofluorescence, using hyperimmune rabbit sera and mouse monoclonal antibodies supplied by J. O'H. Tobin.

For July to September 1980, 64 cases of pneumonia were identified among British tourists who had visited Benidorm; 59 of these individuals had stayed at hotel A. The case definition for LD was met in 25 persons, 23 of whom had stayed at hotel A. Of the 23, 13 were male and 10 female. Sixteen were diagnosed by seroconversion and seven by convalescent titers of >128; one had positive sputum by the direct fluorescent-antibody test. Sera from 27 pneumonia cases were also tested against *L. pneumophila* serogroups 2 to 6; all gave titers less than or equal to those against serogroup 1. In addition, they were tested against *L. micdadei*, *L. bozemanii*, *L. dumoffii*, and *L. gormanii*, but no diagnostic titers were found. Twelve sera were tested against a formolized wash-off of a serogroup 4/5 isolate from the hotel water system, but all gave titers of <32. No cases were found among the hotel staff, but a serological survey found that 4 of 83 staff members had antibody titers of 32 or more (two with titers of 256) against *L. pneumophila* serogroup 1, compared with none among 78 employees in other establishments in the town.

The epidemic curve for the 37 pneumonia cases for which an exact date of onset was given is shown in Fig. 1. The fatality ratio for LD cases was 4.5%, with only 1 death among the 23 cases. The mean interval between day 1 of stay in the hotel and onset of illness in 15 cases was 9.5 days (range, 2 to 15 days). During August and September 4,900 guests stayed at the hotel and 58 developed pneumonia, giving an attack rate of 1.18%; 22 of these were diagnosed as LD cases, giving an attack rate of 0.45%.

Twenty of the cases who had stayed at hotel A were selected for the case-control study; questionnaires were returned from 18 cases (90%) and from 36 controls (86%), 18 of whom were matched for age, sex, and month of stay. The median age of these cases was 52 years (range, 23 to 75 years), and that of the 18 random controls was 39 years (range, 19 to 58 years). The cases were significantly older (exact probability = 0.004, proportions above and below overall median). Smoking was clearly shown to be a risk factor (Table 1), but not alcohol consumption or preexisting respiratory disease. Although no differences were detected in terms of frequency of drinking tap water, use of swimming pool, brushing teeth, bathing, or showering, cases did tend to be the first to use the washing facilities in the morning. Individuals sharing a room with another person were asked whether they were always, usually, sometimes, or never the first to wash in the morning. One of 18 cases and six of 36 controls stated that they were never the first to wash. This difference is not significant, but similar results were found with the age- and sex-matched pairs. As *L. pneumophila* may multiply overnight in water standing in peripheral pipework, perhaps those who wash first in the morning are at a greater risk of developing LD than those who wash

TABLE 1. History of cigarette smoking (age/sex-matched pairs)[a]

Controls	Cases		
	Smoker	Nonsmoker	Total
Smoker	4	0	4
Nonsmoker	11	3	14
Total	15	3	18

[a] Exact binomial probability = 0.0005 (one-tailed test).

later. The most accurate way to test this hypothesis is with matched pairs sharing the same room. Eight case-spouse control pairs were contacted, and seven pairs were analyzed; of these, the members of one pair washed first with the same frequency, leaving six pairs with a predominant "first washer." Of the six, the case washed first in five pairs. With a null hypothesis of equal likelihood for case or control to wash first, the exact binomial probability (five of six) = 0.11. In the one pair where the control washed first, the control always bathed and the case always showered. There were only four pairs where both always used the same facility (three showers, one bath), and in all four pairs the case usually or always washed first [binomial probability = $(1/2)^4 = 0.06$].

Cases occupied bedrooms in the lower part of the building more commonly than controls, although the difference was not significant, perhaps due to the greater likelihood of production of small-particle aerosol at lower levels. The hotel did not have an air-conditioning system, but there was a cooling tower on the grounds of a hotel nearby. However, occupancy of those bedrooms facing the cooling tower, or located in the two wings next to it, was not associated with LD. Furthermore, there were no differences in the frequency or lengths of time spent in the outer reception area of hotel A.

Two water samples from hotel A yielded L. pneumophila serogroup 1; these included a sample from the hairdresser's salon and a pooled sample from 10 room showers on the third floor. A third sample from 10 showers on the 11th floor caused seroconversion in guinea pigs to serogroup 1 formolized yolk sac antigen. The sample from the hairdresser's salon also yielded a second L. pneumophila strain, 034RP, which reacted with both serogroup 4 and serogroup 5 antibodies. L. pneumophila was not isolated from the water samples collected from other hotels in Benidorm.

A review of engineering practices in the hotel revealed that no major work had been carried out on the water system immediately before or during the outbreak, other than the reconnection of a supplementary well water supply which had been out of commission for a month. The person with the first case of LD in this outbreak fell ill 5 days after this reconnnection. An inspection of the hotel water system found that cooling water from the kitchen refrigeration unit was returned to the hotel's cold-water reservoir as a water conservation measure.

Our findings indicated that the hotel's water system was the source of the outbreak, and we found no evidence to implicate a cooling tower. Accordingly, we introduced the following control measures. Initially, the cold-water system was superchlorinated to 60 ppm of free residual chlorine for a minimum of 30 min, the water temperature in the calorifiers (heating cylinders) was raised to 90°C and held at that level for 30 min, and cooling water from the kitchen refrigeration units was diverted to waste. On a continuous basis, the hot- and cold-water systems have been chlorinated so as to achieve 2 to 3 ppm of free residual chlorine as measured at outlets, and the hot-water temperatures have been maintained at 50 to 60°C at all taps and showers in the hotel bedrooms.

No cases of LD have been identified in association with this hotel since the control measures were introduced in the autumn of 1980, although the occupancy has been high during the last 2 years. This provides further circumstantial evidence that the hotel's domestic water system was the source of Legionella infection and clearly shows that legionellosis may be controlled in such situations.

LITERATURE CITED

1. Moss, C. W., R. E. Weaver, S. B. Dees, and W. B. Cherry. 1977. Cellular fatty acid composition of isolates from Legionnaires disease. J. Clin. Microbiol. 6:140–143.
2. Reid, D., N. R. Grist, and R. Najera. 1978. Illness associated with "package tours": a combined Spanish-Scottish study. Bull. W.H.O. 56:117–122.
3. Taylor, A. G., T. G. Harrison, M. W. Dighero, and C. M. P. Bradstreet. 1979. False positive reactions in the indirect fluorescent antibody test for Legionnaires' disease eliminated by use of formolized yolk-sac antigen. Ann. Intern. Med. 90:686–689.
4. Tobin, J. O'H., R. A. Swann, and C. L. R. Bartlett. 1981. Isolation of Legionella pneumophila from water systems: methods and preliminary results. Br. Med. J. 282:515–517.

Legionella pneumophila Contamination of Residential Tap Water

PAUL M. ARNOW AND DIANE WEIL

University of Chicago Hospital, Chicago, Illinois 60637

Ecological studies and investigations of outbreaks of Legionnaires disease have demonstrated species of the genus *Legionella* in many aquatic habitats. Reservoirs include natural bodies of freshwater and water in artificial environments such as cooling towers and evaporative condensers. In addition, potable water systems in many buildings, especially hospitals, have been shown to be contaminated by *Legionella* species (1, 2, 4–8). This contamination has been implicated in cases of nosocomial Legionnaires disease in the United States and Europe (1, 2, 5–8), although in only one instance has the mechanism of transmission been demonstrated (1).

The prevalence and significance of *Legionella* species contamination of residential potable water systems has not been systematically evaluated. Therefore, we examined specimens from selected residences in the community adjacent to our hospital, and we sought evidence by history and serological survey of clinical or subclinical *Legionella* infection in the occupants of those residences.

The residences studied are in 37 apartment buildings in the neighborhood adjacent to the University of Chicago Hospital. Potable water is supplied to the residences by the Chicago Department of Water and Sewers. Fifty-two residences were selected based upon occupancy for at least the previous year by University of Chicago Hospital personnel who volunteered to participate in the study.

During August 1982 through February 1983, the following specimens were collected from a bathroom in each residence: 1 liter of cold tap water, 1 liter of hot tap water, and scrapings obtained separately from the inside of the shower head and sink faucet. Scrapings were suspended in 1 to 2 ml of tap water at the time of collection, and 0.5 ml of each suspension was inoculated onto buffered charcoal-yeast extract agar containing cephalothin (4 μg/ml), colistin (16 μg/ml), and vancomycin (0.5 μg/ml). Hot tap water specimens were concentrated by filtration as previously described (1), and 0.01- to 0.5-ml portions of the concentrate were inoculated onto buffered charcoal-yeast extract agar containing the supplements described above. After incubation under 5% CO_2, colonies resembling legionellae were identified by standard methods, including direct fluorescent-antibody staining (3). Up to 20 colonies from each positive plate were picked for direct fluorescent-antibody staining to permit detection of multiple serogroups.

The temperature of hot-water specimens was recorded at the time of collection, and pH and chlorine levels were measured within 2 h. Also, quantitative cultures of bacteria other than legionellae in hot tap water were performed with sheep blood agar.

Legionella pneumophila was recovered from 19 (37%) of the 52 apartments. The culture-positive apartments were in 14 of the 37 buildings sampled. Specimens from which *L. pneumophila* was isolated were hot tap water only (18 apartments) or hot tap water and shower head scrapings (1 apartment). Most isolates were serogroup 1 (12 specimens), but one to three isolates each of serogroups 2 to 6 (10 specimens) also were recovered. Three specimens contained isolates of two different serogroups. The concentration of *L. pneumophila* in hot tap water ranged from 1 to 5,000 CFU/liter, and the median was 100 CFU/liter.

Tap water systems that were culture positive were almost identical to those that were culture negative in free chlorine content of cold water, and in pH and concentration of bacteria other than *Legionella* species in hot water. The mean temperature of hot water was lower for culture-positive (47.7°C) than for culture-negative (54.9°C) specimens, and the difference in temperature between the two groups was highly significant ($P < 0.01$ by the Mann-Whitney test). An apparent temperature dependence was seen, because *L. pneumophila* was isolated from 19 (50%) of 38 specimens with a temperature of ≤55°C and from none of 14 specimens above 55°C ($P < 0.001$ by Fisher's exact test).

A questionnaire survey showed that the participants in the study had lived in their apartments for a median of 3 years (range, 1 to 19 years). Each individual took 3 to 10 showers in his or her apartment each week (mean, 6.3), and 22 used tap water-filled portable humidifiers during the winter. None of the individuals had had pneumonia during the period of residence in the apartment studied.

Serum specimens were obtained from the study participants and tested for antibodies against *L. pneumophila* serogroups 1 to 6 individually by the indirect fluorescent-antibody method (9). An antibody titer of ≥256 was detected in the serum of only one subject and was directed against serogroup 1. The hot tap water in this subject's apartment contained *L. pneumophila* serogroup 1 at a concentration of 3 × 10³ CFU/liter. The subject had lived in his apartment for 6 years and used a tap water-filled room humidifier for about 20 days each year. The geometric mean titer of antibodies against *L. pneumophila* serogroup 1 was 18 in individuals whose tap water contained the organism, compared with 17 in individuals whose tap water did not yield *L. pneumophila*. The geometric mean titer of the eight individuals whose residential tap water contained *L. pneumophila* serogroup 1 and who used humidifiers was 21.

These data indicate that *L. pneumophila* contamination of hot tap water in apartments occurs commonly in the area of Chicago studied. Contamination was confined to systems in which the temperature of the water at the faucet did not exceed 55°C. This observation is compatible with previous reports of *Legionella* species contamination of hot potable water maintained at a temperature below 55°C but not at higher temperatures (1, 6, 8; J. F. Plouffe, L. R. Webster, B. Hackman, and M. Macynski, Abstr. Annu. Meet. Am. Soc. Microbiol. 1983, L16, p. 103).

Despite a high rate of *L. pneumophila* contamination of residential hot tap water and frequent exposure of residents to aerosols from humidifiers and probably from showers, illness attributable to *L. pneumophila* could not be demonstrated in the healthy, immunocompetent population studied. No cases of pneumonia were reported, and only one subject had unequivocal serological evidence of infection, i.e., an antibody titer of ≥256. Furthermore, the geometric mean titer of antibodies against *L. pneumophila* serogroup 1 was similar in groups whose tap water was culture positive or culture negative. The apparently benign interaction with *L. pneumophila* is most likely attributable to the relatively low concentration of the organism in hot tap water. Low concentrations of *L. pneumophila* in the tap water of our hospital were linked to a small cluster of infections only in immunocompromised patients (1). In contrast, apparent high concentrations of *Legionella* species have been associated with sustained outbreaks of Legionnaires disease (2, 5, 7). An alternative explanation for the absence of infection in the present study is diminished virulence of the residential tap water strains of *L. pneumophila* (8).

L. pneumophila in residential tap water was not shown to be a significant hazard in the immunocompetent population studied, but this reservoir may nonetheless be considered a potential source for winter or other sporadic cases of Legionnaires disease. Until the range of concentrations and role of *Legionella* species in residential tap water is better characterized, it seems advisable to avoid the use of residential hot tap water maintained below 55°C in aerosol-producing devices such as humidifiers and respiratory therapy equipment.

This work was supported in part by a grant from Abbott Laboratories.

LITERATURE CITED

1. **Arnow, P. M., T. Chou, D. Weil, E. N. Shapiro, and C. Kretzschmar.** 1982. Nosocomial Legionnaires' disease caused by aerosolized tap water from respiratory devices. J. Infect. Dis. **146**:460–467.
2. **Brown, A., R. M. Vickers, E. M. Elder, M. Lema, and G. M. Garrity.** 1982. Plasmid and surface antigen markers of endemic and epidemic *Legionella pneumophila* strains. J. Clin. Microbiol. **16**:230–235.
3. **Cherry, W. B., B. Pittman, P. P. Harris, G. A. Hébert, B. M. Thomason, L. Thacker, and R. E. Weaver.** 1978. Detection of Legionnaires disease bacteria by direct immunofluorescent staining. J. Clin. Microbiol. **8**:329–338.
4. **Dennis, P. J., J. A. Taylor, R. B. Fitzgeorge, C. L. R. Bartlett, and G. I. Barrow.** 1982. *Legionella pneumophila* in water plumbing systems. Lancet **i**:949–951.
5. **Edelstein, P. H., J. B. Snitzer, and S. M. Finegold.** 1982. Isolation of *Legionella pneumophila* from hospital potable water specimens: comparison of direct plating with guinea pig inoculation. J. Clin. Microbiol. **15**:1092–1096.
6. **Fisher-Hoch, S. P., C. L. R. Bartlett, J. O'H. Tobin, M. B. Gillett, A. M. Nelson, J. E. Pritchard, M. G. Smith, R. A. Swann, J. M. Talbot, and J. A. Thomas.** 1981. Investigation and control of an outbreak of Legionnaires' disease in a district general hospital. Lancet **i**:932–936.
7. **Stout, J., V. L. Yu, R. M. Vickers, and J. Shonnard.** 1982. Potable water supply as the hospital reservoir for Pittsburgh pneumonia agent. Lancet **i**:471–472.
8. **Wadowsky, R. M., R. B. Yee, L. Mezmar, E. J. Wing, and J. N. Dowling.** 1982. Hot water systems as sources of *Legionella pneumophila* in hospital and nonhospital plumbing fixtures. Appl. Environ. Microbiol. **43**:1104–1110.
9. **Wilkinson, H. W., B. J. Fikes, and D. D. Cruce.** 1979. Indirect immunofluorescent test for serodiagnosis of Legionnaires disease: evidence for serogroup diversity of Legionnaires disease bacterial antigens and for multiple specificity of human antibodies. J. Clin. Microbiol. **9**:379–383.

Epidemic of Community-Acquired Pneumonia Associated with *Legionella pneumophila* Infection in an Iowa County

CHARLES HELMS, LAVERNE WINTERMEYER, RODNEY ZEITLER, RICHARD LAREW, R. MICHAEL MASSANARI, NANCY HALL, WILLIAM HAUSLER, JR., AND WILLIAM JOHNSON

Department of Internal Medicine, University of Iowa Hospitals and Clinics, Department of Microbiology, University of Iowa College of Medicine, and University Hygienic Laboratory, Iowa City, Iowa 52242, and Iowa State Department of Health, Des Moines, Iowa 50308

Legionnaires disease (LD) is a significant cause of epidemic pneumonia in humans. Recent investigations have focused on epidemics and clusters of LD occurring as nosocomial disease (1–3, 4a, 6). Community-acquired epidemics of LD have received less attention.

In late October 1981, physicians affiliated with hospital A in Johnson County, Iowa, noted an abrupt increase in the number of cases of community-acquired pneumonia occurring in adults. A retrospective investigation of this event was carried out in February 1982, and a small epidemic of community-acquired LD was uncovered. We present here the results of this investigation.

Pneumonia frequency in Johnson County in October 1981. Monthly pneumonia rates from hospital A were calculated to determine whether the rate in October 1981 was abnormally high. The rate of primary diagnosis of pneumonia for October (34.1/1,000 admissions) was 7.8 times that for September (4.4/1,000) and 2.6 times that for November (12.9/1,000). The chi-square test was used to compare rates for the months of September, October, and November over the 6-year period 1976 to 1981. The sets of rates for September and November were not different (*P* = 0.148 and *P* = 0.835, respectively). For October, however, the test for no difference was rejected (*P* < 0.0001). Because the set of rates for October 1976 to 1980 was not different (*P* = 0.276), the rate for October 1981 was implicated as an aberrantly high value.

The medical records of patients with primary diagnosis of pneumonia admitted to all Johnson County hospitals in October were reviewed. A case definition was adopted requiring (i) age of ≥18 years, (ii) a physician's diagnosis of pneumonia, and (iii) radiographic pulmonary infiltrates. Forty-three patients with pneumonia were admitted to Johnson County hospitals in October 1981 (Fig. 1). Twenty-five (58%) of these were clustered in onset from 19 to 25 October. Pneumonia cases occurred with higher frequency in this 7-day period than in the rest of the month (*P* < 0.001 by goodness-of-fit chi-square test). Twenty-two (88%) of the 25 clustered patients versus seven (39%) of the 18 nonclustered patients were Johnson County residents (*P* = 0.001 by Fisher's exact test). Based on these data, pneumonia cases in Johnson County residents were selected for further study.

Etiology of pneumonia. Of 21 adequately sampled cases, 12 (57%) showed evidence of *Legionella pneumophila* infection. Eight confirmed cases of *L. pneumophila* infection were identified by significant increases in indirect fluorescent-antibody titer (six cases) and direct fluorescent-antibody examination of lung tissue (two cases). *L. pneumophila* serogroup 1 was isolated from lung tissue in one case. Four suspect cases of *L. pneumophila* infection were identified by a fourfold or greater rise in microagglutination titer (two cases), a standing convalescent microagglutination titer of 1:64 (one case) (4), or a convalescent indirect fluorescent-antibody titer of 1:128 (one case) (5).

Routine microbiological laboratory testing failed to identify pathogens in other pneumonia cases. Paired sera showed no increases in titers of complement-fixing antibodies to influenza A and B viruses, *Mycoplasma pneumoniae* (lipid antigen), or adenovirus (13 cases).

Temporal distribution of LD cases. Eleven of twelve confirmed and suspect LD cases had onset from 19 to 25 October (Fig. 2). Of 17 clustered cases tested, 11 (65%) had laboratory evidence of *L. pneumophila* infection.

Features of LD cases. Males outnumbered females in the LD group by 2:1. Clinical differentiation of LD cases from undiagnosed pneumonias occurring in Johnson County residents was not possible. All LD patients received erythromycin; one died, for a fatality rate of 8%. No significant differences between LD cases and undiagnosed pneumonia cases in Johnson Coun-

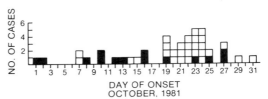

FIG. 1. Temporal distribution of pneumonia cases in Johnson County, Iowa, hospitals in October 1981. Open squares indicate Johnson County cases. Solid squares indicate cases from other Iowa counties.

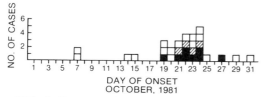

FIG. 2. Temporal distribution of LD cases among cases of pneumonia in Johnson County residents in October 1981. Solid squares indicate confirmed LD cases. Cross-hatched squares indicate suspect LD cases. Open squares indicate pneumonia cases of unclear etiology.

ty residents related to hospital exposure, overnight travel, attendance at conventions, or exposure to excavations were uncovered by questionnaire. Samples of hot and cold water from sinks and showers in homes of four confirmed LD patients and four neighborhood controls were examined in February 1982. No *Legionella* species were isolated.

Discussion. In the present study we uncovered a small epidemic of community-acquired LD pneumonia that occurred in Johnson County, Iowa, in autumn 1981. LD pneumonia has been recognized in Johnson County for several years. Between 1971 and 1980, at least 19 cases of nosocomial or community-acquired disease occurred without obvious temporal clustering. The present epidemic, then, occurred in an endemic setting. A nosocomial epidemic of *L. pneumophila* serogroup 1 pneumonia was in progress at hospital B in Johnson County during this community-acquired epidemic (4a). The possibility of a common source of both epidemics had to be considered, therefore. Nine of ten patients with community-acquired LD questioned denied being a hospital visitor, inpatient, or employee in the 2 weeks before onset of illness. The 10th patient had hospital exposure in a different county. The patient who died of LD had not been an inpatient or outpatient at hospital B in the 2 weeks preceding onset of illness. One LD patient had been an outpatient at hospital B and may have acquired the disease there.

Because the hot water and water delivery system of hospital B had been implicated as reservoirs in the nosocomial epidemic, hot and cold water samples from the homes of four LD patients were examined in February 1982. *Legionella* species were not isolated. The reservoir of this epidemic of community-acquired LD pneumonia, therefore, remains obscure.

LD may be an important cause of epidemic community-acquired pneumonia. Prospective studies of the epidemiology of community-acquired LD pneumonia to determine reservoirs of infection and mechanisms of transmission are in order.

LITERATURE CITED

1. Brown, A., V. L. Yu, E. M. Elder, M. H. Magnussen, and F. Kroboth. 1980. Nosocomial outbreak of Legionnaire's disease at the Pittsburgh Veterans Administration Medical Center. Trans. Assoc. Am. Physicians 93:52–59.
2. Cordes, L. G., A. M. Wiesenthal, G. W. Gorman, J. P. Phair, H. M. Sommers, A. Brown, V. L. Yu, M. H. Magnussen, R. D. Meyer, J. S. Wolf, K. N. Shands, and D. W. Fraser. 1981. Isolation of *Legionella pneumophila* from hospital shower heads. Ann. Intern. Med. 94:195–197.
3. Fisher-Hoch, S. P., C. L. R. Bartlett, J. O'H. Tobin, M. B. Gillett, A. M. Nelson, J. E. Pritchard, M. G. Smith, R. A. Swann, J. M. Talbot, and J. A. Thomas. 1981. Investigation and control of an outbreak of Legionnaires' disease in a district general hospital. Lancet i:932–936.
4. Helms, C. M., W. Johnson, E. D. Renner, W. J. Hierholzer, Jr., L. A. Wintermeyer, and J. P. Viner. 1980. Background prevalence of microagglutination antibodies to *Legionella pneumophila* serogroups 1, 2, 3, and 4. Infect. Immun. 30:612–614.
4a. Helms, C. M., R. M. Massanari, R. Zeitler, S. Streed, M. J. R. Gilchrist, N. Hall, W. J. Hausler, Jr., J. Sywassink, W. Johnson, L. Wintermeyer, and W. J. Hierholzer, Jr. 1983. Legionnaires' disease associated with a hospital water system: a cluster of 24 nosocomial cases. Ann. Intern. Med. 99:172–178.
5. Helms, C. M., E. D. Renner, J. P. Viner, W. J. Hierholzer, Jr., L. A. Wintermeyer, and W. Johnson. 1980. Indirect immunofluorescence antibodies to *Legionella pneumophila*: frequency in a rural community. J. Clin. Microbiol. 12:326–328.
6. Tobin, J. O'H., J. Beare, M. S. Dunnill, S. Fisher-Hoch, M. French, R. G. Mitchell, P. J. Morris, and M. F. Muers. 1980. Legionnaires' disease in a transplant unit: isolation of the causative agent from shower baths. Lancet ii:118–121.

Sporadic Legionellosis in the United States, 1970 to 1982

JOEL N. KURITSKY, ARTHUR L. REINGOLD, ALLEN W. HIGHTOWER, AND CLAIRE V. BROOME

Division of Bacterial Diseases, Center for Infectious Diseases, Centers for Disease Control, Atlanta, Georgia 30333

The Centers for Disease Control has maintained a passive surveillance system for sporadic cases of legionellosis since the recognition of the organism in 1977. A case was included in this report if it met one of the following criteria: fourfold or greater rise in indirect fluorescent-antibody titer to 128 or greater; demonstration of the agent in lung tissue, respiratory secretions, or pleural fluid by direct fluorescent-antibody testing; or isolation of the agent from lung tissue,

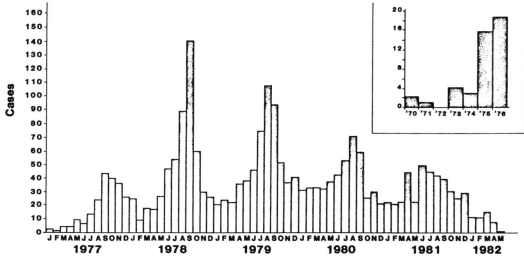

FIG. 1. Sporadic cases of legionellosis, 1970 to 1982. Reports received as of 15 June 1982 are included.

respiratory secretions, pleural fluid, or blood. Cases were excluded if they were known to be associated with an outbreak. Heat-killed antigen for indirect fluorescent-antibody testing to serogroup 1 was available in October 1978 (7). Polyvalent reagents containing heat-killed *Legionella pneumophila* serogroups 1 to 4 became available in July 1979. Antiserum for direct fluorescent-antibody testing was available for serogroups 1 to 4 by November 1978. Antiserum for other *Legionella* species became available shortly after recognition of the organisms (*Legionella bozemanii, Legionella dumoffii,* and *Legionella micdadei* in 1979, and *Legionella longbeachae* in 1980) (2–4, 6). Not all reagents have been available at all laboratories.

Information obtained from the case report form included age, sex, race, state of residence, travel history, history of prior diabetes, renal disease, cancer, prior chronic bronchitis, history of steroid use, smoking history, and outcome. Information regarding method of diagnosis was requested.

A total of 2,470 reported cases occurred between 1970 and the first half of 1982. Of those reported, 2.9% occurred before 1977. The remainder occurred between 1977 and 1982, the largest number (593 [26%]) occurring in 1979 (Fig. 1). By season of onset, significantly more cases (43%) were reported the third quarter of the year than the other three quarters (14, 20, and 23% in the first, second, and fourth quarters, respectively; $P < 0.0001$ by the test of significance of a binomial proportion). Sporadic cases of legionellosis were reported from 49 states and the District of Columbia.

Sixty-nine percent of the subjects were male.

Eighty-five, eleven, one, and less than one percent of the subjects, respectively, were white, black, Hispanic, and Asian or American Indian; the ethnicity of the remaining cases was not reported. A total of 1,241 (52%) of the subjects were between the ages of 50 and 69, 40 subjects (1.6%) were under the age of 20, and 11 subjects were less than 10 years of age (Fig. 2).

A total of 78% (1,941) of the cases were diagnosed by serological tests for antibody, 19% (480) were diagnosed by direct fluorescent-antibody staining, and 8% (201) were diagnosed by culture. Fifty-five serologically confirmed cases were also culture positive, and seventy-four cases positive by direct fluorescent-antibody testing were also culture positive.

Specific species or serogroup identification was available in 146 cases. Of these cases, 126 were *L. pneumophila* (101 serogroup 1, 6 serogroup 2, 4 serogroup 3, 6 serogroup 4, 2 serogroup 5, and 7 serogroup 6), and 20 were not *L. pneumophila* species. Of these, 6 were *L. bozemanii,* 12 were *L. micdadei,* and 2 were *L. longbeachae.*

Frequently recognized risk factors associated with legionellosis, including smoking, underlying malignancy, and steroid use, were similar to what has been previously reported (1). The fatality rate was approximately 20%. This was highest in those subjects under 20 years of age (25%) and those over the age of 70 (27%). Possible environmental associations were first described in a case-control study involving 100 sporadic cases reported between 1976 and 1977 (5). Risk factors identified in that study were noted on many case report forms; these factors included travel overnight (25%), visiting, work-

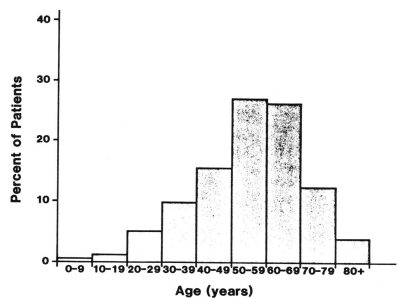

FIG. 2. Sporadic legionellosis by age, 1970 to 1982.

ing, or being in a hospital (27%), and having an excavation site close to one's home (13%).

Many questions regarding risk factors for sporadic legionellosis remain unanswered. The importance of different methods of transmission needs to be determined. Surveillance data can suggest features common to reported cases of legionellosis, but further studies will be needed to assess the relative role of these risk factors.

LITERATURE CITED

1. **England, A. C., D. W. Fraser, B. D. Plikaytis, T. F. Tsai, G. Storch, and C. V. Broome.** 1981. Sporadic legionellosis in the United States: the first thousand cases. Ann. Intern. Med. **94:**164–170.
2. **Hébert, G. A., B. M. Thomason, P. P. Harris, M. D. Hicklin, and R. M. McKinney.** 1980. "Pittsburgh pneumonia agent": a bacterium phenotypically similar to *Legionella pneumophila* and identical to the Tatlock bacterium. Ann. Intern. Med. **92:**53–54.
3. **Lewallen, K. R., R. M. McKinney, D. J. Brenner, C. W. Moss, D. H. Dail, B. M. Thomason, and R. A. Bright.** 1979. A newly identified bacterium phenotypically resembling but genetically distinct from *Legionella pneumophila*: an isolate in a case of pneumonia. Ann. Intern. Med. **91:**831–834.
4. **McKinney, R. M., R. Porschen, P. H. Edelstein, M. L. Bissett, P. P. Harris, S. P. Bondell, A. G. Steigerwalt, R. E. Weaver, M. E. Ein, D. S. Lindquist, R. S. Kops, and D. J. Brenner.** 1981. *Legionella longbeachae* species nova: another etiologic agent of human pneumonia. Ann. Intern. Med. **94:**739–743.
5. **Storch, G., W. B. Baine, D. W. Fraser, C. V. Broome, H. W. Clegg II, M. L. Cohen, S. A. J. Goings, B. D. Politi, W. A. Terranova, T. F. Tsai, B. D. Plikaytis, C. C. Shepard, and J. V. Bennett.** 1979. Sporadic community acquired Legionnaires' disease in the United States: a case control study. Ann. Intern. Med. **90:**596–600.
6. **Thomason, B. M., P. P. Harris, M. D. Hicklin, J. A. Blackmon, C. W. Moss, and F. Matthews.** 1979. A *Legionella*-like bacterium related to WIGA in a fatal case of pneumonia. Ann. Intern. Med. **91:**673–676.
7. **Wilkinson, H. W., B. J. Fikes, and D. D. Cruse.** 1979. Indirect immunofluorescence test for serodiagnosis of Legionnaires disease: evidence for serogroup diversity of Legionnaires disease bacterial antigens and for multiple specificity of human antibodies. J. Clin. Microbiol. **9:**379–383.

Pontiac Fever in an Automobile Engine Assembly Plant

LOREEN A. HERWALDT, GEORGE W. GORMAN, ALLEN W. HIGHTOWER, BONNIE BRAKE, HAZEL WILKINSON, ARTHUR L. REINGOLD, PETER A. BOXER, TERESA McGRATH, DON J. BRENNER, C. WAYNE MOSS, AND CLAIRE V. BROOME

Division of Bacterial Diseases, Center for Infectious Diseases, Centers for Disease Control, Atlanta, Georgia 30333; Hazard Evaluation and Technical Assistance Branch, National Institute for Occupational Safety and Health, Cincinnati, Ohio 45226; and Ministry of Labor, Toronto, Ontario, Canada

From 15 to 21 August 1981, 317 engine assembly plant workers in Windsor, Ontario, Canada, were affected by an explosive outbreak of Pontiac fever. Extensive epidemiological, microbiological, serological, and environmental investigations were undertaken to identify the etiological agent and the source of the outbreak. To facilitate the epidemiological investigation a case was defined as a worker who experienced at least three of the symptoms fever, chills, headache, and myalgia during August 1981. A well person was one who answered "no" to the question, "Were you ill in August?" Individuals who were ill but did not meet the case definition were considered possible cases.

A total of 695 employees (80%) completed the questionnaire. Of these, 317 (46%) met the case definition, 270 (39%) were not ill, and 108 (16%) were ill but did not meet the case definition. The illness had a mean maximum incubation period of 46 h and was characterized by fever ranging from 99.5 to 104°F, severe myalgia, headache, and extreme fatigue. The illness was short (median duration, 3 days) but was severe enough to cause nearly 30% of the workers to miss work (median days of sick leave, 2). There were no fatalities, and only four of the workers reported similar illness in family members within 72 h after the onset of the worker's illness.

Attack rates varied significantly by department, with a progressive decrease from north to south. The departments with the highest attack rates were along a line from the piston, camshaft, and crankshaft lines in the northwest to the cylinder head line in the southeast, which corresponded to the wind direction on 17 and 18 August. The attack rate gradient suggested an association between where an employee worked on 17 and 18 August and the risk of developing illness. The association was confirmed by multivariate analysis ($P = 6.4 \times 10^{-28}$ by chi-square automatic interaction detection).

A *Legionella*-like organism, designated WO-44c, was isolated from a sample of coolant obtained on 19 August from system 17 in the piston department. No other legionellae or *Legionella*-like organisms were isolated from any of the other environmental samples. Like other legionellae, the organism does not grow on blood agar, requires L-cysteine for growth, and produces catalase, but does not produce urease, reduce nitrite to nitrate, or produce acid from carbohydrates (1). Unlike other previously described legionellae, it does not produce gelatinase, and WO-44c and *L. pneumophila* are the only two species that hydrolyze hippurate. WO-44c is antigenically distinct from other legionellae, since it does not stain with direct fluorescent-antibody sera against the nine previously described species. By DNA-DNA hybridization, WO-44c is less than 10% related to all named species. The guanine-cytosine content is 45.7%. Results of cellular fatty acid analysis are similar to those for other legionellae, with the characteristic features of relatively large amounts of branched-chain acids (i-$C_{14:0}$, a-$C_{15:0}$, and a-$C_{17:0}$ acids), the presence of relatively large amounts of $C_{16:0}$ and $C_{16:1}$ straight-chain acids, and the absence of hydroxy acids.

Indirect fluorescent-antibody titers (2) to WO-44c for cases (geometric mean titer [GMT] = 296) were significantly different than those for well persons (GMT = 165, $P = 0.0006$), possible cases (GMT = 128, $P = 0.007$), and controls (GMT = 71, $P < 0.0001$). GMTs for well persons and possible cases were not statistically different from each other, but were significantly higher than the GMT for controls ($P < 0.001$ and $P = 0.0152$, respectively). The well persons and the possible cases had been in the plant at the time of the outbreak and may have been exposed to the etiological agent, whereas the controls had no recent contact with the plant. Since case definition was closely correlated with both location of work and indirect fluorescent-antibody titer, the serological data were stratified by case definition to determine whether indirect fluorescent-antibody titer was merely a function of case definition or varied independently with location of work and therefore with dose. Ill persons (cases) who worked in the three highest-risk areas (departments closest to system 17) had a higher GMT (345) than did cases who worked in the two lowest-risk areas (GMT = 154) ($P = 0.0339$ by Wilcoxon rank sums). There was no statistical difference between the GMTs for well persons who worked in the higher- and the lower-risk areas.

The time course of the outbreak was consistent with an explosive common-source outbreak. Food and waterborne spread were ruled out by the epidemiological survey, which documented no common source of food or water for ill employees. Airborne spread from one or more of the coolant systems, including system 17, was the most likely mode of transmission in this outbreak. System 17, from which WO-44c was isolated, had no coolant circulation during a partial plant shutdown period from 8 to 16 August 1981, because the piston department was not operating. Although the data necessary to prove this hypothesis were not available, one can speculate that during the period of disuse, bacterial overgrowth, which is known to occur more readily when coolants are not circulated, decreased the pH of the coolant and led to separation of the oil-water emulsion. The separation could have been favorable for the growth of WO-44c. With the resumption of production on 17 August, the machines serviced by system 17 may have generated a contaminated aerosol which was spread throughout the plant.

We conclude that this outbreak was caused by a new *Legionella* species. This is the first outbreak of nonpneumonic legionellosis, i.e., Pontiac fever, in which the etiological agent was not *L. pneumophila* serogroup 1.

LITERATURE CITED

1. **Orrison, L. H., W. B. Cherry, R. L. Tyndall, C. B. Fliermans, S. B. Gough, M. A. Lambert, L. K. McDougal, W. F. Bibb, and D. J. Brenner.** 1983. *Legionella oakridgenis*: unusual new species isolated from cooling tower water. Appl. Environ. Microbiol. **45:**536–545.
2. **Wilkinson, H. W., B. J. Fikes, and D. D. Cruce.** 1979. Indirect immunofluorescence test for serodiagnosis of Legionnaires disease: evidence for serogroup diversity of Legionnaires disease bacterial antigens and for multiple specificity of human antibodies. J. Clin. Microbiol. **9:**379–383.

Legionellosis in Southern Africa

ALFRED C. MAUFF AND HENDRIK J. KOORNHOF

South African Institute for Medical Research, Johannesburg, South Africa

A Legionellosis Research Laboratory was established at the South African Institute for Medical Research in 1979 to provide reliable laboratory service for the diagnosis of Legionnaires disease and to determine its prevalence in South Africa. In 1980, 12 cases of Legionnaires disease were confirmed in our laboratory (2). Since then extensive use has been made of the indirect fluorescent-antibody (IFA) test for Legionnaires disease in the Johannesburg area and to a limited extent elsewhere in South Africa. The antigen first used was a heat-killed suspension of *Legionella pneumophila* serogroup 1 in an egg yolk diluent supplied by the Centers for Disease Control, Atlanta, Ga., but this was later replaced by a polyvalent antigen (Centers for Disease Control) consisting of *L. pneumophila* serogroups 1 to 4. Facilities were also available for the direct fluorescent-antibody test, initially for *L. pneumophila* serogroups 1 to 4 only (polyvalent pool A) but subsequently for *L. pneumophila* serogroups 5 and 6, *Legionella dumoffii*, and *Legionella longbeachae* serogroup 1 (polyvalent pool B) and *Legionella micdadei*, *Legionella bozemanii*, *Legionella gormanii*, and *Legionella longbeachae* serogroup 2 (polyvalent pool C). Relatively few specimens were suitable for culture, and initial attempts to isolate legionellae from patients were unsuccessful. We present here an analysis of some of our findings, emphasizing the problems we experienced in the interpretation of IFA results. We also report the isolation of legionellae from human and environmental samples in Johannesburg and Zambia.

Laboratory investigations of clinical cases. During 1982, serum specimens from 2,261 patients were tested for Legionnaires disease, and 225 (9.9%) of these had IFA titers of ≥256. The seropositive individuals were categorized into three groups, consisting of 9 patients who showed at least a fourfold rise in titer with final titers of ≥128 on two or more occasions, 99 patients with repeated titers of ≥256, and 126 patients with a single titer of ≥256. The antigen used in these investigations included *L. pneumophila* serogroups 1 to 4. In contrast to these findings, of 478 patients there were 25 seropositive reactors (5.2%) with titers of ≥256 in 1981 when *L. pneumophila* serogroup 1 (Philadelphia) was used as antigen. Of the 478 patients, 186 (38.9%) had titers of <16, and 83 (17.4%), 90 (18.8%), and 64 (13.4%) had titers of 16, 32, and 64, respectively.

Not infrequently, difficulties were encountered with the interpretation of the IFA results. To illustrate some of the problems, the clinical picture and laboratory results of 48 patients on whom adequate clinical and radiological data were available were analyzed. Based on this information there were 16 cases of Legionnaires disease with incontrovertible laboratory evidence of the disease. *L. pneumophila* serogroups 5 and 6, respectively, were cultured from lung tissue at autopsy from one patient and from a bronchial aspirate of another. The disease was confirmed by direct fluorescent-antibody testing

TABLE 1. Features of positive water samples

Sample	Guinea pig inoculation	Acid treatment and culture on BCYE[a] + antibiotics	Site	Organism
Winter survey				
Sample 8	+	ND[b]	Hotel	*L. pneumophila* serogroup 1
Summer survey				
Sample 1	−	+	Office building	*L. pneumophila* serogroup 1
Sample 5	+	+	Hotel	*L. pneumophila* serogroup 1
Sample 6	+	+	Cinema	*L. pneumophila* serogroup 1
Sample 11	ND	+	Cinema	*L. pneumophila* serogroup 6
Zambia, 1 to 4	ND	+	Water tanks	*L. longbeachae* serogroup 1
Hospital, 1 to 4	ND	+	Hot-water system	*L. pneumophila* serogroup 1

[a] BCYE, Buffered charcoal-yeast extract agars.
[b] ND, Not done.

in another five cases, two at autopsy, while nine patients, as indicated above, showed seroconversion with final titers of ≥128. All of these patients had pneumonia, and the clinical findings were compatible with legionellosis.

In a second category, 18 of the 48 patients were regarded as clinically probable cases. All of these patients had pneumonitis and titers of ≥256 on more than one occasion and met at least two of the following five diagnostic criteria: typical X-ray appearance (pneumonia), fever, leukocytosis of ≥10,000/μl, no confirmed alternative diagnosis, and response to erythromycin or rifampin therapy.

A third group of six patients presented with atypical clinical features. These had X-ray pictures compatible with Legionnaires disease but met only one or none of the above-mentioned criteria, other than radiological evidence of pulmonary consolidation. Five of these patients had chronic illnesses without evidence of acute episodes, and three were confirmed cases of pulmonary tuberculosis. One patient had disease resembling tuberculosis clinically and radiologically, but mycobacteria were not found on microscopy nor were they cultured. The fifth patient had chronic infiltrates of unknown etiology in the bases of both lungs.

The remaining eight patients were unlikely to have had recent legionellosis. Two of these patients had confirmed tuberculosis, while another four had disease resembling pulmonary tuberculosis clinically and radiologically but not confirmed bacteriologically. One of the patients with established tuberculosis had a recent acute episode of pneumonia, but the titers were not compatible with the duration of the illness. This was also the case with another patient who was previously healthy, while a third patient who also suffered from tuberculosis had laboratory-confirmed viral pneumonia.

The association of pulmonary tuberculosis with positive Legionnaires disease serology requires further investigation. Of our 48 patients, 11 had either confirmed tuberculosis (6 patients) or radiological evidence suggestive of tuberculosis (5 patients). Five of the confirmed tuberculosis patients either were considered to be unlikely to have suffered from recent legionellosis (two cases) or had atypical features. All of these patients were in the same hospital, and the possibility of nosocomially acquired Legionnaires disease cannot be excluded. The hospital concerned serves as a referral hospital for patients with tuberculosis and undiagnosed chronic pulmonary infections, and IFA testing for Legionnaires disease is performed routinely in problem cases.

Environmental studies. During the winter of 1981 and summer of 1981 to 1982, water samples from cooling towers in buildings in Johannesburg were examined for legionellae. For both winter and summer surveys, 5-liter samples were filtered under positive pressure through a 0.65-μm membrane filter (Millipore Corp., Bedford, Mass.). The filter was then emulsified in 40 ml of the filtrate and centrifuged at 1,000 rpm for 10 min. The supernatant was subsequently centrifuged for 90 min at $10,000 \times g$, and the pellet was suspended in 4 ml of supernatant. The concentrated specimens were examined by direct fluorescent-antibody testing, and positive samples with microscopic counts of ≥10^5/ml were initially, during the winter survey, inoculated intraperitoneally into guinea pigs and also cultured directly on a variety of media. Among these were a selective medium containing cephalothin, colistin, vancomycin, and amphotericin B (rather than cycloheximide) after low-pH treatment of the samples (1) and a medium containing glycine (3) plus the above-mentioned antibiotics. As a result of a small outbreak of putative legionellosis in a Zambian copper mine, water samples from storage tanks were also investigated. Late in 1982 samples from the hot-water reticulation system of a Johannesburg hospital, including delivery points, were examined for legionellae. The results of these studies are summarized in Table 1.

The studies in Johannesburg suggest that there is a high prevalence of legionellae in the environment, and this is reflected in the high rate of seropositive patients in the region. The relatively large number of individuals with low-titer positive results and others with titers within the diagnostic range but with atypical clinical features suggests that in addition to typical confirmed cases of Legionnaires disease, many subclinical cases occur in the community.

LITERATURE CITED

1. **Bopp, C. A., J. W. Sumner, G. K. Morris, and J. G. Wells.** 1981. Isolation of *Legionella* spp. from environmental water samples by low-pH treatment and use of a selective medium. J. Clin. Microbiol. **13:**714–719.
2. **Koornhof, H. J., and S. Zwi.** 1980. Legionnaires' disease in South Africa (editorial). S. Afr. Med. J. **58:**1–2.
3. **Wadowsky, R. M., and R. B. Yee.** 1981. Glycine-containing selective medium for isolation of *Legionellaceae* from environmental specimens. Appl. Environ. Microbiol. **42:**768–772.

Legionellosis in Italy

DONATO GRECO, FRANCESCO ROSMINI, AND MADDALENA CASTELLANI PASTORIS

Istituto Superiore di Sanità, Rome, Italy

The presence of legionellosis in Italy has been documented since June 1973, when a Dutch citizen became ill after having travelled in Italy (3). Later, other foreign tourists became ill during or immediately after their stay in Italy. Thus, Italian laboratories as well as various foreign laboratories became interested in looking for legionellosis in Italy (1).

A total of 108 cases of legionellosis were reported from January 1979 to September 1982. Laboratory diagnosis was available for 76 (70%) of the 108 cases (Table 1); 32 cases were diagnosed solely on a clinical basis.

The laboratory diagnosis was based on the indirect fluorescent-antibody test in 74 cases and on direct immunoflorescence staining of lung tissue in 2 cases. Sixteen cases diagnosed clinically were suspected to be part of epidemic clusters; 16 other cases were diagnosed abroad, and titer information was unavailable. *Legionella pneumophila* serogroup 1 was the organism most frequently found in tested cases (98%); in only one case was serogroup 3 detected. Fourteen cases showed seroconversions to a titer of ≥128, and in six cases there was a fourfold fall in titer. In 54 cases the single serum titer was ≥128; of these 27 (50%) had a titer of ≥256.

Diagnostic reagents were obtained from the Centers for Disease Control, Atlanta, Ga., and from CDSC, London, England. Diagnostic Italian antigen was available only from January 1983.

Clinical features were known for 86 cases (80%); 83 (96%) of these patients had pulmonary symptomatology. Four patients (4.6%) died. Seventy-seven patients (89%) required hospital-ization, and one patient developed the disease during hospitalization.

Age and sex distributions are shown in Fig. 1. The sex ratio (male/female) was 1.5:1. The relatively high proportion of children under 10 years of age is a result of specific attention to age group (G. Nigro, M. Castellani Pastoris, M. Fantasia Mazzotti, and M. Midulla, Pediatrics, in press, and unpublished data).

A total of 43 cases (40%) were clustered in several epidemic groups with 2 or more cases per group; other cases were classified as sporadic, and 51% of the 108 persons had stayed in an Italian hotel within the incubation time.

From 1978 to 1982, seven epidemic clusters were reported in Italy. The first epidemic of legionellosis in Italy was reported by Danish health authorities concerning 10 cases, serologically confirmed, in a group of Danish tourists who stayed in a hotel on the Lake of Garda (Veneto) in July 1978 (2). In 1979 English health authorities reported on two cases in English citizens who had been in the town of Cattolica (Romagna) and had stayed in the same hotel during August and September 1979. Five other clusters occurred in the town of Lido di Savio (Romagna); four of these occurred in two hotels in two different years, with a total of 41 cases and two deaths (4; F. Rosmini, M. Castellani Pastoris, M. Fantasia Mazzotti, F. Forastiere, A. Gavazzoni, D. Greco, G. Ruckdeschel, E. Tartagni, A. Zampieri, and W. B. Baine, Am. J. Epidemiol., in press). *L. pneumophila* serogroup 1 was isolated from shower heads of one hotel, and the same serogroup was identified by the direct fluorescent-antibody test in samples

TABLE 1. Laboratory evaluations of 108 cases of legionellosis

Type of case	Source of information[a]	No. with:					
		Titer by IFA test[b]:				Positive DFA test[c]	UK[d]
		Seroconversion to ≥128	Fourfold fall	Single or persistent			
				≥256	128		
Sporadic	ISS	11	5	8	21	1	0
	CDSC	0	0	0	0	1	12
	Other	1	1	4	0	0	0
Cluster or epidemic	ISS	1	0	13	6	0	16
	CDSC	1	0	2	0	0	4

[a] ISS, Istituto Superiore di Sanità; CDSC, Communicable Disease Surveillance Centre.
[b] Reciprocal of serum antibody titer to *L. pneumophila*. IFA, Indirect fluorescent antibody.
[c] DFA, Direct fluorescent antibody.
[d] Unknown or not performed.

taken from different points of the hotel water system, including a cistern receiving water from a local well. The presence of legionellae in the water system was documented after intensive chlorination of the system.

To date not one of the 21 Italian regions can be considered free from legionellosis; in fact, reports on cases were obtained from all regions. Two human isolates of *L. pneumophila* and *Legionella micdadei* were obtained by D. Fumarola in Bari from bronchial aspirates of two patients.

In January 1983 a national surveillance program on legionellosis was implemented, with the following objectives: (i) to define the clinical and epidemiological features of the disease in Italy, (ii) to define the role of legionellosis in acute pulmonary pathology, (iii) to define the environmental behavior of the microorganism and test effective control measures, and (iv) to develop knowledge of the disease within the national health service. The program is coordinated by the Istituto Superiore di Sanità, Rome. The main tools of the surveillance activity are (i) a network of about 50 laboratories receiving the indirect fluorescent-antibody test antigen prepared and distributed by the national reference laboratory, (ii) the availability of a national epidemiological team for investigations of clusters, and (iii) a notification system based on a national legionellosis form and on periodic data feedback through the national communicable disease surveillance system.

LITERATURE CITED

1. Castellani Pastoris, M., M. Fantasia Mazzotti, F. Mondello, L. Bonazzi, F. Ingrao, P. Serra, and G. Stagni. 1981. Antibody reacting with *Legionella pneumophila* in sera of Italian patients with respiratory illness of unknown cause. Microbiologica 4:205–214.
2. Jorgensen, K. A., B. Korsager, G. Johannsen, L. G. Freund, and H. W. Wilkinson. 1981. Legionnaires' disease imported to Denmark from Italy. Scand. J. Infect. Dis. 13:133–136.
3. Meenhorst, P. L., J. W. M. van der Meer, and J. Borst. 1979. Sporadic cases of Legionnaires' disease in the Netherlands. Ann. Intern. Med. 90:529–532.
4. Rosmini, F., D. Greco, M. Castellani Pastoris, and A. Zampieri. 1983. Legionellosi a Lido di Savio: materiali da 18 mesi di sorveglianza (agosto 1981–febbraio 1983). Rapp. ISTISAN 1983/7.

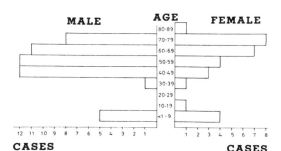

FIG. 1. Age and sex distribution of cases.

Distribution of *Legionella* Species in Japan and Cases of Legionnaires Disease

ATSUSHI SAITO, NAOMI ITO, KIYO FUJITA, YOJI SUZUYAMA, KEIZO YAMAGUCHI, AND KOHEI HARA

Second Department of Internal Medicine, Nagasaki University School of Medicine, Nagasaki 852, Japan

To study the distribution of *Legionella* species from environmental sources in Japan, we studied 408 water samples, including water from 343 cooling towers from sites widely distributed over Japan and 25 samples of paddy field water, 29 samples of river water, and 11 samples of lake, fountain, puddle, and shower water from the Nagasaki area. A 500-ml sample was centrifuged at 10,000 rpm for 20 min. The sediments were cultured on a selective medium (1) for isolation of *Legionella* species after low-pH treatment. In some negative samples, 3 ml of the sediments was inoculated intraperitoneally into guinea pigs, which were sacrificed 4 days later. A peritoneal swab and homogenate of spleen were inoculated onto buffered charcoal-yeast extract agar. In addition, 200-g samples of soil were suspended in 400 ml of distilled water with 0.5% of Tween 60 for 20 min. The supernatant was centrifuged at 1,000 rpm for 10 min, removed, and centrifuged again at 3,000 rpm for 10 min, and the sediments were used for isolation of *Legionella* species.

A total of 166 strains of *Legionella* spp. were isolated from 126 (36.7%) of 343 cooling-tower samples. The origin of positive samples ranged from Hokkaido (northern Japan) to Okinawa (southern Japan) (Table 1). *L. pneumophila* serogroup 1 was the most frequent isolate (110 strains), and *Legionella*-like organisms were second in frequency (27 strains). We used direct fluorescent-antibody (DFA) reagents for *L. pneumophila* serogroups 1 to 6, *L. micdadei*, *L. bozemanii*, *L. dumoffii*, *L. gormanii*, and *L.*

longbeachae serogroup 1; the reagents were obtained from the Centers for Disease Control. *L. pneumophila* serogroups 7 and 8, *L. longbeachae* serogroup 2, *L. jordanis*, *L. wadsworthii*, and *L. oakridgensis* were not examined.

No *Legionella* species were isolated from water samples of paddy fields (25 samples), rivers (29 samples), lakes (3 samples), fountains (1 sample), wells (4 samples), or showers (1 sample). In the soil, 3 of 13 samples from construction sites and 2 of 3 samples from river banks were positive. All eight strains isolated were *L. pneumophila*, including serogroup 1 (five strains), serogroup 4 (one strain), and serogroup 5 (two strains). *L. pneumophila* and *L. bozemanii* were widely distributed over Japan, but we could not obtain *L. micdadei*, *L. dumoffii*, *L. gormanii*, and *L. longbeachae* serogroup 1. Yield was highest in urban locations (Tokyo, Osaka, etc.) and lowest in rural areas. There was no relationship between ambient temperature and culture yield. The main reason for not isolating *Legionella* species was considered to be overgrowth of other bacteria or fungi.

In a study of Legionnaires disease in Japan, five cases of pneumonia were confirmed by isolation of *L. pneumophila* from pleural effusions of the patients. The four cases diagnosed by us in the Nagasaki area were all caused by *L. pneumophila* serogroup 1, and the remaining case, in Tokyo Metropolitan Hospital, was caused by serogroup 3.

We studied nosocomial Legionnaires disease by DFA staining. A total of 362 of 1,488 cases

TABLE 1. Distribution of *Legionella* species from cooling towers in Japan

Area	No. of samples	No. (%) positive	*L. pneumophila* serogroup:						*L. bozemanii*	LLO[a]
			1	2	3	4	5	6		
Hokkaido	12	3 (25.0)	2	0	0	1	0	0	0	0
Tohoku	30	4 (13.3)	4	0	0	0	0	0	0	0
Kanto	43	22 (51.2)	19	0	0	3	0	0	3	8
Chubu	29	11 (37.9)	10	0	0	2	0	0	1	0
Kinki	34	14 (41.2)	13	0	0	0	0	0	0	3
Chugoku	38	9 (23.7)	8	0	0	0	0	0	0	1
Shikoku	26	11 (42.3)	9	0	0	1	0	0	1	6
Kyushu	131	52 (39.7)	45	0	2	5	0	3	7	9
Total	343	126 (36.7)	110	0	2	12	0	3	12	27

[a] LLO, *Legionella*-like organisms.

autopsied in six general hospitals had bacterial pneumonia by histological examination, and 196 of these cases were diagnosed as fatal bacterial pneumonia by both histological and clinical findings. These cases were studied retrospectively by DFA staining of Formalin-fixed lung tissues (7 to 10 different specimens in each case). Forty-two cases (22%) were positive in DFA staining with reagents for *L. pneumophila*, serogroups 1 to 6. The ratio of male to female subjects was 4:1, and the average age was 62.3 years. In the remaining 154 subjects without Legionnaires disease, the ratio of male to female was 3:1, and the average age was 62.2 years. There was no seasonal trend in occurrence of either Legionnaires disease or the other fatal bacterial pneumonias. The distribution of *L. pneumophila* serogroups in these 42 cases is shown in Table 2. *L. pneumophila* serogroup 1 was the most frequent pathogen (29 cases, 69.0%), followed by serogroup 4 (9 cases, 21.4%). There were no lung

TABLE 2. Serogroup distribution of *L. pneumophila* in 42 cases of nosocomial Legionnaires disease

L. pneumophila serogroup	No. of cases (%)
1	29 (69.0)
2	2 (4.8)
3	1 (2.4)
4	9 (21.4)
5	0 (0)
6	1 (2.4)

tissue specimens which had positive DFA staining to multiple serogroups, but four of nine cases with positive DFA staining for serogroup 4 showed positive DFA staining for *L. micdadei*. We could not confirm that these cases involved two kinds of organisms.

Cough, sputum, chest pain, and dyspnea were commonly observed in these patients, but high fever (over 40°C) and relative bradycardia were not characteristic signs (6.4 and 10.6%, respectively). Hypoxemia with hyperventilation and respiratory alkalosis were seen in early stages of the disease despite small shadows on chest X ray. The main underlying diseases of the 42 patients were cancer (eight cases of primary lung cancer, five cases of stomach cancer, three cases of gall bladder cancer, and eight cases of other cancers), followed by lymphoma (four cases), cerebrovascular disorder (three cases), heart failure (two cases), and diabetes mellitus (two cases). Chest X-ray findings were analyzed in 25 of 42 cases; a noncircumscribed homogenous shadow was seen in 17 cases. Pleural effusion was seen in 13 (75%) cases; this was on the left in 8 cases, on the right in 5 cases, and on both sides in 5 cases.

LITERATURE CITED

1. **Bopp, C. A., J. W. Sumner, G. K. Morris, and J. G. Wells.** 1981. Isolation of *Legionella* spp. from environmental water samples by low-pH treatment and use of a selective medium. J. Clin. Microbiol. **13**:714–719.
2. **Edelstein, P. H., and S. M. Finegold.** 1979. Use of a semiselective medium to culture *Legionella pneumophila* from contaminated lung specimens. J. Clin. Microbiol. **10**:141–143.

Seroepidemiological Survey of Irrigation Workers in Israel and Isolation of *Legionella* spp. in the Environment

HERVÉ BERCOVIER, MICHELE DERAI-COCHIN, JONI SHERMAN, BADRI FATTAL, AND HILLEL SHUVAL

Department of Clinical Microbiology and Environmental Health, Hebrew University Hadassah Medical School, Jerusalem, Israel

Previously we showed that patients with waterborne leptospirosis (i.e., *Leptospira icterohemorrhagiae*) had a higher prevalence of *Legionella* antibodies than did patients with pneumonic leptospirosis (1a). This finding suggested that *Legionella* infections linked to occupational activities could be frequent. In the present study, *Legionella* antibodies in the sera of irrigation workers in Israel were measured to evaluate the potential risk for legionellosis in this population. The irrigation workers in this investigation were exposed to water aerosolized from sprinkler irrigation and were selected from rural populations living in 30 different agricultural communi-

ties in various regions of Israel. The irrigation workers could be separated into two groups; the first group was exposed to aerosolized fresh water or potable water (population C, 81 individuals), whereas the second group was exposed to aerosolized wastewater effluents from oxidation ponds (population A + B, 240 individuals). From the population living in the 30 rural communities, a matched control population (72 individuals) of agricultural workers who did not work with irrigation was also investigated. Fifteen Formalin-killed antigens, suspended in egg yolk sac (3), representing *Legionella pneumophila* serogroups 1 to 8, *Legionella bozemanii*,

Legionella gormanii, Legionella micdadei, Legionella jordanis, Legionella dumoffii, Legionella longbeachae serogroup 1, and *Legionella oakridgensis*, were used to perform the serological tests by the indirect immunofluorescent-antibody technique. With our procedure, a reciprocal titer of 32, equivalent to a titer of 128 to 256 by the procedure of the Centers for Disease Control, was chosen as the cutoff between positive and negative results. The serological survey (Table 1) showed that 12.5% of the irrigation workers exposed to wastewater aerosols (population A + B) had *Legionella* antibodies at a titer of ≥32, whereas 7.2% of the irrigation workers exposed to aerosols of clean water (population C) and 1.4% of the matched control population had antibodies at this titer. The difference in percentages between population A + B and the control population was statistically significant.

Two-thirds of the positive individuals had antibodies directed against more than one antigen of *Legionella* spp. However, if we consider only the antigen against which each individual had the highest antibody titer, 9.5% of population A + B reacted with *L. pneumophila* (serogroups 1 to 8), and 3% reacted with one of the seven other *Legionella* species investigated. In population C, the corresponding percentages were, respectively, 1.2 and 6.1%. Unfortunately, the small size of population C (82 individuals) does not allow any conclusion about the differences of specific antibody prevalence between the two populations. If we combine the serological data on these two populations, 70% of the positive individuals had their highest antibody titer directed against *L. pneumophila*, whereas 30% were directed against the seven other *Legionella* species. These data, which were obtained in a nonepidemic situation from a young population (predominantly under 40 years of age) exposed to aerosols, are unique and cannot be compared with previously published data. The serologically positive individuals living in the same community had *Legionella* antibodies directed against the same antigen(s), whether it

was a single antigen or a combination of two or more antigens.

These clusters of positive individuals suggested a common source of contamination, unique for each community. Therefore, 28 samples of water used for irrigation and 28 samples of drinkable water were collected and concentrated by filtration when necessary. These water samples were first examined by direct fluorescence assay (DFA), using labeled antibodies (pools A, B, and C) provided by the Centers for Disease Control. Of 28 samples of water used for irrigation, 6 were DFA positive whereas only 4 samples of drinkable water were DFA positive. The water samples were further treated by the acid treatment described by Bopp et al. (1), the antibiotic treatment described by Thorpe and Miller (4), and a combination of the two treatments. The treated samples, diluted or not, were plated on buffered charcoal-yeast extract agar and on glycine-vancomycin-polymyxin agar (4a). Suspected *Legionella* colonies growing on these media were picked and further examined according to recognized procedures (2, 5). No presumptive *Legionella* colonies grew from DFA-negative water samples or from the four DFA-positive drinkable-water samples. One strain of *L. pneumophila* serogroup 4 and five organisms resembling *Legionella* spp. were isolated from one oxidation pond of the six DFA-positive irrigation water samples. This is the first isolation and direct proof of the presence of *L. pneumophila* in Israel. A retrospective study based on medical files showed that 2 of 37 individuals with *Legionella* antibodies had developed in the past 2 years a pneumonia that responded to erythromycin treatment without hospitalization. The results of this serological survey and the isolation of *Legionella* spp. from oxidation ponds rich in cyanobacteria indicate the potential risk of spreading *Legionella* spp. by land application of wastewater. Because of their structures (well-defined environment, closed communities), these rural communities might be the perfect setting for study to better understand the full clinical spectrum of legionellosis in a nonepidemic situation.

This research was funded in part by the U.S. Environmental Protection Agency under cooperative agreement no. CR806416 to the Hebrew University.

TABLE 1. *Legionella* antibody titers in a rural population[a]

Reciprocal titer	% of population:		
	A + B (n = 240)	C (n = 81)	Control (n = 72)
≥32	12.5	7.2	1.4
≥64	9.1	4.8	1.4
≥128	2.5	2.4	0

[a] Fifteen antigens representing eight *Legionella* species were used for the indirect fluorescence assay. These antigens were Formalin killed and suspended in normal egg yolk sac.

LITERATURE CITED

1. **Bopp, C. A., J. W. Sumner, G. K. Morris, and J. G. Wells.** 1981. Isolation of *Legionella* spp. from environmental water samples by low-pH treatment and use of a selective medium. J. Clin. Microbiol. **13:**714–719.
1a. **Dournon, E., and H. Bercovier.** 1980. Maladie des légionnaires et leptospirose. Parentés épidémiologiques ou antigéniques. Med. Mal. Infect. **10:**418–421.
2. **Hébert, G. A.** 1981. Hippurate hydrolysis by *Legionella pneumophila*. J. Clin. Microbiol. **13:**240–242.
3. **Lattimer, G. L., and B. A. Cepil.** 1980. Effect of an antigen

preparation on specificity and sensitivity of the indirect fluorescent antibody test. J. Clin. Pathol. **33**:585–590.

4. **Thorpe, T. C., and R. D. Miller.** 1980. Negative enrichment procedure for isolation of *Legionella pneumophila* from seeded cooling tower water. Appl. Environ. Microbiol. **40**:849–851.

4a.**Wadowski, R. M., and R. B. Yee.** 1981. Glycine-containing

selective medium for the isolation of *Legionellaceae* from environmental specimens. Appl. Environ. Microbiol. **42**:768–772.

5. **Weaver, R. E., and J. C. Feeley.** 1979. Cultural and biochemical characterization of Legionnaires' disease bacterium, p. 19–26. *In* G. L. Jones and G. A. Hébert (ed.), "Legionnaires' ": the disease, the bacterium and methodology. Centers for Disease Control, Atlanta, Ga.

Prevalence of Antibodies to *Legionella* Species in Young Men in Taiwan

MELISSA M. GUO CHEN, YUN RU YANG, AND H. T. SHEN

National Yang Ming Medical College and National Defense Medical College, Taipei, Taiwan, Republic of China

Since the discovery of Legionnaires disease in Philadelphia, Pa., in 1976, knowledge has increased about its causative agents. Owing to the fastidious nature of these organisms, their isolation from clinical specimens is not always successful. Many serological methods have been developed for detection of antibodies to *Legionella pneumophila* to aid in the diagnosis of the disease (2, 6, 10). Although the indirect immunofluorescence test has proved valuable in diagnosing legionellosis (9), it is nevertheless technically difficult and has drawbacks. A microagglutination test was developed in response to the need for a relatively simple and dependable serological test to detect antibodies to the Legionnaires disease organism (2).

Prevalence studies of antibodies to *L. pneumophila* have been carried out in many parts of the world in association with investigation of particular age groups or communities (3, 7, 8). However, antibodies to *Legionella* species have not been reported in Taiwan. Therefore, the present study was undertaken to determine the level of antibodies to various *Legionella* species in apparently normal young men.

Sera were collected aseptically from 256 men, 20 to 30 years old, from different localities in Taiwan. The organisms used to prepare the antigens for the microagglutination test included *L. pneumophila* serogroups 1 to 6 (strains Philadelphia 1, Atlanta 1, Bloomington 2, Los Angeles 1, Cambridge 2, Houston 2), *Legionella micdadei* TATLOCK, *Legionella dumoffii* NY-23, and *Legionella bozemanii* WIGA, all of which were obtained from the Centers for Disease Control, Atlanta, Ga. *Bacteroides fragilis* YM-65 and *Pseudomonas pseudomallei* YM-1 were clinical isolates. The microagglutination technique used was as described by Farshy et al. (2).

The serum specimens were twofold serially diluted by using a 0.025-ml microdiluter (Dynatech Laboratories, Inc.), with dilutions from 1:4

to 1:1,024. U-shaped 96-well microtitration plates were used. Positive and negative control sera were included and diluted as for the test specimens. A control with only antigen and diluent was also included. To the serially diluted serum (0.025 ml in each well), 0.025 ml of properly prepared bacterial antigen was added. The mixture was mixed well, incubated at room temperature (22°C) overnight, and refrigerated at 4°C for 2 h before the result was read. Reading of microagglutination was aided by using a reading mirror. The highest dilution of serum which gave no definite button after reaction with the antigen was taken as the endpoint and the titer of that serum.

To test the specificity of these agglutinins, serum specimens were absorbed with antigens prepared from heat-killed, washed, and lyophilized *B. fragilis* YM-65 and *P. pseudomallei* YM-1. The antibody titers after absorption were then determined as described above.

The distributions of serum microagglutination titers against different *Legionella* species were found to be quite similar, except for some minor differences. Antibodies to six serogroups of *L. pneumophila* were found in titers of ≥1:16 in 15.6 to 44.0% of the sera, and titers of ≥1:32 were found in 1.9 to 8.2%. Antibodies to *L. micdadei* TATLOCK, *L. dumoffii* NY-23, and *L. bozemanii* WIGA were found in titers of ≥1:16 in 19.3 to 43.1% of the sera, and titers of ≥1:32 were found in 2.0 to 12.2%. According to these results, the upper limit of normal titer determined by the microagglutination test in young men in Taiwan may be set at 1:16, since 85% of the normal sera examined did not exceed this titer (5).

It has been suggested that *L. pneumophila* and *P. pseudomallei* share an antigen when tested by a microagglutination technique (4). It has also been reported that *L. pneumophila* and some strains of *B. fragilis* have cross-reactions among 324 isolates of anaerobic and microaerophilic organisms examined (1). Hence, in this study some sera were absorbed with antigens prepared from *P. pseudomallei* and *B. fragilis* to demonstrate the specificity of antibodies to *L. pneumophila*. The results of the serum absorption test with the homologous antigen (*L. pneumophila*) and with the heterologous antigens (*P. pseudomallei* and *B. fragilis*) indicated that the microagglutinating antibodies against *L. pneumophila* Philadelphia 1 (serogroup 1) observed in this investigation are relatively specific. However, since *L. pneumophila* was reported to cross-react with many other infectious agents, the possibility of the existence of other cross-reacting antibodies cannot be ruled out, and the presence of antibody in the absence of disease may be difficult to interpret.

LITERATURE CITED

1. Edelstein, P. H., R. M. McKinney, R. D. Meyer, M. A. C. Edelstein, C. J. Krause, and S. M. Finegold. 1980. Immunologic diagnosis of Legionnaires' disease: cross-reactions with aerobic and microaerophilic organisms and infections caused by them. J. Infect. Dis. 141:652–655.
2. Farshy, C. E., G. C. Klein, and J. C. Feeley. 1978. Detection of antibodies to Legionnaires disease organism by microagglutination and micro-enzyme-linked immunosorbent assay tests. J. Clin. Microbiol. 7:327–331.
3. Helms, C. M., E. D. Renner, J. P. Viner, W. J. Hierholzer, Jr., L. A. Wintermeyer, and W. Johnson. 1980. Indirect immunofluorescence antibodies to *Legionella pneumophila*: frequency in a rural community. J. Clin. Microbiol. 12:326–328.
4. Klein, G. C. 1980. Cross-reaction to *Legionella pneumophila* antigen in sera with elevated titers to *Pseudomonas pseudomallei*. J. Clin. Microbiol. 11:27–29.
5. Klein, G. C., W. L. Jones, and J. C. Feeley. 1979. Upper limit of normal titer for detection of antibodies to *Legionella pneumophila* by the microagglutination test. J. Clin. Microbiol. 10:754–755.
6. Lennette, D. A., E. T. Lennette, B. B. Wentworth, M. L. V. French, and G. L. Lattimer. 1979. Serology of Legionnaires disease: comparison of indirect fluorescent antibody, immune adherence hemagglutination, and indirect hemagglutination tests. J. Clin. Microbiol. 10:876–879.
7. Macrae, A. D., P. N. Appleton, and A. Laverick. 1979. Legionnaires' disease in Nottingham, England. Ann. Intern. Med. 90:580–583.
8. Storch, G., P. S. Hayes, D. L. Hill, and W. B. Baine. 1979. Prevalence of antibody to *Legionella pneumophila* in middle-aged and elderly Americans. J. Infect. Dis. 140:784–787.
9. Wilkinson, H. W., D. D. Cruce, and C. V. Broome. 1981. Validation of *Legionella pneumophila* indirect fluorescence assay with epidemic sera. J. Clin. Microbiol. 13:139–146.
10. Wilkinson, H. W., C. E. Farshy, B. J. Fikes, D. D. Cruce, and L. P. Yealy. 1979. Measure of immunoglobulin G-, M-, and A-specific titers against *Legionella pneumophila* and inhibition of titers against nonspecific, gram-negative bacterial antigens in the indirect immunofluorescence test for legionellosis. J. Clin. Microbiol. 10:685–689.

Prevalence of Antibody to Various *Legionella* Species in Ill and Healthy Populations in Michigan

BERTTINA B. WENTWORTH, WILLIAM A. CHADWICK, HARLAN E. STIEFEL, and DALICE S. BENGE

Michigan Department of Public Health, Lansing, Michigan 48909, and Michigan State University, East Lansing, Michigan 48824

An indirect fluorescent-antibody technique using Formalin-fixed antigen (3, 10) was used to detect antibody to *Legionella* species in sera submitted for the diagnosis of pneumonia during the period September 1980 through March 1983. Sera from a total of 5,682 patients were tested for antibody to *L. micdadei*, *L. bozemanii*, and all six serogroups of *L. pneumophila*. *L. dumoffii* and *L. gormanii* antigens were added in January 1981 and used to test 4,827 of these sera. Only 3,159 sera were tested with *L. longbeachae* serogroups 1 and 2, added in January 1982, and only 1,611 were tested with *L. jordanis* antigen, added in October 1982. The results were analyzed for changes in antibody prevalence over time and for relationship to cases diagnosed by either isolation of an agent or rises in antibody titer. Only 304 cases were detected among these patients, so that antibody prevalence was based almost entirely upon reactivity in early, acute-phase sera, indicative of past contact with these antigens. Overall, sera from 42.6% of patients showed reactivity to one or more *Legionella* antigens at a dilution equal to or greater than 1:8. Studies of the specificity of this technique did not demonstrate the same cross-reactivity between species that has been found among serogroups (8).

Antibody to *L. bozemanii* was most prevalent (21.1%), and *L. bozemanii* infection was second most frequent among cases (Table 1). Otherwise, there was little relationship between prevalence of antibody and frequency of cases by species. *L. pneumophila* serogroup 1 (S1), which was responsible for more than one-third of the cases, had a prevalence rate of only 7.9%. It appeared that many low-grade or inapparent infections may occur with various *Legionella* species without a corresponding number of diagnosed cases of more severe illness.

Antibody prevalence tended to peak in either fall or spring months, except for antibodies to *L.*

TABLE 1. Prevalence of *Legionella* antibodies and distribution of 304 cases

Antigen	No. (%) of:		
	Sera	Reactive sera	Cases
L. pneumophila	5,682		
S1		447 (7.9)	116 (39.2)
S2		8 (0.1)	1 (0.3)
S3		69 (1.2)	16 (5.4)
S4		127 (2.2)	1 (0.3)
S5		123 (2.2)	4 (1.3)
S6		403 (7.1)	15 (5.1)
L. bozemanii	5,682	1,197 (21.1)	81 (26.7)
L. micdadei	5,682	633 (11.1)	20 (6.4)
L. dumoffii	4,827	661 (13.7)	10 (3.0)
L. gormanii	4,827	857 (17.8)	13 (3.7)
L. longbeachae	3,159		
S1		314 (9.9)	6 (2.0)
S2		553 (17.5)	21 (6.9)
L. jordanis	1,611	88 (5.5)	0

pneumophila S2, S3, and S4, which were found at low prevalence with no particular peak activity. *L. pneumophila* S1 showed peak prevalence in November 1980, as did *L. bozemanii*. *L. pneumophila* S5 and S6 both showed peaks in March 1981, while *L. dumoffii* and *L. gormanii* peaked in the winter of 1981 to 1982.

An unusually high prevalence of antibody to *L. bozemanii* was seen in November 1980, and 15 cases of *L. bozemanii* infection were detected that month. Both the prevalence and the number of cases were significantly higher than for any other month during the study period ($P < 0.001$) (7). Antibody prevalence rose from 12% (23 of 191) in September to 20% (50 of 248) in October and peaked at 52.8% (67 of 127) in November. Thereafter, prevalence declined to 27% (46 of

222) in January 1981 and continued at a level between 25 and 35% through June.

Since these sera were all from patients with pneumonia, they represented a biased sample in relation to the general population. To determine whether this peak of activity would occur at the same time in subsequent years and be reflected in sera from presumably healthy individuals, we tested 1,159 sera in 1981 by an indirect hemagglutination procedure (3, 10) for antibodies to *L. bozemanii* and *L. pneumophila* S1. For this study, premarital and preemployment sera were selected from among nonreactive sera submitted for syphilis serology. These sera were tested at an initial dilution of 1:16.

There was a significant increase in prevalence to *L. bozemanii* among sera collected in the fall (October and November) of 1981 (53.4%) versus sera collected earlier (January and February) that year (44.8%). This was a higher prevalence than that seen in sera from pneumonia patients for that year, which varied from 20 to 35% during the year.

Among the 1,159 survey sera, antibody prevalence to *L. bozemanii* was significantly higher in age groups from 15 to 49 years (53 to 59%) than in sera from older groups (31 to 39%). Prevalence was higher in females aged 20 to 49 years (63 to 71%) than in males of the same age range (41 to 56%) for fall sera but not for the sera from the previous winter months.

Immunofluorescence (5–8) and other immunological studies (2, 4, 9) of *Legionella* species have found these species to be antigenically distinct, except for a partial serological relationship between *L. bozemanii* and *L. jordanis* (1). Although previous cross-absorption studies had not shown any cross-reactivity between *L. boze-*

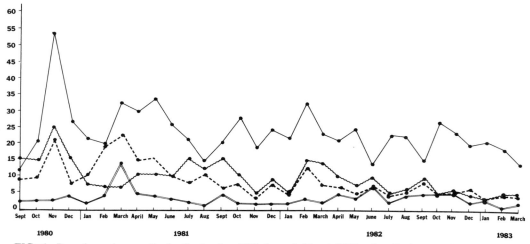

FIG. 1. Prevalence by month, for September 1980 through March 1983, of antibody to *L. bozemanii* (——) and to *L. pneumophila* S1 (∿), S5 (═), and S6 (------).

manii and other *Legionella* species (8), further cross-absorptions were done to determine the extent, if any, of cross-reactivity between *L. bozemanii* and *L. pneumophila* S1. There appeared to be some cross-reactivity between these two species, but not a cross which would influence *L. bozemanii* antibody prevalence. A one-way cross was found, in that absorption with S1 antigen of sera reactive to both species removed only homologous S1 activity and did not affect *L. bozemanii* titers, while absorption of the same sera with *L. bozemanii* antigen removed *L. pneumophila* S1 antibody as well as homologous antibody from two of six sera absorbed. The same one-way cross was shown by Cherry et al. (1) for *L. bozemanii* and *L. jordanis*, where absorption with *L. bozemanii* reduced *L. jordanis* titers.

Since 21.1% of these sera showed antibody to *L. bozemanii*, cross-reactivity may account for some of the reactivity to *L. jordanis* (5.5%) or to *L. pneumophila* S1 (7.9%), particularly in 1980. However, the curves for prevalence of reactivity to *L. bozemanii* and *L. pneumophila* S1 do not coincide after January 1981 (Fig. 1). The *L. bozemanii* prevalence appears to be specific for this species, since 85% of the sera reactive with this antigen did not react with *L. pneumophila* S1, and only 19.5% of the sera reactive with *L. bozemanii* were also reactive with *L. jordanis*.

Among the sera tested in November 1980, 85% (48 of 67) of the sera reactive with *L. bozemanii* were not reactive with *L. pneumophila* S1 antigen. These 48 sera alone constitute a significant increase over the prevalence of sera reactive to *L. bozemanii* in other months of the study. Therefore, we concluded that an outbreak of infection due to *L. bozemanii* occurred in Michigan during the fall of 1980 and was not duplicated during the succeeding 2 years, although sufficient infections occurred in 1981 to increase antibody prevalence. This increase occurred predominantly in females aged 20 to 49 years. There appears to be no obvious explanation for the occurrence of antibody predominantly in this segment of the population. However, there was a change in the male/female ratio among cases due to *L. bozemanii* during 1981. Cases in 1980 had a male/female ratio of 1.58:1, and those in 1982 had a ratio of 1.42:1. But this ratio was 1:1.10 for cases occurring in 1981, suggesting that the increased prevalence of antibody seen in female sera for that year was accompanied by an increase in more serious disease in females. This reversal in male/female ratio among 1981 cases was not seen for cases due to *L. pneumophila* S1, where the ratio for all three years was equal to or greater than 1.25:1.

There was no geographical concentration of either female or male cases for any year to suggest the occurrence of a localized outbreak which might account for an increase in antibody prevalence.

LITERATURE CITED

1. **Cherry, W. B., G. W. Gorman, L. H. Orrison, C. W. Moss, A. G. Steigerwalt, H. W. Wilkinson, S. E. Johnson, R. M. McKinney, and D. J. Brenner.** 1982. *Legionella jordanis*: a new species of *Legionella* isolated from water and sewage. J. Clin. Microbiol. **15**:290–297.
2. **Cordes, L. G., R. L. Myerwitz, A. W. Pasculle, L. Corcoran, T. A. Thompson, G. W. Gorman, and C. M. Patton.** 1981. *Legionella micdadei* (Pittsburgh pneumonia agent): direct fluorescent-antibody examination of infected human lung tissue and characterization of clinical isolates. J. Clin. Microbiol. **13**:720–722.
3. **Edson, D. C., H. E. Stiefel, B. B. Wentworth, and D. L. Wilson.** 1979. Prevalence of antibodies to Legionnaires' disease. A seroepidemiologic survey of Michigan residents using the hemagglutination test. Ann. Intern. Med. **90**:691–693.
4. **Lewellan, K. R., R. M. McKinney, D. J. Brenner, C. W. Moss, D. H. Dail, B. M. Thomason, and R. A. Bright.** 1979. A newly identified bacterium phenotypically resembling, but genetically distinct from, *Legionella pneumophila*: an isolate in a case of pneumonia. Ann. Intern. Med. **91**:831–834.
5. **McKinney, R. M., R. K. Porschen, P. H. Edelstein, M. L. Bissett, P. P. Harris, S. P. Bondell, A. G. Steigerwalt, R. E. Weaver, M. E. Ein, D. S. Lindquist, R. S. Kops, and D. J. Brenner.** 1981. *Legionella longbeachae* species nova, another etiologic agent of human pneumonia. Ann. Intern. Med. **94**:739–743.
6. **Morris, G. K., A. Steigerwalt, J. C. Feeley, E. S. Wong, W. T. Martin, C. M. Patton, and D. J. Brenner.** 1980. *Legionella gormanii* sp. nov. J. Clin. Microbiol. **12**:718–721.
7. **Remington, R. D., and M. A. Schork.** 1970. Statistics with applications to the biological and health sciences. Prentice-Hall, Inc., Englewood Cliffs, N.J.
8. **Wentworth, B. B., and H. E. Stiefel.** 1982. Studies of the specificity of *Legionella* serology. J. Clin. Microbiol. **15**:961–963.
9. **Wilkinson, H. W., and B. J. Fikes.** 1981. Detection of cell-associated or soluble antigens of *Legionella pneumophila* serogroups 1 to 6, *Legionella bozemanii*, *Legionella dumoffii*, *Legionella gormanii*, and *Legionella micdadei* by staphylococcal coagglutination tests. J. Clin. Microbiol. **14**:322–325.
10. **Yonke, C. A., H. E. Stiefel, D. L. Wilson, and B. B. Wentworth.** 1981. Evaluation of an indirect hemagglutination test for *Legionella pneumophila* serogroups 1 to 4. J. Clin. Microbiol. **13**:1040–1045.

Frequency of Seroreactors to *Legionella* spp. Among Pneumonic Patients in a Danish Epidemic Ward

A. FRIIS-MØLLER, C. RECHNITZER, F. T. BLACK,† M. T. COLLINS, AND O. AALUND

Statens Seruminstitut, Department of Clinical Microbiology, Rigshospitalet, Department of Infectious Diseases, Rigshospitalet, and Royal Veterinary and Agriculture College, Laboratory of Preventive Medicine, Copenhagen, Denmark, and Department of Microbiology, Colorado State University, Fort Collins, Colorado 80523

To determine the incidence and severity of *Legionella* infections in pneumonic patients, a prospective study was conducted with patients with a diagnosis of pneumonia who were consecutively admitted to the University Clinic of Infectious Diseases, Rigshospitalet, Copenhagen, Denmark, from 1 January 1982 to 1 March 1983.

Bacteriological investigations were done on tracheal aspirates and pleural fluids. Antibodies to *Legionella pneumophila* serogroups (SG) 1 to 6, *Fluoribacter (Legionella) bozemanae*, *Fluoribacter (Legionella) dumoffii*, *Fluoribacter (Legionella) gormanii*, and *Tatlockia (Legionella) micdadei* were measured by the microagglutination technique and the indirect fluorescent-antibody test. Cutoff values for microagglutination positivity were the 2% upper limits of normal titers measured in 200 healthy Danish adults, i.e., a single titer of or seroconversion to ≥64 to *L. pneumophila* SG 1, 3, 4, 5, or 6, *F. gormanii*, or *T. micdadei*, and a single titer of ≥256 or seroconversion to ≥128 to *L. pneumophila* SG 2, *F. bozemanae*, or *F. dumoffii*. Indirect fluorescent-antibody titers were considered positive when there was a single titer of ≥256 or seroconversion to ≥128. Serological testing for antibodies to *Mycoplasma pneumoniae*, *Chlamydia psittaci*, influenza viruses A and B, adenovirus, and respiratory syncytial virus was performed simultaneously. Laboratory parameters and antibody titers were followed for at least 1 month.

During the 14-month study period, 45 male and 47 female patients were investigated. The laboratory diagnosis of legionellosis was based on positive serology, as all cultures for legionellae on buffered charcoal-yeast extract agar were negative. Legionnaires disease (LD) was diagnosed in 25 patients, including 14 men aged 16 to 76 years and 11 women aged 19 to 73 years. Twenty of these patients showed seroconversion, one patient had a single indirect fluorescent-antibody titer of 512 to *L. pneumophila* SG 5 and died before a second investigation, and four patients had standing titers higher than or equal to the cutoff values at several investiga-

tions. By these serological parameters, *Legionella* infection was the most common cause of pneumonia (26%), followed by *Haemophilus influenzae* infection (25%) and *Streptococcus pneumoniae* infection (20%) (Table 1). *Legionella* spp. appeared to be the only agent of infection in seven cases, while they were associated with *S. pneumoniae*, *H. influenzae*, *M. pneumoniae*, *C. psittaci*, or a virus in 16 cases. Diagnosis of *M. pneumoniae*, *C. psittaci*, or viral infection was based on positive serology, i.e., a fourfold rise or fall in titer. Two patients had concomitant *Salmonella* infection caused by *S. typhimurium* and *S. paratyphi*, respectively.

A total of 15 of the 92 pneumonic infections were hospital acquired; these included four cases of LD, one of *L. pneumophila* SG 5, one of *F. bozemanae*, one of *L. pneumophila* SG 5 plus *F. bozemanae* plus *T. micdadei*, and one of *L. pneumophila* SG 1 plus SG 5 plus *F. dumoffii* plus *T. micdadei*. Five patients had probably contracted LD in the Far East; one, in Italy. Of the 36 patients who were immunocompromised because of underlying disease, 10 had LD. Six of the 25 patients with LD showed respiratory failure and received assisted ventilation. Only

TABLE 1. Pathogens associated with pneumonia in 92 patients

Organism	No. of cases	No. of associated infections
L. pneumophila (SG 1 to 6), *F. bozemanae*, *F. dumoffii*, *F. gormanii*, or *T. micdadei*	25	18
H. influenzae	23	11
S. pneumoniae	19	10
Staphylococcus aureus	5	1
Pseudomonas aeruginosa	3	0
Mycobacterium tuberculosis	2	1
Peptostreptococcus sp.	1	0
M. pneumoniae	2	2
C. psittaci	4	4
Influenza A virus	4	2
Adenovirus	3	2
Respiratory syncytial virus	2	2
Unknown	20	

† Present address: Department of Infectious Diseases, Marselisborg Hospital, Aarhus, Denmark.

TABLE 2. Positive *Legionella* serology in 25 patients

No. of cases	Serology
1	L. pneumophila SG 1, T. micdadei
1	L. pneumophila SG 1 and SG 5, F. dumoffii, T. micdadei
1	L. pneumophila SG 1, SG 4, and SG 6
1	L. pneumophila SG 1 and SG 6
1	L. pneumophila SG 3
3	L. pneumophila SG 5
1	L. pneumophila SG 5, F. bozemanae
1	L. pneumophila SG 5, F. bozemanae, T. micdadei
1	L. pneumophila SG 6
1	L. pneumophila SG 6, F. bozemanae, F. dumoffii
4	F. bozemanae
2	F. bozemanae, F. dumoffii
1	F. bozemanae, T. micdadei
1	F. dumoffii, T. micdadei
5	T. micdadei

one of them, a 55-year-old man with *L. pneumophila* SG 5 infection, died. A 32-year-old pregnant woman with *F. bozemanae* infection gave birth to a healthy child (with negative microagglutination titers) and recovered fully.

The distribution of cases of LD (Table 2) indicates cross-reactivity among the genera *Legionella*, *Fluoribacter*, and *Tatlockia*, as has been shown in vitro by crossed immunoelectrophoresis (1). We found no case of *L. pneumophila* SG 2 or *F. gormanii*. By our parameters for diagnosis, our results show that LD occurs more often as a mixed infection than as a monoinfection.

Symptoms such as diarrhea, confusion, and hallucinations, and biochemical abnormalities such as elevated serum levels of creatinine, lactate dehydrogenase, and alanine aminotransferase, were slightly but not significantly more frequent in LD than in other types of pneumonia. Serum levels of natrium and creatine kinase and leukocyte counts varied independently of the cause of pneumonia. We therefore conclude that clinical features and laboratory parameters cannot establish the diagnosis of LD, nor can the severity of the illness distinguish LD from other types of pneumonia.

LITERATURE CITED

1. Collins, M. T., F. Espersen, N. Høiby, S.-N. Cho, A. Friis-Møller, and J. S. Reif. 1983. Cross-reactions between *Legionella pneumophila* (serogroup 1) and twenty-eight other bacterial species, including other members of the family *Legionellaceae*. Infect. Immun. 39:1441–1456.

Studies with Monoclonal Antibodies to *Legionella* Species

I. D. WATKINS AND J. O'H. TOBIN

Sir William Dunn School of Pathology, Oxford University, Oxford, England

Serological identification and grouping of *Legionella* species have been done by the use of antisera from immunized rabbits, guinea pigs, or mice; these studies have indicated that strain differences exist between members of serogroups, as do crosses between groups, sometimes leading to difficulty in allocation of isolates to a specific group. An alternative approach is to use monoclonal antibodies for testing; this offers benefits of clarity and reproducibility and the prospect of better standard serological reagents. A set of monoclonal antibodies to *Legionella* has been produced and applied to the work of grouping and subgrouping and also to preliminary work on the recognition of strains in the environment and on the antigenic structure of the organism.

BALB/c mice were immunized with whole *Legionella* antigen (heat killed by treatment at 65°C for 1 h); the immunization schedule involved an intraperitoneal injection of about 0.2 mg (dry weight) of bacteria, followed after about 6 weeks by intravenous injections of similar dose at 4 days and, in some cases, 2 days before splenic cell fusion with mouse myeloma cells (P2-NSI-1-Ag4-1), using polyethylene glycol 1500. Stable hybrid lines producing a single antibody were obtained after recloning.

Assays using whole bacteria were performed by indirect immunofluorescence with a minimum incubation period of 3 h at 37°C for antibody-antigen interaction and also by an enzyme-linked immunosorbent assay (4). In the latter the wells of the assay plates are precoated with poly-L-lysine, which adsorbs to the plastic. A suspension of whole bacteria is pelleted in the wells by centrifugation of the plates, and then the bacteria are covalently cross-linked to the poly-L-lysine, and hence bound to the plate, by the addition of glutaraldehyde. This method retains the antigenic integrity of the bacteria without the requirement for extraction, and there is no need to let the antigenic material adsorb to the plate by itself, which may need a long incubation period and may not be reliable.

Monoclonal antibodies with different specificities were used for typing. The set comprises antibodies derived from fusions involving 19 different strains of *Legionella*: five *L. pneumophila* serogroup 1 strains; one strain each from *L. pneumophila* serogroups 2, 3, and 6; two strains from *L. pneumophila* serogroup 4; three *L. pneumophila* serogroup 5 strains; two *L. pneumophila* strains of uncertain serotype; and four strains from other species. Very clear all-or-none results were usually obtained, without the difficulties posed by animal sera variability and background effects. Of 104 *L. pneumophila* strains tested with monoclonal antibodies, 99 could be clearly placed in one of the conventional serogroups 1 through 6.

Monoclonal antibodies specific for *L. pneumophila* serogroup 3 (9 strains, including Bloomington 2), serogroup 6 (11 strains, including Chicago 2), and serogroup 2 (Togus 1 strain)

have been obtained. It is not possible at present to subdivide these groups, probably because of the limited number of monoclonal antibodies and strains available. In contrast, it has been possible to produce subgrouping of varying complexity with serogroups 1 and 5.

As one antibody (Wash. 29) reacts with 46 of 50 strains, all of the serogroup 1 strains tested can be covered by the combination of this monoclonal antibody and another, F7-9C3, which recognizes 17 of 50 strains, including the 4 strains not detected by Wash. 29 (Table 1). Preliminary results from antigen extraction work and immunodiffusion and crossed immunoelectrophoresis studies suggest that these two monoclonal antibodies recognize determinants on the serogroup-specific antigen described elsewhere (1, 3).

The reactions of eight other serogroup 1-

TABLE 1. Monoclonal subgrouping, *L. pneumophila* serogroup 1

Strains[a]	Subgroup (no. of strains)	Reaction[b] with monoclonal antibody:									
		Wash. 29	Pontiac 7-4C3	Pontiac 7-9C3	Pontiac C13-15	Wash/36	Wash/32 11	K1/G6 22	JR5/25 16/53	4032/4 19	Wash/39/7
Pontiac E	A (6)	+	++	++	++	++	−	++	++	−	−
Philadelphia 1		+	++	++	++	++	−	++	++	−	−
Kingston P129 E		+	++	++	++	++	−	++	++	−	−
3 E strains		+	−	++	++	+	−	++	++	−	−
Nottingham GF	B (4)	+	−	++	+	++	++	−	−	−	++
Benidorm 030 E		+	−	++	+	+/−	++	−	−	−	++
Wadsworth F 264 and F 265		−	−	++	+	+/−	++	−	−	−	++
Corby E, 6842 E	C (3)	+	−	++	++	++	+	+	−	−	−
Kingston P15 E		++	−	++	++	+	+	−	−	−	−
10979 E, JR6 E	D (5)	++	++	−	++	++	+	+	++	++	−
3 E strains		++	++	−	++	+	+	++	++	++	−
Kingston 1 and 2 E, 4032 E	E (12)	++	+	−	−	++	−	++	++	++	−
Wadsworth E402 E		++	+/−	−	−	++	−	++	++	++	−
8 E strains		++	+	−	−	++	−	++	++	++	−
Bellingham	F (7)	+	−	−	−	++	++	++	−	−	++
Washington, Cairo-B E		++	−	−	−	++	++	++	−	−	++
Nottingham RH		++	−	−	−	++	++	++	+/−	−	++
3 E strains		++	−	−	−	++	++	+	−	−	++
		++	++	+/−		++	++	+	++	−	−
Kingston P64 E, 8141 E	G (6)										
6507 E, 3 E strains		++	+/−	−	−	++	++	++	++	−	−
74/81 E	O (5)	++	−	++	++	−	−	−	++	−	−
105/81 E		++	++	+/−	++	−	−	−	++	++	−
Porton 1093 E		++	−	−	−	++	−	−	++	−	−
7384 E		−	+	++	+/−	−	++	−	−	−	−
Heysham E		++	−	−	−	+/−	−	+/−	−	−	−

[a] E, Environmental strain.
[b] Symbols: ++, maximum fluorescence; +, good fluorescence; +/−, some fluorescence; −, no fluorescence.

TABLE 2. Monoclonal subgrouping, *L. pneumophila* serogroups (SG) 4 and 5[a]

SG 5 subgroup (no. of strains)	Strains	SG 4		SG 5									
				Dallas			P188 E				Benidorm 034 E		Cambridge 2
		LA-10	K4-8	0	1	12	1	4	11	6L2	9-5B1	9-5G7	
A (4)	Dallas 1 E	−	−	++	++	++	+	−	−	−	++	++	−
	Pittsburgh U7W E	−	−	++	++	++	++	+/−	+/−	−	+	++	−
	Pittsburgh U8W E	−	−	++	++	++	++	−	−	−	++	+	−
	Pittsburgh MICU-B E	−	−	++	++	++	+	−	−	−	+	++	−
B (4)	4 E strains	+	−	+/−	++	++	+	−	+/−	−	+	++	+/−
C (3)	P188 E, P157 E	+	+	−	−	−	++	++	++	++	+/−	++	−
	8078 E	−	−	−	−	−	++	−	−	+	+	++	−
D (14)	Leiden 1, Chow E Pittsburgh 684 and	+	+	−	−	++	+	−	+/−	−	++	++	−
	687 E	+	+	−	−	++	++	−	+/−	−	++	+	−
	10 E strains	+	+	−	−	++	+	−	+/−	−	+	++	−
(E) (1)	Cambridge 2	+/−	+/−	−	−	−	−	−	−	+/−	+/−	−	++
SG 4	Los Angeles 1	++	++	−	−	−	−	−	−	−	−	−	+/−
	Kingston 4 E	++	++	−	−	−	−	−	−	−	−	−	+/−

[a] See footnotes to Table 1.

specific monoclonal antibodies have been used to divide the serogroup 1 strains into subgroups by their reaction patterns. The number of subgroups needed to give epidemiologically relevant divisions has yet to be determined, as the patterns produced can be varied by the number of antibodies used and their choice. Clear relationships, in terms of shared antigenic determinants, can be recognized within the serogroup. The example scheme given in Table 1 places 45 of 50 strains in seven subgroups, the remaining 5 strains (group O for others) having their own distinctive antibody reactions. The few isolates obtained from patients were in subgroups 1A, 1B, and 1F, which also included environmental strains.

By using monoclonal antibodies a similar picture of subgrouping has been obtained with serogroup 5 (Table 2). Subgroups 5B through 5D cross-react with serogroup 4 monoclonal antibodies (but not the reverse), suggesting that these serogroups are not distinct. Recent work suggests that there are subgroups within serogroup 4 and that Dallas 1E-like strains (subgroup 5A) share minor antigenic determinants with some serogroup 4 strains.

That the five strains of serogroup 5 from Pittsburgh described by Garrity et al. (2) fall into two subgroups has been confirmed. Three strains (U7W, U8W, and MICU-B) are in subgroup 5A with Dallas 1E, whereas strains 684 and 687 fall into subgroup 5D with Leiden 1 and

11 environmental strains. The Cambridge 2 isolate, which is considered a serogroup 5 strain, seems distinct from the others and would not appear to be a typical serogroup 5 strain.

The five strains not serogrouped by our monoclonal antibodies, including Cambridge 2, are Chicago 8 (the serogroup 7 reference strain) and three environmental strains from the Thames Valley (P183, JR4, and 4037) which react specifically with monoclonal antibodies to JR4 and P183 and appear to form a separate group of *L. pneumophila*.

Another monoclonal antibody recognizes 53 of 104 *L. pneumophila* strains across the serogroups and shows a weak cross-reaction with strain TEX-KL (*L. dumoffii*). Otherwise, the *L. pneumophila*, *L. bozemanii*, *L. longbeachae* serogroup 1, and *L. jordanis* monoclonal antibodies obtained thus far do not cross-react with each other or with other *Legionella* species (*L. gormanii*, *L. oakridgensis*, *L. wadsworthii*, *L. micdadei*, *L. dumoffii*, and strain E327F, provisionally designated *L. morrisii*). The *L. longbeachae* monoclonal antibodies tested recognize specifically the Long Beach 4 serogroup 1 strain, but not the Tucker 1 serogroup 2 strain.

A set of seven antibodies (one specific for *L. longbeachae* serogroup 1 and six specific for *L. pneumophila* serogroups 1, 2, 3, and 5) have been tested with non-*Legionella* bacteria. These were 3 strains of *Staphylococcus* sp., 2 *Streptococcus* sp., and 18 strains from 11 species of gram-negative organisms (*Escherichia* sp., five *Pseudomonas* spp., *Achromobacter* sp., *Flavobacterium* sp., *Bacteroides* sp., *Salmonella* sp., and *Serratia* sp.). There was no evidence of any recognition of these bacteria by the monoclonal antibodies; the antigenic determinants involved appear to be specific to legionellae. It should be noted that four of the strains used (one *Escherichia coli*, three *Bacteroides fragilis*) were clinical isolates from the Wadsworth Medical Center, which have been reported to give weak cross-reactions with *Legionella* subgroup 1 antiserum (P. H. Edelstein, personal communication).

Monoclonal antibodies appear to be superior to whole sera for the accurate and reproducible serogrouping of *Legionella* strains and should be useful for diagnosis and environmental studies, as well as for analysis of structure, and investigation of other problems. At some point centers working with monoclonal antibodies to *Legionella* species should exchange antibody from adequately cloned and tested cultures so that an orderly system of grouping and subgrouping can be developed by their use.

We thank P. H. Edelstein, J. B. Kurtz, R. A. Swann, R. Newnham, F. G. Rodgers, P. J. Dennis, R. J. Fallon, G. M. Garrity, S. Fisher-Hoch, M. Smith, C. Anand, and P. Meenhorst for supplying *Legionella* and other strains.

LITERATURE CITED

1. **Flesher, R. A., H. J. Jennings, C. Lugowski, and D. L. Kaspar.** 1982. Isolation of a serogroup 1-specific antigen from *Legionella pneumophila*. J. Infect. Dis. **145**:224–233.
2. **Garrity, G. M., E. M. Elder, B. Davis, R. M. Vickers, and A. Brown.** 1982. Serological and genotypic diversity among serogroup 5-reacting environmental *Legionella* isolates. J. Clin. Microbiol. **15**:646–653.
3. **Joly, J. R., and G. E. Kenny.** 1982. Antigenic analysis of *Legionella pneumophila* and *Tatlockia micdadei* (*Legionella micdadei*) by two-dimensional (crossed) immunoelectrophoresis. Infect. Immun. **35**:721–729.
4. **Kennett, R. H.** 1980. Enzyme-linked antibody assay with cells attached to polyvinyl chloride plates. *In* R. H. Kennett, T. J. McKearn, and K. B. Bechtol (ed.), Monoclonal antibodies. Hybridomas: a new dimension in biological analyses. Plenum Publishing Corp., New York.

Production of Monoclonal Antibodies to *Legionella pneumophila* and Relationship of Monoclonal Binding to Plasmid Content

MICHAEL F. PARA, J. F. PLOUFFE, AND W. E. MAHER

The Ohio State University, Columbus, Ohio 43210

Isolates of *Legionella pneumophila* are divided into eight serogroups based upon reactivity with antisera from rabbits immunized with the prototype strains (4). Additional antigenic complexity within the serogroups has been demonstrated by studies with absorbed sera (1).

To better define the serological relationship among *L. pneumophila* strains for epidemiological and taxonomic purposes and to develop tools for isolation of *Legionella* antigens, monoclonal antibodies were produced to *L. pneumophila* serogroup 1 (SG-1) and *L. pneumophila* serogroup 6 (SG-6).

BALB/c mice were immunized with either Formalin-fixed Knoxville strain (SG-1) or Chicago strain (SG-6) which had been mixed with incomplete Freund adjuvant. Hybridoma cells were produced by fusion of the mouse spleen cells and SP 2/0 myeloma cells according to the method of McKearn (3). Hybridomas producing antibody to the SG-1 or SG-6 strains were selected by indirect immunofluorescent assay with goat anti-mouse immunoglobulin M and immunoglobulin G. The positive wells were cloned by limiting dilution.

Two antibodies to SG-1 were produced, LP-I-17 and LP-I-81. LP-I-81 recognized only SG-1 strains. The medium from this hybridoma was capable of microagglutinating a bacterial cell suspension of either the Philadelphia or the Knoxville strain, the prototype SG-1 isolates (Fig. 1). LP-I-17 recognized the six *L. pneumophila* reference strains supplied by the Centers for Disease Control for SG-1 through SG-6, but the fluorescence in the indirect immunofluorescent assay was very weak, except with SG-1. This antibody did not agglutinate bacterial cell suspensions of any serogroup.

Three monoclonal antibodies were produced to SG-6. These were LP-VI-2, LP-VI-5, and LP-VI-6. All three antibodies were specific for SG-6. In addition, each could microagglutinate a bacterial cell suspension of the SG-6 Chicago strain.

Isolates of SG-1 from the environment and from Legionnaires disease patients were tested for microagglutination with LP-I-81 (Table 1). Of the 20 SG-1 isolates from Columbus nosocomial pneumonia cases, 19 were agglutinated by LP-I-81. Environmental isolates from the potable water in the rooms of Ohio State University Hospitals and other Columbus, Ohio, hospitals were similarly tested. Approximately 50% of the SG-1 strains isolated from separate sites were agglutinated. The organisms which did agglutinate had a skewed distribution within the university medical complex (J. F. Plouffe, M. F. Para, B. Hackman, L. Webster, and W. Maher, this volume).

The patient and environmental isolates were further analyzed for plasmid content by alkaline sodium dodecyl sulfate treatment (2) and agarose gel electrophoresis (W. E. Maher, J. F. Plouffe, and M. F. Para, this volume). Table 1 lists the plasmid content of these isolates. All isolates from the potable water in one university hospital building (RH) had two plasmids (85 and 45 megadaltons [Mdal]). Culture of the potable water in another university hospital building (UH) yielded both a plasmidless and a 45-Mdal-plasmid-containing organism. Two other Columbus hospitals had an SG-1 isolate that contained a single 85-Mdal plasmid.

The presence of the 85-Mdal plasmid alone or in combination with the 45-Mdal plasmid correlated with the lack of microagglutination by the monoclonal antibody LP-I-81. The presence of the 45-Mdal plasmid alone did not interfere with the microagglutination by this antibody. Occasional strains were found that did not agglutinate with LP-I-81 antibody and did not contain a plasmid.

The availability of monoclonal antibodies which recognize subsets of *L. pneumophila* and the analysis of the plasmid content of these organisms provide markers for strains of SG-1. These markers are useful in epidemiological

TABLE 1. Microagglutination by LP-I-81 and plasmid content of *L. pneumophila* SG-1

Isolate and origin	Agglutination	Plasmid
Patient isolates		
Philadelphia 1, CDC[a]	+	None
Knoxville 1, CDC	+	None
Nosocomial cases	+	None ($n = 16$)
($n = 20$), Columbus	+	45 Mdal ($n = 3$)
	−	85 and 45 Mdal ($n = 1$)
Environmental isolates		
University hospital,	+	None
Columbus	+	45 Mdal
	−	85 and 45 Mdal
Community hospital 1,	−	85 Mdal
Columbus		
Community hospital 2,	+	None
Columbus	−	85 and 45 Mdal
Community hospital 3,	+	None
Columbus		
Community hospital 4,	−	85 Mdal
Columbus	−	None

[a] CDC, Centers for Disease Control.

studies of Legionnaires disease and in understanding the environmental colonization by legionellae (4).

Monoclonal antibodies to the surface components will help define the nature of the antigens. Preliminary studies of the antigen recognized by LP-I-81 have been performed (M. F. Para and J. F. Plouffe, Abstr. Annu. Meet. Am. Soc. Microbiol. 1983, D23, p. 62). Antigenicity is retained after heating to 100°C for 5 min and after treatment with proteinase K, neuraminidase, and mercaptoethanol. On the other hand, treatment with sodium dodecyl sulfate appears to dissociate the antigen, and periodate destroys it. These results are consistent with this antigen being an endotoxin-like molecule. This hypothesis is also supported by the bacterial cell microagglutination by a monoclonal antibody.

The presence of the 85-Mdal plasmid in two strains from Columbus correlates with the lack of microagglutination by LP-I-81. Whether this plasmid is responsible for this interference is not known. However, in other bacterial systems, plasmids code for substances which inhibit agglutination by antibodies to endotoxin (e.g., K-88 of *Escherichia coli*). Experiments are under way to cure the nonagglutinable strains of SG-1

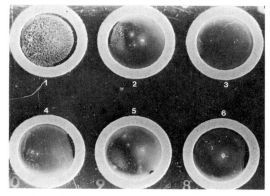

FIG. 1. SG-1-specific microagglutination of *L. pneumophila* by LP-I-81. Wells contain hybridoma medium plus a bacterial cell suspension of SG-1 strain Philadelphia 1 (well 1), SG-2 strain Togus (well 2), SG-3 strain Bloomington (well 3), SG-4 strain Los Angeles (well 4), SG-5 strain Dallas (well 5), and SG-6 strain Chicago (well 6).

containing the 85-Mdal plasmid and then reevaluate the reactivity with the monoclonal antibody LP-I-81.

LITERATURE CITED

1. **Brown, A., R. M. Vickers, E. M. Elder, M. Lema, and G. M. Garrity.** 1982. Plasmid and surface antigen markers of endemic and epidemic *Legionella pneumophila* strains. J. Clin. Microbiol. **16:**230–235.
2. **Kado, C. I., and S. T. Liu.** 1981. Rapid procedure for detection of large and small plasmids. J. Bacteriol. **145:**1365–1373.
3. **McKearn, T. J.** 1980. Methods for growing hybridomas in rats and mice, p. 403–404. *In* R. H. Kennette, R. J. McKearn, and K. B. Bechtol (ed.), Monoclonal antibodies: a new dimension in biological analysis. Plenum Publishing Corp., New York.
4. **McKinney, R. M., L. Thacker, P. P. Harris, K. R. Lewallen, G. A. Hébert, P. H. Edelstein, and B. M. Thomason.** 1979. Four serogroups of Legionnaires' disease bacteria defined by direct immunofluorescence. Ann. Intern. Med. **90:**621–624.

Subtyping of *Legionella pneumophila* Serogroup 1 by Using Monoclonal Antibodies

JEAN R. JOLY, YUAN YAO CHEN,† AND DANIELLE RAMSAY

Département de Microbiologie, Faculté de Médecine, Université Laval, Québec G1K 7P4, Canada

Legionella pneumophila serogroup 1 is the most frequent cause of legionellosis (4, 8, 13); it is also the most frequently isolated member of *Legionella* in natural environments (5) and has frequently been observed in man-made aquatic environments (3, 9). Because of this ubiquity in the environment, epidemiological studies on the origin of strains causing either epidemic or sporadic cases of Legionnaires disease have been extremely difficult. Antigenic differences between different isolates of a bacterial species are frequently used as epidemiological markers (11), and such differences could also be useful in the study of Legionnaires disease.

The introduction by Köhler and Milstein (6) of cell fusion for the production of monoclonal antibodies of defined specificity has opened new fields in antigenic analysis of microorganisms. These antibodies have already allowed the serogrouping of organisms that were previously unseparable with polyclonal animal antisera (2, 7). The use of monoclonal antibodies has allowed us to define five subtypes of *L. pneumophila* serogroup 1.

Monoclonal antibodies were produced by fusing spleen cells of mice immunized with *L. pneumophila* serogroup 1 (Philadelphia 1 strain) with the Sp2-0/Ag14 mouse myeloma cell line. After cloning, eight stable hybrids were obtained. Culture fluids from these hybrids, at optimal dilution, were all specific for *L. pneumophila* serogroup 1 when tested by an indirect fluorescent-antibody assay. Using these eight monoclonal antibodies, 47 clinical and environmental isolates of *L. pneumophila* serogroup 1 were stained by an indirect microimmuno-fluorescent-antibody technique. For this, bacteria were grown for 48 to 72 h on buffered charcoal-yeast extract medium, harvested with a glass rod, and suspended in 1% (vol/vol) Formalin in 0.85% (wt/vol) saline. This suspension was then diluted 1:10 in 3% normal yolk sac and was placed with a pen on multiple-well microscope slides according to the procedure of Wang and Grayston (10). After drying and fixing in acetone for 10 min, 50 µl of monoclonal antibodies was applied and slides were incubated at 37°C for 30 min; they were then washed in phosphate-buffered saline and stained for 30 min with fluorescein-labeled anti-mouse immunoglobulin sera. Washing was performed as described above, and slides were mounted and examined with an epiillumination fluorescence microscope at 400× magnification. Positive and negative controls were included in all experiments.

Because the staining intensity of different isolates varied markedly with a single monoclonal antibody, strains were considered as either negative (no fluorescence) or positive (1+ to 4+ fluorescence). At least three different epitopes were recognized by these eight antibodies, and five different subtypes of *L. pneumophila* serogroup 1 could be defined (Table 1).

Antigenic differences among different isolates of *L. pneumophila* serogroup 1 were previously noted (1, 12). However, the production of strain-specific antisera required adsorption, and these antisera were not generally available. These monoclonal antibodies provide standardized reagents that can be used in a simple technique for subtyping purposes. The exact significance of the antigenic differences in *L. pneumophila* serogroup 1 is unknown. Additional studies on the possible relationships between antigenic subtypes and the different clinical presentations of Legionnaires disease are warranted.

† Present address: Department of Pathology, Bethune Medical University, Changchun, People's Republic of China.

TABLE 1. Subtypes of *L. pneumophila* identified with monoclonal antibodies

No. of strains belonging to each subtype	Staining by the following monoclonal antibody:							
	33G2	31C7	32D10	31G5	34G8	31G4	32E6	32A12
10 (21.3%)	+	+	+	+	+	+	+	+
25 (53.2%)	−	+	+	+	+	+	+	+
3 (6.4%)	+	+	+	+	+	+	+	−
1 (2.1%)	−	+	+	+	+	+	+	−
8 (17%)	−	−	−	−	−	−	−	+
Total no. stained	13 (28%)	39 (83%)	39 (83%)	39 (83%)	39 (83%)	39 (83%)	39 (83%)	43 (91%)

This study was supported by grant MA-7689 from the Medical Research Council of Canada.

LITERATURE CITED

1. **Brown, A., R. M. Vickers, E. M. Elder, M. Lema, and G. M. Garrity.** 1982. Plasmid and surface antigen markers of endemic and epidemic *Legionella pneumophila* strains. J. Clin. Microbiol. **16:**230–235.
2. **Coates, A. R. M., B. W. Allen, J. Hewitt, J. Ivanyi, and D. A. Mitchison.** 1981. Antigenic diversity of *Mycobacterium tuberculosis* and *Mycobacterium bovis* detected by means of monoclonal antibodies. Lancet **ii:**167–169.
3. **Cordes, L. G., A. M. Wiesenthal, G. W. Gorman, J. P. Phair, H. M. Sommers, A. Brown, L. Yu, M. H. Magnussen, R. D. Meyer, J. S. Wolf, K. N. Shands, and D. W. Fraser.** 1981. Isolation of *Legionella pneumophila* from hospital shower heads. Ann. Intern. Med. **94:**195–197.
4. **England, A. C., D. W. Fraser, B. D. Plikaytis, T. F. Tsai, G. Storch, and C. V. Broome.** 1981. Sporadic legionellosis in the United States: the first thousand cases. Ann. Intern. Med. **94:**164–170.
5. **Fliermans, C. B., W. B. Cherry, L. H. Orrison, S. J. Smith, D. L. Tison, and D. H. Pope.** 1981. Ecological distribution of *Legionella pneumophila*. Appl. Environ. Microbiol. **41:**9–16.
6. **Köhler, G., and C. Milstein.** 1975. Continuous cultures of fused cells secreting antibody of predefined specificity. Nature (London) **256:**495–497.
7. **Koprowski, H., W. Gerhard, T. Wiktor, J. Martinis, M. Shander, and C. M. Croce.** 1978. Anti-viral and anti-tumor

antibodies produced by somatic cell hybrids. Curr. Top. Microbiol. Immunol. **81:**8–19.
8. **Macfarlane, J. T., M. J. Ward, R. G. Finch, and A. D. Macrae.** 1982. Hospital study of adult community-acquired pneumonia. Lancet **ii:**255–258.
9. **Stout, J., V. L. Yu, R. M. Vickers, J. Zuravleff, M. Best, A. Brown, R. B. Yee, and R. Wadowsky.** 1982. Ubiquitousness of *Legionella pneumophila* in water supply of a hospital with endemic Legionnaires' disease. N. Engl. J. Med. **306:**466–468.
10. **Wang, S. P., and J. T. Grayston.** 1970. Immunologic relationship between genital TRIC, lymphogranuloma venereum, and related organisms in a new microtiter indirect immunofluorescence test. Am. J. Ophthalmol. **70:**367–374.
11. **Wang, S. P., K. K. Holmes, J. S. Knapp, S. Ott, and D. D. Kyser.** 1977. Immunologic classification of *Neisseria gonorrhoeae* with microimmunofluorescence. J. Immunol. **119:**795–803.
12. **Wilkinson, H. W., B. J. Fikes, and D. D. Cruce.** 1979. Indirect immunofluorescence test for serodiagnosis of Legionnaires disease: evidence for serogroup diversity of Legionnaires disease bacterial antigens and for multiple specificity of human antibodies. J. Clin. Microbiol. **9:**379–383.
13. **Wilkinson, H. W., A. L. Reingold, B. J. Brake, D. L. McGiboney, G. W. Gorman, and C. V. Broome.** 1983. Reactivity of serum from patients with suspected legionellosis against 29 antigens of *Legionellaceae* and *Legionella*-like organisms by indirect immunofluorescence assay. J. Infect. Dis. **147:**23–31.

Characterization of Membrane Proteins from Various Strains and Serogroups of *Legionella pneumophila* and Other *Legionella* Species

W. EHRET, G. ANDING, AND G. RUCKDESCHEL

Max von Pettenkofer Institute of Hygiene and Medical Microbiology, University of Munich, Munich, West Germany

Since the discovery of *Legionella pneumophila*, seven serogroups of this species and nine related species, one with two serogroups, have been found. Accordingly, 17 serologically different types of legionellae can be distinguished, most of them possible etiological agents of respiratory tract diseases. This group of antigenically distinct pathogens and their multiple cross-relations greatly complicate the serological diagnosis of legionelloses. Therefore, it would be most interesting to find alternative ways of identifying and differentiating these organisms.

Considerable work has been done in studies related to the protein composition of the outer membranes of gram-negative bacteria (5). However, little is known about the membrane proteins of legionellae (1). In this investigation, outer and total membrane protein preparations from various *Legionella* strains were compared to find new tests for answering taxonomic, epidemiological, and clinical questions.

Twenty-one *Legionella* strains were studied, including strains of *L. pneumophila* serogroups 1 (Philadelphia 1, Knoxville 1, München 1, München 2, München 3, Augsburg 1, Berlin 1, Zürich 1), 2 (Togus 1), 3 (Bloomington 2), 4 (Los Angeles 1), 5 (Dallas 1E), 6 (Chicago 2), and 7 (ATCC 33823); *L. bozemanii* (ATCC 33217); *L. dumoffii* (ATCC 33279); *L. gormanii* (ATCC 33297); *L. jordanis* (ATCC 33623); *L. micdadei* (ATCC 33204); and *L. longbeachae* serogroups 1 (Long Beach 4, ATCC 33462) and 2 (Tucker 1, ATCC 33484). The other bacterial species compared with legionellae were clinical isolates from our diagnostic laboratory.

The legionellae were grown on buffered charcoal-yeast extract agar and, after harvesting, were killed by formaldehyde, diethyl ether, or heat. A preparation of the total bacterial membrane was obtained by ultracentrifugation after disintegration of the cells by sonication. As known from studies with other gram-negative bacteria, sodium lauryl sarcosinate dissolves the inner membrane from the detergent-insoluble cell envelope (2). The detergent-insoluble membrane fraction was therefore considered to represent the outer membrane. Alternatively, the outer membrane was prepared by isopycnic sucrose gradient ultracentrifugation as described by Osborn and Munson (6).

After the different inactivation procedures, membrane fractions isolated by these methods were analyzed by sodium dodecyl sulfate-polyacrylamide gel electrophoresis, using 11% polyacrylamide gels prepared as described by Laemmli (3). In special cases, 7.5% gels or 4 to 30% gradient gels were also used. Commercial molecular weight markers served as references. Gels were stained with Coomassie brilliant blue and, when necessary, with the more sensitive silver stain.

The results demonstrate that all strains of *L. pneumophila* contain a major outer membrane protein (MOMP) of 29,000 (±500) daltons (29 kd), regardless of serogroup (Fig. 1). This protein was particularly predominant in formaldehyde-inactivated preparations. The 29-kd proteins could also be observed when 7.5% or 4 to 30% gradient polyacrylamide gels were used. Analyses of the outer membrane isolated by isopycnic sucrose gradient centrifugation gave the same result.

Investigation of clinical isolates of some other gram-negative rods (*Escherichia coli, Citrobacter freundii, Klebsiella pneumoniae, Enterobacter cloacae, Serratia marcescens, Proteus mirabilis, Pseudomonas aeruginosa,* and *Acinetobacter calcoaceticus*) indicated the 29-kd protein to be a distinguishing feature of *L. pneumophila*. Similarly, bacterial species with known cross-reactivity in indirect immunofluorescence (*Pseudomonas fluorescens, Pseudo-*

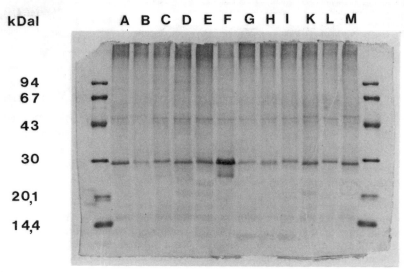

FIG. 1. Sodium dodecyl sulfate-polyacrylamide gel electrophoresis of lauryl sarcosinate-insoluble membrane proteins from formaldehyde-inactivated strains of *L. pneumophila*. Lanes: A, serogroup 1, Philadelphia 1; B, serogroup 2, Togus 1; C, serogroup 3, Bloomington 2; D, serogroup 4, Los Angeles 1; E, serogroup 5, Dallas 1E; F, serogroup 6, Chicago 2; G, serogroup 7, ATCC 33823; H, serogroup 1, Berlin 1; I, serogroup 1, Zürich 1; K, serogroup 1, München 1; L, serogroup 1, München 2; M, serogroup 1, München 3.

monas pseudomallei, Bacteroides fragilis, and Salmonella typhi) did not contain a protein component similar to the 29-kd protein (MOMP) of L. pneumophila.

Only one more weakly staining protein band, with an apparent mass of about 46 kd, was regularly seen in formaldehyde-inactivated strains of L. pneumophila. A more complex protein pattern was obtained after inactivation by ether or heat. Although the 29-kd protein was the most intensely staining component after these alternative inactivation procedures, additional weakly staining protein bands of higher molecular weight were observed. These five to eight components were identical in all 14 strains of L. pneumophila investigated.

Unlike in some other gram-negative bacteria (4), disulfide bridges seem to play an important role in maintenance of the structural integrity of L. pneumophila envelopes. The 29-kd protein was only obtained after heating to a minimum of 80°C in the presence of 2-mercaptoethanol. Lower temperature or omission of the reducing agent resulted in a complete loss of this main component. Even with polyacrylamide gels of lower concentration, no equivalent to the 29-kd component could be found in a higher-molecular-weight range after omission of 2-mercaptoethanol.

Analyses of outer membrane preparations of Legionella species other than L. pneumophila inactivated by formaldehyde showed no similarity in the characteristics described above. Only L. micdadei exhibited a prominent protein band of about 39 kd; this feature could be observed after all three killing methods and seems to represent the MOMP of L. micdadei. The other Legionella species did not have any characteristic protein pattern after formaldehyde treatment. Preparations of the remaining Legionella species were therefore investigated after killing by ether or heat.

After ether inactivation, both serogroups of L. longbeachae showed identical protein patterns with a characteristic doublet band at 43 and 51 kd. Except for L. pneumophila and L. micdadei, no intensely staining protein band corresponding to a MOMP could be detected. L. bozemanii, L. dumoffii, and L. gormanii had a common characteristic protein of about 48 kd. Additional weaker bands occurred at 43 kd (L. bozemanii), 35 kd (L. dumoffii), and 33 kd (L. gormanii). L. jordanis was characterized by a protein band at 45 kd.

After heat inactivation, both serogroups of L. longbeachae yielded an identical pattern, again with a protein band at 45 kd. The most heavily staining protein of L. bozemanii, L. dumoffii, and L. gormanii had a mass of 44 kd. L. bozemanii and L. dumoffii had a second component

at 64 kd, corresponding to a similar band in L. pneumophila.

The analysis of total membrane fractions after sonication only also revealed the main 29-kd component in all 14 strains of L. pneumophila, but the resulting protein pattern was much more complex and enabled the detection of minor differences between single strains. More differentiated protein patterns were obtained also with the other Legionella species; these patterns cannot be described here in detail but seem to be very specific for each single species (Fig. 2).

Preliminary results obtained by use of the Western blot technique (7) and autoradiographic detection of immune complexes with ^{125}I-labeled protein A indicate that a rabbit antiserum against L. pneumophila serogroup 1 (strain Philadelphia 1) is able to react with the 29-kd MOMP of all serogroups of L. pneumophila. By the same method the development of strongly reacting antibodies against the 29-kd MOMP of L. pneumophila in human convalescent-phase sera could be demonstrated.

The 29-kd protein should be a suitable antigen for simpler and more specific tests for the diagnosis of legionellosis. Further investigations of cell wall components may provide methods to support taxonomic, epidemiological, and clinical studies.

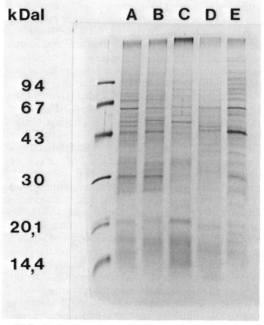

FIG. 2. Sodium dodecyl sulfate-polyacrylamide gel electrophoresis of membrane proteins from different Legionella species. Lanes: A, L. bozemanii; B, L. dumoffii; C, L. gormanii; D, L. jordanis; E, L. longbeachae Tucker 1.

LITERATURE CITED

1. **Amano, K. I., and J. C. Williams.** 1983. Peptidoglycan of *Legionella pneumophila*: apparent resistance to lysozyme hydrolysis correlates with a high degree of peptide cross-linking. J. Bacteriol. **153**:520–526.
2. **Filip, C., G. Fletcher, J. L. Wulff, and C. F. Earhart.** 1973. Solubilization of the cytoplasmic membrane of *Escherichia coli* by the ionic detergent sodium-lauryl sarcosinate. J. Bacteriol. **115**:717–722.
3. **Laemmli, U. K.** 1970. Cleavage of structural proteins during the assembly of the head of bacteriophage T4. Nature (London) **227**:680–685.
4. **Logan, S. M., and T. J. Trust.** 1982. Outer membrane characteristics of *Campylobacter jejuni.* Infect. Immun. **38**:898–906.
5. **Lugtenberg, B., and L. Van Alphen.** 1983. Molecular architecture and functioning of the outer membrane of *Escherichia coli* and other gram-negative bacteria. Biochim. Biophys. Acta **737**:51–115.
6. **Osborn, M. J., and R. Munson.** 1974. Separation of the inner (cytoplasmic) and outer membranes of gram-negative bacteria. Methods Enzymol. **31**:642–653.
7. **Towbin, H., T. Staehelin, and J. Gordon.** 1979. Electrophoretic transfer of proteins from polyacrylamide gels to nitrocellulose sheets: procedure and some applications. Proc. Natl. Acad. Sci. U.S.A. **76**:4350–4354.

Immunochemical Analysis of Cell Envelopes of *Legionella pneumophila* Serogroup 1 Strains Isolated from Patients and Water During an Epidemic in Amsterdam

O. G. ZANEN-LIM, N. J. VAN DEN BROEK, P. J. G. M. RIETRA, R. J. VAN KETEL, AND H. C. ZANEN

Laboratorium voor de Gezondheidsleer, Mauritskade 57, Universiteit van Amsterdam, 1092 AD Amsterdam, The Netherlands

Contaminated tap water has been incriminated as a source of *Legionella pneumophila* in clinical outbreaks of Legionnaires disease. Environmental and clinical isolates belonging to the same serogroup are generally considered to be epidemiologically related. As the majority of infection is caused by *L. pneumophila* serogroup 1, more refined analytical methods for epidemiological studies might be useful. We therefore applied a combination of biochemical and immunochemical methods to the analysis of strains isolated during an outbreak of legionellosis in a newly built hospital in Amsterdam.

The investigated strains included isolates from eight patients (strains P1 to P8), 12 hot water taps (strains T1 to T12), two cooling towers (strains A and B), and an energy building (strain C). In addition the reference Philadelphia 1 and Philadelphia 2 strains (Phi1 and Phi2) were examined. The 23 isolated strains were all *L. pneumophila* serogroup 1 by direct immunofluorescence with reagents from the Centers for Disease Control. In slide agglutination with rabbit antiserum raised against the reference Phi2 strain (Phi2-Ab), all of the strains except strain C agglutinated strongly.

Rabbits were immunized with heated suspensions of the strains in complete Freund adjuvant. Adsorbed and unadsorbed antisera were reacted with heated cell suspensions of all 25 strains in the double diffusion test. Two main precipitation lines were seen. The line near the antibody well was common to all the strains with Phi2-Ab. With P1-Ab and T1-Ab this line was absent with strains A, B, and C as antigen. After adsorption of Phi2-Ab with P1 or T1 and of P1-Ab or T1-Ab

with Phi2, the line disappeared. The other line near the antigen well remained after adsorption and showed precipitation patterns which are summarized in Table 1. It appears that serogroup-specific antigen is adsorbed, leaving the serotype-specific antigen to react.

In immunoelectrophoresis (3) the reaction of strains from patients and from water with Phi2-Ab showed similar patterns, whereas those of strains A, B, and C were different. Moreover the antigens near the antigen wells of strains A, B, and C were negatively charged, and those of strains Phi1, Phi2, P1 to P8, and T1 to T12 were positively charged (Fig. 1).

Cell envelopes obtained by sonication of strains P1 to P8, T1 to T12, A, B, C, Phi1, Phi2, and *L. pneumophila* serogroups 2 to 6 were subjected to sodium dodecyl sulfate-polyacrylamide gel electrophoresis (2). The gel slabs were

TABLE 1. Results of double diffusion test in which representative rabbit antisera reacted with heated cell suspensions of 25 *L. pneumophila* serogroup 1 strains

Rabbit antibody	Reaction[a] with the following antigen:						
	Phi1	Phi2	P1 to P8	T1 to T12	A	B	C
Phi2-Ab	+	+	+/−	+/−	+/−	+/−	w
P1-Ab	+/−	+/−	+	+	0	0	w
T1-Ab	+/−	+/−	+	+	0	0	w
A-Ab	0	0	0	0	+	+	+
C-Ab	0	0	0	0	+/−	+	+

[a] Reaction for the precipitation line near the antigen well. Symbols: +, complete identity; +/−, partial identity; w, weak precipitation; 0, no visible reaction.

Phi2   T VI   ————————————patients strains————————————

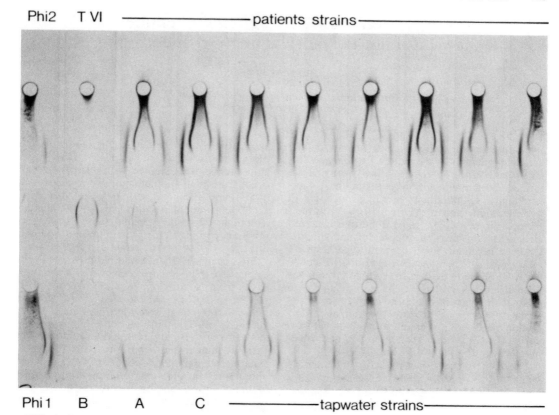

Phi 1   B   A   C   ————————tapwater strains————————

FIG. 1. Immunoelectrophoresis of strains of *L. pneumophila* from patients, tap water strains, and control strains reacted with adsorbed Phi2 antiserum.

stained with Coomassie brilliant blue (1) for proteins and with silver for lipopolysaccharides (4). Two major proteins were present in all strains. They had molecular weights of approximately 25,000 and 47,000; the 25,000 molecular-weight protein was heat modifiable. Strains Phi1 and Phi2 and the strains of serogroups 2 to 6 showed clear differences in protein and lipopolysaccharide patterns. The two major proteins of strains P1 to P8 and T1 to T12 had similar mobilities, whereas those of strains A, B, and C showed small differences. Silver staining revealed quantitative differences between strains P1 to P8 and between strains T1 to T12, the meaning of which remains to be elucidated. The lipopolysaccharide patterns of strains A, B, and C not only differed from the other strains quantitatively but also differed in mobility.

These data indicate immunological and biochemical differences among the 23 strains isolat-ed at the hospital in Amsterdam, despite the fact that they all belonged to the same serogroup. Strains isolated from tap water and from patients showed greater similarity with each other than with strains isolated from the cooling towers or from the energy building or with the Phi1 and Phi2 reference strains. Thus, these methods enable a further differentiation in subtypes.

LITERATURE CITED

1. **Fairbanks, G., T. L. Steck, and D. F. H. Wallach.** 1971. Electrophoretic analysis of the major polypeptides of the human erythrocyte membrane. Biochemistry **10:**2606–2617.
2. **Laemmli, U. K.** 1970. Cleavage of structural proteins during the assembly of the head of bacteriophage T4. Nature (London) **227:**680–685.
3. **Scheidegger, J. J.** 1955. Une micro-méthode de l'immuno-électrophorése. Int. Arch. Allergy Appl. Immunol. **7:**103–110.
4. **Tsai, C.-M., and C. E. Frasch.** 1982. A sensitive silver stain for detecting lipopolysaccharides in polyacrylamide gels. Anal. Biochem. **119:**115–119.

Analysis of Detergent-Soluble *Legionella* Peptides by Sodium Dodecyl Sulfate-Polyacrylamide Gel Electrophoresis

M. LEMA AND A. BROWN

William Jennings Bryan Dorn Veterans Hospital and University of South Carolina School of Medicine, Columbia, South Carolina 29201

Legionella pneumophila and phenotypically similar organisms of the genera *Legionella, Tatlockia*, and *Fluoribacter* (1, 3) have been characterized by antigen-antibody techniques, fatty acid analysis, ubiquinone profiling, nucleic acid hybridization, and a relatively few "routine" phenotypic tests. The most definitive test, nucleic acid hybridization, is complex and tedious. On the other hand, standard phenotypic tests are often not sufficiently discriminating. Electrophoresis of crude extracts of soluble protein has been used either alone or as a supplement to other techniques in the study of many bacterial groups. This paper provides preliminary information about the protein patterns obtained with various strains of *L. pneumophila* and other members of the family *Legionellaceae*.

The bacteria used in this study included 21 strains of *L. pneumophila* representing serogroups 1 through 6, 9 strains of *Tatlockia (Legionella) micdadei*, 4 *Fluoribacter* strains representing three species, and 1 *Legionella longbeachae* strain. The confluent (72-h, 37°C) bacterial growth on buffered charcoal-yeast extract agar was harvested and suspended in 2 ml of a solution of 0.1 M Tris-hydrochloride (pH 6.8)–15% glycerol–2 mM phenylmethylsulfonyl fluoride. To this, 10% sodium dodecyl sulfate (SDS) was added to yield a final concentration of 2.0%. The suspension was vigorously mixed, placed in a boiling water bath for 5 min, again mixed, and then centrifuged at room temperature for 10 min at 10,000 rpm with a JA-20 rotor in a Beckman J21B centrifuge. The supernatant was removed, and the protein content was determined by the method of Lowry et al. (4). Extracts were stored at −20°C.

The stacking gel was 4.5% polyacrylamide–0.125 M Tris-hydrochloride–0.1% SDS (pH 6.8); the running gel was 8.5% polyacrylamide–0.375 M Tris-hydrochloride–0.1% SDS (pH 8.8). The running buffer was 0.025 M Tris-hydrochloride–0.2 M glycine–0.1% SDS (pH 8.8). Equal volumes of sample and 2× sample buffer (10% glycerol, 4.0% SDS, 0.02% bromthymol blue, 10% [vol/vol] 2-mercaptoethanol, 0.125 M Tris-hydrochloride [pH 6.8]) were mixed and placed in a boiling water bath for 5 min. Samples contained 150 to 200 μg of protein in Coomassie blue-stained gels and 10 μg of protein per sample in gels which were to be silver stained. Electrophoresis was performed at 100 V (constant volt-

age) until the tracking dye reached the bottom of the gel. Gels were fixed in 12.5% (wt/vol) trichloroacetic acid, stained with 0.018% (wt/vol) Coomassie blue in methanol-water-acetic acid (5:5:1), and destained with 10% (vol/vol) acetic acid. Gels were also stained by a silver staining technique (Gelcode; Upjohn Co., Kalamazoo, Mich.) according to the protocol of the manufacturer.

Figures 1 and 2 show the patterns seen with Coomassie blue-stained gels with protein extracts from isolates representing five of the six serogroups of *L. pneumophila, T. micdadei, Fluoribacter (Legionella) bozemanae*, and *L. longbeachae*. The patterns of the *L. pneumophila* strains are visually very similar and could be easily distinguished from those of the other genera. *L. longbeachae*, by its DNA homology, ubiquinone profile, and carbohydrate pattern (data not shown) and by its protein pattern (Fig. 1), probably is a nonfluorescing *Fluoribacter* species. In the gels stained by the silver staining technique (data not shown), more than 50 to 60 distinct bands of various colors could be seen in each lane, even though 20-fold-less material was used.

Figure 2 shows the Coomassie blue staining pattern of extracts obtained from serogroup 1 *L. pneumophila* strains isolated at the Veterans Administration Medical Center, Pittsburgh, Pa. The characteristics of these strains have been reported previously (2); the strains include environmental isolates containing a unique 80-megadalton plasmid (lanes 1 through 5), environmental isolates without a plasmid (lanes 6 and 7), and plasmidless clinical isolates (lanes 8 and 9). The patterns are similar to those of the *L. pneumophila* strains shown in Fig. 1. The most obvious difference seen between strains in Fig. 2 was due to the presence of a dense low-molecular-weight band in the extracts of the five plasmid-containing environmental isolates (lanes 1 through 5). Other more subtle differences may be seen; however, the major band patterns are almost identical. A band with identical mobility and appearance to that seen in the plasmid-containing isolates was also seen in some plasmidless strains from other outbreaks.

When the densitometer tracings of Coomassie blue-stained gels were compared, very high correlation coefficients were obtained for the following comparisons: (i) different preparations of

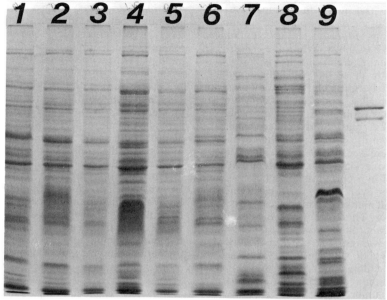

FIG. 1. Coomassie blue-stained gel patterns of extracts of strains representing various serogroups of *L. pneumophila*, with one representative each of *T. micdadei*, *L. longbeachae*, and *F. bozemanae*. Strains: lane 1, Philadelphia 1 *(L. pneumophila)*; lane 2, Togus 1 *(L. pneumophila)*; lane 3, Bloomington 2 *(L. pneumophila)*; lane 4, Los Angeles 1 *(L. pneumophila)*; lane 5, Dallas 1E *(L. pneumophila)*; lane 6, 687 *(L. pneumophila)*; lane 7, TATLOCK *(T. micdadei)*; lane 8, Long Beach 4 *(L. longbeachae)*; lane 9, WIGA *(F. bozemanae)*. The last lane contains bovine serum albumin and ovalbumin.

the same strain of *L. pneumophila* (Philadelphia 1), $r = 0.983 \pm 0.014$ ($n = 7$); (ii) epidemiologically related plasmid-containing environmental

L. pneumophila strains, $r = 0.985 \pm 0.004$ ($n = 7$); and (iii) *T. micdadei* isolates (TATLOCK and four different clinical PPA strains), $r = 0.983 \pm$

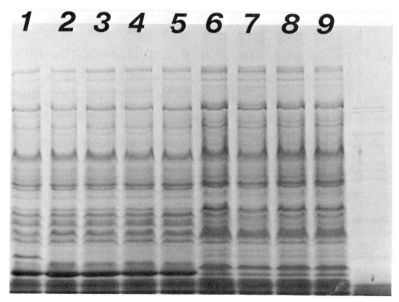

FIG. 2. Coomassie blue-stained gel patterns of *L. pneumophila* strains from the Pittsburgh Veterans Administration Medical Center, including plasmid-containing environmental isolates (lanes 1 through 5), plasmidless environmental isolates (lanes 6 and 7), and plasmidless clinical isolates (lanes 8 and 9). The specific isolates are as follows: lane 1, 1CMH; lane 2, 11EJ; lane 3, 10E36; lane 4, 10E115; lane 5, HWT3; lane 6, 11ES; lane 7, 7W; lane 8, FOS; lane 9, CAR. The last lane contains a mixture of protein standards.

0.010 ($n = 5$). The intergroup comparisons between strains of *L. pneumophila, T. micdadei,* and the various *Fluoribacter* species ranged from $r = 0.593$ to 0.898. When the patterns of plasmid-containing environmental isolates were compared with those of plasmidless isolates from the Pittsburgh Veterans Administration Medical Center, a correlation coefficient of 0.710 ± 0.009 was obtained; however, when the most prominent plasmid-associated band was eliminated from the comparison, the correlation coefficient increased to 0.966 ± 0.005. Thus, a difference in one major band between two patterns resulted in a correlation coefficient which was as low, or even lower, than that found in a comparison of patterns which were not at all visually similar. Therefore, while this test could detect significant relatedness with values greater than 0.96, it was not useful to distinguish between degrees of difference, at least as we performed the data analysis. With automated digitization of the scans and with the comparison of a larger number of smaller intervals, the useful range of values would probably be greater.

The data obtained from the Coomassie blue staining patterns are restricted to band position and density. With the silver stain used in this study another parameter, color, was added, thus significantly increasing the amount of available information. The clarity and distinctiveness of these patterns make data analysis simple. Characteristic patterns were observed for the *L. pneumophila* and *T. micdadei* strains. The *Fluoribacter* species, being a genetically heterogeneous group, had more variable peptide patterns. A good correlation has been observed by others between similarities found by DNA homology studies and those found through the examination of protein electrophoretic patterns. A comparison of the peptide patterns of the *Legionellaceae* with the results of DNA homology studies (1, 3) shows that the methods are in agreement with respect to the grouping of organisms tested. Other techniques, such as carbohydrate profiling, also support these taxonomic distinctions among the genera *Legionella, Tatlockia,* and *Fluoribacter* (A. Fox, P. Y. Lau, A. Brown, S. L. Morgan, Z.-T. Zhu, M. Lema, and M. Walla, this volume). Because of its sensitivity and ability to discriminate between proteins with color differences, the silver staining technique in particular appears to be a powerful tool for both taxonomic and epidemiological studies.

LITERATURE CITED

1. **Brown, A., G. M. Garrity, and R. M. Vickers.** 1981. *Fluoribacter dumoffii* (Brenner et al.) comb. nov. and *Fluoribacter gormanii* (Morris et al.) comb. nov. Int. J. Syst. Bacteriol. **31:**111–115.
2. **Brown, A., R. M. Vickers, E. M. Elder, M. Lema, and G. M. Garrity.** 1982. Plasmid and surface antigen markers of endemic and epidemic *Legionella pneumophila* strains. J. Clin. Microbiol. **16:**230–235.
3. **Garrity, G. M., A. Brown, and R. M. Vickers.** 1980. *Tatlockia* and *Fluoribacter:* two new genera of organisms resembling *Legionella pneumophila.* Int. J. Syst. Bacteriol. **30:**609–620.
4. **Lowry, O. H., N. J. Rosebrough, A. L. Farr, and R. J. Randall.** 1951. Protein measurement with the Folin phenol reagent. J. Biol. Chem. **193:**265–275.

Plasmid Profiles of Clinical and Environmental Isolates of *Legionella pneumophila* Serogroup 1

WILLIAM E. MAHER, JOSEPH F. PLOUFFE, AND MICHAEL F. PARA

Division of Infectious Diseases, Department of Medicine, The Ohio State University, Columbus, Ohio 43210

Several groups have observed plasmids in clinical and environmental *Legionella pneumophila* isolates (1, 3, 5, 7, 8). Plasmid analysis of *L. pneumophila* serogroup 1 (SG-1) isolates from diverse geographic locations has revealed 80-, 84-, 60-, and 43-megadalton (Mdal) plasmids (3, 5, 8). Studies conducted by one group (3) have correlated plasmid content with differential virulence and surface antigen composition, but, overall, specific gene products or functions have yet to be assigned to any of these plasmids. We report here the results of plasmid analysis of clinical and environmental SG-1 isolates from an area in Columbus, Ohio, serviced by the same main cold-water line.

Clinical and environmental specimens were cultured on buffered charcoal-yeast extract agar (4) after selective acid washing (2). Organisms were identified as SG-1 by characteristic colonial morphology and direct fluorescent-antibody staining with reagents supplied by the Centers for Disease Control, Atlanta, Ga. Plasmid analysis was conducted by the alkaline sodium dodecyl sulfate lysis method of Kado and Liu (6). Molecular weights of the crude plasmid DNA were estimated from their electrophoretic mobilities on horizontal agarose gels relative to those of the four plasmids of *Legionella bozemanii* (35, 40, 45, and 80 Mdal) (1). The *L. bozemanii* isolate was obtained from the Centers for Disease Control.

Environmental samples were obtained from

the hot-water tanks and plumbing fixtures of three buildings (UH, RH, and UPH) within the Ohio State University medical complex and from one neighboring community hospital (CH). The same city water services all of these sites, but each has an independent hot-water system. Clinical SG-1 isolates were obtained from patients with nosocomially acquired Legionnaires disease seen in RH and UH during the period of the environmental survey (January 1982 to February 1983). No culture-positive cases of Legionnaires disease have been reported from UPH or CH.

Analysis of environmental isolates from these four sites revealed five plasmid profiles: (i) plasmidless (UH-1); (ii) a single 45-Mdal plasmid (UH-2); (iii) a 45-Mdal and a 85-Mdal plasmid (RH-1); (iv) a single 85-Mdal plasmid (CH-1); and (v) a single 33-Mdal plasmid (UPH-1). The plasmid profiles are shown in Fig. 1 and are listed according to their environmental origins in Table 1. Each plasmid profile was a stable characteristic of its respective isolate, and no spontaneous curing of plasmids was noted de-

TABLE 1. Distribution of clinical (C) and environmental (E) SG-1 isolates among different hospital buildings

No. of plasmids (Mdal)	SG-1 isolates in:			
	UH	RH	CH	UPH
0	E and C	—	—	—
1 (45)	E and C	—	—	—
2 (45 and 85)	E	E and C	—	—
1 (85)	—	—	E	—
1 (33)	—	—	—	E

spite many plasmid analyses and several passages on buffered charcoal-yeast extract agar.

Twenty clinical SG-1 isolates were available for analysis. Of these, 16 (80%) were plasmidless (UH-1), 3 (15%) possessed a single 45-Mdal plasmid (UH-2), and 1 (5%) contained 45- and 85-Mdal plasmids (RH-1). The patient whose SG-1 isolate contained two plasmids was housed in RH. The remaining 19 cases were seen in UH.

Johnson and Schalla (5) have observed an 84-Mdal plasmid in SG-1 isolates Albuquerque 1, Flint 1, and OLDA. These plasmids displayed the same EcoRI cleavage pattern, implying identity. The same group has detected 43- and 85-Mdal plasmids in SG-1 isolate Allentown 1. We have found SG-1 in Columbus, Ohio, with different combinations of 45- and 85-Mdal plasmids. These data support the idea that certain *Legionella* plasmids have been disseminated over a broad geographic area. Whether this is a consequence of frequent plasmid transfer (conjugation) between various legionellae, acquisition of plasmids by legionellae from ubiquitous environmental commensals, or a selective advantage imparted by specific plasmid gene products is unknown. We are currently investigating the possibility that plasmid content influences the colonization of hot-water systems and are attempting to define the selective pressures involved.

The presence of the same plasmid profiles in clinical and environmental SG-1 isolates further supports the concept of acquisition of Legionnaires disease from the local hospital environment. Similarly, this suggests that plasmid analysis will be useful in the epidemiological study of nosocomial Legionnaires disease.

The distribution of plasmid profiles among clinical SG-1 isolates was skewed. Most clinical isolates were plasmidless, and 19 of 20 cases of nosocomial Legionnaires disease were associated with building UH. During the same time period, only one case of Legionnaires disease was seen in RH, which houses similar numbers and types of patients. The characteristics of SG-1 isolates bearing specific plasmids which may

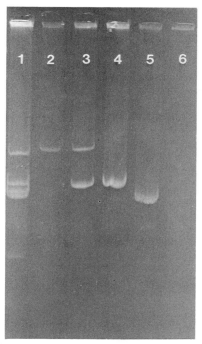

FIG. 1. Agarose gel (0.6%) electrophoresis of alkaline sodium dodecyl sulfate lysates of *Legionellaceae*. Lanes: 1, *L. bozemanii* (WIGA); 2, community hospital *L. pneumophila* isolate (CH-1); 3, 1979 annex *L. pneumophila* isolate (RH-1); 4, main hospital *L. pneumophila* isolate (UH-2); 5, psychiatric hospital *L. pneumophila* isolate (UPH-1); 6, plasmidless main hospital *L. pneumophila* isolate (UH-1).

be responsible for these observations are currently under study (J. F. Plouffe, M. F. Para, B. Hackman, L. Webster, and W. E. Maher, this volume).

LITERATURE CITED

1. Aye, T., K. Wachsmuth, J. C. Feeley, R. J. Gibson, and S. R. Johnson. 1981. Plasmid profiles of *Legionella* species. Curr. Microbiol. 6:389–394.
2. Bopp, C. A., J. W. Sumner, G. K. Morris, and J. G. Wells. 1981. Isolation of *Legionella* spp. from environmental water samples by low-pH treatment and use of a selective medium. J. Clin. Microbiol. 13:714–719.
3. Brown, A., R. M. Vickers, E. M. Elder, M. Lema, and G. M. Garrity. 1982. Plasmid and surface antigen markers of endemic and epidemic *Legionella pneumophila* strains. J. Clin. Microbiol. 16:230–235.
4. Feeley, J. C., R. J. Gibson, G. W. Gorman, N. C. Langford, J. K. Rasheed, D. C. Mackel, and W. B. Baine. 1979. Charcoal-yeast extract agar: primary isolation medium for *Legionella pneumophila*. J. Clin. Microbiol. 10:437–441.
5. Johnson, S. R., and W. O. Schalla. 1982. Plasmids of serogroup 1 strains of *Legionella pneumophila*. Curr. Microbiol. 7:143–146.
6. Kado, C. I., and S. T. Liu. 1981. Rapid procedure for detection of large and small plasmids. J. Bacteriol. 145:1365–1373.
7. Knudson, G. B., and P. Mikesell. 1980. A plasmid in *Legionella pneumophila*. Infect. Immun. 29:1092–1095.
8. Mikesell, P., J. W. Ezzell, and G. B. Knudson. 1981. Isolation of plasmids in *Legionella pneumophila* and *Legionella*-like organisms. Infect. Immun. 31:1270–1272.

Application of Capillary Gas-Liquid Chromatography for Subtyping *Legionella pneumophila*

N. P. MOYER, A. F. LANGE, N. H. HALL, and W. J. HAUSLER, JR.

Hygienic Laboratory, University of Iowa, Iowa City, Iowa 52242

Members of the genus *Legionella* contain large amounts of branched-chain fatty acids, with both quantitative and compositional differences occurring between species. The only reported quantitative differences in cellular fatty acids within a species are small (2 to 4%) variations in the amount of *dl-cis*-9,10-methylhexadecanoate (17.0Δ) in the Pontiac and Philadelphia 2, Philadelphia 3, and Philadelphia 4 strains (2, 4) and in environmental and control strains of *Legionella pneumophila* (Philadelphia 1 and Dallas 1 E) (1). Relative differences in fatty acids within a species have been attributed to variations in growth medium. The effect of culture incubation time before extraction and methylation upon the gas-liquid chromatography

(GLC) profile is controversial (2, 3). No major differences in cellular fatty acid composition have been shown between serogroups of *L. pneumophila* or within a serogroup for either environmental or patient strains.

We examined 20 patient and environmental cultures isolated during a nosocomial outbreak and subsequent environmental surveillance in Iowa City, Iowa, by analyzing fatty acid profiles produced by capillary GLC. Both patient and environmental cultures were biotypically and serotypically identified as *L. pneumophila* serogroup 1. Cultures were grown on buffered charcoal-yeast extract agar at pH 6.9 and incubated in 5% CO_2 at 95% relative humidity and 35°C for 48 h. After extraction and methylation (6) the

TABLE 1. Specimens and sources

Source	Group A[a]		Group B		Group C	
	Patient	Room[b]	Patient	Room[b]	Patient	Room[b]
University Hospital	D144 D159/D160[c] D184	7044 (7066, 7068) (2089, 3094) 7067	D176	(2048, 2095) 3066, W225	D337	(7083) 7093, 7045, 3059, W316
VA Hospital	D157[d]	(4W14)			D234 D238	(7W14) (4W14) 8W14, 1W35
Community acquired					D172	(7095)

[a] Includes control strains Philadelphia 1 and Philadelphia 2.
[b] Parentheses indicate patient room where no environmental samples were cultured.
[c] Two isolates obtained from one patient.
[d] Patient received radiation therapy at University Hospital.

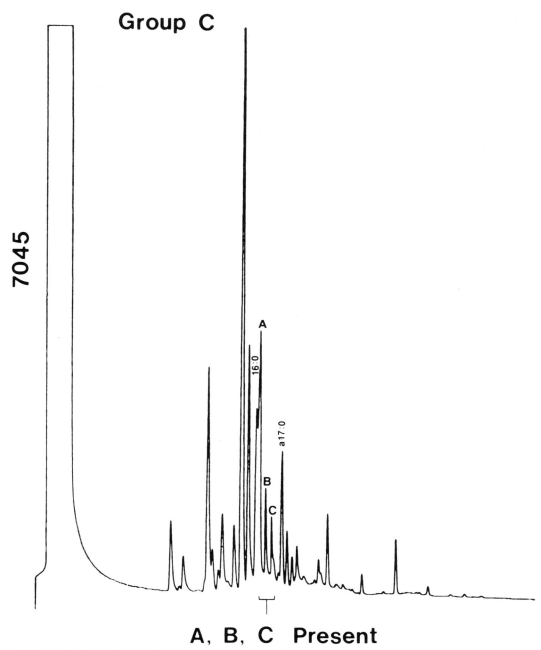

FIG. 1. Capillary GLC profile of a representative group C strain of *L. pneumophila* serogroup 1 isolated from tap water in a patient room (room 7045) in University Hospital, Iowa City. Peaks A, B, and C are unidentified cyclic, unsaturated, and branched-chain C_{16} to C_{18} fatty acids.

samples were analyzed on a Varian 3700 GLC equipped with a flame ionization detector, the Hewlett-Packard HP 3388A data analysis system, and a 30-m SE-54 silica capillary column. Peak identification was done by comparing re-

tention times of standards with those of the unknowns. Selected peaks were confirmed by mass spectrographic analysis.

During analysis of the chromatograms from the 20 Iowa City isolates, three distinct patterns emerged. These patterns were arbitrarily classified into groups labeled A, B, and C; group C is shown in Fig. 1. Group A and control cultures of *L. pneumophila* Philadelphia 1 and Philadelphia 2 were closely related, but minor differences were detected between strains Philadelphia 1 and Philadelphia 2. Groups B and C were closely related but differed in the number and relative amounts of unidentified cyclic, unsaturated, and branched-chain C_{16} to C_{18} fatty acids (Fig. 1, peaks A, B, and C). Definitive identification of these components by mass spectroscopy is under way.

The patient and hospital room association for the Iowa City isolates of *L. pneumophila* is presented in Table 1. A patient culture with a corresponding environmental culture was available only for group A (patient D144, room 7044); however, the similarity between patient isolate GLC profiles and those from nearby rooms reinforced the hypothesis that capillary GLC profiles could be used to determine relationships between patient and room culture pairs in support of the epidemiological investigation of nosocomial outbreaks. Patient D157 could have been exposed at Veterans Hospital or University Hospital, since environmental surveillance identified contaminated sites to which the patient was exposed. Chromatographic evidence supports infection resulting from exposure while receiving radiation therapy at University Hospital. While these findings are encouraging, further refinement of this chromatographic technique is necessary for the production of consistent epidemiologically useful information.

ADDENDUM IN PROOF

The legend to Fig. 1 is slightly misleading and should read ". . . branched-chain C-16 to C-18 fatty acid methyl esters."

LITERATURE CITED

1. **Garrity, G. M., E. M. Elder, B. Davis, R. M. Vickers, and A. Brown.** 1982. Serological and genotypic diversity among serogroup 5-reacting environmental *Legionella* isolates. J. Clin. Microbiol. **15**:646–653.
2. **Moss, C. W., and S. B. Dees.** 1979. Further studies of the cellular fatty acid composition of Legionnaires disease bacteria. J. Clin. Microbiol. **9**:648–649.
3. **Moss, C. W., D. E. Karr, and S. D. Dees.** 1981. Cellular fatty acid composition of *Legionella longbeachae* sp. nov. J. Clin. Microbiol. **14**:692–694.
4. **Moss, C. W., R. E. Weaver, S. B. Dees, and W. B. Cherry.** 1977. Cellular fatty acid composition of isolates from Legionnaires disease. J. Clin. Microbiol. **6**:140–143.
5. **Orrison, L. H., W. B. Cherry, R. L. Tyndall, C. B. Fliermans, S. B. Gough, M. A. Lambert, L. K. McDougal, W. F. Bibb, and D. J. Brenner.** 1983. *Legionella oakridgensis*: unusual new species isolated from cooling tower water. Appl. Environ. Microbiol. **45**:536–545.
6. **Tisdall, P. A., G. D. Roberts, and J. P. Anhalt.** 1979. Identification of clinical isolates of mycobacteria with gas-liquid chromatography alone. J. Clin. Microbiol. **10**:506–514.

SUMMARY

Sources of Legionellosis

DAVID W. FRASER

Swarthmore College, Swarthmore, Pennsylvania 19081

Since the First International Symposium on Legionnaires Disease there has been remarkable progress in the search for the environmental source of *Legionella* spp. and delineation of conditions under which they spread to infect humans. Much of the work presented at the present Symposium focused on this issue of environmental sources and on ways that this spread could be controlled. In summarizing these new findings, it may be useful to look carefully at what it means for an environmental site to be a "source" of legionellosis. One such conceptual scheme has the following six links in the chain of causation that leads from environmental site to infection in humans: (i) that there exist an environmental reservoir where legionellae live; (ii) that there be one or more amplifying factors which allow legionellae to grow from low concentrations to high ones; (iii) that there be some mechanism for dissemination of legionellae from the reservoir so as to expose people; (iv) that the strain of *Legionella* that is disseminated be virulent for humans; (v) that the organism be inoculated at an appropriate site on the human host; and (vi) that the host be susceptible to *Legionella* infection. In reviewing the progress of research as presented at the Symposium, it may be of use both to outline how epidemiological methods can be used to document each link in this chain and to see how each element of the reported research fits into the scheme.

Reservoir. To show that an environmental site is in fact a reservoir of *Legionella* spp. that infect humans, it is sufficient to show that there is an association between the presence of legionellae at that site and the occurrence of legionellosis. Alternatively, it can be shown that control methods that eradicate legionellae from the site lead to the disappearance of disease.

Several studies described at this Symposium add to the body of information that implicates potable water, especially hot water, as a major reservoir of *Legionella* spp. Potable water supplies in hospitals have been implicated by Morse and colleagues, Jones and colleagues, and Plouffe and colleagues. Water in the home has been shown by Arnow to be contaminated with legionellae, and some serological data suggest

that this contamination may be associated with infection of residents. Jones and colleagues showed that water used in a whirlpool bath could be the source of Pontiac fever. Additional data from Benidorm, Spain, and Lido de Savio, Italy, suggest that drinking water in hotels can be implicated in legionellosis. Finally, Beam and colleagues suggest that drinking water in a developmental center in New York may have been the reservoir for outbreaks of legionellosis. Of considerable interest is the report by Ruden and colleagues, who suggest that potable water may *not* be the source of an outbreak of *Legionella micdadei* infection at Montefiore Hospital. They observed that *Legionella pneumophila* could be recovered in much larger numbers than *L. micdadei* from their hospital water supply, during a time when nosocomial disease caused by *L. micdadei* was more common than that caused by *L. pneumophila*. Raising the temperature of the water led to a marked drop in the concentration of *L. pneumophila* in water but was not associated with a decrease in clinical cases due to *L. micdadei*. These observations might still be consistent with a potable water source of *L. micdadei* infection if heat treatment were not as effective in decreasing environmental contamination with *L. micdadei* as it is with *L. pneumophila* and if *L. micdadei* were considerably more difficult than *L. pneumophila* to recover from potable water, but the authors' suggestion that alternative reservoirs of *L. micdadei* should be looked for certainly seems warranted.

The report by Herwaldt and colleagues that industrial coolants can be epidemiologically important reservoirs is of great interest. Coolants are used in a wide variety of manufacturing processes and may expose large numbers of workers. An unpublished study by the Centers for Disease Control and the New York State Health Department has shown *L. pneumophila* contamination of industrial coolants at a plant in Jamestown, N.Y., but in that case the epidemiological link to cases of legionellosis was not nearly as convincingly demonstrated as in the engine assembly plant in Ontario.

The study of kibbutzim in Israel by Bercovier and colleagues raises the intriguing thought that wastewater might be a reservoir for legionellosis. In this case wastewater is used for drip or spray irrigation of fields. The finding of *L. pneumophila* in the water and some evidence of increased seropositivity in kibbutz workers where wastewater is used bears further investigation.

Amplifying factors. The recognition that some environmental factors may lead to marked increases in the concentration of *L. pneumophila* opens an exciting area for future work. The presence of such amplifying factors can be shown epidemiologically by demonstrating an association of the presence of that factor with markedly increased concentrations of *Legionella*.

Relatively low temperatures in hot water systems have been shown by several workers to be associated with increasing concentrations of *L. pneumophila* or increasing probabilities of positive cultures. The study by Arnow and colleagues in Chicago, Ill., homes suggest that water with temperatures of less than 55°C is particularly likely to be heavily positive. It is not clear, however, that temperatures of less than 55°C are sufficient to lead to large concentrations of legionellae. The presence of dead ends in piping or obstruction, such as aerators on sinktaps, seems to play a role. The fascinating observations by Colburne and colleagues that certain resins on rubber gaskets may play a critical role could point to relatively simple ways to control *Legionella* concentrations in water systems. Active investigations along those lines are likely to pay handsome dividends.

There was considerable talk at the Symposium about the roles of sediment, location of the heating element within the hot water system, and the metals used in pipes on the concentration of legionellae in water systems, but few interpretable data were presented. The observation that high copper levels may inhibit *Pseudomonas* spp. much more than *Legionella* spp. may be of some importance, but any suggestion that this is relevant to the fact that *L. pneumophila* was recovered from water from a Zambian copper mine would be purely speculative.

There is a growing body of information that *Legionella* spp. can grow in association with green algae, ciliated protozoa, and amoebae, in addition to the earlier reports of growth with cyanobacteria. However, there is little evidence that these more complex life forms are important in increasing the risk of legionellosis by amplifying legionellae in the environment. Ecological studies that demonstrated the epidemiological importance of these observations would be of great interest.

Disseminators. The importance of a particular disseminator of legionellae can be shown physically and microbiologically by demonstration of the dissemination of legionellae from a particular site. By using epidemiological techniques one would expect to be able to show that infected people had exposure to the putative dissemination side that corresponds closely with the demonstrated pattern of distribution of organisms.

The demonstration by Morse and colleagues that hospital rooms of legionellosis patients were strikingly near communal showers comes closer than most studies to suggesting a way that potable water might be disseminated to infect people. The critic may have a bit of residual concern about the confounding effect of length of hospital stay in the Morse study, but the possibility that showers may infect those nearby should be evaluated in future investigations. Bartlett's report that in a Benidorm hotel the person who bathed earlier in the morning was more likely to acquire legionellosis fits with the concept that *Legionella* spp. may increase in concentrations when water stagnates in pipes.

Arnow's observation of the association of the use of room humidifiers with seropositivity to legionellae is tantalizing, although based on few data. The association of the use of respiratory equipment and acquisition of legionellosis suggests that the equipment acted as a disseminator in the Connecticut hospital studied by Jones, but the problem of the possible confounding effect of other factors must lead to some caution in interpreting that association.

Additional observations of interest include the role of the industrial grinding procedures in disseminating organisms in the engine assembly plant (Herwaldt and colleagues) and the observation that 2 ft (ca. 0.6m) above the surface of the water in an implicated whirlpool particles could be found that were 3 to 5 μm in diameter, a size that would permit deposition of bacteria in alveolae (Jones and colleagues). Vogt and colleagues presented additional information on the role of cooling towers as disseminators of legionellae in their studies of community outbreaks in Vermont.

It would be of great interest to focus more work in the future on direct demonstration of the dissemination of *Legionella* spp. from mechanisms that might act in this crucial way to spread the organism. It is only with this direct demonstration that one can separate the role of the disseminator from that of the relative virulence of the strain.

Virulence of strains. To demonstrate that strains are of different virulence to humans, it is necessary to show that given equal exposure of groups to a disseminator of organisms, there is an association of a strain of particular characteristics with a risk of illness or infection. In this Symposium several intriguing marker systems have been reported for identifying strains. These include the use of monoclonal antibodies or absorbed sera to distinguish strains immunologically, the use of plasmid analysis, and the use of gas-liquid chromatography. Although a great deal of enthusiasm about the use of these markers for epidemiological purposes is appropriate, much more work is needed to show that the markers are in fact stable and reproducible. One runs the risk of a circular argument if one observes that an environmental strain and a clinical strain have the same markers, deduces from that that the site from which the environmental strain was recovered was in fact the source of infection for the human, and concludes therefrom that markers are useful epidemiologically. A more proper sequence would be first to show epidemiologically that a particular site was the likely source of infection of humans by showing a particular pattern of exposure of affected persons to that site that differed from that of controls and, only then, to demonstrate that *Legionella* isolates from the site had the same marker characteristics as those from cases.

The observation by Plouffe and colleagues that the risk of legionellosis was strikingly different in two hospital buildings that had similar concentrations of legionellae of two different strain types in their water supplies may well prove to be a signal that the two strains were of markedly different virulence. However, it is also possible that an efficient disseminator of legionellae was present in only one of the two buildings, so that effective exposure of patients was markedly different despite similar concentrations of the organism in the water supplies. Until the disseminators and precise modes of exposure of patients to *Legionella* spp. have been identified it will not be possible to conclude with certainty that strains are of different virulence for humans.

The observation that passage of water through a recirculating cooling system of a power plant is associated with increased infectivity of legionellae for guinea pigs despite a slight decrease in observable numbers of organisms suggests another direction that might be explored in looking for factors that affect the virulence of strains.

In addition to the work that has been presented on variations in virulence among strains of *L. pneumophila*, it will be exceedingly interesting to see whether different species of *Legionella* vary in their virulence for humans.

Site of inoculation. To show that a particular site of inoculation is important in the spread of legionellosis to humans, it is of value to show an association of the type of exposure with the occurrence of illness or to show an association of a particular breach of local defenses with exposure and subsequent infection. Several investigations reported in this Symposium confirmed the importance of inhalation as a mode of infection for both Legionnaires disease and Pontiac fever. This inhalation appears to have occurred through the use of respiratory therapy equipment, exposure to a cloud of coolant droplets, proximity to showers, or exposure to 3- to 5-μm particles emitted by whirlpools. Little evidence was presented about alternative methods of exposure, although the possibility of aspiration is raised by the observation by direct immunofluorescence that there may be oropharyngeal colonization by *L. pneumophila* and *L. micdadei*. Also, the observation by Jones and colleagues that the use of antacids was associated with increased risk of legionellosis in the outbreak in Connecticut suggests the possibility that ingestion of legionellae may be of some importance, since the antacids could be expected to permit the breaching of the barrier normally presented by gastric acid.

Susceptibility of host. To demonstrate that variations in host susceptibility may be of importance in the spread of legionellosis, it is necessary to show that given equal exposure to the organisms there is an association of illness with selected host characteristics. Despite considerable sophisticated laboratory work on mechanisms of host defense, little progress on the issue of host susceptibility was seen in the epidemiologically oriented work presented at this conference. Essentially no information has been obtained on the host susceptibility factors that are important for *Legionella* spp. other than *L. pneumophila* and *L. micdadei*.

Testing the chain of causation. Once the chain of causation has been hypothesized for an outbreak of legionellosis, there are two important ways to test whether that chain is plausible. The first is to seek an explanation for the start of an outbreak. Herwaldt and colleagues showed that the outbreak of Pontiac fever in the engine assembly plant began immediately after a 9-day shutdown in which the incriminated industrial coolants were not in use. Herwaldt postulated that this interval may have permitted a growth of *Legionella feeleii*, although this growth was not directly demonstrated. The study by Bartlett et al. suggested that the reconnection of the supplemental well water supply was associated with one of the clusters of disease in Benidorm. In each instance the association of a particular plausible event with a subsequent outbreak adds some confidence to the interpretation of causation offered by the authors.

Of great practical importance is the testing of the hypothesized sequence of events by a specific intervention designed to stop the outbreak. Although one is always faced with the possibility of coincidence unless one undertakes a controlled intervention, one is more confident that one understands how an outbreak occurs if a rational intervention strategy is followed by the cessation of disease. It is important to realize that control can occur at any link in the chain of causation and does not necessarily involve elimination of legionellae from the reservoir. In fact, it may be far more efficient to focus control efforts on amplifiers and disseminators, especially with organisms that are as widely distributed as *Legionella* spp. Methods that appeared to have been effective in halting outbreaks included the turning off of the disseminator in the cases of the implicated whirlpool and the industrial grinding operation and the use of chlorine or high temperature to decrease numbers of *L. pneumophila* in potable water systems.

Areas for future epidemiological study. The aspect of epidemiological studies of legionellosis that I find most disquieting is that it has been exceedingly hard to document that the configuration of human exposure (as shown in most instances by case-control studies) fits neatly with the distribution of legionellae from the putative site of dissemination. Although airborne infections are notoriously tricky in this regard, one is left with the nagging concern that either we do not know the disseminator of *Legionella* spp. or we are missing some important modes of spread. In our enthusiasm for associating legionellosis with particular reservoirs such as potable water, we should not slacken our efforts to prove the modes of spread of legionellae in a rigorous fashion. It may be that our efforts to pinpoint sites of dissemination have been hampered by heterogeneity of virulence of *Legionella* strains, with only a small subpopulation of organisms being dangerous for humans. If so, the newly described epidemiological markers may provide invaluable help.

In looking for specific dangerous sites of *Legionella* concentration, it will be increasingly important to look for factors that amplify the numbers of virulent organisms and to document in a quantitative fashion the dissemination of organisms from those sites, using physical and bacteriological measures of shedding.

Well designed epidemiological studies can be of great value in confirming the utility of markers that distinguish strains of *Legionella*. This confirmation in turn is likely to increase the power of epidemiological investigations considerably.

The consistently low attack rate of Legionnaires disease suggests that most people are immune to *Legionella* spp. In many infectious diseases for which immunity is widespread, epidemiological investigations are made much more powerful by excluding those exposed persons who are already immune before using the pattern of ill and well exposed people to determine how exposure occurred. For legionellosis this is not possible, as serology provides at best an incomplete view of immunity in a population. It would be of great value to develop an immunological method such as the skin test to permit clear separation of immune from susceptible people.

Work is especially needed on the sources and modes of spread of *Legionella* spp. other than *L. pneumophila*. This is highlighted by the observation of an *L. micdadei* cluster that suggests that water may not be its reservoir.

Finally, there is a major question left from the last symposium regarding the factors that determine whether a particular outbreak of legionellosis will be in the form of Legionnaires disease or Pontiac fever. In investigations of future outbreaks of Pontiac fever it would be of great interest to determine the sizes of particles emitted from the putative disseminator, the ratio of live versus dead organisms disseminated, and the presence of other chemical and biological agents in the medium. In addition, it would be of value to assess urine of Pontiac fever patients for *Legionella* antigen, as this may provide a clue as to the size of the challenging dose and the occurrence of multiplication of organisms in the host.

ECOLOGY AND ENVIRONMENTAL CONTROL

STATE OF THE ART LECTURE

Current Microbiological Methods Used in the Analysis of Environmental Specimens for *Legionella* spp.

JAMES C. FEELEY

Respiratory and Special Pathogens Laboratory Branch, Center for Infectious Diseases, Centers for Disease Control, Atlanta, Georgia 30333

Only a few short years ago, the Legionnaires disease bacterium was detected and isolated (15). Although this organism initially seemed to defy detection, the rate that advances have been made in delineating its microbiology is probably higher than that for any other microbe in history. The program of this Symposium is ample testimony to this.

In depicting the state of the art for environmental methods, I will first address the types of specimens that should be collected and the methods that can be used to detect legionellae in them. I will then briefly discuss quantitation and marker systems. Since the methodologies for sampling and analyzing water and air are very different, I will discuss each separately.

Water. Legionellae have been isolated from a wide variety of types of water. Examples are potable water of hospitals and hotels, cooling-tower water, and ground- and surface waters. For the purpose of bacteriological analysis, water specimens can be classified into two types, "low count" and "high count." This is determined according to the concentration of non-*Legionella* microorganisms rather than the numbers of legionellae present. Low-count water specimens, such as potable water, must be concentrated. Consequently a large volume of water, usually 10 liters, is collected and concentrated before media are inoculated. Various methods can be used for this concentration, namely, high-speed continuous centrifugation, pressure filtration through membrane filters, and filtration through polycarbonate filters. In contrast, high-count water specimens, like cooling-tower waters, usually can be analyzed without concentration, and only 1 liter of water is usually collected. Both types of water samples either can be inoculated directly to media or can be exposed to low-pH (2.2) treatment for 10 min (1, 12) or heated at 60°C for 1 to 2 min (5) and then inoculated to media.

The state of the art for media and reagents for isolating and identifying legionellae is well portrayed by the commercial products currently available, as exhibited at the Symposium. Their worldwide availability should enable trained microbiologists in well-equipped laboratories to isolate legionellae from environmental water specimens.

Several types of media have been reported in the literature (8, 9, 13). However, the basal medium that appears to be the most commonly employed is a modification of charcoal-yeast extract agar (7). I believe that charcoal-yeast extract agar supplemented with ACES buffer (16) and alpha-ketoglutarate (2) is the best basal medium that presently is available for growing all *Legionella* species. This medium, termed BCYEα agar, supplemented with effective selective agents (1–4, 6, 17, 18), has enabled major breakthroughs in the recovery of legionellae from environmental specimens. I consider this one of the most important advances for the study of legionellae. Recovery of legionellae from water specimens can be better accomplished by direct plating than by inoculation of guinea pigs. Direct plating has allowed the processing of large numbers of specimens in a very short time, which is a great advantage during investigations of epidemics.

Currently there is no single selective medium or pretreatment method that can be used exclusively. A combination of pretreatment techniques and selective media is recommended, since some media will work for some specimens and species but not for others. Variation of microbial flora between geographical locations may be a partial explanation for this.

Air. Collection of air samples is considerably more difficult than collection of water samples. The most efficient air samplers for legionellae, unfortunately, have been and remain the individuals who become ill. Sentinel guinea pigs are second in efficiency and have been employed very successfully in an investigation of Pontiac fever (11, 14). Air sampling can also be accomplished by mechanical samplers like the Anderson multistage sampler and the all-glass impinger (19), but only a few isolations have been made with this type of equipment. Even so, the fact remains that better documentation of legionellae can be done by mechanical samplers than by passage through animals. The reason for poor recovery with mechanical samplers may be the injury they inflict on legionellae, or any microorganism, through dehydration and air turbulence. For this reason, another detection system for documenting the presence of legionellae should also be used. I recommend direct fluorescent-antibody staining, keeping in mind that extreme caution should be used in interpreting the results because of the possible serological cross-reactions with non-*Legionella* bacteria.

Quantitation. The current methods for quantitation only provide estimates of the numbers of legionellae in environmental specimens. The reasons for this is that legionellae are being injured in the collection and processing of the specimens. Pretreatment of specimens with acid and heat and the use of selective agents in the media are extremely detrimental to the survival of legionellae and inhibit the growth of some of the legionellae that are present in a sample. For best recovery no selective agent should be used; however, this currently is not possible. At present, I recommend direct fluorescent-antibody and vital staining (10). In any case improved methods must be developed to quantitate the legionellae in a specimen.

Marker systems. A number of different typing systems, such as monoclonal antibody, outer membrane protein, and plasmids, have been recently developed. They should enable more precise characterization of a *Legionella* isolate and therefore have immense epidemiological value, permitting specific tracing of an etiological agent to its source. Another equally important development that needs to be made is a simple test for the determination of the virulence of *Legionella* strains.

In conclusion, the current methods for isolating legionellae from environmental specimens are much improved since the First International Symposium, but many improvements need to be made before we reach a technical level comparable to that used for isolation of most other bacteria from environmental sources known to harbor the organism in question. The best techniques to use obviously will vary with the environmental specimen to be collected and tested.

LITERATURE CITED

1. **Bopp, C. A., J. W. Sumner, G. K. Morris, and J. G. Wells.** 1980. Isolation of *Legionella* spp. from environmental water samples by low-pH treatment and use of a selective medium. J. Clin. Microbiol. **13**:714–719.
2. **Edelstein, P. H.** 1981. Improved semiselective medium for isolation of *Legionella pneumophila* from contaminated clinical and environmental specimens. J. Clin. Microbiol. **14**:298–303.
3. **Edelstein, P. H.** 1982. Comparative study of selective media for isolation of *Legionella pneumophila* from potable water. J. Clin. Microbiol. **16**:697–699.
4. **Edelstein, P. H., and S. M. Finegold.** 1979. Use of a semiselective medium to culture *Legionella pneumophila* from contaminated lung specimens. J. Clin. Microbiol. **10**:141–143.
5. **Edelstein, P. H., J. B. Snitzer, and J. A. Bridge.** 1982. Enhancement of recovery of *Legionella pneumophila* from contaminated respiratory tract specimens by heat. J. Clin. Microbiol. **16**:1061–1065.
6. **Edelstein, P. H., J. B. Snitzer, and S. M. Finegold.** 1982. Isolation of *Legionella pneumophila* from hospital potable water specimens: comparison of direct plating with guinea pig inoculation. J. Clin. Microbiol. **15**:1092–1096.
7. **Feeley, J. C., R. J. Gibson, G. W. Gorman, N. C. Langford, J. K. Rasheed, D. C. Mackel, and W. B. Baine.** 1979. Charcoal-yeast extract agar: a primary isolation medium for *Legionella pneumophila*. J. Clin. Microbiol. **10**:436–441.
8. **Feeley, J. C., G. W. Gorman, and R. J. Gibson.** 1979. Primary isolation media and methods, p. 77–84. *In* G. L. Jones and G. A. Hébert (ed.), "Legionnaires' ": the disease, the bacterium and methodology. Centers for Disease Control, Atlanta, Ga.
9. **Feeley, J. C., G. W. Gorman, R. E. Weaver, D. C. Mackel, and H. W. Smith.** 1978. Primary isolation media for Legionnaires disease bacterium. J. Clin. Microbiol. **8**:320–325.
10. **Fliermans, C. B., R. J. Soracco, and D. J. Pope.** 1981. Measure of *Legionella pneumophila* activity in situ. Curr. Microbiol. **6**:89–94.
11. **Fraser, D. W.** 1980. Legionellosis: evidence of airborne transmission. Ann. N.Y. Acad. Sci. **353**:61–66.
12. **Gorman, G. W., and J. C. Feeley.** 1982. Procedures for the isolation of *Legionella* species from environmental samples. Centers for Disease Control, Atlanta, Ga.
13. **Greaves, P. W.** 1980. New methods for the isolation of *Legionella pneumophila*. J. Clin. Pathol. **33**:581–584.
14. **Kaufmann, A. F., J. E. McDade, C. M. Patton, J. V. Bennett, P. Skaliy, J. C. Feeley, D. C. Anderson, M. E. Potter, V. F. Newhouse, M. B. Gregg, and P. S. Brachman.** 1981. Pontiac fever: isolation of the etiologic agent (*Legionella pneumophila*) and demonstration of its mode of transmission. Am. J. Epidemiol. **114**:337–347.
15. **McDade, J. E., C. C. Shepard, D. W. Fraser, T. F. Tsai, M. A. Redus, W. R. Dowdle, and the Laboratory Investigation Team.** 1977. Legionnaires' disease. Isolation of a bacterium and demonstration of its role in other respiratory diseases. N. Engl. J. Med. **297**:1197–1203.
16. **Pasculle, S. W., J. C. Feeley, R. J. Gibson, L. G. Cordes, R. L. Myerowitz, C. M. Patton, G. W. Gorman, C. L. Carmack, J. W. Ezzell, and J. N. Dowling.** 1980. Pittsburgh pneumonia agent: direct isolation from human lung tissue. J. Infect. Dis. **141**:727–732.
17. **Vickers, R. M., A. Brown, and G. M. Garrity.** 1981. Dye-containing medium for differentiation of members of the family *Legionellaceae*. J. Clin. Microbiol. **13**:380–382.
18. **Wadowsky, R. M., and R. B. Yee.** 1981. Glycine-containing selective medium for isolation of *Legionellaceae* from environmental specimens. Appl. Environ. Microbiol. **42**:768–772.
19. **Wolf, H. W., P. Skaliy, L. B. Hall, M. M. Harris, H. M. Decker, L. M. Buchanan, and C. M. Dahlgren.** 1964. Sampling microbial aerosols. Public Health Service publication no. 686. U.S. Public Health Service, Washington, D.C.

Philosophical Ecology: *Legionella* in Historical Perspective

C. B. FLIERMANS

Ecological Microbes, Augusta, Georgia 30909

Since the initial findings that *Legionella* spp. cause serious infection in humans, a large number of studies have centered on the clinical, epidemiological, physiological, and ecological importance of the disease and the organism. The purpose of this short ecological overview is to put into perspective the current status of ecological research on *Legionella*, how the data have and will continue to be applied, and where one goes from here.

At the First International Symposium on Legionnaires Disease in 1978 our knowledge of the ecological importance of *Legionella* was virtually nonexistent, since isolates had not been obtained from habitats that were not associated with outbreaks of the disease. In fact, environmental habitats were being redefined by epidemiologists as any location that was not clinical or not associated with a patient. These habitats were cooling systems, i.e., towers, exchangers, and air wash systems; subsequently, potable water systems and hot and cold water systems in both domestic and institutional settings were added. Most of these habitats are highly artificial in that they are heavily influenced by humans and may only be niches for *Legionella* spp., rather than normal habitats. Nevertheless, the redefinition of terms began because of the training and disciplines of the majority of the researchers associated with study of *Legionella*.

HISTORICAL HINDSIGHT

Aquatic microbiologists recognize that aquatic bacteria follow a seasonal cycle with respect to their population densities. The seasonality of Legionnaires disease cases appeared to be cyclic (4), which stimulated us to look for legionellae in aquatic habitats. Also, the fact that the isolates studied had a very unusual fatty acid composition triggered rememberances of past research that seemed to fit the puzzle.

Since 1969, I have been conducting research on microorganisms associated with thermal habitats, including natural thermal habitats like those in Yellowstone National Park and man-made thermal habitats created by heat rejection from electrical power and nuclear generation facilities. The microorganisms associated with these habitats were often mesophilic and thermophilic with regard to their physiological response, in that they had temperature optima for growth between 30 and 90°C. A second characteristic that was unusual for these thermophiles was the large number of branched-chained fatty acids that they contained (13), just like the clinical isolates of *Legionella*. Armed with this information, I began looking in aquatic habitats, both natural and man-made, that were both ambient and thermally altered for the presence of legionellae.

Our initial findings (8) that *Legionella pneumophila* could be isolated from natural habitats that were not associated with an outbreak of the disease, opened a new arena for thinking, experimentation, and understanding. The lid on the proverbial "Pandora's box" would not be shut. Pieces of the puzzle were beginning to be found in the aquatic habitats. The question now became, "How and where do the legionellae fit?"

Through the course of the various studies on legionellae a wide spectrum of media had been developed to isolate the bacterium from human hosts and subsequently from environmental samples as well as from natural habitats. As would be expected, these habitats differ enormously in their bacterial densities with regard to both legionellae and other "contaminating" microorganisms. It should not be surprising that media which worked for one habitat had relatively little use in a different habitat. This was the case for media developed for legionellae cultured from human patients and legionellae from the natural environment. It was not that the legionellae were different, but the types and densities of the associated flora and fauna differed dramatically. Thus, no media (other than guinea pigs, with all of their drawbacks) were effective in selectively enriching the *Legionella* population from habitats where legionellae made up less than 1% of the total bacterial population. We still find that guinea pigs are a useful "selective enrichment medium" for culturing legionellae from natural aquatic habitats.

Microbial ecologists look for and welcome a

selective medium which allows the ready isolation of specific microorganisms from natural habitats that contain great microbial diversity. Such media, however, are practically nonexistent. I still remember my surprise at finding a variety of bacteria present in a sample of acidic thermophilic soil (pH 0.5, 80°C) from Yellowstone National Park. The medium for growth and enrichment was highly restrictive, i.e., sulfur as the sole energy source in a minimal salts medium, carbon dioxide from the air as the sole carbon source, pH 1.5, and temperature of 65°C, and yet several different organisms grew. Even under these extreme conditions a selective medium was not achieved.

By using guinea pigs as a selective enrichment medium, isolates of *Legionella* have been obtained from natural habitats that have a tremendous range of environmental parameters (7). These habitats have included temperatures of 63°C and frozen rivers, and lakes with an oxygen content below 0.2 ppm and algae-clogged lakes with a dissolved oxygen content greater than 15.0 ppm. The pH values of the habitats from which legionellae have been isolated have ranged from 5.0 to 8.5, and conductivity has ranged from 10 to 120 μS/cm. Thus, *Legionella* has the ability to survive a great diversity of environmental characteristics, although the initial data from the laboratory suggested that *Legionella* could only grow under a very restrictive range of temperature and pH. Such discrepancies clearly indicate to the microbial ecologist that the media used for *Legionella* do not allow it to express all of its physiological characteristics. Statistical analyses with chi-square values indicated that habitats with temperatures from 36 to 70°C yielded significantly greater number of isolates than lower-temperature habitats, at the $P < 0.05$ level.

The distribution studies of legionellae have developed to the point that the organism is isolatable from virtually every aquatic habitat conceivable, including potable hot and cold water systems. On an equal-effort sampling basis it appears that *Legionella* is more readily isolated from naturally warm or thermal habitats than from ambient temperature systems. The various pieces of the puzzle seemed to indicate that temperature is a significant key to understanding the ecological distribution and role of *Legionella*. Initially, we believed that many of the data on *Legionella* in heat rejection systems could be explained by the association with temperature and that in fact temperature would probably explain, either directly or indirectly, many of the reasons for the outbreaks that were occurring. Although temperature is probably very important, it is not an overriding paradigm. Such data have been expanded by Christensen et al. (5)

and Hazen and Fliermans (unpublished data) and indicate that no single environmental factor thus far identified is a useful predictor of environmental densities of legionellae, by a variety of statistical tests.

Distribution studies of legionellae in natural habitats have been confined primarily to freshwater systems. Data collected in a river continuum system along the Savannah River to the coast at Savannah, Ga., and into the open ocean indicated that once salinity levels increased to saltwater concentrations, neither isolates of *Legionella* nor samples positive by the direct fluorescent-antibody test could be obtained (C. M. Fliermans, *in* S. M. Katz, ed., *Legionellosis*, in press). Ortiz-Roque and Hazen (Abstr. Annu. Meet. Am. Soc. Microbiol. 1982, N15, p. 180) have isolated *L. pneumophila* from rain forest areas of Puerto Rico and observed *Legionella*-like organisms in the Caribbean Ocean. Thus, the distribution of legionellae in a tropical habitat may well be distinctively different from that in the temperate habitats that have thus far been studied.

Associations among organisms play a dynamic and important role in aquatic ecosystems. Such observations are not unusual (T. C. Hazen, Ph.D. thesis, Wake Forest University, Winston-Salem, N.C., 1978), but techniques for observing such relationships are specialized. To my knowledge only the direct fluorescent-antibody technique allows one to assess the presence of a particular organism in its environment, identify it to serogroup if necessary, and measure its density. Such a technique combined with a viability test such as autoradiography or electron transport system activity (9, 10) provides information on the viability of an organism under in situ conditions. Early in the ecological research on *Legionella*, Tison et al. (19) demonstrated that *L. pneumophila* could derive all of its nutritional requirements for growth through its association with *Fisherella* sp., a blue-green alga, both in situ and in vitro. These initial data have been expanded by Pope et al. (15) to include other species of algae which are primary producers of organic material. These findings indicate a reason why cooling towers may play a significant role in the amplification of legionellae. Because the towers are generally exposed to either direct or indirect lighting, are warm, moist habitats, often receive makeup water from sources that contain algae, and are efficient scrubbers of the microorganisms from the air (of which algae are a part), these systems are logically important in the amplification of *Legionella*. Towers simply provide the right ecological niche for both the amplification and the dissemination of all types of microorganisms, including legionellae (1).

Before a wide variety of algae was known to be able to support the growth of legionellae, it was hypothesized that *Fisherella* spp. may produce a particular kind of nutrient that is specific for legionellae. Since a wide variety of algae can play a role in the production and transfer of nutrients to legionellae, it is not reasonable to assume that the algae are producing a particular kind of photosynthate just for legionellae.

Not only do heat rejection systems such as cooling towers provide the right ecological niche for the amplification of legionellae, but the dissemination of the organism through aerosol drift is likewise achieved. The work of Berendt (2) is quite interesting on this point. He demonstrated that the survivability of *L. pneumophila* in an aerosolized form was enhanced when the bacterium was associated with *Fisherella* sp. Such findings suggest that the dissemination of legionellae away from a source such as a cooling tower could be enhanced by the association with an algal component. This enhancement may be due to a physical protection of legionellae simply from desiccation, since *Fisherella* sp. will produce a mucilagenous matrix. If such protection occurs in heat rejection systems with the formation of aerosols, then the dissemination of viable *Legionella* particles away from a point source may be quite significant; this would increase considerably the job of epidemiologists, since the source of infection may be "miles" away. Simple modeling using wind trajectories from meteorological conditions established for plume chasing suggests that particles the size of *Legionella* cells may be carried 20 to 50 km downwind before they are removed by gravitational settling (20). This says nothing about viability of the *Legionella* cells once they travel those distances. The data from aerosol studies done by Berendt et al. (3) suggest that a much lower dose of *L. pneumophila* is necessary for guinea pig infection via the aerosol route than is required via the interperitoneal route. If such data were to apply to human infection, then the increased viability and dissemination from cooling towers would be quite significant in affecting a large portion of the human population. Such thinking may have merit in explaining why areas of the country have a higher incidence of human sera positive for *Legionella* than do other areas of the country.

The association of legionellae with newly developed habitats such as lakes formed by the eruption of Mount St. Helens is noteworthy. During the time of outblast, numerous lakes were formed, some of which had thermal inputs. These lakes provided nutrients as well as appropriate temperatures for *Legionella* spp. It is of interest that the lakes contained large amounts of burned timber so that an in situ "charcoal"

medium was readily available.

During a U.S. Department of Energy workshop on aquatic ecology in March 1981, I learned that researchers sampling these lakes were often contracting flulike symptoms after the retrieval of samples. Access to the area was by helicopter, and the actual sampling was often done with direct body contact with the lake water being sampled. We initially obtained water samples from the area and isolated *L. pneumophila* serogroup 2. Tison et al. (Abstr. Annu. Meet. Am. Soc. Microbiol. 1982, Q16, p. 212) did a much more extensive study, demonstrating the presence of *L. pneumophila* serogroups 2 and 6 as well as assessing the densities of legionellae present in these habitats.

Associations have been observed with *L. pneumophila* and free-living amoebae such as *Acanthamoeba* and *Naegleria* spp. (17, 21). The significance of these findings is yet to be fully determined, but it appears that the amoeba can use legionellae as a food source and at times legionellae can use the amoeba as a host in which to replicate. Rowbotham (17) has suggested that the different manifestations of legionellosis may be due to the association of legionellae and the respective amoeba. It is significant that both of these organisms are associated with thermal habitats (11).

A less-studied association is that observed between *L. pneumophila* and *Myriophillum spicatum*, a submerged aquatic macrophyte, in Par Pond, S.C. It was observed that when this plant dies back each season in the fall of the year, large populations of *L. pneumophila* become associated with the degradation of the plant material (Fliermans, in press). Such associations are not observed microscopically during the time of the year when the plant is actively growing. Microscopic examinations have been followed by use of direct fluorescent antibodies with subsequent culturing in which *L. pneumophila* serogroup 1 has been isolated. This association appears to be somewhat specific in that other aquatic plants, i.e., *Typha latifolia*, *Potamogeton crispus*, *Najas* spp., and *Eichhornia crassipes*, did not show any relationship with *L. pneumophila* during any of the seasons studied. This association between aquatic plants and aquatic bacteria is not uncommon. Hazen (Ph.D. thesis) demonstrated a similar association between *Aeromonas hydrophila* and *M. spicatum*. Because of the global ubiquity of *M. spicatum*, the relationship between it and legionellae may be quite important for explaining the growth of legionellae in a variety of habitats and the subsequent role of those habitats as inoculum for amplifiers. This could readily be achieved even for potable water systems, since legionellae appear to survive much of the water treatment process now in use across the United States.

How does one evaluate the ecological evidence in light of the fact that Legionnaires disease has an apparent low frequency in humans, yet the organism appears to be ubiquitous? That question does not have a ready-made answer, yet one still searches. Since this is a philosophical paper on the ecology of *Legionella*, I propose the following rationale: all of the physical and biological processes that are currently operational in the world are controlled by the First and Second Laws of Thermodynamics. That is to say, very briefly, that energy may be transformed, but it is always conserved, and the ability to utilize that energy is never 100%. This simply means that things are running down, they get old, they decay and wear out. For all the knowledge and skill that medicine brings to the marketplace of human affairs, the average human lifespan has not increased significantly beyond the proverbial "threescore years and ten." It is true that many more people have reached that average age, but the overall life expectancy has not increased. Humanity has always had a "cultural" mandate to be a steward of the earth. Its response has at times been quite poor, to say the least, and often the resources have been squandered. Nevertheless, humanity seeks to utilize the resources at its disposal to bring order from a system that continually increases in entropy, or disorder (16).

Part of the way to achieve order from disorder is through the use of enormous amounts of energy to perform useful work. Such is the case with the heat rejection systems discussed earlier. I believe that these are the largest single amplifiers and disseminators of legionellae. It is also evident that the greatest densities of legionellae appear to be associated with the thermal portions of these systems. Heat rejection systems serve as the acceptor of the increase in entropy in that they receive dissipated heat from the production of energy used to perform work, which increases the order in a localized habitat while increasing the disorder over a much larger region. The loss of energy to perform useful work is the system's response to the Second Law of Thermodynamics. The energy is dissipated into the environment, where the temperature of the local habitat is raised. Such habitats become conducive to the growth and proliferation of a large number of organisms that respond positively to the added heat. I believe that *Legionella* is one of those organisms.

Since 1947, the consumption of electricity has gone from 0.033×10^{15} BTUs to a peak in 1979 of 78.9×10^{15} BTUs (14), a tremendous increase in the amount of energy used to perform useful work. That dissipated heat has to go somewhere and it does—to the environment. The ubiquity of *Legionella* assured that a suitable habitat would have a ready inoculum. Humanity, through its technology, has gotten in the way of the Second Law of Thermodynamics.

APPLICATION OF KNOWLEDGE

Legionella spp. are aquatic bacteria found in lakes, streams, and ponds, and it is not reasonable to have amplifiers, which receive water from these habitats, free from aquatic bacteria. Thus, virtually all cooling towers will contain legionellae at some concentration. Furthermore, it is unreasonable to attempt to erradicate legionellae from all of the aquatic habitats in which it is found. Yet I believe it is reasonable to place restrictions on ecological niches which amplify and disseminate *Legionella*, particularly if these niches are near a susceptible population.

Because the mere presence of legionellae in an amplifier is nothing unusual, the quantity of legionellae in the habitat of question ought to be determined. *L. pneumophila* serogroups 1 to 4 can be quantified by direct fluorescent antibodies, since much is known about their reaction with water samples, their use in both clinical and environmental systems, and their levels and span of cross-reactions. Studies should center on identifying what is an acceptable density of legionellae for a given niche. In 1981 it was suggested that 10^5 *Legionella* organisms per ml in a cooling tower was cause for concern (6). Recent studies by Christensen et al. (5) suggested that this level may be too high and that legionellae from environmental samples may be infective at much lower concentrations.

Additionally, all studies on the effectiveness of biocides against *Legionella* should be tested in situ, in the field. Such studies have been conducted with chlorine (6), potassium *n*-methyldithiocarbamate with disodium cyanodithioimidocarbonate, bis-tributyltin oxide with dimethylbenzyl ammonium chlorides (E. B. Braun, Ph.D. thesis, Rensselaer Polytechnic Institute, Troy, N.Y., 1982), and Bromicide (Great Lakes Chemical Corp., Lafayette, Ind., 1-bromo-3-chloro-5,5-dimethylhydantoin) (Fliermans, submitted for publication). Only chlorine at the recommended dosage and duration described has been effective in removing legionellae from the studied cooling towers and air wash systems.

The thermal physiology data appear to fit well with what we know about *Legionella*. It is a eurythermal organism able to maintain itself over a wide range of temperatures. It appears to favor thermal habitats, as it is more readily isolated at elevated temperatures and its fatty acid composition is very similar to that of previously described thermophiles. In situ measurements of electron transport activity indicate that

L. pneumophila demonstrates greatest activity in habitats of 45°C, with very little if any activity above 70°C (10). Such findings are relevant to hot water systems lowered to 43–45°C due to energy conservation measure for hospitals, in that these systems provide niches for the proliferation of legionellae. Elevated temperature as a means of controlling legionellae in hospital systems makes sense from the ecological physiology of the organism. Raising the temperature of hot water systems above 63°C, or even into the low 60°C range, has a marked effect on reducing the electron transport system activity of legionellae. Such means of decontaminating the hot water system appear to have promise (14a).

FUTURISTIC FORESIGHT

Where do we go from here? The cream has been skimmed from the ecological, clinical, and bacteriological studies of *Legionella*. We in the field of ecology are ready for a major breakthrough; the research has plateaued. Maybe that break will come through the use of cytofluorographic tools to assess virulence. It is clear that aerosol studies are needed. The paucity of information in this area is of concern. How far does *Legionella* travel and how healthy is it when it gets there? How do the operations of electrical power plants, the largest operators of cooling towers in the world, affect the densities of *Legionella* in ambient habitats and thermally altered water? Is there any other reservoir besides aquatic systems? Soil was suggested early as a reservoir for *Legionella*, but no isolates have been obtained. It certainly houses many other pathogens, why not *Legionella*?

I believe that *Legionella* is not strictly a pathogenic organism but that the bacterium has a unique role to play in the natural environment as a consumer. It may indeed have transformation powers for organic matter of which we currently have no knowledge. Its association with a variety of algae, amoebae, and aquatic plants suggests to me that *Legionella* "knows" exactly what it is doing; we simply do not yet have the understanding.

Unless greater strides are made in all areas of *Legionella* research than have been made in the past 2 years, a 3rd International Symposium on *Legionella* would be a waste of time and money. The entropy principle is alive and well.

LITERATURE CITED

1. **Adams, A. P., M. Garbett, H. B. Rees, and B. G. Lewis.** 1980. Bacterial aerosols produced from a cooling tower using wastewater effluent as makeup water. J. Water Pollut. Control Fed. **52:**498–501.
2. **Berendt, R. F.** 1981. Influence of blue-green algae (cyanobacteria) on survival of *Legionella pneumophila* in aerosols. Infect. Immun. **32:**690–692.
3. **Berendt, R. F., H. W. Young, R. G. Allen, and G. L. Knutsen.** 1980. Dose-response of guinea pigs experimentally infected with aerosols of *Legionella pneumophila*. J. Infect. Dis. **141:**186–192.
4. **Center for Disease Control.** 1978. Legionnaires' disease—United States. Morbid. Mortal. Weekly Rep. **27:**439.
5. **Christensen, S. W., R. L. Tyndall, J. A. Solomon, and C. B. Fliermans.** 1982. Legionnaires' disease bacterium in power plant cooling systems. Electr. Power Res. Inst. J. **1:**1–97.
6. **Fliermans, C. B., G. E. Bettinger, and A. W. Fynsk.** 1981. Treatment of cooling systems containing high levels of *Legionella pneumophila*. Water Res. **16:**903–909.
7. **Fliermans, C. B., W. B. Cherry, L. H. Orrison, S. J. Smith, D. L. Tison, and D. H. Pope.** 1981. Ecological distribution of *Legionella pneumophila*. Appl. Environ. Microbiol. **41:**9–16.
8. **Fliermans, C. B., W. B. Cherry, L. H. Orrison, and L. Thaker.** 1979. Isolation of *Legionella pneumophila* from nonepidemic-related aquatic habitats. Appl. Environ. Microbiol. **37:**1239–1242.
9. **Fliermans, C. B., and E. L. Schmidt.** 1975. Autoradiography and immunofluorescence combined for autecological study of single cell activity with *Nitrobacter* as a model system. Appl. Microbiol. **30:**676–684.
10. **Fliermans, C. B., R. J. Soracco, and D. H. Pope.** 1981. Measure of *Legionella pneumophila* activity *in situ*. Curr. Microbiol. **6:**89–94.
11. **Fliermans, C. B., R. L. Tyndall, E. L. Domingue, and E. J. P. Willaert.** 1979. Isolation of *Naegleria fowleri* from artificially heated water. J. Therm. Biol. **4:**303–305.
12. **Hazen, T. C.** 1979. Ecology of *Aeromonas hydrophila* in a South Carolina reservoir. Microb. Ecol. **5:**179–195.
13. **Heinen, W.** 1970. Extreme thermophilic bacteria: fatty acids and pigments. Antonie van Leeuwenhoek J. Microbiol. Serol. **36:**582–584.
14. **Monthly Energy Review.** 1983. Consumption of energy by type. DOE/EIA-0035 (83-06).
14a. **Plouffe, J. F., L. R. Webster, and B. Hackman.** 1983. Relationship between colonization of hospital buildings with *Legionella pneumophila* and hot water temperatures. Appl. Environ. Microbiol. **46:**769–770.
15. **Pope, D. H., R. J. Soracco, H. K. Gill, and C. B. Fliermans.** 1982. Growth of *Legionella pneumophila* in two membered cultures with green algae and cyanobacteria. Curr. Microbiol. **7:**319–322.
16. **Rifkin, J.** 1980. Entropy, p. 303. Viking Press, New York.
17. **Rowbotham, T. J.** 1980. Pontiac fever explained? Lancet **ii:**969.
18. **Tansey, M. R., C. B. Fliermans, and C. D. Kern.** 1979. Aerosol dissemination of veterinary pathogenic and human opportunistic thermophilic and thermotolerant fungi from thermal effluents of nuclear production reactors. Mycopathologia **69:**91–115.
19. **Tison, D. L., D. H. Pope, W. B. Cherry, and C. B. Fliermans.** 1980. Growth of *Legionella pneumophila* in association with blue-green algae (cyanobacteria). Appl. Environ. Microbiol. **39:**456–459.
20. **Tyndall, R. J.** 1982. Concentrations, serotypic profiles and infectivity of Legionnaires' disease bacteria populations in cooling towers. J. Cooling Tower Inst. **3:**25–33.
21. **Wadowsky, R. M., R. B. Yee, L. Mezmar, E. J. Wing, and J. N. Dowling.** 1982. Hot water systems as sources of *Legionella pneumophila* in hospital and nonhospital plumbing fixtures. Appl. Environ. Microbiol. **43:**1104–1110.

Media for Detection of *Legionella* spp. in Environmental Water Samples

R. L. CALDERON AND A. P. DUFOUR

Department of Epidemiology and Public Health, Yale University, New Haven, Connecticut 06520, and Health Effects Research Laboratory, U.S. Environmental Protection Agency, Cincinnati, Ohio 45268

Recent outbreaks of legionellosis have been attributed to water aerosols created by shower heads and heat exchange systems, such as cooling towers or air-conditioning units (1, 3, 4; J. Stout, V. L. Yu, R. M. Vickers, J. Zuraleff, and A. Brown, Program Abstr. Intersci. Conf. Antimicrob. Agents Chemother. 21st, Chicago, Ill., 1981, abstr. no. 293). There is speculation that *Legionella* spp. are ubiquitous in aquatic environments and that these organisms are able to survive water treatment barriers and enter distribution systems, probably in very low densities. In plumbing systems, especially the hot water portions, *Legionella* spp. are thought to multiply to high enough concentrations to present a health hazard (5, 7). The extent of the occurrence of *Legionella* spp. in raw surface waters and in treated water distribution systems has not, however, been fully investigated. This research has been hampered by the lack of a simple, facile technique for quantifying *Legionella* spp. in environmental waters.

There are presently three selective media which have appeared in the published literature that have been used in limited environmental surveys. However, these media were developed for use in hospital settings and have not been extensively evaluated to determine their usefulness for measuring *Legionella* spp. in the presence of autochthonous freshwater bacteria. All three media contain a buffered charcoal-yeast extract base supplemented with iron and cysteine, but each contains different inhibitors. The first medium, CCVC (2), contains cephalothin (4 μg/ml), colistin (16 μg/ml), cycloheximide (80 μg/ml), and vancomycin (0.5 μg/ml). The second medium, WY (8), contains glycine (0.3%), polymyxin B (100 U/ml), and vancomycin (5 μg/ml). The third medium, MWY (6), was a modification of WY containing glycine (0.3%), polymyxin B (50 U/ml), vancomycin (1 μg/ml), anisomycin (80 μg/ml), bromothymol blue (10 μg/ml), and bromocresol purple (10 μg/ml). A second base in which proteose peptone no. 3 (Difco) was substituted for yeast extract (EPA basal medium) was tested with each of the above inhibitor combinations and designated EPA-CCVC, EPA-WY, and EPA-MWY, respectively.

Ohio River water was seeded with seven stressed *Legionella* spp. The effect of acid treatment and the ability of various media to recover *Legionella* spp. from the seeded samples and to reduce background bacteria and fungi were all examined. The media were also evaluated against 11 environmental water samples previously found positive for *Legionella* spp.

The average heterotrophic bacterial concentration in the seeded samples was approximately 950 CFU/ml, determined with standard plate count agar, and the average numbers of legionellae seeded in the raw Ohio River water samples ranged from 300 to 8,000 CFU/ml. Samples were directly plated onto duplicate plates. The six media were graded in descending order by ability to recover *Legionella* spp., as follows: EPA-MWY > MWY > EPA-CCVC > EPA-WY > WY > CCVC (Table 1). EPA basal medium appeared to improve recovery of some of the *Legionella* spp., especially *L. gormanii*. However, glycine appeared to be inhibitory to *L. gormanii* because this species grew well on EPA-CCVC medium, which does not contain glycine, but grew poorly on EPA media containing glycine.

One-half of each seeded sample was treated with acid (pH 2) for 5 min, neutralized, and then plated. The recovery of various *Legionella* spp. was similar to the results obtained with untreated samples. The ranking of media for recovery of *Legionella* spp. on the various acid-treated samples was: EPA-MWY > MWY > EPA-CCVC > EPA-WY > CCVC > WY. It appears that acid treatment had a negative effect on recovery of *L. gormanii*, *L. dumoffii*, and *L. wadsworthii* (data not shown). However, only the decrease in recovery of *L. wadsworthii* was statistically significant ($P < 0.05$).

The efficiency of each medium for reducing background organisms was evaluated by comparing numbers of non-*Legionella* colonies that developed on the test medium relative to the numbers that developed on noninhibitory medium (i.e., standard plate count agar). The media were ranked for efficiency of reduction of background flora in descending order, as follows: EPA-WY > WY > EPA-MWY > MWY >

TABLE 1. Recovery of *Legionella* spp. on six isolation media from surface water[a]

Species	CFU/0.1 ml of medium:					
	CCVC	MWY	WY	EPA-CCVC	EPA-MWY	EPA-WY
L. pneumophila	207	216	116	232	232	162
L. micdadei	151	354	220	270	321	200
L. gormanii	10	3	0	347	1	0
L. longbeachae	423	693	330	808	879	511
L. bozemanii	266	515	423	335	407	490
L. dumoffii	81	224	90	83	336	134
L. wadsworthii	165	184	190	120	345	340

[a] All strains were stressed in potassium phosphate buffer for 3 days before being seeded.

CCVC > EPA-CCVC. Recovery of *Legionella* spp. was particularly reduced by the presence of fungal or spreading colonies that often covered the entire plate.

Environmental water samples were divided arbitrarily into two groups, those with high densities and those with low densities of *Legionella* spp. Low-density samples were concentrated by centrifugation before being plated to media. We found that the EPA-CCVC and MWY media had the best overall recoveries of *Legionella* spp. from the samples examined which we classified as high density (Table 2). The low-density samples were also graded for *Legionella* spp. isolation capacity. The CCVC and EPA-CCVC media had the highest frequency of *Legionella* isolations.

The reduction of background flora in environmental samples was similar to that observed in seeded water samples. EPA-WY and WY media were most effective in reducing non-*Legionella* heterotrophic bacterial populations. However, these two media showed the lowest recoveries of *Legionella* spp. from unseeded natural samples. In contrast, the two media showing the poorest reduction of background flora had the best detection rate of *Legionella* spp.

It appears that an inverse trend in the relationship of a medium's ability to isolate *Legionella* spp. and its ability to reduce background flora exists in the three medium formulations examined; correspondingly, the greater the capacity of a medium for the reduction of non-*Legionella* spp., the lower is its capacity for the recovery of legionellae. We do not consider that any of the media evaluated was very effective for both reducing background flora and simultaneously recovering *Legionella* spp.

Many questions remain regarding the significance of *Legionella* spp. in potable source waters and treated water distribution systems. To answer some of these questions, it is imperative that an isolation and enumeration method tailored to the requirements for measuring *Legionella* spp. in such waters be developed, because the ones currently available, in our opinion, are not sufficient.

LITERATURE CITED

1. **Band, J. D., M. La Venture, J. P. Davis, G. F. Mallison, P. Skaliy, P. S. Hayes, W. L. Schell, H. Weiss, D. J. Green-**

TABLE 2. Enumeration of *Legionella* spp. from environmental water samples on six isolation media

Source	CFU/0.1 ml on medium:					
	CCVC	MWY	WY	EPA-CCVC	EPA-MWY	EPA-WY
High-density specimens						
Hospital hot water tank	148	404	55	290	496	100
Hospital hot water tap	10	19	14	20	12	7
Hospital shower	179	175	82	130	42	17
Hospital shower	5	3	2	26	5	0
Low-density specimens						
River	2	1	1	2	1	1
River	1	0	0	2	0	0
Lake[a]	1	0	0	2	0	1
Lake[b]	3	3	0	6	7	2
Lake[a]	1	0	2	3	2	0
Lake[a]	5	6	0	6	1	1
Lake[a]	1	2	1	1	2	2

[a] Concentrated 100-fold before plating.
[b] Concentrated 10-fold before plating.

berg, and D. W. Fraser. 1981. Epidemic Legionnaires' disease. Airborne transmission down a chimney. J. Am. Med. Assoc. **245**:2404–2407.

2. **Bopp, C. A., J. W. Sumner, G. K. Morris, and J. G. Wells.** 1981. Isolation of *Legionella* spp. from environmental water samples by low-pH treatment and use of a selective medium. J. Clin. Microbiol. **13**:714–719.

3. **Cordes, L. G., A. M. Wiesenthal, G. W. Gorman, J. P. Phair, H. M. Sommer, A. Brown, V. Yu, M. H. Magnussen, R. D. Meyer, S. S. Wolf, K. N. Shands, and D. W. Fraser.** 1981. Isolation of *Legionella pneumophila* from hospital shower heads. Ann. Intern. Med. **94**:195–197.

4. **Dondero, T. J., Jr., R. C. Rendtorff, G. F. Mallison, R. M. Weeks, J. S. Levy, E. W. Wong, and W. Schaffner.** 1980. An outbreak of Legionnaires' disease associated with a contaminated air-conditioning cooling tower. N. Engl. J. Med. **302**:365–370.

5. **Dufour, A. P., and W. Jakubowski.** 1982. Drinking water and Legionnaires' disease. Am. Water Works Assoc. **74**:631–637.

6. **Edelstein, P. H.** 1982. Comparative study of selective media for isolation of *Legionella pneumophila* from potable water. J. Clin. Microbiol. **16**:697–699.

7. **Stout, J., V. L. Yu, J. Zuraleff, M. Best, A. Brown, R. B. Yee, and R. Wadowsky.** 1982. Ubiquitousness of *Legionella pneumophila* in the water supply of a hospital with endemic Legionnaires' disease. N. Engl. J. Med. **306**:466–468.

8. **Wadowsky, R., and R. B. Yee.** 1981. Glycine-containing selective medium for isolation of *Legionellaceae* from environmental specimens. Appl. Environ. Microbiol. **42**:768–772.

Sampling Methodology for Enumeration of *Legionella* spp. in Water Distribution Systems

LANCE VOSS, KENNETH S. BUTTON, MELVIN S. RHEINS, AND OLLI H. TUOVINEN

Department of Microbiology, The Ohio State University, Columbus, Ohio 43210, and Columbus Division of Water, Water Research Laboratory, Columbus, Ohio 43215

Legionella spp. contamination has been repeatedly reported for hot water systems and associated plumbing fixtures (1, 2, 5). The present paper describes work that was undertaken to seek evidence for the presence of *Legionella* spp. in a metropolitan water distribution system. The two treatment plants in this system utilize surface water and incorporate coagulation, lime softening, sand filtration, and chlorination in the treatment process. The finished water has a free chlorine residual of 1.0 to 1.5 mg/liter before it is pumped to the distribution system at a rate of 100×10^6 gallons a day. The system has a mainline length of about 2,494 miles (ca. 4,015 km) with an average residence time of 18 h.

First, it was established that several hot water recirculating systems in the study area were contaminated with *Legionella* spp. The viable counts of *Legionella* spp. varied depending on the location and sampling site, but the contamination was so persistent that several sites could be used as a continuous source of environmental isolates of *Legionella* spp. The contamination problem was particularly associated with drainage valves and nipples where small volumes of water remain relatively stagnant but tempered for long periods of time. With a buffered charcoal-yeast extract medium with or without antibiotics (3, 4), viable *Legionella* counts of up to several thousand cells per milliliter could be readily detected. Elevated temperature of the water in the recirculating line was not a prohibitive factor to the *Legionella* spp. contamination because a temperature gradient resulting from heat loss was evident in valves and nipples. To date no evidence has been obtained for the presence of *Legionella* spp. in noncirculating hot water systems, including water heater deposits, in private residences.

Attempts to detect *Legionella* spp. in very large cold water trunk lines were unsuccessful. A few samples from customers' lines were positive for *Legionella* spp. in spite of associated free chlorine residual (Table 1). In extreme cases, several thousand viable *Legionella* cells per milliliter were detected in the cold water system in sites that were stagnant and had no or low chlorine residual. These sites included drainage valves and hose bibs with no physical contact with hot water. Periodical flushing of water from a plumbing system in one building transiently reduced the bacterial counts below the level of detection, but *Legionella* spp. counts increased again within a matter of days (Table 2). Whether this was due to regrowth in the contaminated fixture or to ingress of *Legionella* spp. in the main-line water could not be resolved with the data available. In this instance, repeated sampling of the main-line service water (66-in. [ca. 167.6-cm] trunk line) failed to demonstrate *Legionella* spp. by direct plating. Therefore, it would appear that regrowth of *Legionella* spp. was occurring in the contaminated fixture, although ingress by dormant or culturally nonrecoverable organisms could not be ruled out.

As a sampling method, conventional membrane filtration was of limited potential because of overgrowth by the total bacterial population in many cases. It could be feasibly used for 2 to 3 liters of finished water depending on the water turbidity, the type of membrane filter, and the

TABLE 1. Detection of *Legionella pneumophila* in cold water distribution system samples[a]

Sampling date	Sample code[b]	Turbidity (NTU)[c]	Chlorine: free/total (mg/liter)	Concn	Viable counts per 0.1 ml (*L. pneumophila*/total)		Recovery[d] (%)
					BCY[e]	PAV[e]	
1/31/83	CL-26	22	0.9/1.4	1,000	106/109	94/96	
3/4/83	CL-26	ND[f]	0.8/1.1	0	0/1	2/6	100[g]
				100	33/35	32/35	32.5
				1,000	476/482	622/626	54.9
2/2/83	CL-29	0.35	0/0.3	1,000	70/71	91/91	
3/4/83	CL-29	ND	0/0.15	0	0/8	3/23	100[g]
				100	165/165	189/189	118
				1,000	TNTC/TNTC[h]	TNTC/TNTC	ND
2/17/83	DCW	ND	ND	0	67/108	78/78	
4/15/83	DCW	ND	ND	0	80/80	83/83	

[a] Thirty-four samples were collected in sterile carboys containing sodium thiosulfate to neutralize the chlorine residual. The samples were collected at arbitrarily selected sites throughout the distribution system and were concentrated by either batch centrifugation (100-fold) or continuous-flow centrifugation (1,000-fold). Three of the 34 samples examined were positive for *L. pneumophila* (samples CL-26, CL-29, and DCW). They were later retested and again found to be positive. Chlorine residuals were determined on site after sample collection and duplicate determinations were made.

[b] Sample CL-26 was taken from a 0.5-in. (ca. 1.27-cm)-diameter hose bib connected to a 3-in. (ca. 7.5-cm) main cold water supply pipe through an extension (48 in.; ca. 122 cm) of copper pipe. The location is in the utility area of a large public building, and the environment is relatively temperate. Sample CL-29 was taken from a 0.5-in. hose bib connected to the main water supply of a large warehouse complex through approximately 200 feet (ca. 61 m) of copper piping. This sample site was also indoors in a relatively temperate environment. Sample DCW was taken from a 3-in. cold water main in the physical service tunnel of a large building complex.

[c] NTU, Nephelometric turbidity units (Hach 18900 Turbidimeter).

[d] Percent ratio of *L. pneumophila* counts from concentrated and unconcentrated samples.

[e] BCY, Buffered charcoal-yeast extract agar medium; PAV, BCY medium supplemented with polymyxin B (4 mg/liter), anisomycin (80 mg/liter), and vancomycin (0.5 mg/liter). Both media also contained 0.1% (wt/vol) α-ketoglutarate and 0.3% (wt/vol) glycine (3, 4).

[f] ND, Not determined.

[g] Arbitrarily assigned as 100%.

[h] TNTC, Too numerous to count.

prevalence of *Legionella* spp. Obviously this technique is only as effective as the medium employed.

Batch centrifugation (10,000 × *g*) was tested for concentrating low-bacterial-density samples for direct spread plating. Experiments with seeded samples indicated that *Legionella* spp. recoveries were consistently low (<20%), suggesting loss of bacterial viability or irreversible cell aggregation upon centrifugation. The cell counts in supernatants were negligible and did not account for the poor recovery of *Legionella* spp. from the sedimented fraction.

Similarly, continuous-flow centrifugation as a concentration method resulted in low recovery of *Legionella* spp. from seeded samples. In a few cases involving very low bacterial densities, the presence of *Legionella* spp. in distribution water remained undetected by direct plating but could be demonstrated if the samples were first concentrated 1,000-fold by continuous-flow centrifugation. Loss of *Legionella* spp. in the effluent of continuous-flow centrifugation averaged about 20% and did not alone account for the poor *Legionella* spp. recovery.

Tangential filtration was evaluated as a sample concentration method by using a Millipore Pellicon cassette assembly with Durapore, Nuclepore, and molecular-weight 100,000 filters. Again, low recoveries (usually <10%) were obtained from water samples seeded with *Legionella* spp. regardless of the type of filter material. The transmembrane flux was increased by using a high-volume pump, but it was still unable to prevent membrane surface polarization and adhesion of bacterial cells to the membrane surface. Linear-path gaskets were installed to improve the hydraulic flow pattern, but they could only be used with the molecular-weight filter packets because of retentate leakage associated with the use of filter sheets when the in-house-designed linear-path gaskets were in place. Quantitation of *Legionella* spp. retained on the surface of molecular-weight filters was achieved by removing the bacteria with a swab and plating on buffered charcoal-yeast extract agar. Results showed that the retention of bacteria on the membrane surface was sizable, which explained why tangential filtration produced only low recoveries of *Legionella* spp. However, this tech-

TABLE 2. Sequential recovery of *L. pneumophila* from a contaminated combination hot and cold water faucet[a]

Date	Flow time (min)	Chlorine: free/total (mg/liter)	Viable *L. pneumophila* counts per 0.1 ml		Date	Flow time (min)	Chlorine: free/total (mg/liter)	Viable *L. pneumophila* counts per 0.1 ml	
			Unconcentrated (BCY/PAV[b])	Concentrated 100× (BCY/PAV)				Unconcentrated (BCY/PAV[b])	Concentrated 100× (BCY/PAV)
2/24/83	0	0/0	108/80	854/1036	3/19/83	0	0/0	22/19	375/454
	1	0/0.05	7/5	72/69		1	0/0.25	0/0	7/5
	5	0/0.05	0/1	5/11		5	0/0.25	1/0	1/0
	10	0/0.05	0/0	ND[c]		10	0/0.30	0/0	1/1
	60	0/0.05	0/0	ND		60	0/0.30	0/0	0/0
3/4/83	0	ND	100/83	973/934	3/25/83	0	0/0	14/21	72/97
	1	ND	8/4	66/48		1	0/0.10	0/3	16/13
	5	ND	0/0	2/1		5	0/0.15	1/0	1/1
	10	ND	0/0	0/0		10	0/0.10	0/0	1/3
	60	ND	0/0	0/0		60	0/0.15	0/0	0/0
3/11/83	0	0/0.05	26/19	252/247	4/1/83	0	0/0	222/94	825/761
	1	0.2/0.35	0/0	80/68		1	0/0.10	7/6	78/65
	5	0.25/0.40	1/0	7/4		5	0/0.10	1/0	8/12
	10	0.25/0.50	0/0	2/1		10	0/0.10	0/0	5/11
	60	0.25/0.50	0/0	5/5		60	0/0.10	0/0	1/3

[a] Cold water samples only were taken. The flow rate was approximately 2 liters/min, and the faucet was not used for other purposes during the study period. The faucet was equipped with a back-flow preventer. Samples were quenched with thiosulfate to neutralize the residual chlorine and concentrated 100-fold by batch centrifugation. The sampling site was indoors, and the ambient temperature was about 22 to 24°C.
[b] Media as described in Table 1, footnote *e*.
[c] ND, Not determined.

nique may, if it is further developed, prove useful for examining large volumes of water. If so, its precision and reproducibility need to be evaluated.

LITERATURE CITED

1. **Brown, A., V. L. Yu, M. H. Magnussen, R. M. Vickers, G. M. Garrity, and E. M. Elder.** 1982. Isolation of Pittsburgh pneumonia agent from a hospital shower. Appl. Environ. Microbiol. **43:**725–726.
2. **Cordes, L. G., A. M. Wiesenthal, G. W. Gorman, J. P. Phair, H. M. Sommers, A. Brown, V. L. Yu, M. H. Magnussen, R. D. Mayer, J. S. Wolf, K. N. Shands, and D. W. Fraser.** 1980. Isolation of *Legionella pneumophila* from hospital shower heads. Ann. Intern. Med. **94:**195–197.
3. **Edelstein, P. H.** 1982. Comparative study of selective media for isolation of *Legionella pneumophila* from potable water. J. Clin. Microbiol. **16:**697–699.
4. **Wadowsky, R. M., and R. B. Yee.** 1981. Glycine-containing selective medium for isolation of *Legionella pneumophila* from environmental specimens. Appl. Environ. Microbiol. **42:**768–772.
5. **Wadowsky, R. M., R. B. Yee, L. Mezmar, E. J. Wing, and J. N. Dowling.** 1982. Hot water systems as sources of *Legionella pneumophila* in hospital and nonhospital plumbing fixtures. Appl. Environ. Microbiol. **43:**1104–1110.

Comparison of Isolation Methods for *Legionella* spp.

P. JULIAN DENNIS, CHRISTOPHER L. R. BARTLETT, AND ARTHUR E. WRIGHT

Environmental Microbiology & Safety Reference Laboratory, Public Health Laboratory Service, Centre for Applied Microbiology and Research, Porton Down, Salisbury, Wiltshire, and Public Health Laboratory Service, Communicable Disease Surveillance Centre, Colindale, London NW9, England

Organisms of the genus *Legionella* can now be isolated as successfully by direct culture as by guinea pig inoculation (3). A number of formulae for the preparation of suitable culture media are available, the best of these being based on a charcoal-yeast extract agar (CYE) with ACES buffer and α-ketoglutarate (BCYE; 2). Selective media for the examination of contaminated samples include those of Edelstein (2), containing antibiotics, and Wadowsky and Yee (4), who added glycine and antibiotics. Other methods to enhance isolation from contaminated environmental specimens include pretreatment with acid (1) or the method used in this laboratory, which is based on the ability of *Legionella* sp. organisms to withstand heat.

The work presented here had four aims: first, to ascertain which of the two selective media described was the most effective for the isolation of legionellae from environmental water samples; second, to determine whether pretreatment of water samples by the methods suggested would enhance recovery of the organisms, and which of the two methods was the most effective; third, to find whether the sample source, e.g., hot water systems, cooling towers, etc., would influence the choice of medium or pretreatment method; and finally, to determine whether the isolation of legionellae by guinea pig inoculation is still required regardless of source or whether guinea pigs are only required for samples taken from specific sources, e.g., cooling towers.

The selective media used in the study were obtained by replacing anisomycin with cycloheximide (80 mg/liter) in the medium described by Edelstein (given the abbreviation CCP), and by adding α-ketoglutarate (1 g/liter) and cyclo-

heximide (80 mg/liter) to the selective medium described by Wadowsky and Yee (given the abbreviation GVPC).

Pretreatments to reduce unwanted microbial flora were based on those described by Bopp et al. (1) and on a heat treatment developed by us. The heat treatment relies on the greater tolerance *Legionella pneumophila* has to a temperature of 50°C in comparison to other organisms isolated from environmental water samples (Fig. 1). *D* values were calculated for *L. pneumophila* and other organisms, and it was found that if samples of water were held at 50°C for 30 min the counts of pseudomonads, coliforms, and micrococci were significantly reduced whereas the *L. pneumophila* count was little affected.

A total of 148 water samples were taken from different sources and cultured for *Legionella* spp. Of these, 74 samples were positive by at least one method. The samples were grouped according to origin: hot water (domestic), cold

FIG 2

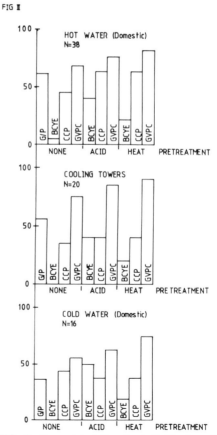

FIG I

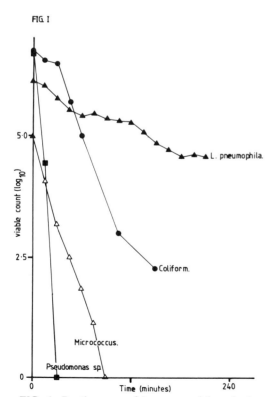

FIG. 1. Death curves of *L. pneumophila* and other environmental organisms when held at 50°C.

FIG. 2. Percent isolation for individual media and methods as compared with the total isolation given by all the methods (*N* = total number positive).

water (domestic), and cooling towers. This was done to ascertain whether any one or combination of culture methods could be used for all samples or whether the sample source influenced the choice of medium or method.

The most effective selective medium was the GVPC medium ($P < 0.01$), but when compared with guinea pig inoculation there was no significant difference ($P > 0.2$; Fig. 2). However, when heat or acid pretreatment was used on the samples before inoculation onto selective medium, a significantly greater ($P < 0.01$) number of the samples were shown to contain Legionella spp. If then compared again with guinea pig inoculation, the isolation rate with GVPC and heat or acid pretreatment was significantly better ($P < 0.001$).

The results also show that sample source did not affect the choice of medium or test procedure and that guinea pigs were no longer required even for cooling tower samples. This work clearly shows that legionellae can be suc-cessfully isolated from environmental samples by using the media and methods described. The procedures used were simple and inexpensive both in cost of materials and time. We therefore recommend these media and methods for the routine examination of water for Legionella spp.

LITERATURE CITED

1. Bopp, C. A., J. W. Sumner, G. K. Morris, and J. G. Well. 1981. Isolation of Legionella spp. from environmental water samples by low-pH treatment and use of a selective medium. J. Clin. Microbiol. 13:714–719.
2. Edelstein, P. H. 1981. Improved semiselective medium for isolation of Legionella pneumophila from contaminated clinical and environmental specimens. J. Clin. Microbiol. 14:298–303.
3. Edelstein, P. H., J. B. Snitzer, and S. M. Finegold. 1982. Isolation of Legionella pneumophila from hospital potable water specimens: comparison of direct plating with guinea pig inoculation. J. Clin. Microbiol. 15:1092–1096.
4. Wadowsky, R. M., and R. B. Yee. 1981. A glycine-containing selective medium for isolation of Legionellaceae from environmental specimens. Appl. Environ. Microbiol. 42:768–772.

Detection of *Legionella* spp. in Water Samples by Electron-Capture Gas Chromatography

JAMES GILBART

Public Health Laboratory Service, Bacterial Metabolism Research Laboratory, Centre for Applied Microbiology and Research, Porton Down, Salisbury SP4 OJG, England

A consistent property of species throughout the genus Legionella is the possession of unusually abundant branched-chain fatty acids. Analysis of these fatty acids by capillary gas chromatography gives rise to specific profiles which are routinely used for the identification of isolates and in the description of new species. Although the lipids of the legionellae have been studied by various workers (2, 6, 7), little assessment of the specificity of fatty acid chromatograms compared to those of other commonly encountered bacterial species has yet been published. In this study, therefore, comparisons were made between fatty acid profiles of some of the Legionella spp. and a range of other bacteria. Once this specificity was established, the possibility of using the characteristic fatty acids of the legionellae as markers of contamination in water samples was investigated.

As part of a taxonomic study, the fatty acid profiles of the legionellae were compared with those of 55 commonly isolated species from 23 different genera. Many of these were pathogens isolated from respiratory and other clinical specimens; some were of environmental origin. None of these other bacteria were found to produce chromatographic profiles similar to the Legionella spp., although several produced diagnostic patterns. The results of this study indicate that the fatty acid profiles of the Legionella spp. are unique among bacteria, although some similar profiles have been observed among the rickettsiae (8).

The fatty acid profiles of most of the bacteria examined were sufficiently different to be distinguishable by eye. The chromatograms of some of the Legionella spp., however, were less easy to distinguish on this basis and were therefore subjected to computer discriminant analysis. This type of statistical analysis has enabled workers on various bacterial genera to differentiate between apparently similar chromatograms (4, 5). Each chromatogram was plotted as a coordinate in a multidimensional space so that similar chromatograms form clusters whereas dissimilar patterns are plotted farther apart. The resultant plot for five species of Legionella is shown in Fig. 1. Each point represents a single isolate analyzed. Once this data base containing mostly well-separated clusters is established, chromatograms from unknown isolates can be entered, and the program determines a "best-fit" identification.

To detect Legionella sp. fatty acids at very

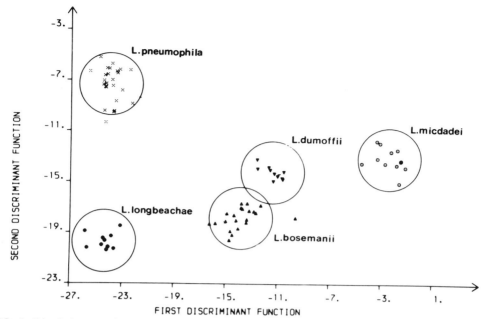

FIG. 1. Discriminant analysis of fatty acids from five species of *Legionella*. Circles around species clusters represent 95% confidence limits. Data from 89 chromatograms were analyzed in this study.

low concentrations as markers of contamination in deposits derived from filtered water samples (3), much greater sensitivity than that provided by a flame ionization detector was clearly needed. For this reason, a chromatograph equipped with an electron-capture detector was used. This type of detector is highly sensitive to many halogenated compounds, particularly pesticides, at concentrations often below 10^{-10} g/ml. Various types of electron-capture derivatives were prepared, the most successful of which were found to be trichloroethyl esters produced by a modified method of Alley et al. (1).

In an initial study, duplicate cultures of *Legionella pneumophila*, *L. bozemanii*, *L. dumoffii*, and *L. micdadei* were suspended in distilled water at a concentration of approximately 10^7 bacteria per ml. The suspensions were then serially diluted. Extraction of the fatty acids from each of the dilutions and electron-capture detector analysis of the trichloroethyl ester derivatives revealed that between 10^3 and 10^4 cells per ml could be detected. This approximated the lowest concentration of *Legionella* cells normally found in filter deposits which are culture positive for the organism.

In a subsequent series of experiments, nine water samples taken from the freshwater system of an ocean-going vessel were pressure filtered through 0.2-μm nylon membrane filters. The retained deposits were saponified in 50% methanol–5% sodium hydroxide, acidified, extracted into hexane, and dried. After derivatization with tricholoroethanol and haptafluorobutyric anhydride, samples were washed, dried under nitrogen, and suspended in hexane. Analysis was made on a Carlo Erba 4160 chromatograph fitted with a 25-m nonpolar, bonded-phase, vitreous silica capillary column (see Fig. 2). Mass spectra of derivatized fatty acids from bacterial cultures were obtained by using a Dupont 21-491 mass spectrometer coupled to a Varian 2700 series gas chromatograph. The concentration of fatty acids in the water samples was too low for mass spectrometry, so identification was made by retention time against standards.

In eight of the nine water samples, characteristic fatty acids of *L. pneumophila* were detected. No such compounds were observed in control waters which were culture negative for *Legionella* sp. In some cases, other contaminating materials gave additional peaks on the chromatograms, but these did not seriously alter the pattern or conflict with the analysis. The results indicate that electron-capture detector analysis of derivatized deposits concentrated from water samples could provide a rapid method for screening large numbers of water samples to determine which areas are worthy of further investigation by using the described culture

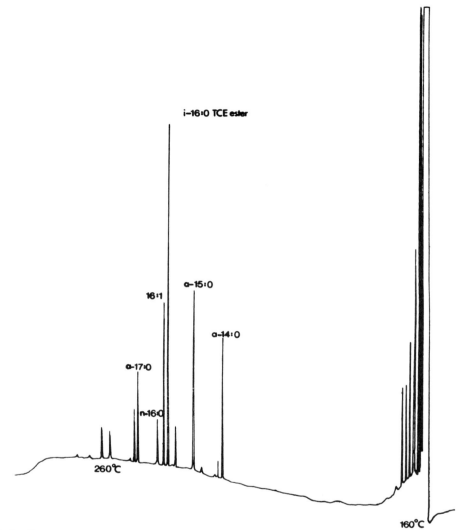

FIG. 2. Chromatogram of fatty acid trichloroethyl esters of *L. pneumophila* detected in water filtration deposits. Column: 25-m, nonpolar, bonded-phase, vitreous silica. Temperature: 160°C for 4 min, then raised at 4°C/min to 260°C and held for 10 min. Carrier: Helium (1 cm³/min). Detector: Frequency-pulsed electron capture at constant current.

techniques (3). Further evaluation of this technique, using waters taken on a much larger scale, is planned.

LITERATURE CITED

1. **Alley, C. C., J. B. Brooks, and D. S. Kellogg.** 1979. Electron capture gas-liquid chromatographic-mass spectral identification of acids produced by *Neisseria meningitidis* in a defined medium. J. Clin. Microbiol. **9:**97–102.
2. **Collins, M. D., and J. Gilbart.** 1983. New members of the coenzyme Q series from the *Legionellaceae.* FEMS Microbiol. Lett. **16:**251–255.
3. **Dennis, P. J., J. A. Taylor, R. B. Fitzgeorge, C. L. R. Bartlett, and G. I. Barrow.** 1982. *Legionella pneumophila* in water plumbing systems. Lancet **i:**949–951.
4. **French, G. L., I. Phillips, and S. Chinn.** 1981. Reproducible pyrolysis-gas chromatography of microorganisms with solid stationary phases and isothermal oven temperatures. J. Gen. Microbiol. **125:**347–355.
5. **Macfie, H. J. H., C. S. Gutteridge, and J. R. Norris.** 1978. Use of canonical variates analysis in differentiation of bacteria by pyrolysis gas liquid chromatography. J. Gen. Microbiol. **104:**67–74.
6. **Moss, C. W., and S. B. Dees.** 1979. Further studies of the cellular fatty acid composition of Legionnaires disease bacteria. J. Clin. Microbiol. **9:**648–649.
7. **Moss, C. W., R. E. Weaver, S. B. Dees, and W. B. Cherry.** 1977. Cellular fatty acid composition of isolates from Legionnaires disease. J. Clin. Microbiol. **6:**140–143.
8. **Tzianabos, T., C. W. Moss, and J. E. McDade.** 1981. Fatty acid composition of rickettsiae. J. Clin. Microbiol. **13:**603–605.

Detection and Quantitation of *Legionella pneumophila* by Immune Autoradiography

WILLIAM T. MARTIN, JAMES M. BARBAREE, AND JAMES C. FEELEY

Respiratory and Special Pathogens Laboratory Branch, Division of Bacterial Diseases, Center for Infectious Diseases, Centers for Disease Control, Atlanta, Georgia 30333

Detection and isolation of *Legionella* spp. in water specimens have primarily been qualitative. However, most investigators feel that this is no longer adequate and that quantitative analysis is required. Two procedures that are frequently used for quantitation are direct fluorescent-antibody staining and plating on selective media. Unfortunately, both have problems and are, at best, only semiquantitative for *Legionella* spp. Direct fluorescent-antibody staining restricts the volume of the sample which can be examined and does not distinguish viable from nonviable cells. Plating on selective media often partially or totally inhibits legionellae.

To achieve effective quantitation, specimens must be examined in duplicate or triplicate at more than one concentration, and the identity of all *Legionella* colonies should be confirmed by antigenic characterization and determination of L-cysteine requirement. Because the effort required to do this by current methodology is prohibitive, we investigated semiautomated ways to accomplish the task. One of the methods investigated was immune autoradiography (IR); the following preliminary report is presented.

Specificity and sensitivity of the IR test were determined by preparing suspensions of *L. pneumophila* serogroups 1 through 6, *Escherichia coli, Staphylococcus aureus*, and *Pseudomonas aeruginosa* in sterile distilled water to an optical density of a McFarland no. 1 standard. Portions (5 µl) of each suspension were inoculated to 82-mm circular nitrocellulose papers (NCPs) (Schleicher & Schuell, Inc.,), allowed to air dry, and heat fixed (1). Nonspecific receptor sites were blocked by putting the NCPs in a solution of 1% bovine fraction V albumin in TSGAN (50 mM Tris, pH 7.5; 0.15 M NaCl; 0.25% gelatin; 0.15% azide; and 0.1% Nonidet P-40 in distilled water) and placing on a rotary shaker (10 rpm) for 45 min. After rinsing in TSGAN, NCPs were immersed in polyvalent or monovalent antiserum diluted 1:1,500 in TSGAN and incubated at room temperature for various lengths of time (15 min, 1 h, and 4 h). The NCPs were then rinsed with TSGAN, and each was placed in 15 ml of TSGAN containing 1.5 µl of ^{125}I-labeled staphylococcal protein A (New England Nuclear Corp.) for 2 h at room temperature, then washed with TSGAN for 45 min. Autoradiography was performed by exposing X-Omat-R X-ray film (Kodak) for 24 h at −70°C (2). The best exposure was for 15 min. Although a few cross-reactions occurred, they were expected because of shared antigens between some serogroups of *L. pneumophila*. Increased cross-reactions were observed at 1 h, and nonspecific staining of non-legionellae was evident at 4 h.

Portions of deliberately or naturally contaminated environmental water samples (either concentrated or diluted) were inoculated onto selective and nonselective buffered charcoal-yeast extract agar plates (15 by 100 mm) with and without L-cysteine, streaked to permit growth of isolated colonies, and incubated at 35°C in moist air supplemented with 2.5% CO_2. After incubation for 72 h, a black-and-white photograph was

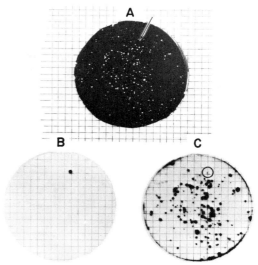

FIG. 1. IR analysis of a natural water sample for *L. pneumophila* serogroups 1 and 5. (A) Buffered charcoal-yeast extract agar plate incubated for 72 h, from which duplicate filter impressions were made. Arrow indicates serogroup 1 colony. (B) Radiograph overlaid on filter grid, showing a single colony reacting with *L. pneumophila* serogroup 1. (C) Radiograph showing many colonies of *L. pneumophila* serogroup 5. Circled area is location of serogroup 1 colony.

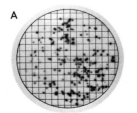 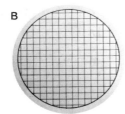

FIG. 2. Determination of L-cysteine requirement of *L. pneumophila* by autoradiography. Buffered charcoal-yeast extract agar plates with (A) and without (B) L-cysteine. Portions (0.1 ml) of a hospital potable-water sample were inoculated onto plates A and B. A total of 190 *L. pneumophila* serogroup 1 colonies grew and radiographed on (A) but not on (B), which indicates an L-cysteine requirement.

made of the plates. NCPs were then carefully placed onto the growth of each plate so that antigen imprints were obtained for each colony. After the impressions were made, the original plate was reincubated for 24 h and then held at 4°C. NCPs were dried, heat fixed for 2 h in an 80°C vacuum oven, reacted with polyvalent or monospecific antiserum for 15 min, treated with ^{125}I-labeled protein A, and exposed to X-ray film as previously described. The photograph of the original plate provided a permanent record and was used to count colonies. The subsequent radiograph was used as a transparent template to locate specific *L. pneumophila* colonies of a specific serogroup for further work-up. By this procedure, a single colony of *L. pneumophila* serogroup 1 was detected, located, and isolated from a plate inoculated with 0.1 ml of a water specimen collected from a hospital intensive

care unit. This plate also had 151 colonies of *L. pneumophila* serogroup 5 and 396 other non-*Legionella* colonies (Fig. 1).

Since it is extremely difficult to detect and quantitate legionellae in environmental water specimens containing large numbers of bacteria that produce colonies morphologically indistinguishable from *Legionella* spp., we seeded such a specimen with *L. pneumophila* serogroup 1 and evaluated the ability of the IR procedure to separate *Legionella* from non-*Legionella* colonies. To insure that radiographic reactions were not false-positive cross-reactions, the L-cysteine requirement of the colonies was determined by examining duplicate samples plated on buffered charcoal-yeast extract agar with and without L-cysteine. Antigenic imprints, characterization with polyvalent *Legionella* antiserum and ^{125}I-labeled protein A, and autoradiography were performed as previously described. We found that the IR procedure easily detected and characterized the antigenic types and L-cysteine requirement of *Legionella* colonies among non-*Legionella* colonies (Fig. 2).

The preliminary results of this study suggest that the IR technique may be used as a means of detecting and quantitating *L. pneumophila* in water specimens.

LITERATURE CITED

1. **Renart, J., J. Reiser, and G. R. Stark.** 1979. Transfer of proteins from gels to diazobenzyloxymethyl paper and detection with antisera: a method for studying antibody specificity and antigen structure. Proc. Natl. Acad. Sci. U.S.A. 76:3116–3120.
2. **Symington, J., M. Green, and K. Brackman.** 1981. Immunoautoradiographic detection of proteins after electrophoretic transfer from gels to diazo-paper: analysis of adenovirus encoded proteins. Proc. Natl. Acad. Sci. U.S.A. 78:177–181.

Survival of Airborne *Legionella pneumophila*

PETER HAMBLETON, P. JULIAN DENNIS, AND ROY FITZGEORGE

Vaccine Research and Production Laboratory, Environmental Microbiology & Safety Reference Laboratory, and Experimental Pathology Laboratory, Public Health Laboratory Service, Centre for Applied Microbiology and Research, Porton Down, Salisbury, Wiltshire, England

Since the earliest recognized outbreaks of legionellosis, strong epidemiological evidence has associated spread of the disease with certain types of domestic and water cooling systems, and it is thought that many outbreaks of *Legionella pneumophila* disease may have resulted from inhalation of droplet clouds generated by air-conditioning cooling towers, evaporative condensers, and domestic plumbing systems containing water contaminated with *Legionella* spp. Supportive evidence for this was obtained by Baskerville et al. (1a), who successfully induced experimental respiratory infection, similar to human legionellosis, in guinea pigs and rhesus monkeys.

In this work the survival of *Legionella pneumophila* serogroup 1 strain 74/81, aerosolized from water and held at different relative humid-ity (r.h.) values, was investigated. The organisms were grown on a modified charcoal-yeast extract (CYE) agar for 4 days at 37°C in air and, for some experiments, in a liquid medium. Suspensions of the organism grown in liquid medium or washed from CYE agar with distilled water were washed twice with and suspended in distilled water; washed *Bacillus subtilis* spores were added to the suspensions to serve as a tracer of physical losses. Aerosols were generated with a three-jet Collison nebulizer into a Henderson apparatus and stored in a 55-liter rotating drum. Samples were recovered from aerosols for 1 min by using Porton raised impingers.

CYE agar-grown *L. pneumophila* 74/81 was

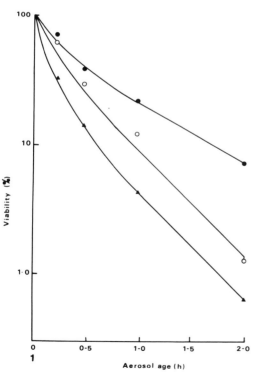

FIG. 1. Survival of aerosolized *L. pneumophila* 74/81 grown on solid CYE medium, sprayed from water, and held at different r.h. values: (▲) 30% r.h.; (●) 65% r.h.; (○) 90% r.h.

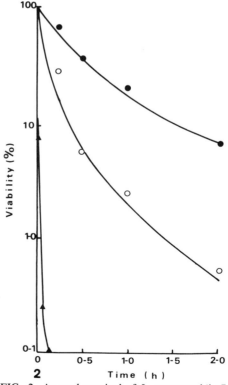

FIG. 2. Aerosol survival of *L. pneumophila* 74/81 sprayed from water and held at 65% r.h.: (●) grown on solid medium for 4 days; (○) liquid culture, stationary phase; (▲) exponential phase.

most stable at intermediate humidity (65% r.h.) and showed less stability in drier air (30% r.h.) or more humid air (90% r.h.). Significant numbers of viable organisms were recovered from aerosols stored at 65% r.h. even after 2 h.

The organisms survived poorly at 55% r.h. compared with r.h. values slightly higher (65%) or lower (40%), suggesting that r.h. zones of instability may exist for *L. pneumophila*.

Compared with CYE agar-grown organisms, aerosols generated from suspensions of broth-grown, stationary-phase cells survived less well at 65% r.h., the rate loss of viability being similar to that of aerosols of solid medium-grown bacteria held at 30% r.h. Exponential-phase, broth-grown organisms were particularly unstable; the viability fell to 1 to 2% within 1 to 2 s and to about 0.1% within 5 min, and thereafter, viable organisms were not recovered.

The ability of aerosolized legionellae to cause disease is likely to depend on the prevailing atmospheric conditions, since the survival and infectivity of airborne bacteria are influenced by many environmental factors (3). Of these the r.h. of the atmosphere is perhaps most important, although its effect is often complicated by other factors, including oxygen tension, temperature, and even the method of sampling the microbial cloud.

The first laboratory studies on aerosols of *L. pneumophila* were described by Berendt in 1980 (2). A direct comparison of his data with ours is difficult, since he used suspensions of bacteria in tryptose saline rather than distilled water and did not use a stable tracer to compensate for physical losses. Since sodium chloride may be lethal to airborne bacteria (1), its inclusion in the spray fluid could have diminished the aerosol stability of the organism.

The work presented here demonstrates that the metabolic state of the organism may determine the survival of airborne legionellae since stationary-phase, broth-grown organisms are markedly more stable in aerosols than exponential-phase, broth-grown organisms, which presumably have a higher metabolic activity.

The survival of airborne organisms with low metabolic activity is such that significant numbers of viable organisms can be recovered from bacterial clouds even after 2 h. Contaminated evaporative condensers, for example, might be anticipated continuously to generate aerosols containing viable *L. pneumophila* organisms, so that exposed susceptible humans might readily inhale and retain sufficient viable organisms to acquire an infection.

LITERATURE CITED

1. **Anderson, J. D., and C. S. Cox.** 1967. Microbial survival. Symp. Soc. Gen. Microbiol. **17:**203–206.
1a. **Baskerville, A., M. Broster, R. B. Fitzgeorge, P. Hambleton, and P. J. Dennis.** 1981. Experimental transmission of Legionnaires' Disease by exposure to aerosols of *Legionella pneumophila*. Lancet **ii:**1389–1390.
2. **Berendt, R. F.** 1980. Survival of *Legionella pneumophila* in aerosols: effect of relative humidity. J. Infect. Dis. **141:**689.
3. **Strange, R. E., and C. S. Cox.** 1976. Survival of dried and airborne bacteria. Symp. Soc. Gen. Microbiol. **26:**111–154.

Source of *Legionella pneumophila* in a Hospital Hot Water System

S. P. FISHER-HOCH,† M. G. SMITH, D. HARPER,‡ AND J. COLBOURNE

St. George's Hospital Medical School, London, SW17; Kingston Hospital, Kingston-on-Thames; and Thames Water Authority, New River Head Laboratories, London, U.K.

In 1980 an outbreak of hospital-acquired Legionnaires disease was reported in Kingston Hospital in a new building opened in 1976 (3). This building comprises operating theaters, an intensive care unit, a plant room, various services, and 12 wards on the top three floors to which about 1,000 patients a month are admit-

† Present address: Special Pathogens Reference Laboratory, Public Health Laboratory Service, Centre for Applied Microbiology and Research, Porton Down, Salisbury, Wiltshire SP4 0JG, U.K.
‡ Present address: Communicable Disease Surveillance Centre, Public Health Laboratory Service, London NW9 5EQ, U.K.

ted. *Legionella pneumophila* serogroup 1 was grown from the cooling tower pond water and from the hot and cold tap water systems in the building. Hospital-acquired cases occurring at a rate of two per month stopped abruptly only when control measures involving chlorination of the tap water to 2 ppm and raising the hot water temperature above 55°C were fully implemented.

The hospital was subjected to intense public, press, and Union interest, with pressure to close the building. Comprehensive surveillance procedures were set up by engineers, epidemiologists, microbiologists, nurses, and clinicians. Stringent procedures were laid down for such routine processes as water tank cleaning, plumbing maintenance, and daily monitoring of chlorine levels and water temperatures. Regular samples were taken for isolation of *L. pneumophila*. All procedures were exhaustively reviewed and recorded at weekly meetings. Surveillance of the wards to identify cases continued on the basis of weekly reports by nursing officers on lower respiratory tract infections. This was supplemented by a nursing research assistant who visited each ward at least two to three times a week and examined medical and nursing notes, talked to staff and took appropriate specimens from all possible cases if not already available, kept detailed records, and followed up on all hospital-acquired respiratory disease.

For 13 months no cases of hospital-acquired *L. pneumophila* pneumonia were identified, despite the fact that *L. pneumophila* could be isolated periodically from certain outlets. Then in late August 1981, two cases developed. Both had been in-patients in the new building 2 to 10 days before the development of pneumonia. Case 1 was a 38-year-old male of previous good health, admitted for routine investigations, whose antibody titer to *L. pneumophila* serogroup 1 rose during his illness from <16 to 128. Case 2 was a 71-year-old female who developed a postoperative pneumonia. Her antibody titer in two specimens taken after the development of clinical pneumonia was 256, and radiographs taken during the illness were compatible with pneumonia due to *L. pneumophila*. Both patients made a full recovery.

Control measures were immediately reviewed. Chlorine levels in the cold water system had not dropped below 2 ppm at the outlets, and there was no longer a cooling tower. The temperature of the hot water had, however, dropped briefly due to increased demand, and a second hot water cylinder, which had been stagnant throughout the summer months, had immediately been brought into use. Brown discoloration of hot water was then noticed on the wards. *L. pneumophila* serogroup 1 was grown from samples of tap water from hot outlets and the brown deposit found at the base of a third, empty, hot water cylinder; in the latter, a titer of 10^9/liter was found. Metal analysis showed that the brownish discoloration at the outlets was due to contamination with similar deposit (4).

Since despite cleaning and heat sterilization of the cylinders *L. pneumophila* could still be grown sporadically from outlets, the cylinders were examined to see whether they might also play a part in infection of patients. A mixer tap and shower unit of the type used in the bathrooms in the wards were removed and taken apart, and each component was examined for *L. pneumophila*. Larger components were thoroughly swabbed with moist swabs. Small components were shaken in sterile distilled water and removed, the water was centrifuged, and the deposit and the swabs were inoculated onto an improved semiselective medium described by Edelstein (2, 5). Only the black rubber washers in the base of the hose and the head of the shower yielded *L. pneumophila* serogroup 1 (Table 1). Next, 25-ml water samples were taken from 20 showers throughout the wards. Each shower was then dismantled, the washers were removed and shaken in 5 ml of sterile distilled water, and *L. pneumophila* isolation was carried out as described above. Two water samples grew *L. pneumophila* serogroup 1, but black rubber washers from these two showers and from three others were positive. Finally, a single outlet was studied by taking serial 25-ml samples after overnight stagnation. Table 2 shows that the numbers of organisms were highest in the initial samples and tailed off as the flow of water washed through the outlet. As a result of these findings, all black rubber washers were immediately removed and replaced with Proteus 80 compound washers (approved by the National Water Council). Since then we have been unable to isolate *L. pneumophila* from any outlet.

Since 1980 close surveillance revealed 93 cases of hospital-acquired pneumonia during a study period of 1 year, of which only 2 were *L. pneumophila* pneumonia. The incubation periods of both patients coincided with the reintroduction into use of a stagnant hot water cylinder. Brown stains at outlets were shown to be due to material similar to the deposits found in hot water cylinders in the building which grew *L. pneumophila* to high titer. It is possible that this was the source of infection of these two cases.

Despite sterilization of the hot water cylinders, hot water at a minimum temperature of 55°C, and chlorination of cold water to 2 ppm, we were still able to isolate *L. pneumophila* from outlets in the building. We therefore examined the outlet fittings and found that the black rub-

TABLE 1. Isolation of *L. pneumophila* serogroup 1 from dismantled shower unit components

Component	Sample	Result
Shower head	Water from head	Negative
	Swab inside head	Negative
	Swab outside head	Negative
	Black rubber washer	Positive
Shower hose	Water from hose	Negative
	Swab inside barrel of hose	Negative
	Black rubber washer from base of hose	Positive

TABLE 2. Isolation of *L. pneumophila* from serial 25-ml water samples from a single outlet

Sample no.	Vol of sample (ml)	*L. pneumophila* CFU per sample
1	"Morning drop"	0
2	25	20
3	25	14
4	25	20
5	25	5
6	25	0
7	25	4
8	25	1
9	25	2
10	25	4
11	25	1
12	25	2

ber tap washers in general use were a source of the organism in the outlets. Black rubber of this type is known not only to deviate chlorine chemically, but to be capable of supplying sufficient nutrients for microbial growth (1). After changing all washers to materials approved by the National Water Council as not being able to support microbial growth, we were unable to reisolate the organism.

A new building such as the one at Kingston Hospital may be seeded with *L. pneumophila* during construction when pipes and tanks lie around building sites collecting dust and rain water for many months. When installed they are filled for pressure testing and left stagnant for long periods. After commissioning, the system provides convenient amplification reservoirs, such as the type of hot water cylinder in Kingston Hospital, in which constant temperature gradients are maintained and a high metal content is provided. During routine inspection these cylinders are left open to the atmosphere, sometimes for months, with a large residual pool of water and metallic deposit. Contaminated deposit can be discharged to the outlets when major changes in flow patterns are introduced by

engineering maneuvers, such as bringing into line stagnant cylinders. Materials in the plumbing system are thus regularly reseeded, and where they provide both nutrients and a sheltered niche further multiplication takes place. Since both hot and cold water supplies in the building are pressurized to 60 lb/in^2, sufficient energy is theoretically available at the taps to generate 5-μm particles, small enough to penetrate to the alveoli and thus infect (G. Harper, personal communication). It is possible that pressurization is one of the features which determines those buildings liable to outbreaks.

In an epidemic situation such as at Kingston Hospital, the information gained has been limited by the need at each stage to take immediate and public steps to eradicate the organism. Our experience nevertheless illustrates the importance of close understanding of water supplies in a building in which cases have occurred to identify possible amplification reservoirs and to select control measures appropriate to that particular building. Replacement of the type of black rubber tap washers we described was a very cheap, but in our experience effective, method of finally ridding the outlets of *L. pneumophila*.

LITERATURE CITED

1. Colbourne, J. S., and D. Brown. 1979. Dissolved oxygen utilisation as an indicator of total microbial activity on non-metallic materials in contact with potable water. J. Appl. Bacteriol. 47:223–228.
2. Edelstein, P. H. 1981. Improved semiselective medium for isolation of *Legionella pneumophila* from contaminated and environmental specimens. J. Clin. Microbiol. 14:298–303.
3. Fisher-Hoch, S. P., C. L. R. Bartlett, J. O'H. Tobin, M. B. Gillett, A. M. Nelson, J. E. Pritchard, M. G. Smith, R. A. Swann, J. M. Talbot, and J. A. Thomas. 1981. Investigation and control of an outbreak of Legionnaires' Disease in a district general hospital. Lancet i:932–936.
4. Fisher-Hoch, S. P., M. G. Smith, and J. S. Colbourne. 1982. *Legionella pneumophila* in hospital hot water cylinders. Lancet i:1073.
5. Smith, M. G. 1982. A simple disc technique for the presumptive identification of *L. pneumophila*. J. Clin. Pathol. 35:1353–1355.

Source of *Legionella pneumophila* Infection in a Hospital Hot Water System: Materials Used in Water Fittings Capable of Supporting *L. pneumophila* Growth

J. S. COLBOURNE, M. G. SMITH, S. P. FISHER-HOCH,[†] AND D. HARPER[‡]

Thames Water Authority, New River Head Laboratories, London EC1R 4TP, Kingston Hospital, Kingston-upon-Thames, and St. George's Hospital Medical School, Cranmer Terrace, London, England

During the 1970s, Burman and Colbourne (2, 4, 6) demonstrated that certain types of material used in direct contact with drinking water in consumers' premises can contaminate water by leaching compounds that either impart a taste, odor, color, or turbidity to water or are considered to be hazardous to health. In particular, they demonstrated that some materials are capable of supporting the growth of microorganisms to such a degree that water quality is impaired by the release of unpleasant-tasting by-products of microbial multiplication, by the appearance in water of slimes and particulate matter, or by the presence of bacteria such as *Pseudomonas aeruginosa* and coliform organisms. Methods for assessing the ability of materials to contaminate water were developed at Thames Water Authority (3, 5) and adopted by the National Water Council (NWC) (10), the body responsible in the United Kingdom for assessing the compliance of water fittings with Model Water Byelaws (7). These methods, now standardized by the British Standards Institute (1), are internationally recognized and are applied in areas such as Hong Kong, Singapore, and the Middle East.

During the 1980 outbreak of Legionnaires disease at Kingston Hospital, Thames Water (the supplying Authority) assisted in an advisory capacity with the chlorination of the hospital's internal water systems (8). As described in a companion paper (S. P. Fisher-Hoch, M. J. Smith, J. S. Colbourne, and D. Harper, this volume), further cases occurred in 1981, and contaminated hot water cylinders were then thought to be the primary source of the organism (9). After sterilization of the cylinders, *Legionella pneumophila* persisted in the periphery of the hot water system, being isolated from water samples and rubber gaskets taken from shower units on the wards. During this investigation, components of one fitting were assessed by the NWC procedure for their ability to support microbial growth (5). The rubber gasket in the shower head gave a positive microbial growth

result, but the plastic shower head and hose gave negative microbial growth results. (Microbial growth is measured by the difference in dissolved oxygen uptake between the test system containing the component and a negative glass control system; values greater than 2.3 mg/liter are recorded as positive responses; Table 1.) These findings led us to question whether *L. pneumophila* may be capable of multiplication in water fittings in association with certain types of component material. We therefore proceeded to a comprehensive survey of all water fittings in the hospital building. We identified 31 nonmetallic components of seven representative water fittings including two shower units, three bath and sink mixer units, a stopcock, and a drinking-water tap. All received both hot and cold water with the exception of the stopcock and drinking water tap, which received only mains water. Each component was tested for the presence of *L. pneumophila* (8) and for its ability to support microbial growth (5). The stopcock and drinking-water tap components gave negative microbial growth responses, but five other components gave positive responses (Table 1). Three of the five were rubber components, and *L. pneumophila* was isolated from one of these.

This survey was carried out when full control procedures were in operation (hot water maintained at 60°C; chlorine residual in the cold water, 2 mg/liter) and the hot water cylinders had been sterilized. The isolation rate of *L. pneumophila* was, therefore, considerably lower than at the time of the 1980 and 1981 cases (Fisher-Hoch et al., this volume) when *L. pneumophila* was isolated from several components. For this reason we turned to a study of the capacity of *L. pneumophila* to multiply in direct association with components of water fittings. We adapted the NWC microbial growth procedure to permit a more direct assessment of the ability of *L. pneumophila* to multiply on rubber components. An environmental isolate of *L. pneumophila* serogroup 1 from Kingston Hospital was used as the inoculum at 10^6 cells per liter. Two parallel series of three tests were performed at 30 ± 1°C, each using negative glass controls, positive paraffin wax controls, and test samples of rubber. One test series (system A) was inoculated with both a mixture of natural aquatic microorga-

† Present address: Public Health Laboratory Service Centre for Applied Microbiology and Research, Porton Down, Salisbury, England.

‡ Present address: Communicable Disease Surveillance Centre, Public Health Laboratory Service, London, England.

TABLE 1. Microbiological testing of components removed from water fittings within Kingston Hospital

Fitting	Component	Material	L. pneumophila isolated	MDOD[a] (mg/liter)
December 1981				
Shower	Gasket	Rubber	+	4.0
	Hose	Plastic	−	0.5
	Head	Plastic	−	2.0
February 1982				
Sink mixer tap	Hot washer	Rubber	−	1.7
	Hot washer	Fiber	−	1.2
	Cold washer	Rubber	−	1.1
	Cold washer	Fiber	−	0.0
Drinking-water tap	Cold washer	Rubber	−	0.3
	Cold washer	Fiber	−	0.1
Stopcock	Cold washer	Rubber	−	0.0
	Cold washer	Fiber	−	0.6
Bath mixer tap	Hot washer	Fiber	−	0.0
	Hot washer	Rubber	−	3.8
	Hot washer	Fiber	−	1.6
	Cold washer	Fiber	−	0.6
	Cold washer	Fiber	−	0.4
	Cold washer	Rubber	−	2.1
Shower unit	Hose washer	Rubber	−	1.2
	Hose washer	Rubber	−	0.2
	Head body	Plastic	−	0.6
	Head washer	Rubber	+	3.2[b]
Bath mixer tap	Cold washer	Rubber	−	0.7
	Cold washer	Fiber	−	0.1
	Hot washer	Rubber	−	0.5
	Hot washer	Rubber	−	0.5
	Hot 'O' ring	Rubber	−	0.4
Shower unit	Hose washer	Plastic	−	3.2
	Hose connector	Plastic	−	0.3
	Valve washer	Rubber	−	0.1
	Valve washer	Rubber	−	0.4
	Head gasket	Rubber	−	4.1
	Head body	Plastic	−	0.9
	Head body	Plastic	−	0.2
	Head washer	Fiber	−	2.2

[a] MDOD, Mean dissolved oxygen difference (5). Values of ≥2.3 mg/liter are considered positive.
[b] Positive for *P. aeruginosa* growth.

nisms (River Thames; 10%) (3) and *L. pneumophila* (10^6/liter), and the water was changed with ordinary dechlorinated, unsterilized tap water. The second test series (system B) was inoculated with *L. pneumophila* (10^6/liter) alone, and the water was changed with dechlorinated, sterilized tap water.

The levels of microbial activity were similar in both test series (Table 2). *L. pneumophila* multiplied rapidly in the positive controls and the

TABLE 2. Ability of *L. pneumophila* to multiply in tap water in association with rubber materials

Test system	Estimated no. of *L. pneumophila*[a] at:					MDOD[b] (mg/liter)
	Week 4	Week 5	Week 6	Week 10	Week 12	
A						
Glass	+++	++	+	+	−	
Paraffin wax	++++		++	+++	+++	5.8
Rubber	++++	+++	+++	++++	++++	3.9
B						
Glass	++++	++	+	+	−	
Paraffin wax	++++		+	++	+++	6.7
Rubber	++	+++	++	+++	+++	4.2

[a] ++++, 10^6 to 10^5/liter; +++, 10^5 to 10^4/liter; ++, 10^4 to 10^3/liter; +, 10^3 to 10^2/liter; −, 10^2/liter or less.
[b] MDOD, Mean dissolved oxygen difference.

tests containing the rubber material, but did not multiply in the negative controls. As the water changing procedure diluted the microbial content in each test by 99% every 3 days, these increases in counts were indicative of rapid multiplication as opposed to survival of the organism. No differences were observed between the test system containing natural aquatic flora and that containing only *L. pneumophila*. While these experiments were being performed, the hospital authorities replaced all the rubber components with a type of rubber known to give a negative microbial growth response and hence approved for use in contact with potable water by the NWC. Since the rubber components were replaced it has not been possible to detect *L. pneumophila* from any outlets in the hospital.

It is our opinion that throughout the life of the building the hot water system was periodically contaminated with *L. pneumophila* from the hot water cylinder deposits. Even after effective treatment of the cylinders, *L. pneumophila* persisted within certain water fittings in the hospital. These contained rubber components which we have shown to be capable of supporting the growth of microorganisms including *L. pneumophila*. The microenvironment within fittings was such that chlorine was unable to exert a local biocidal effect and raised water temperatures were selective for *L. pneumophila*. Thus fittings that contained certain types of rubber components provided an ecological niche for *L. pneumophila* within the plumbing system.

LITERATURE CITED

1. **British Standards Institution.** 1982. Specification of requirements for suitability of materials for use in contact with water for human consumption with regard to their effect on the quality of the water. Draft for development. DD82:1982. British Standards Institution, London.
2. **Burman, N. P., and J. S. Colbourne.** 1976. The effect of plumbing materials on water quality. J. Inst. Plumbing 3:12–13.
3. **Burman, N. P., and J. S. Colbourne.** 1977. Techniques for the assessment of growth of micro-organisms on plumbing materials used in contact with potable water. J. Appl. Bacteriol. 43:137–144.
4. **Burman, N. P., and J. S. Colbourne.** 1979. Effect of non-metallic materials on water quality. J. Inst. Water Eng. Scientists. 1:11–18.
5. **Colbourne, J. S., and D. Brown.** 1979. Dissolved oxygen utilisation as an indicator of total microbial activity on non-metallic materials in contact with potable water. J. Appl. Bacteriol. 47:223–231.
6. **Colbourne, J. S.** 1981. The influence of non-metallic materials and water fittings on the microbial quality of water within plumbing systems. Proceedings of the Symposium on Hospital Water Supplies. Institute of Public Health Engineering and Department of Health and Social Security, Loughborough, England.
7. **Department of the Environment.** 1966. Model water bye-laws. Her Majesty's Stationery Office, London.
8. **Fisher-Hoch, S. P., C. L. R. Bartlett, J. O'H. Tobin, et al.** 1981. Investigation and control of an outbreak of Legionnaire's Disease in a district general hospital. Lancet i:932–936.
9. **Fisher-Hoch, S. P., M. G. Smith, and J. S. Colbourne.** 1982. Legionella pneumophila in hospital hot water cylinders. Lancet i:1073.
10. **National Water Council.** 1982. Requirements for the testing of non-metallic materials for use in contact with potable water. Document 1O8D01, issue 2. National Water Council, London.

Role of Stagnation and Obstruction of Water Flow in Isolation of *Legionella pneumophila* from Hospital Plumbing

CAROL A. CIESIELSKI, MARTIN J. BLASER, F. MARC LaFORCE, AND WEN-LAN L. WANG

Medical Service and Microbiology Laboratory, Denver Veterans Administration Medical Center, Denver, Colorado 80220, and Division of Infectious Diseases, Department of Medicine, and Department of Pathology, University of Colorado School of Medicine, Denver, Colorado 80262

Nosocomial infections due to *Legionella pneumophila* are being recognized more frequently as appropriate diagnostic techniques are becoming generally available. Often these infections have occurred in association with isolation of *L. pneumophila* from potable water supplies. The organism has been found in naturally occurring aquatic habitats as well as in the plumbing systems of many hotels, hospitals, and other large buildings (3).

After three nosocomial infections due to *L. pneumophila* were detected at the Denver Veterans Administration Medical Center (DVAMC) in late 1981, ongoing environmental surveillance of the hospital's potable water system was initiated. The DVAMC water system includes five hot water storage tanks that are heated by central steam coils. Two tanks, used on a rotating basis, supply hot (114 to 118°F; ca. 45.5 to 47.5°C) water to the hospital at any given time. Tanks not in use are kept filled with water which remains at the ambient boiler room temperatures.

At monthly intervals, water from the hot water storage tanks was collected from a valve near the bottom of the tanks. Immediately after collection, we noted the temperature of the water and whether the tank was in use (on-line). We then plated 0.2 to 0.5 ml in duplicate onto semiselective medium. The medium contained

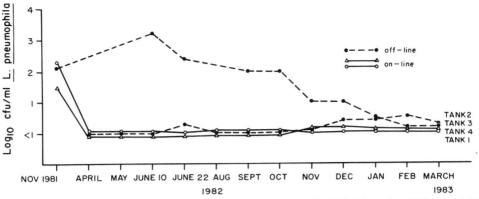

FIG. 1. Isolation of *L. pneumophila* from hot water storage tanks, DVAMC, November 1981–March 1983.

buffered charcoal-yeast extract agar with vancomycin, colistin, cephalothin, and cycloheximide (1). Plates were incubated at 35°C and were examined daily for 10 days. Colonies with typical morphology and growth characteristics were confirmed to be *L. pneumophila* and serogrouped by direct fluorescent-antibody staining.

Shower heads were also cultured at monthly intervals by removing the shower head, dislodging sediment with a sterile swab, and plating onto the semiselective agar. During a survey to determine environmental contamination from the hospital's faucets, *L. pneumophila* was isolated from 3 of the 12 sampled. Each of these three positive faucets was found to contain an aerator, a device that obstructs water flow. To further investigate the role of obstruction in the isolation of *L. pneumophila* from plumbing fixtures, a survey of all faucets in the DVAMC to identify outflow obstructions indicated that three types were present: aerators, vacuum breakers, and backflow preventers. The latter two devices contain one-way valves designed to prevent the backflow of water in the event of a pressure drop.

Seventy-two faucets were then cultured. From each, tap water was obtained before any manipulation and 0.5 ml of this water was plated onto semiselective medium. The faucets were

then disassembled, and cultures taken as follows from the indicated sites were plated onto the medium: from the unobstructed faucets, the spout sediment was collected with a sterile swab; from the aerated faucets, the aerator screen and the pipe proximal to the aerator were cultured; from faucets with backflow preventers or vacuum breakers, swabs of the one-way valves and the pipes proximal and distal, respectively, to the obstruction were cultured.

All five hot water storage tanks were positive for *L. pneumophila* on initial culture in November 1981 (Fig. 1). Tank 5 was subsequently drained dry. Two of the tanks (1 and 4) have been kept in continual use since that time, thereby preventing stagnation of water. Colony counts in these two tanks fell quickly to very low levels, and no subsequent multiplication of *L. pneumophila* was observed. *L. pneumophila* persisted in the other two tanks (2 and 3) which were not on-line. The temperature of the water in these two tanks was at the ambient temperature of the boiler room (84 to 88°F; ca. 28.8 to 31.1°C), which is below the temperature at which *L. pneumophila* has been shown to multiply (4).

Six of the eight shower heads cultured in November 1981 were positive for *L. pneumophila*. Since then, shower head culture positivity has ranged from 50 to 100%.

L. pneumophila was isolated from 22 of the 30 aerated faucets versus none of the 26 nonobstructed faucets (Table 1). Two of the 16 vacuum breakers, but none of the six backflow preventers, grew *L. pneumophila*. From the aerated faucets, legionellae were most frequently isolated from the aerator screen itself, being positive in 17 of the 30 sampled, compared with 11 positive of 30 faucet pipes and 9 positive of 30 water samples. All environmental isolates from the DVAMC plumbing system have been sero-

TABLE 1. Isolation of *L. pneumophila* from faucet fixtures, DVAMC, August 1982

Type of faucet	No. sampled	No. positive	P value[a]
Aerator	30	22	
Nonobstructed	26	0	<0.001
Vacuum breakers	16	2	<0.001
Backflow preventers	6	0	<0.005

[a] Comparison with results from faucets with aerators; Fisher's exact test.

group 1, as were isolates from the three affected patients.

To further characterize the role of aerators as environmental reservoirs, nine previously positive aerators were recultured, then sterilized overnight in 70% alcohol. They were replaced on the faucets, and the aerator screens were recultured monthly. Within 1 month, one aerator became recolonized with *L. pneumophila*. Several other aerators have been intermittently positive during the 7 months of this ongoing study.

Infections due to *L. pneumophila* have been associated with its isolation from potable water systems, but *L. pneumophila* also has been found in many water systems unassociated with clinical illness. When deemed necessary on the basis of epidemiological considerations, eradication of this organism from potable water supplies has been difficult and costly.

Conditions known to be favorable for the multiplication of *L. pneumophila* include stagnating warm water. After a small nosocomial outbreak, we eliminated stagnation of warm water in our potable water system. We found that colony counts of *L. pneumophila* dropped quickly to low levels in the two tanks that were continuously supplying hot water to the hospital. There have been no further cases of Legionnaires disease recognized at the DVAMC since that time. Where feasible, use of this simple technique may keep *L. pneumophila* concentrations at low levels in water systems serving high-risk populations.

Faucet aerators have not been permitted to be used in hospitals since they were shown to be a source of *Pseudomonas* sp. infection (2). Our data suggest that faucet aerators, devices which obstruct water flow, become colonized by *L. pneumophila* and may constitute a secondary reservoir for the organisms in the water system. This study provides further evidence that these devices should be eliminated from hospital plumbing fixtures. The role of obstruction as shown in the faucet study may provide a model to explain contamination of shower heads, another clinically relevant site.

LITERATURE CITED

1. **Bopp, C. A., J. W. Sumner, G. K. Morris, and J. G. Wells.** 1980. Isolation of *Legionella* spp. from environmental water samples by low pH treatment and use of a selective medium. J. Clin. Microbiol. **40:**714–719.
2. **Cross, D. F., A. Benchimol, and S. E. G. Dimond.** 1966. The faucet aerator—a source of *Pseudomonas* infection. N. Engl. J. Med. **274:**1430–1431.
3. **Wadowsky, R. M., R. B. Yee, L. Mezmar, E. P. Wing, and J. N. Dowling.** 1982. The hot water system as a source of *Legionella pneumophila* in plumbing fixtures in hospital and non-hospital environments. Appl. Environ. Microbiol. **43:**1104–1110.
4. **Yee, R. B., and R. M. Wadowsky.** 1982. Multiplication of *Legionella pneumophila* in unsterilized tap water. Appl. Environ. Microbiol. **43:**1330–1334.

Distribution of *Legionella pneumophila* in Power Plant Environments

JEAN A. SOLOMON, SIGURD W. CHRISTENSEN, RICHARD L. TYNDALL, CARL B. FLIERMANS, AND STEPHEN B. GOUGH†

Oak Ridge National Laboratory, Oak Ridge, Tennessee 37830; University of Tennessee, Knoxville, Tennessee 37916; Savannah River Laboratory, Aiken, South Carolina 29801

Legionnaires disease bacteria (*Legionella* spp.) are a natural component of freshwater aquatic communities; when aerosolized, however, they can be pathogenic to humans. *Legionella* spp. in air-conditioning cooling towers have been tentatively linked with outbreaks of legionellosis. Because utilities operate many large cooling towers, the Electric Power Research Institute judged it desirable to support a study of the distribution, abundance, viability, and infectivity of *Legionella* spp. from power plant cooling systems. This paper discusses abundance and viability; distribution of *Legionella* species and serogroups and infectivity results are presented later in this volume (R. L. Tyndall, S. W. Christensen, J. A. Solomon, C. B. Fliermans, and S. B. Gough, this volume; S. W. Christensen, R. L. Tyndall, J. A. Solomon, C. B. Fliermans, and S. B. Gough, this volume).

Water samples were collected during each of the four seasons (1981–1982) at various plant-affected locations within each of nine power plants (e.g., precondenser, postcondenser, cooling tower basin, discharge canal) and from source waters at each site. Geographic diversity was achieved by including four plants from the northern midwest, four from the southeast, and one from the east. Six of the plants were coal-fired, and three were nuclear. The type of cooling system (once-through versus closed-cycle) was of particular interest; five of the plants used once-through cooling and the remaining four were predominantly closed-cycle. Among the closed-cycle plants, both mechanical-draft and natural-draft cooling towers were represented.

† Present address: System Development Corp., Fredericksburg, VA 22401.

TABLE 1. Mean cell densities of *L. pneumophila* serogroups 1 to 4

Season	Sample location	No. of cells per ml[a]								
		Plants operating in once-through mode					Plants operating in closed-cycle mode			
		A	B	C	D	E	F	G	H	I
Spring	Ambient	35	48	22	227	15,733 *	3,967 *	8,767	10,133 *	11,333 *
	Plant-exposed	33	35	28	233	343	508	19,400	300	248
Summer	Ambient	520	140	71	56	300[b]	263[c]	183	650 *	327 *
	Plant-exposed	235	225	105	128	—[b]	4,081[c]	368	146	95
Fall	Ambient	63	80	33	21	93	40	45	49	372 *
	Plant-exposed	70	59	45	22	42	36	112	30	55
Winter	Ambient	665	185	320	114	1,205	1,360	957	4,450 *	45
	Plant-exposed	337	352	710	158	1,070	2,050	1,640	315	56

[a] Mean cell densities separated by an asterisk represent a significant ($P < 0.05$) decrease with passage through a power plant.
[b] No samples taken. Plant was not operating.
[c] Plant operating system could not be classified.

Measurements of temperature, dissolved oxygen, pH, conductivity, alkalinity, phosphate, nitrate, ammonia, and inorganic and total dissolved carbon were obtained from each sample. In addition, the samples were concentrated 500-fold by centrifugation and processed to determine the density of *Legionella pneumophila* serogroups 1 to 4 by using direct fluorescent-antibody staining (1). Viability, defined as possession of a functional electron transport apparatus, was estimated by using a tetrazolium dye (2).

The ubiquitous occurrence of the organism was demonstrated in this study. *L. pneumophila* was detected by direct fluorescent-antibody staining in 265 of 270 samples, in source water as well as in plant-affected waters, and from all nine plants during all seasons. The minimum cell density detectable by this method was approximately 3 organisms per ml, and the maximum density found was 2.3×10^4 cells per ml. The highest average densities (7.8×10^3 cells per ml) were found in spring in the source waters of closed-cycle sites; these values were significantly higher than the total cell densities in the source waters of the once-through plants in spring (Table 1). This difference in source waters is not attributed to the plants themselves, but could be due to different siting requirements for the two cooling system types (e.g., a larger water volume is required for once-through systems). The high springtime cell densities in the source water of plant E, a variable-mode plant operated in once-through configuration, are probably an artifact of sampling immediately

upstream from a skimmer wall. On the average, source-water cell densities tended to be higher during winter and spring than during summer and fall, with numerous exceptions at individual plants. The effect of plant passage (difference between source and plant-affected water), when detectable, was to decrease the cell density.

In addition to differing cell densities, the source waters of closed-cycle and once-through sites were also found to have *L. pneumophila* populations with differing levels of viability. In both the spring and summer samples, the proportion of viable cells in source waters was significantly lower at the closed-cycle sites than at the once-through sites ($P < 0.05$). The effect of retention in a closed-cycle plant was to increase viability levels in spring and decrease them in the summer. However, viability proved to be a highly variable parameter in our study, and statistically significant results were few.

Analysis of the density of viable *L. pneumophila* cells (product of density and viability) showed a difference in the source waters of once-through and closed-cycle sites only in the spring, with closed-cycle sites having an average of 30 times as many viable cells (Table 2). At closed-cycle sites, densities of viable cells decreased significantly with retention in the plant in both spring and summer.

Both correlation analysis and stepwise multiple regression were used to identify potential relationships between the physical-chemical variables and the density and viability of *L. pneumophila*. Due to extensive multicolinearity, the results were inconclusive, but several obser-

TABLE 2. Mean densities of viable *L. pneumophila* cells before and after plant passage for the two operating modes

Season	Operating mode	No. of cells per ml at location[a]:	
		Source	Plant
Spring	Once-through	21	17
		*	*
	Closed-cycle	645 *	162
Summer	Once-through	20	32
			*
	Closed-cycle	14 *	2
Fall	Once-through	7	8
	Closed-cycle	7	6
Winter	Once-through	83	103
	Closed-cycle	243	126

[a] An asterisk between two numbers indicates that these two means are significantly different from one another ($P < 0.05$).

vations can be made. Density of viable cells was correlated more strongly with many physical-chemical variables than either total cell density or percent viability alone. A positive correlation ($n = 60$, $P < 0.001$) among the closed-cycle plant-affected samples was noted between viable cell density and pH ($r = +0.42$) and shock temperature ($r = +0.43$), whereas negative relationships ($n \geq 50$, $P < 0.02$) existed with organic carbon ($r = -0.46$), total carbon ($r = -0.41$), alkalinity ($r = -0.44$), P_i ($r = -0.32$), and growth temperature ($r = -0.32$). Stepwise multiple regression, using the same data, indicated a negative relationship between viable cell density and both organic carbon and conductivity ($R^2 = 0.77$). Caution is needed in interpreting statistically significant correlation and regression results. In neither case is causation implied; for example, two variables may be changing in response to a third, unmeasured, factor.

In summary, our results demonstrate the ubiquitous distribution of *L. pneumophila* serogroups 1 to 4. No effect of passage through a once-through cooling system was found, but retention within a recirculating (closed-cycle) system resulted in some observable effects: total cell density and the density of viable cells were lower in the plant water than in source water in both spring and summer. The *L. pneumophila* populations from source waters of once-through plants were observed to differ from those of closed-cycle plants; this may be a reflection of differing siting requirements for the two system types. Attempts to relate density and viability with water quality parameters in these field samples were inconclusive due to multiple correlations among the parameters.

This research was sponsored by the Electric Power Research Institute (EPRI Project RP1909-1) pursuant to a Participation Agreement dated 29 September 1981 with the U.S. Department of Energy under Contract W-7405-eng-26 with Union Carbide Corp.

LITERATURE CITED

1. Cherry, W. B., B. Pittman, P. P. Harris, G. A. Hebert, B. M. Thomason, L. Thacker, and R. E. Weaver. 1978. Detection of Legionnaires disease bacteria by direct immunofluorescent staining. J. Clin. Microbiol. 8:329–338.
2. Fliermans, C. B., R. J. Soracco, and D. H. Pope. 1981. Measure of *Legionella pneumophila* activity in situ. Curr. Microbiol. 6:89–94.

Thermally Altered Habitats as a Source of Known and New *Legionella* Species

R. L. TYNDALL, S. W. CHRISTENSEN, J. A. SOLOMON, C. B. FLIERMANS, AND S. B. GOUGH

Oak Ridge National Laboratory, Oak Ridge, Tennessee 37830; University of Tennessee, Knoxville, Tennessee 37916; Savannah River Laboratory, Aiken, South Carolina 29801; and System Development Corporation, Fredericksburg, Virginia 22401

Thermally altered waters of electric power plants were analyzed for the presence of a variety of human microbial pathogens including *Legionella* spp. Water samples from both ambient (source) (Table 1) and plant-affected (Table 2) locations at nine geographically disparate sites (designated A through I) in the continental United States were concentrated by centrifugation, examined by the fluorescent-antibody technique, and injected into guinea pigs for infectivity and isolation studies. The test sites are described in a companion paper (S. W. Christensen, R. L. Tyndall, J. A. Solomon, C. B. Fliermans, and S. B. Gough, this volume). In the 51 samples that proved infectious for guinea pigs and from which *Legionella* spp. were isolated, *Legionella pneumophila* serogroup 1 was the most prevalent. This serogroup was isolated from a total of 14 plant-affected and 5 ambient samples from six of the nine sites. The second most prevalent *Legionella* isolate was *L. pneumophila* serogroup 4, which was isolated from nine plant-affected and six ambient samples and was found at seven of the nine sites. A newly

TABLE 1. Types of pathogens isolated from ambient water samples

Site	Pathogen[a]			
	Spring	Summer	Fall	Winter
A	—	K	K, LA, GO	—
B	—	—	—	—
C	—	K	"Vibrio"	K
D	—	—	LA, CH	—
E	—	—	OR	—
F	—[b]	—[b]	LA, J19	—
G	—[b]	K, CH	—	BOZ
H	—	—[b]	LA	LA
I	—	—	—	LA

[a] Reactive with fluorescent-antibody conjugates prepared against the following *L. pneumophila* serogroups: K, Knoxville 1 (serogroup 1); LA, Los Angeles 1 (serogroup 4); CH, Chicago 2 (serogroup 6); BL, Bloomington 2 (serogroup 3); TO, Togus 1 (serogroup 2). Other *Legionella* spp. tested were as follows. BOZ, *L. bozemanii*. GO, *L. gormanii*. J19, *Legionella* sp. not typable with antiserum prepared against known species or serogroups of *Legionella*, but reacting with antiserum made against a "J19" (Jamestown) strain of *Legionella*. OR, *L. oakridgensis*. L?, Species of *Legionella* not typable with antiserum prepared against known species or serogroups of *Legionella*. "Vibrio," Vibrio-like organism, at present unidentifiable by the Centers for Disease Control. —, *Legionella* spp. were not isolated.

[b] One or more samples were injected. Samples either were toxic to the guinea pigs or resulted in contaminated plates upon subsequent plating of tissue; therefore, *Legionella* spp. could not have been isolated if present.

discovered species of *Legionella* (*L. oakridgensis*; see below) was isolated from 12 plant-affected samples at three sites but was rare or absent in ambient water samples. Other less frequently isolated *Legionella* spp. included serogroups 2, 3, and 6 of *L. pneumophila* as well as *L. gormanii* and *L. bozemanii*.

L. oakridgensis (OR10 isolate) was first obtained from water sample concentrates taken during the spring sampling from the cooling-tower basins and water boxes at site G (northeastern). Guinea pigs inoculated with these samples became ill and were killed. Late-appearing (>5 days) colonies from spleen and liver tissue plated on charcoal-yeast extract agar had the cultural and microscopic characteristics of *Legionella* sp. and were not reactive in fluorescent-antibody analysis with known *Legionella* antisera. Two separate samples of cooling tower water concentrates from site F (northern midwest) in the spring also yielded *Legionella* isolates not typable with known *Legionella* antisera.

None of the isolates from either site could be grown on blood or brain heart infusion agar. Weakly staining gram-negative rods of various lengths were apparent upon microscopic examination of all the isolates. All 11 isolates were weakly catalase positive. Whereas none of the isolates reacted with conjugated antibodies specific for known *Legionella* species and serotypes, they all reacted maximally with conjugate prepared against the OR10 isolate (supplied through the courtesy of W. B. Cherry, Centers for Disease Control). Detailed studies of the Oak Ridge isolates by Orrison et al. (1) have confirmed these isolates as a new species of *Legionella*, i.e., *Legionella oakridgensis*.

L. oakridgensis was isolated from these same two sites (F and G) in other seasons. In addition, it was isolated from site E (southern) in the fall. In other related studies, analyses of power plant-affected and ambient water concentrates from various locations in the continental United States showed a wide distribution and variable concentration of the Oak Ridge species of *Legionella*, not unlike those of the four major serogroups of *L. pneumophila* combined (2).

In contrast to *L. pneumophila*, the newly discovered *L. oakridgensis* has yet to be isolated from clinical specimens. It was, however, the second most prevalent species isolated in this study. Fluorescent-antibody analysis of various ambient and thermally altered waters indicates a wide distribution of this *Legionella* sp., but the

TABLE 2. Types of pathogens isolated from plant-affected water samples

Site	Pathogen[a]			
	Spring	Summer	Fall	Winter
A	—	K	CH	GO
B	—	—	—	—
C	—	CH	"Vibrio"	—
D	—	LA	—	LA, L?
E	K	—[b]	—	LA
F	OR	K	K, BL, LA, OR	K, CH, OR
G	K, BL, LA, OR	CH, OR	K, OR	K, CH, OR
H	—[b]	LA	LA, CH	LA
I	K	K, CH	K, TO	—

[a] Abbreviations given in Table 1, footnote *a*.
[b] See Table 1, footnote *b*.

isolation of infectious *L. oakridgensis* was more site specific. Of 13 isolates, 12 were obtained from only two of the nine sites. The apparent site dependency for the presence of infectious *L. oakridgensis* may explain in part why it has yet to be detected in clinical material.

In addition to *L. oakridgensis*, other unidentified *Legionella* spp. and another currently unidentified microbial pathogen were isolated during the course of this study. A *Legionella* species not reactive with antisera against known species of *Legionella* was isolated from site F. Analysis of this isolate by personnel at Centers for Disease Control showed it to be identical to a previously isolated but currently unnamed new *Legionella* species. Yet another untypable *Legionella* sp. was isolated from a winter discharge sample at plant D (northern midwest). This isolate is being examined to determine whether it too is a new *Legionella* species.

In addition to the *Legionella* isolates, another microbial pathogen has been isolated from fall samples of intake and discharge water at plant C. Guinea pigs inoculated with discharge samples developed high fevers within 3 days after inoculation, and on autopsy, pure cultures of a vibrio-like (i.e., weakly gram-negative, comma-shaped bacteria capable of growth at 25 to 37°C) organism grew from the plated tissue. Personnel at Centers for Disease Control have examined this isolate but have been unable to classify the microbe as belonging to any known genus.

Although the major goal of the ongoing study is to delineate those ecological variables important in the propagation of infectious *Legionella* spp., the isolation of a new species of *Legionella* illustrates the value of environmental information as input to clinical studies. Specifically, many clinical specimens are treated with Formalin before analysis, obliterating any possibility of isolating new species of pathogens. Environmental material serves not only as a source of clinical infection but also as a reservoir from which the isolation of previously undiscovered pathogens of clinical importance is possible.

This research was sponsored in part by the Electric Power Research Institute (EPRI Project RP1909-1) pursuant to a Participation Agreement dated 29 September 1981 with the U.S. Dept. of Energy, and in part by the Office of Nuclear Regulatory Research, U.S. Nuclear Regulatory Commission, under Interagency Agreement DOE-40-550-75 with the U.S. Dept. of Energy under contract W-7405-eng-26 with the Union Carbide Corp.

LITERATURE CITED

1. **Orrison, L. H., W. B. Cherry, R. L. Tyndall, C. B. Fliermans, S. B. Gough, M. A. Lambert, L. K. McDouglas, W. F. Bibb, and D. J. Brenner.** 1983. *Legionella oakridgensis*: an unusual new species isolated from cooling tower water. Appl. Environ. Microbiol. 45:536–545.
2. **Tyndall, R. L., S. B. Gough, C. B. Fliermans, E. L. Dominigue, and C. B. Duncan.** 1983. Isolation of a new *Legionella* species from thermally altered waters. Curr. Microbiol. 9:77–80.

Patterns of *Legionella* spp. Infectivity in Power Plant Environments and Implications for Control

SIGURD W. CHRISTENSEN, RICHARD L. TYNDALL, JEAN A. SOLOMON, CARL B. FLIERMANS, AND STEPHEN B. GOUGH†

Oak Ridge National Laboratory, Oak Ridge, Tennessee 37830; University of Tennessee, Knoxville, Tennessee 37916; and Savannah River Laboratory, Aiken, South Carolina 29801

Water samples were collected in 1981–1982, during each of the four seasons, from source water and plant-affected locations at nine power plant sites. Further information about the test sites is provided in a companion paper (J. A. Solomon, S. W. Christensen, R. L. Tyndall, C. B. Fliermans, and S. B. Gough, this volume) and a report by Christensen et al. for the Electric Power Research Institute, Palo Alto, Calif. (1). *Legionella pneumophila* was found in nearly all samples. Infectivity of some of the water samples was determined by injecting two uncompromised guinea pigs intraperitoneally with 2 ml of a 500-fold concentration of each sample. Animals

† Present address: System Development Corp., Fredericksburg, VA 22401.

were observed over a 10-day period for signs of fever and overt illness. Those with signs of illness were killed, and peritoneal swabs and organ tissues were cultured on charcoal-yeast extract agar at 35°C. Colonies typical of *Legionella* spp. were then analyzed morphologically and physiologically and were serologically identified with respect to species, serogroup, or both by the fluorescent-antibody technique. The water sample was classified "noninfectious" if the animals showed no signs of illness and "infectious" if any *Legionella* spp. were isolated from either of the two guinea pigs after appearance of typical legionellosis symptoms. Samples causing at least one of the two guinea pigs to become ill, but from which no organisms could be isolated, were classified as noninfectious. Samples that were toxic (causing rapid death of both guinea pigs) or contaminated (yielding only non-*Legionella* spp. bacteria in cultures from both guinea pigs) were excluded as uninterpretable with respect to *Legionella* infectivity. Infectivity was not confirmed by reinoculation of guinea pigs. For a number of reasons, it is not possible to directly relate infectivity in this study to human risk. Contingency table analysis was used to analyze the data statistically.

Of the 143 samples tested for infectivity, 51 were infectious, 70 were noninfectious, and 22 were unsuccessful tests (infectivity could not be determined due to toxic reactions or presence of bacterial contaminants). Such interference was most acute in the spring, with 28% of the tested samples having inconclusive results. Overall, the infectivity rate was lowest in spring, with 32% of the samples tested (both source and plant-affected water) yielding positive results. Somewhat higher infectivity rates were seen in the other three seasons: 48, 43, and 45% of samples yielded positive results in the summer, fall, and winter, respectively. These percentages, including spring, are not significantly different from each other ($P < 0.05$).

Examination of infectivity levels at individual plant sites revealed that site G, a northeastern closed-cycle plant, had the highest level of infectivity, with 79% (15 of 19) of all the tested samples throughout the entire year yielding positive results (Table 1). Plant F, a north-midwestern variable-mode plant, had the second highest overall level of infectivity, with 62% (8 of 13) of successfully tested samples producing positive results. Toxicity, as mentioned above, proved to be a problem. Plant B, a southern once-through plant, produced no infectious samples. Thus, there is an association of infectious *Legionella* spp. with certain power plant sites; infectivity is not distributed evenly among the sites ($P < 0.01$).

TABLE 1. Results of interpretable tests for infectivity at the nine power plant sites[a]

Plant	Operating mode[b]	No. of interpretable tests	Annual % infectious
A	OT	14	43
B	OT	14	0
C	OT	14	29
D	OT	15	20
E	OT	10	30
F	CC	13	62
G	CC	19	79
H	CC	11	55
I	CC	11	55
Total		121	42

[a] Intake, precondenser, postcondenser, and outfall samples were combined over all seasons.
[b] OT, Once through; CC, closed cycle.

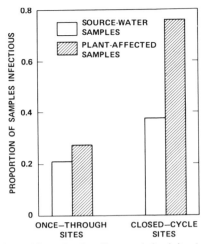

FIG. 1. Mean *Legionella* spp. infectivity (proportion of samples infectious) by type of cooling system and location of sample. Plant-affected samples at closed-cycle plants were infectious more frequently than source water or once-through samples ($P < 0.05$).

Attempts were made to relate infectivity to water quality parameters by using the statistical techniques of logistic regression, discriminant analysis, and the Wilcoxon rank sum test. Interpretation of these results was inconclusive, but conductivity and, to a lesser degree, growth temperature and dissolved inorganic carbon bore a consistently positive relationship with infectivity, while there were indications of a negative relationship with dissolved oxygen. Caution is needed in interpreting statistically significant correlation and regression results. In neither case is causation implied; for example, two variables may be changing in response to a third, unmeasured factor.

Student's *t* test was used to test several reasonable hypotheses about the relationship between infectivity and *L. pneumophila*, namely, that the infectivity of *L. pneumophila* for guinea pigs is a function of (i) density (number per milliliter) of *L. pneumophila*, (ii) viability (proportion alive) of *L. pneumophila*, or (iii) the product of density and viability (i.e., the density of live *L. pneumophila* cells). No significant differences in density, viability, or the concentration of viable cells were detected between the infectious and the noninfectious groups of samples, indicating that these quantities are not predictors of infectivity in our study. This suggests that the current practice (2) of using densities of *L. pneumophila* greater than 10^8 per liter as the sole "trigger" for instituting control measures may not be appropriate for all systems.

There was more infectivity associated with all closed-cycle power plant sites (i.e., source water and plant-affected samples combined) than with once-through power plant sites ($P < 0.01$): 64% of the samples from closed-cycle sites were infectious to guinea pigs, compared with 24% at the once-through sites. The data were examined in more detail to detect patterns of distribution of infectivity at these two plant types. It was found that a disproportionately large number of the plant-affected samples from closed-cycle plants were infectious ($P < 0.05$; see Fig. 1). These analyses were repeated with a different protocol for judging infectivity, as follows: cases in which a guinea pig developed fever but where no organisms could be isolated (normally classified noninfectious with respect to *Legionella* spp.) were now classified infectious under the presumption that *Legionella* spp. are often difficult to isolate and could have been responsible for the symptoms. This resulted in the shifting of 20 samples from the noninfectious to the infectious category. Results of the analyses were the same with respect to both level of significance and direction of effect; in particular, the proportion of samples designated infectious was highest in the plant-affected samples of closed-cycle sites ($P < 0.05$). It is apparent that some attribute of the closed-cycle systems tends to cause an increase in infectivity over ambient conditions.

This research was sponsored by the Electric Power Research Institute (EPRI project RP1909-1) pursuant to a Participation Agreement dated 29 September 1981 with the U.S. Department of Energy under Contract W-7405-eng-26 with Union Carbide Corp.

LITERATURE CITED

1. **Christensen, S. W., R. L. Tyndall, J. A. Solomon, C. B. Fliermans, and S. B. Gough.** 1983. Legionnaires' disease bacterium in power plant cooling systems: phase I final report. EPRI EA 3153, Electric Power Research Institute, Palo Alto, Calif.
2. **Fliermans, C. B., G. E. Bettinger, and A. W. Fynsk.** 1982. Treatment of cooling systems containing high levels of *Legionella pneumophila*. Water Res. **16:**903–909.

Legionella pneumophila in Vermont Cooling Towers

LINDEN E. WITHERELL, LLOYD F. NOVICK, KENNETH M. STONE, ROBERT W. DUNCAN, LILLIAN A. ORCIARI, DAVID A. JILLSON, RICHARD B. MYERS, AND RICHARD L. VOGT

Vermont State Department of Health, Burlington, Vermont 05401

Legionella pneumophila may be contained in aerosols ejected from contaminated cooling towers. There have been several legionellosis outbreaks associated with aerosols emitted from contaminated cooling towers, including two outbreaks in Vermont (3, 4).

A cooling tower is a wet-type heat rejection unit (WTHRU) used to dissipate unwanted heat from air conditioning, materials processing, or manufacturing into the atmosphere (6). The heat exchange is accomplished by passing heated water through an airstream with cooling resulting from evaporation. The cooled water is collected and passed through the process again. Depending upon design and operation, approximately 5% of the water in the system is continuously lost by a combination of evaporation, drainage of water from the unit to control the buildup of solids, and ejection of aerosols from the unit in the form of fine water droplets which become entrained in the airstream.

Because of two outbreaks of Legionnaires disease in Vermont in 1980 and reports from other states of outbreaks of legionellosis associated with WTHRUs (2; D. N. Klaucke, R. L. Vogt, D. LaRue, L. E. Witherell, L. Orciari, K. C. Spitalny, R. Pelletier, W. B. Cherry, and L. F. Novick, this volume), it was decided to inventory, inspect, and sample all units in Vermont.

Vermont Department of Health staff contacted WTHRU manufacturers, engineers, architects, and suppliers of air-conditioning and refrigeration equipment in Vermont and completed a statewide roster of this equipment.

On-site inspections of all units were conducted by Vermont Department of Health, Division of Environmental Health personnel, using a standard six-page form covering 151 attributes of design, construction, location, and operation of the units.

Water in each WTHRU was sampled for pH, temperature, turbidity, microbiocide content, standard plate count (1), and *L. pneumophila* (5). WTHRUs in Vermont were sampled once per cooling season. Sixteen units located within 150 m of hospitals or nursing homes in Vermont were sampled once every 2 weeks. Ten units located on the Burlington university campus, where two previous outbreaks of legionellosis had occurred, were sampled once per week. Twenty-eight other WTHRUs in Burlington were sampled once each month.

Because owners or operators of WTHRUs in Burlington or at medical care facilities had been advised to institute microbiological treatment of their units after the 1980 Legionnaires disease outbreaks, these units were not included in statistical analyses. In addition, 11 units were excluded from statistical analyses because of incomplete analyses of samples from these units.

Pertinent descriptive variables gathered on WTHRUs were examined using chi-square contingency tables for possible association with the presence of *L. pneumophila*. Stepwise logistic regression analysis of the Biomedical Data Program software (7) was used to identify variables significantly associated with *L. pneumophila*. Presence or absence of *L. pneumophila* served as the dependent variable.

It was found that 185 WTHRUs were operating in Vermont during the study period of April 1981 to April 1982. The water for the majority of the units came from surface water systems (134 of the 185) rather than groundwater sources (38) or a combination of surface and groundwater sources (13). The categorical variable "clustered" was used to indicate whether a WTHRU was located within 1.6 km of another unit. The majority of units were clustered: 159 versus 26 nonclustered.

Statewide, *L. pneumophila* was isolated from 18 of the 185 WTHRUs during the study period. Overall, 11.7% (14 of 120) of the WTHRUs in the statistical study group were positive for *L. pneumophila*.

Initial examination by use of chi-square contingency tables found possible association between recovery of *L. pneumophila* and temperature, pH, turbidity, nonclustered location, and water source. The results of the stepwise logistic regression showed water source, turbidity, and pH to be significantly associated with *L. pneumophila* recovery. There were no significant interaction terms among independent variables. Only 6.3% (5 of 79) of WTHRUs utilizing surface water sources for make-up water were positive, in contrast to 25% (7 of 28) of units utilizing ground water sources. Turbidity tended to be lower in positive units (mean = 5.49, standard deviation = 10.93) than in negative units (mean = 8.87, standard deviation = 12.46). pH was higher in positive units (mean = 8.29, standard deviation = 0.44) than in negative units (mean = 7.83, standard deviation = 0.65). The analysis gave an excellent chi-square goodness-of-fit statistic (chi-square = 66.60, 116 df, $P > 0.995$).

LITERATURE CITED

1. **American Public Health Association.** 1980. Standard methods for the examination of water and wastewater, 15th ed. American Public Health Association, Washington, D.C.
2. **Broome, C. V., S. A. J. Goings, S. B. Thacker, R. L. Vogt, H. N. Beaty, D. W. Fraser, and the Field Investigation Team.** 1979. The Vermont epidemic of Legionnaire's disease. Ann. Intern. Med. **90:**573–577.
3. **Cordes, L. G., D. W. Fraser, P. Skaliy, C. A. Perline, W. R. Elsea, G. F. Mallison, and P. S. Hayes.** 1980. Legionnaire's disease outbreak at an Atlanta, Georgia, country club: evidence for spread from an evaporative condensor. Am. J. Epidemiol. **111:**425–431.
4. **Dondero, T. J., Jr., R. C. Rendtoroff, G. F. Mallison, R. M. Weeks, J. S. Levy, E. W. Wong, and W. Schaffner.** 1980. An outbreak of Legionnaire's disease associated with a contaminated air-conditioning cooling tower. N. Engl. J. Med. **302:**365–370.
5. **Gorman, G. W., and J. C. Feeley.** 1982. Procedures for the recovery of *Legionella* from water. Centers for Disease Control, Atlanta, Ga.
6. **Miller, R. P.** 1979. Cooling towers and evaporative condensors. Ann. Intern. Med. **90:**667–670.
7. **University of California.** 1981. BMDP statistical software. University of California Press, Berkeley.

Environmental Factors Influencing Growth of *Legionella pneumophila* in Operating, Biocide-Treated Cooling Towers

KARLA M. GROW, DAVID O. WOOD, JOSEPH H. COGGIN, JR., AND EDWIN D. LEINBACH

Department of Microbiology and Immunology, University of South Alabama College of Medicine, Mobile, Alabama 36688

Although cooling towers and similar aerosol-generating heat rejection devices have been implicated in several outbreaks of legionellosis, the importance of these units as reservoirs for and disseminators of *Legionella pneumophila* has remained controversial. Published extensive surveys of *L. pneumophila* in cooling towers (2, 5, 6) have generally been limited to single time-point samplings. Likewise, the techniques for analysis have generally involved immunofluorescence and guinea pig inoculation. The former method suffers from questionable specificity and inability to distinguish between viable and non-viable organisms, and the latter is not quantitative.

We have attempted to overcome these problems by following levels of viable *L. pneumophila* in two cooling towers at a single site over a 9-month period to examine the environmental and operating parameters most likely to influence levels of *L. pneumophila* contamination.

Water samples collected at 2-week intervals from the basins of two operating, biocide-treated Marley cooling towers (towers 1 and 3) were concentrated and examined microscopically for the presence of "viable" legionella-like bacteria by combined immunofluorescence (DFA) and staining with *p*-iodonitrotetrazolium violet (INT) as described by Fliermans et al. (3). Samples of the concentrates were also diluted and cultured on GVP medium (7) containing cycloheximide. Other samples were acid-washed and then cultured on CCVC medium by the procedures of Bopp et al. (1) and Gorman and Feeley (4). No single culture method allowed optimal detection of *L. pneumophila* at all time points; therefore, we performed all three procedures for each sample and used the highest level of *L. pneumophila* detected for all data analyses.

Both DFA-INT and culture showed significant fluctuations in levels of viable *L. pneumophila* which did not correlate with the seasons of the year or the mean ambient air temperature. Although viable *L. pneumophila* levels as determined by DFA-INT were generally at least 10-fold higher than those measured by culture, the two methods nonetheless showed similar trends.

In mid-March, both cooling towers were cleaned with a high-pressure water spray followed by resumption of the usual twice-weekly addition of thiocarbamate-based biocides. This procedure resulted in a sharp drop in viable *L. pneumophila* to barely detectable levels. By late May, legionella levels had begun to increase slowly in both towers.

A multiple linear regression analysis was performed on the data collected from each tower, using the number of culturable *L. pneumophila* as the dependent variable. As independent variables we included the environmental parameters of mean, maximum, and minimum air temperature, relative humidity, and barometric pressure. Operating parameters included were basin water conductivity, pH, and temperature as well as dissolved oxygen, free chlorine, and the level of total viable bacteria determined by culture.

In tower 3, only conductivity significantly predicted levels of culturable *L. pneumophila* at the 95% confidence level. At the 85% confidence level, pH, water temperature, and relative humidity also had significant predictive value. A regression model constructed with these four variables showed good agreement with actual levels of culturable *L. pneumophila* in tower 3 up to the point at which the tower was cleaned, after which the fit was less satisfactory (Fig. 1). When "viable *L. pneumophila*" levels determined by DFA-INT were used as the dependent variable, only relative humidity was found to enter the regression model at the 85% confidence level. This discrepancy could be interpreted as suggesting that conductivity and pH influenced our ability to culture *L. pneumophila* as opposed to the actual level of viable organisms in the tower. Alternatively, many of the DFA-INT-positive bacteria may have been nonlegion-

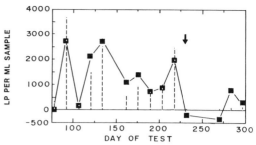

FIG. 1. *L. pneumophila* levels in tower 3 as determined by culture (— — —) or as predicted by a multiple linear regression model based on conductivity, pH, water temperature, and relative humidity (■). The arrow indicates the point at which the cooling tower was cleaned.

ellae whose growth was dependent on factors different from those determining *L. pneumophila* growth.

Levels of culturable *L. pneumophila* in tower 1 were generally quite low, making calculation of a meaningful regression model difficult. In contrast, levels of DFA-INT-positive organisms in tower 1 were frequently higher than those in tower 3. In this instance, relative humidity, pH, minimum air temperature, and dissolved oxygen levels in the basin water were the most significant factors in the regression model at the 85% confidence level.

The data gathered in these studies suggest that as long as a cooling tower is in virtually continuous operation only relative humidity, among the common environmental variables, has a significant impact on levels of viable *L. pneumophila*. Whether this is a direct effect or an indirect influence exerting its effect through a change in the temperature drop across the cooling tower (ΔT) or some other parameter cannot be determined from these studies. However, the difference in *L. pneumophila* levels and in the pattern of fluctuation in those levels in these two cooling towers, despite their common siting and shared make-up water supplies, implies that operating parameters characteristic of a given system are more important determinants of *L. pneumophila* growth.

The results of these studies also point out the danger of using single time-point samplings to assess either the suitability of a particular maintenance procedure or the potential for a given system to serve as a source of an outbreak of legionellosis. There is a clear need for improvements in surveillance techniques which will allow for simpler and yet more reliable quantitation of *L. pneumophila* levels as well as provide an estimate of the virulence potential of the bacteria present.

Refinement of the data presented here is exceedingly difficult in the open system of a "real world" cooling tower, yet laboratory studies of *L. pneumophila* growth have thus far shown generally low applicability to operating systems. Plans are under way to develop an experimental cooling tower system which can be contaminated with *L. pneumophila* against a variety of types of microbial backgrounds and in which the operating and environmental parameters can be controlled and varied singly or in combination to simulate "worst-case" conditions encountered in open, operating cooling systems. Only then can we determine what procedures will be most effective in controlling *L. pneumophila* growth under the wide variety of conditions which prevail in the real world.

LITERATURE CITED

1. **Bopp, C. A., J. W. Sumner, G. K. Morris, and J. G. Wells.** 1981. Isolation of *Legionella* spp. from environmental samples by low pH treatment and use of a selective medium. J. Clin. Microbiol. **13:**714–719.
2. **England, A. C., III, D. W. Fraser, G. F. Mallison, D. C. Mackel, P. Skaliy, and G. W. Gorman.** 1982. Failure of *Legionella pneumophila* sensitivities to predict culture results from disinfectant-treated air-conditioning cooling towers. Appl. Environ. Microbiol. **43:**240–244.
3. **Fliermans, C. B., R. J. Soracco, and D. H. Pope.** 1981. Measure of *Legionella pneumophila* activity in situ. Curr. Microbiol. **6:**89–94.
4. **Gorman, G. W., and J. C. Feeley.** 1982. Procedures for the recovery of *Legionella* from water. Centers for Disease Control, Atlanta, Ga.
5. **Kurtz, J. B., C. L. R. Bartlett, U. A. Newton, R. A. White, and N. L. Jones.** 1982. *Legionella pneumophila* in cooling water systems. J. Hyg. **88:**369–381.
6. **Tyndall, R. L.** 1982. Concentration, serotypic profiles, and infectivity of Legionnaires' disease bacteria populations in cooling towers. J. Cooling Tower Inst. **3:**25–33.
7. **Wadowsky, R. M., and R. B. Yee.** 1981. Glycine-containing selective medium for isolation of *Legionellaceae* from environmental specimens. Appl. Environ. Microbiol. **42:**768–772.

Occurrence of *Legionella* spp. and Other Aquatic Bacteria in Chemically Contaminated Groundwater Treated by Aeration

DONALD F. SPINO, EUGENE W. RICE, AND EDWIN E. GELDREICH

Microbiological Treatment Branch, Drinking Water Research Division, Municipal Environmental Research Laboratory, U.S. Environmental Protection Agency, Cincinnati, Ohio 45268

Groundwater contaminated with volatile organic chemicals poses a major challenge to utilities which rely upon these waters as drinking water sources. Many of these volatile organic compounds, such as trichloroethylene, are suspected carcinogens. The inclusion of induced-

TABLE 1. Numbers of heterotrophic bacteria isolated from raw groundwater and aerated water from two New England aeration towers

Date of sample	Water temp (°C)		Plate count agar			R2A medium		
			CFU/ml		Ratio, aerated/raw	CFU/ml		Ratio, aerated/raw
	Raw	Aerated	Raw	Aerated		Raw	Aerated	
10-27-81	11.5	11.6	11	26	2.4	40	84	2.1
11-3-81	11.5	11.5	29	17	0.58	390	880	2.3
12-2-81	11.5	11.6	25	3,500	140.0	128	2,200	17.2
12-8-81	11.9	12.0	28	460	16.4	27	200	7.4
12-15-81	11.3	11.3	19	1,050	55.3	42	1,150	27.4
1-5-82	9.1	8.8	9	12	1.3	74	700	9.5
1-20-82	9.9	10.8	9	18,100	2,000.0	115	14,800	129.0
5-11-82	9.2	9.7	103	7	0.07	128	168	1.3
7-21-82	11.3	12.3	23	230	10.0	119	1,800	15.0
12-21-82	11.4	11.4	1	2	2.0	10	178	17.8
1-11-83	11.3	11.3	1	4	4.0	9	34	3.8
1-25-83	11.3	11.4	1	3	3.0	53	120	2.3
2-9-83	11.3	11.3	1	33	33.0	186	190	1.0
2-23-83	11.2	11.2	89	24	0.25	1,300	316	0.24
2-25-83	11.4	11.4	10	20	2.0	53	120	2.3
3-29-83[a]	11.4	11.5	9	44	4.9	45	298	6.6
3-29-83	6.0	6.0	9	19	2.1	57	108	1.9
4-12-83[a]	6.0	6.0	2,810	4,830	1.7	2,690	6,500	2.4
4-19-83[a]	9.0	9.0	26	26	1.0	86	72	0.83

[a] Outdoor tower.

TABLE 2. Assays for total coliform bacteria and *Legionella* spp. from raw groundwater and aerated water from two New England aeration towers

Date of sample	*Legionella* spp.[a] (CFU/liter)		Total coliforms (CFU/100 ml)		Species identification (no. of isolates)	
	Raw	Aerated	Raw	Aerated	Raw	Aerated
10-27-81	<1	<1	<1	<1		
11-3-81	<1	<1	<1	<1		
12-2-81	<1	<1	<1	<1		
12-8-81	<1	<1	<1	<1		
12-15-81	<1	<1	1	1	*Enterobacter agglomerans*	*E. agglomerans*
1-5-82	<1	<1	<1	<1		
1-20-82	<1	<1	<1	<1		
5-11-82	<1	<1	<1	1		*Citrobacter freundii*
7-21-82	<1	<1	<1	10		*Klebsiella pneumoniae* (5)
						Klebsiella oxytoca (1)
						C. freundii (1)
						Escherichia coli (1)
						Aeromonas hydrophila (2)
12-21-82	<1	<1	<1	<1		
1-11-83	<1	<1	<1	<1		
1-25-83	<1	<1	NA[b]	NA		
2-9-83	<1	<1	NA	NA		
2-23-83	<1	<1	<1	<1		
2-25-83	<1	<1	1	<1	*E. agglomerans*	
3-19-83	<1	<1	<1	1		*K. pneumoniae*
3-29-83	<1	<1	<1	<1		
4-12-83	<1	<1	<1	<1		
4-19-83	NA	NA	<1	<1		

[a] Positive controls: lyophilized *L. pneumophila* serogroup 1 and environmental samples containing *L. pneumophila* serogroups 1 and 5.

[b] NA, Not available.

draft, redwood slat aerators in a water treatment train has been found to be an efficient and cost-effective method for the removal of these contaminants. This study was undertaken to evaluate the effect of the aeration process on the microbiological quality of the water. The similarity in design between the aeration towers and cooling towers, which have previously been implicated in outbreaks of Legionnaires disease, provided the impetus for assaying for *Legionella* spp.

Water samples from an indoor aeration tower and an outdoor tower in New England were analyzed over a period of 1.5 years. Preaeration (raw groundwater) and aerated samples were examined for the occurrence of *Legionella* spp., coliform bacteria, and heterotrophic bacteria. Water temperatures for each sample were also recorded. Various culturing procedures, including direct plating, and acid pretreatment were utilized in assaying for *Legionella* spp. (4) with either buffered charcoal-yeast extract agar (BCYE; 5), a glycine-containing selective medium (7), or Edelstein's semiselective medium, MWY (3). All plates were incubated at 36°C and 3% CO_2. The inclusion of negative and positive control samples in each assay confirmed the adequacy of the methods used. Positive controls included spiked samples containing laboratory cultures of *L. pneumophila* serogroup 1 and environmental samples from a hospital hot water tank known to contain *L. pneumophila* serogroups 1 and 5. Suspicious colonies which morphologically resembled *Legionella* spp. were isolated from the assays, and these and positive controls were subcultured on BCYE, BCYE minus cysteine, and blood agar plates. Subcultures which grew only on BCYE were verified by the direct immunofluorescent technique (2) using fluorescein-labeled antibody conjugates of *L. pneumophila*, polyvalent serogroups 1 to 4, and monovalent conjugates 1 through 6. Conjugates for *L. micdadei*, *L. gormanii*, *L. bozemanii*, and *L. dumoffii* were also utilized.

Heterotrophic bacteria were enumerated by the pour plate technique using standard plate count agar (1) and a low-nutrient medium designated R2A (D. J. Reasoner and E. E. Geldreich, Abstr. Annu. Meet. Am. Soc. Microbiol. 1979, N7, p. 180). The R2A medium has been found to provide more representative profiles of the heterotrophic bacteria in water than those provided by using conventional methodology. Analyses for total coliform bacteria were conducted by membrane filtration with M-Endo agar (1). All sheen colony isolates from M-Endo were identified to species with the API 20E system (Analytab Products, Plainview, N.Y.).

Levels of heterotrophic bacteria recovered during the various seasonal time periods of the study are shown in Table 1. The numbers of heterotrophic bacteria, with few exceptions, were consistently higher in the postaeration samples, indicating possible colonization of the redwood slats.

Results from some individual samples showed dramatic increases in the number of bacteria in the postaeration samples, suggesting perhaps a sloughing off of the biomass from the redwood slats. A higher number of organisms were generally recovered with the R2A medium.

Total coliform bacterial levels (Table 2) were consistently low, with the majority of the isolates coming from the postaeration samples. These results give further evidence that microbial colonization may be taking place in the aerators. The frequent isolation of *Klebsiella* spp. in the postaeration samples is in keeping with previous studies reporting the association of this organism with redwood (6).

All cultural attempts to isolate *Legionella* spp. (Table 2) both from the water samples and from scrapings of the biomass from the redwood slats were negative. Analyses of the water samples by direct immunofluorescence also failed to elucidate any *Legionella* spp.

These results indicate that the use of aeration towers in drinking water treatment technology may be responsible for significantly higher levels of bacterial flora in the aerated water than that encountered in the raw water, thus indicating the need for a good postaeration disinfection program. These results further suggest that aeration towers may not pose a substantial risk for the introduction of *Legionella* spp. into the water supply. If *Legionella* organisms are present, their numbers must be below the levels for

detection by accepted procedures. It has been reported that *Legionella* spp. fail to multiply in aquatic environments below 25°C (8). The lower temperatures characteristic of the groundwaters studied (≤12°C) may have been a major hindrance in the detection of this organism.

Further investigations are planned to study the microbial quality of water taken from the aeration tower located outdoors. These additional studies will investigate the influence of water hydraulics as a major cause of heterotrophic bacterial releases in the effluents.

We gratefully acknowledge the technical assistance of Janet C. Blannon and Carol Ann Fronk-Leist.

LITERATURE CITED

1. **American Public Health Association.** 1981. Standard methods for the examination of water and wastewater, 15th ed., p. 789 sect. 907, p. 806 sect. 909. American Public Health Association, New York.
2. **Cherry, W. B., and R. M. McKinney.** 1979. Detection of Legionnaires' disease bacteria by direct immunofluorescence, p. 130–145. *In* G. L. Jones and G. A. Hébert (ed.), "Legionnaires' ": the disease, the bacterium and methodology. Centers for Disease Control, Atlanta, Ga.
3. **Edelstein, P. H.** 1982. Comparative study of selective media for isolation of *Legionella pneumophila* from potable water. J. Clin. Microbiol. **16**:697–699.
4. **Gorman, G. W., and J. C. Feeley.** 1982. Procedures for the recovery of *Legionella* from waters. Developmental manual. Centers for Disease Control, Atlanta, Ga.
5. **Pasculle, A. W., J. C. Feeley, R. J. Gibson, L. G. Cordes, R. L. Myerowitz, C. M. Patton, G. W. Gorman, L. L. Carmack, J. W. Ezzell, and J. N. Dowling.** 1980. Pittsburgh pneumonia agent: direct isolation from human lung tissue. J. Infect. Dis. **141**:727–732.
6. **Seidler, R. J., J. E. Morrow, and S. T. Bagley.** 1977. Klebsielleae in drinking water emanating from redwood tanks. Appl. Environ. Microbiol. **33**:893–900.
7. **Wadowsky, R. M., and R. B. Yee.** 1981. Glycine-containing medium for the isolation of *Legionellaceae* from environmental specimens. Appl. Environ. Microbiol. **42**:768–772.
8. **Yee, R. B., and R. W. Wadowsky.** 1982. Multiplication of *Legionella pneumophila* in unsterilized tap water. Appl. Environ. Microbiol. **43**:1330–1334.

Legionella spp. in Environmental Water Samples in Paris

NICOLE DESPLACES, ANNE BURE, AND ERIC DOURNON

Hygiene Laboratory City of Paris, 75004 Paris, and Central Laboratory Claude Bernard Hospital, 75019 Paris, France

In the United States and United Kingdom many investigations have shown that water from plumbing systems and from cooling towers is frequently contaminated by *Legionella* spp. The presence of *Legionella* spp. in environmental water samples has not yet been systematically studied in France.

Between January 1982 and April 1983, we examined 190 water samples from the Paris area

for the presence of legionellae, unrelated to any Legionnaires disease outbreak. A total of 54 water samples were obtained from 11 different air-conditioning systems (18 from cooling towers, 36 from different airwash systems). A further 128 water samples from plumbing systems in 20 buildings (88 cold water samples: 19 distribution water, 42 cold tap water, 27 water fountains; 40 hot water samples: 15 hot water tanks,

25 hot tap water) and 8 samples from the main water supplies of the municipality were obtained. One liter per sample (10 liters for the main water supplies) was filtered through one or several polycarbonate filters (pore size, 0.2 μm) to concentrate the bacteria. The residues were washed off the filters and suspended in 10 ml of sterile tap water. A portion of this suspension was treated with low pH (pH 2.2) for 10 min (1) at room temperature to reduce the number of non-*Legionella* organisms. Volumes of 0.1 ml of this suspension and 0.1 ml of the suspension not treated with acid were plated after neutralization onto two plates each of buffered charcoal-yeast extract agar (BCYE) (3) with α-ketoglutarate (0.1%) (2) and glycine (0.3%) (4) and on BCYE supplemented with antibiotics (cephalothin, 4 μg/ml; colistin, 16 μg/ml; cycloheximide, 80 μg/ml; vancomycin, 0.5 μg/ml) (1). Direct fluorescent assay was performed on all the suspensions with the three *Legionella* spp. polyvalent fluorescent-antibody pools (A, B, C) supplied by the Biological Products Division, Centers for Disease Control, Atlanta, Ga. When the sample was culture negative but positive by direct fluorescent assay, 10-fold dilutions to 10^{-2} and 10^{-4} of the suspension were plated on the same media with the same procedures. All of the plates were incubated in a moist atmosphere in 2.5% CO_2 at 35°C for 14 days and were examined daily for growth under a magnifying glass. Any colony that grew only on BCYE agar and not on sheep blood agar and on BCYE agar without L-cysteine was identified by direct fluorescent assay as recommended by the Centers for Disease Control. The cellular fatty acids of some strains were analyzed by gas-liquid chromatography.

Sixty-six (34.7%) of the 190 water samples contained at least one *Legionella* strain. Thirty (45.5%) of the 66 positive specimens yielded two or three *Legionella* strains. Nine of the 11 air-conditioning systems were contaminated: 15 of the 18 cooling tower samples and 6 of the 36 airwash system samples were positive. Ten of the 20 buildings investigated yielded at least one *Legionella*-positive water sample: 17 (24.6%) of the 69 cold water samples, 9 (21.4%) of the 42 cold tap water specimens, 8 (29.6%) of 27 water fountain specimens, 18 (72%) of the 25 hot tap water specimens, and 10 (66.6%) of the 15 hot water tanks sampled. In some tap water samples the number of legionellae was as high as 10^5 CFU/liter. No *Legionella* strains were isolated, until now, from either the 8 main water supplies or 19 building distribution entry points.

Ninety-two strains of *Legionella* were isolated: 46 (50%) were of *L. pneumophila* serogroup 1, 11 were serogroups 3 and 6, 8 were serogroup 2, 5 were serogroup 5, 3 were serogroup 4–5, and 2 were serogroup 4. One strain was *L. longbeachae* serogroup 2. Five *Legionella*-like organisms are being further characterized.

Our study showed that *Legionella* spp. seems to be a common environmental bacterium in the Paris area and that *L. pneumophila* serogroup 1 is the predominant organism.

LITERATURE CITED

1. **Bopp, C. A., J. W. Sumner, G. K. Morris, and J. G. Wells.** 1981. Isolation of *Legionella* spp. from environmental water samples by low-pH treatment and use of a selective medium. J. Clin. Microbiol. **13**:714–719.
2. **Edelstein, P. H.** 1981. Improved semiselective medium for isolation of *Legionella pneumophila* from contaminated clinical and environmental specimens. J. Clin. Microbiol. **14**:298–303.
3. **Pasculle, A. W., J. C. Feeley, R. J. Gibson, L. G. Cordes, R. L. Myerowitz, C. M. Patton, G. W. Gorman, C. L. Carmack, J. W. Ezzel, and J. N. Dowling.** 1980. Pittsburgh pneumonia agent: direct isolation from human lung tissue. J. Infect. Dis. **141**:727–732.
4. **Wadowsky, R. M., and R. B. Yee.** 1981. A glycine-containing selective medium for isolation of *Legionellaceae* from environmental specimens. Appl. Environ. Microbiol. **42**:768–772.

Detection of *Legionella pneumophila* and *Legionella micdadei* in Water Collected from a Rainfall in Southeastern Pennsylvania

WALLACE E. TURNER, PHILIP NASH, GEORGE F. HARADA, AND LARRY M. IAMPIETRO

Bureau of Laboratories, Pennsylvania Department of Health, Lionville, Pennsylvania 19353

The isolations of *Legionella* spp. from a variety of water sources—lakes, rivers, streams (2, 3, 5), and water-associated equipment such as air-conditioning cooling towers (1)—suggest that water may be the natural reservoir for those organisms. Consequently, freshly collected rainwater was investigated as another water source for *Legionella* spp. Water was collected during rainfalls that occurred in southeastern Pennsylvania in June, July, and August 1982. The collections were made at two location points (one in Lancaster County and one in Delaware County), one approximately 60 miles (96 km) east of the other.

TABLE 1. Rainwaters collected in 1982 and 1983 that were positive by direct fluorescent-antibody assay for *Legionella* spp.[a]

Date of collection	County location	*Legionella* spp.
6/12/82	Lancaster	*L. pneumophila* serogroup 1
		L. micdadei
6/12/82	Delaware	*L. pneumophila* serogroup 1
		L. micdadei
4/15/83	Lancaster	*L. gormanii*
4/16/83	Delaware	*L. gormanii*
4/18/83	Berks	*L. gormanii*
4/24/83	Lancaster	*L. micdadei*
		L. gormanii
5/22/83	Lancaster	*L. micdadei*
5/23/83	Lancaster	*L. gormanii*
5/24/83	Delaware	*L. gormanii*
5/27/83	Berks	*L. gormanii*
5/29/83	Delaware	*L. gormanii*

[a] Smears made from plate growth.

At each location point, falling rain was caught in a sterile 1-liter beaker and later transferred to a sterile bottle for storage at 4°C until tested. The pH of each sample was measured, and the samples were filtered through a sterile 0.45-μm membrane filter. The filter was then divided; half was placed on a buffered charcoal-yeast extract agar plate containing antibiotics, and the other half was set on a buffered charcoal-yeast extract agar plate containing dye. All plates were incubated at 35°C in 3% CO_2 and observed for growth at 48 h and daily thereafter.

The direct fluorescent-antibody procedure was used to examine smears from plate growth. The fluorescent conjugates used were: *Legionella pneumophila* (polyvalent groups 1, 2, 3, and 4), *L. pneumophila* group-specific conjugates (1, 2, 3, and 4), and *L. micdadei*.

The investigation of rainwater was continued in 1983. Water samples were collected as described during March, April, May, and June. One more location point (Berks County) was added. The point in Berks County is approximately 35 miles (56 km) northeast of the point in Lancaster County and approximately 28 miles (44 km) northwest of the one in Delaware County. Filtration and culturing were done as described, with a modification to include a second membrane filtration for each sample. The conjugates used in the direct fluorescent-antibody assay were as mentioned, plus *L. gormanii* and *L. longbeachae*.

A total of 58 water samples were tested (1982 and 1983). Eleven of these were positive by direct fluorescent-antibody assay for one or more *Legionella* spp. (Table 1).

Growth on and surrounding the membrane filter usually showed varied amounts of mixed colonial forms. Although these bacteria stained brightly (3 to 4+) with the conjugates of either *L. pneumophila*, *L. micdadei*, or *L. gormanii*, none has been isolated, even after numerous attempts. Active efforts have been made to determine whether the positive staining results were due to cross-reactions, but no such cross-reacting bacteria have been found. This does not exclude the possibility of cross-reactions with some unknown or untested organisms. Future studies of rainwater might reveal more positive reactions by including additional conjugates to the study.

The identification of *Legionella* spp. from rainwater collected at widely separated locations indicates that rain may act as a disseminator of these organisms. In this connection, it is interesting that in at least three reports of outbreaks of legionellosis (1, 4, 6) the occurrence of heavy rains was mentioned as one of the events preceding the outbreaks.

LITERATURE CITED

1. **Dondero, T. J., Jr., R. C. Rendtorff, G. F. Mallison, R. M. Weeks, J. S. Levy, E. W. Wong, and W. Schaffner.** 1980. An outbreak of Legionnaires' disease associated with a contaminated air-conditioning cooling tower. N. Engl. J. Med. **302:**365–370.
2. **Fliermans, C. B., W. B. Cherry, L. H. Orrison, and L. Thacker.** 1979. Isolation of *Legionella pneumophila* from nonepidemic related aquatic habitats. Appl. Environ. Microbiol. **37:**1239–1242.
3. **Fliermans, C. B., W. B. Cherry, L. H. Orrison, S. J. Smith, D. L. Tison, and D. H. Pope.** 1981. Ecological distribution of *Legionella pneumophila*. Appl. Environ. Microbiol. **41:**9–16.
4. **Glick, T. H., M. B. Gregg, B. Berman, G. Mallison, W. W. Rhodes, Jr., and I. Kassanoff.** 1978. Pontiac fever—an epidemic of unknown etiology in a health department. 1. Clinical and epidemiologic aspects. Am. J. Epidemiol. **107:**149–160.
5. **Morris, G. K., C. M. Patton, J. C. Feeley, S. E. Johnson, G. Gorman, W. T. Martin, P. Skaliy, G. F. Mallison, B. D. Politi, and D. C. Mackel.** 1979. Isolation of the Legionnaires' disease bacterium from environmental samples. Ann. Intern. Med. **90:**664–666.
6. **Thacker, S. B., J. V. Bennett, T. F. Tsai, D. W. Fraser, J. E. McDade, C. C. Shepard, K. H. Williams, Jr., W. H. Stuart, H. B. Dull, and T. C. Eickhoff.** 1978. An outbreak in 1965 of severe respiratory illness caused by the Legionnaires' disease bacterium. J. Infect. Dis. **138:**512–519.

Growth Relationships of *Legionella pneumophila* with Green Algae (Chlorophyta)

R. DOUGLAS HUME AND WILLIAM D. HANN

Bowling Green State University, Bowling Green, Ohio 43403

Although *Legionella* spp. appear to be ubiquitous, little is known about their relationship with other species. The growth of *Legionella pneumophila* ATCC 33152 in association with green algae (chlorophyta) was examined. This investigation indicates that the association of *L. pneumophila* with other aquatic species may be more extensive than previously thought since growth was minimal in the absence of algae.

Growth of *L. pneumophila* when cultured with *Scenedesmus* sp., *Chlorella* sp., and *Gloeocystis* sp. in Bold basal salts (BBS) medium at 20 to 22°C was compared with the growth of *Escherichia coli* ATCC 8739 and *Pseudomonas aeruginosa* ATCC 27853 under similar conditions. All three bacterial species grew better in mixed cultures with green algae than when grown alone. A 2-log increase in cell number was seen during the first 48 h of incubation when *L. pneumophila* was grown with *Scenedesmus* sp. and *Chlorella* sp. (Fig. 1). In contrast, *Gloeocystis* sp. only supported a 1.5-log increase in legionellae during this time period. The doubling time was 5, 6.9, and 13 h, respectively, for *Chlorella* sp., *Scenedesmus*, sp., and *Gloeocystis* sp.

The elemental composition of all the organisms used, as well as the various growth media, was determined by energy-dispersive X-ray analysis and atomic absorption spectrophotometry. Elemental analysis of *L. pneumophila* showed evidence of concentrating phosphorus and sulfur. No other detectable elements appeared to be concentrated above the levels found in charcoal-yeast extract-BBS agar. *Chlorella* sp. showed increased concentrations of phosphorus above those found in BBS agar. *Gloeocystis* sp. had an increased concentration of magnesium and phosphorus, whereas *Scenedesmus* sp. did not have any increased concentration of any detectable elements. The analysis of elements lends support to the idea of a symbiotic or competitive relationship between *L. pneumophila* and *Chlorella* sp. or *Gloeocystis* sp. based upon their mutual accumulation of phosphorus. The accumulation of Mg and P may possibly influence the symbiotic relationships of *P. aeruginosa* and *Gloeocystis* sp. Similarly, phosphorus accumulation may influence the relationship between *P. aeruginosa* and *Chlorella* sp. These relationships would be of increased importance in an aquatic ecosystem where nutrient compounds may be rapidly diluted. The highest concentrations of extracellular metabolic by-products of the algal metabolism exist close to the cell surface, and it would be of

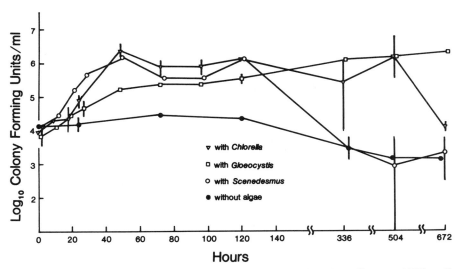

FIG. 1. Growth of *L. pneumophila* with green algae at 20 to 22°C. Growth medium was BBS medium. Bars represent ± 1 standard error of the mean of three experiments. Standard errors less than the span of a data point are not noted.

benefit for the bacterial cell to be in this area so as to utilize the metabolites before dilution in the aquatic environment. Nitrogen was supplied by the medium used for the algal cultures. The carbon sources were not examined in detail, but the carbon compounds used by the bacteria ultimately derive from algal photosynthesis. Close contact would also be important if the bacterial cell received carbon or nitrogen compounds from the algae, or O_2 from photosynthesis. Alternatively, the bacterial cells may contribute CO_2 for utilization by the algal cells.

All experiments were performed at 20 to 22°C to more closely approximate normal environmental growth temperatures. As *L. pneumophila* appears to be a common environmental organism, it may be more appropriate to examine its metabolic and growth characteristics at 20 to 22°C than at 37°C. Growth characteristics and metabolic behavior of *L. pneumophila* when grown at 20 to 22°C may more closely approximate those actually seen in nature.

Algal cultures which contained *L. pneumophila*, *E. coli*, or *P. aeruginosa* were observed by scanning electron microscopy. When grown in yeast extract broth-BBS medium, the typical cell morphology of *L. pneumophila* resembled that of a coccobacillus.

L. pneumophila, *E. coli*, and *P. aeruginosa* grew among the cells of mats of *Chlorella* sp., associated with the areas between individual algal cells. *L. pneumophila* formed small colonies in these interstitial areas. *E. coli* was more widely distributed as individual cells; *P. aeruginosa* was seen in extremely heavy concentrations in the interstitial areas as well as over the surface of the individual algal cells.

Unlike *Chlorella* sp., *Gloeocystis* sp. did not support growth of colonies of bacteria; instead, bacterial cells of all three bacterial species were distributed evenly over the surface of the mucilage coat (Fig. 2). Bacterial cells appear to become partially embedded in the mucilage, and it seems reasonable to assume that additional bacterial cells could be completely engulfed in the mucilage layer and thus not visible.

This relationship of *L. pneumophila* with green algae (chlorophyta) may parallel a similar symbiosis with blue-green algae (cyanobacteria) as reported by Tison et al. (2), who suggested that production of carbon compounds by blue-green algae supports the growth of *L. pneumophila*. The growth of *P. aeruginosa* is also

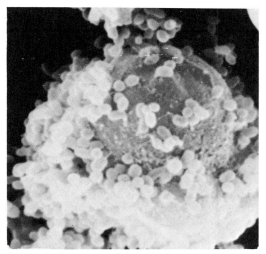

FIG. 2. *L. pneumophila* and *Gloeocystis* sp. mixed culture grown in BBS medium. Individual *L. pneumophila* cells can be seen on the surface of the algal cells. Magnification, ×2,600.

apparently stimulated by algal blooms. Pellett et al. (1) reported that *P. aeruginosa* is an epibacterium more frequently found in sediment and solid-water interfaces than in the water column itself. The natural habitat of *L. pneumophila* may parallel that of *P. aeruginosa*, making surveillance of water systems most effective at interfaces, in sediment, and near algal blooms. *Chlorella* sp. and *Gloeocystis* sp. are common contaminants in water systems which create aerosols. Inhalation of a single algal cell carrying *L. pneumophila* may be an infective dose sufficient to cause disease. The proper use of algicides could help alleviate any potential health problems.

Knowledge of the metabolic pathways responsible for the symbiotic relationships of *L. pneumophila* with other organisms in the environment will lead to a better understanding of the ecological distribution of this ubiquitous organism.

LITERATURE CITED

1. Pellet, S., D. V. Bigley, and D. J. Grimes. 1983. Distribution of *Pseudomonas aeruginosa* in a riverine ecosystem. Appl. Environ. Microbiol. 45:328–332.
2. Tison, D. L., D. H. Pope, W. B. Cherry, and C. B. Fliermans. 1980. Growth of *Legionella pneumophila* in association with blue-green algae (cyanobacteria). Appl. Environ. Microbiol. 39:456–459.

Legionellae and Amoebae

TIMOTHY J. ROWBOTHAM

Leeds Regional Public Health Laboratory, Leeds LS15 7TR, England

Amoebae appear to be important to our understanding of the ecology of legionellae and the epidemiology of Legionnaires disease (4, 5). This is a report of the interaction of *Legionella* spp. and amoebae.

For aquatic bacteria, legionellae are surprisingly fastidious, which perhaps implies a close relationship with other organisms. *Legionella pneumophila* may grow in association with axenic algae and cyanobacteria (3), but *L. pneumophila* has been isolated from many dark habitats. Some cyanobacteria excrete amino acids and peptides into the surrounding environment, and as *L. pneumophila* has been grown in defined amino acid media (6), such compounds might be utilized by legionellae for growth. Cyanobacterial filaments are usually colonized by bacteria, e.g., *Bacillus polymyxa*, source of the polymyxin used in selective media for legionellae. Other environmental bacteria, e.g., *Pseudomonas aeruginosa*, are very inhibitory to *L. pneumophila* in vitro (4). In natural habitats, inhibition or competition for nutrients or both may perhaps limit, or prevent, growth of legionellae in association with algae and cyanobacteria. Amoebae which can be infected with legionellae in the laboratory (4, 7) can feed on green algae (7), on cyanobacteria (7), and probably on the bacteria associated with the latter.

Macrophages, the main host cells for legionellae in the lung, are bacteriovorous amoeboid cells. Bacteriovorous amoebae are ubiquitous; e.g., *Acanthamoeba polyphaga*, whose cysts are resistant to the concentrations of chlorine used in the water supply industry, can be isolated from potable water, water tanks, showers, humidifiers, aquaria, and halogenated swimming pools. Cyanobacterial communities and cooling tower ponds can also contain acanthamoebae. *A. polyphaga* (AP L1501/3A) trophozoites are readily attacked by non-laboratory-adapted *L. pneumophila* serogroup 1 (Fig. 1), as well as by *L. pneumophila* serogroup 6, *L. bozemanii*, and *L. jordanis* (5a). There appears to be a degree of host-parasite specificity (4). Virulent legionellae may be ingested by nonhost amoebae, but are then usually egested or digested. Laboratory-adapted legionellae may be digested by potential host amoebae. As an aid to isolation, the number of *L. pneumophila* serogroup 1 organisms in homogenized and washed clinical specimens can be greatly increased by incubation with *A. polyphage* trophozoites (5a). This effect was not achieved with *L. pneumophila* serogroup 5, as large numbers of this serogroup are needed to

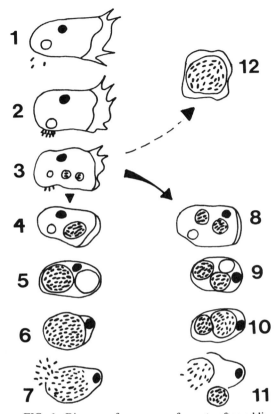

FIG. 1. Diagram of sequence of events after addition of *L. pneumophila* (Leeds-1A SAP) to *A. polyphaga* (AP L1501/3A) trophozoites. The infective process can be divided into 12 stages. (1) Legionellae are attracted to extended trophozoites. (2) They cluster around the rear end of the amoebae, close to where the contractile vacuole empties. (3) The legionellae are phagocytosed, several in each vesicle. (4) The legionellae multiply inside the vesicle and are nonmotile at this stage. (5) The vesicle of legionellae fills most of the cytoplasm, and about this time the contractile vacuole may cease to function properly. (6) A single large vesicle of motile legionellae almost fills the amoeba. (7) The vesicle and amoeba rupture, liberating motile legionellae. (8 and 9) Sometimes two, rarely three (more with attenuated strains), vesicles of legionellae develop. (10) The legionellae inside them may not become motile at the same time, leading to (11) release of vesicles (some <5 μm) of motile or nonmotile legionellae. (12) Production of a mature cyst containing motile bacteria has so far only been accomplished with a legionella-like amoebal pathogen (LLAP-1; no growth on buffered charcoal-yeast extract medium) and *A. polyphaga* and *Acanthamoeba palestinensis* strains (grown on dead klebsiellae) from the same water tank (5a).

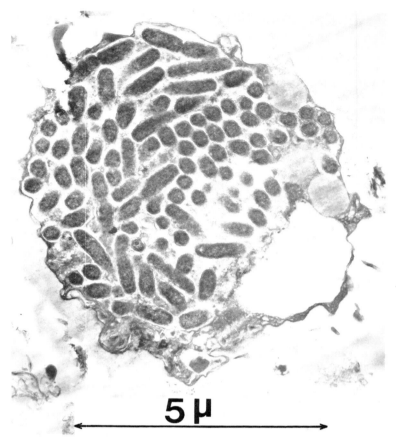

FIG. 2. Electron micrograph of a section through a vesicle full of *L. pneumophila* (see Fig. 1). The vesicle is probably still inside a trophozoite. Photo courtesy of G. E. Bellamy (Leeds Public Health Laboratory, Leeds, U.K.).

infect *A. polyphaga*, and motile organisms of this serogroup are not attracted to this amoeba.

The results diagrammed in Fig. 1 were obtained after addition of serogroup 1 *L. pneumophila* (Leeds-1A SAP) to *A. polyphaga* (AP L1501/3A) trophozoites. The amoebae were grown axenically in peptone-yeast extract broth and washed three times with amoeba saline before inoculation. Leeds-1A SAP was originally isolated from a single infected amoeba from an amoebal enrichment inoculated with the sputum of a mild case of Legionnaires disease (5a). At 35°C the first amoebae full of motile legionellae were seen after 22 h of incubation. Material from a second passage of Leeds-1A SAP in washed axenic *A. polyphaga* trophozoites was used for Fig. 2. The 12 stages proposed for the infective process are outlined in Fig. 1.

Inside amoebae, *L. pneumophila* serogroup 1 divides by pinching division (Fig. 2). The dividing legionellae, with their convoluted outer surface and lack of vacuoles, develop into single small bacilli with smooth walls, numerous vacuoles, and a single subpolar-polar flagellum. Pairs of extracellular legionellae which could be considered as dividing bacteria were rare in secondary or later passages of *L. pneumophila* serogroup 1 in amoebae, as they were in amoebal enrichments.

Most of the *L. pneumophila* organisms contained in the material for Fig. 2 were 0.32 to 0.42 μm in diameter and 0.6 to 1.3 μm long (dividing bacilli considered as two bacilli). If the legionellae occupied 80% of the available space inside a 5-μm vesicle (or amoeba), this particle (which if inhaled could reach the alveoli of the lung) could contain about 326 of the larger or 1,320 of the smaller legionellae (no allowance made for fixation shrinkage or for evaporative shrinkage of an aerosolized particle). Such particles could be the main infective particles for humans (4). When a concentrated water sample associated with a case of Legionnaires disease was incubated on a lawn of UV-killed *Klebsiella aerogenes*, tiny, unidentified cyst-forming amoebae (rounded-up trophozoites about 5 to 7 μm in diameter) were the apparent host for tiny coccobacilli which stained with *L. pneumophila* antiserum (serogroups 1 through 4). *Filamoeba nolandi*, a second amoeba in this material, was not infected, and a single cyst-derived clone could not be infected with the *L. pneumophila* serogroup 1 strain supplied by workers at Oxford who had previously isolated it from the same material by guinea pig inoculation. Numerous attempts to purify the slow-growing amoeba, which grew at 30 but not 35°C, failed.

The original investigators of Pontiac fever (1) noted the similarities of the disease to some types of hypersensitivity disease. *A. polyphaga* and *Naegleria gruberi*, potential hosts for legionellae, have been implicated in a form of hypersensitivity pneumonitis known as "humidifier fever." I have previously suggested that Pontiac fever may be a hypersensitivity pneumonitis to amoebae, coupled with a mild infection by *L. pneumophila* in most cases (5). The hypersensitivity reaction accounts for the rapid onset of symptoms seen in some cases. Activated lung macrophages are presumably better able to counter infection by legionellae, as has now been demonstrated for activated blood monocytes (2); this, with other factors such as age and fitness, might explain the lack of pneumonia seen in Pontiac fever.

In summary, there is strong circumstantial evidence to implicate amoebae as natural hosts of legionellae. This thesis is strongly supported by microscopic observation and explains much of the epidemiology of Legionnaires disease.

LITERATURE CITED

1. **Glick, T. H., M. B. Gregg, B. Berman, G. Mallison, W. W. Rhodes, Jr., and I. Kassanoff.** 1978. An epidemic of unknown etiology in a health department. 1. Clinical and epidemiologic aspects. Am. J. Epidemiol. **107:**149–160.

2. **Horwitz, M. A., and S. C. Silverstein.** 1981. Activated human monocytes inhibit the intracellular multiplication of Legionnaires' disease bacteria. J. Exp. Med. **154:**1618–1635.

3. **Pope, D. H., R. J. Soracco, H. K. Gill, and C. B. Fliermans.** 1982. Growth of *Legionella pneumophila* in two-membered cultures with green algae and cyanobacteria. Curr. Microbiol. **7:**319–322.

4. **Rowbotham, T. J.** 1980. Preliminary report on the pathogenicity of *Legionella pneumophila* for freshwater and soil amoebae. J. Clin. Pathol. **33:**1179–1183.

5. **Rowbotham, T. J.** 1980. Pontiac fever explained? Lancet **ii:**969.

5a. **Rowbotham, T. J.** 1983. Isolation of *Legionella pneumophila* from clinical specimens via amoebae, and the interaction of those and other isolates with amoebae. J. Clin. Pathol. **36:**978–986.

6. **Tesh, M. J., and R. D. Miller.** 1981. Amino acid requirements for *Legionella pneumophila* growth. J. Clin. Microbiol. **13:**865–869.

7. **Wright, S. J. L., K. Redhead, and H. Maudsley.** 1981. *Acanthamoeba castellanii*, a predator of cyanobacteria. J. Gen. Microbiol. **125:**293–300.

Proliferation of *Legionella pneumophila* as an Intracellular Parasite of the Ciliated Protozoan *Tetrahymena pyriformis*

BARRY S. FIELDS, EMMETT B. SHOTTS, JR., JAMES C. FEELEY, GEORGE W. GORMAN, AND WILLIAM T. MARTIN

Department of Medical Microbiology, College of Veterinary Medicine, The University of Georgia, Athens, Georgia 30602, and Division of Bacterial Diseases, Center for Infectious Diseases, Centers for Disease Control, Atlanta, Georgia 30333

Legionella pneumophila has been shown to infect amoebae of the genera *Acanthamoeba* and *Naegleria* on an agar surface (6). Our study was initiated to determine whether *L. penumophila* would infect and multiply within a freshwater protozoan in an aquatic environment.

This study consisted of a series of five experiments. The *L. pneumophila* strain used in these tests was a Philadelphia 1 strain (serogroup 1) which had not been previously passed on laboratory media. The protozoan selected for this study was the ciliated holotrich *Tetrahymena pyriformis*, a University of Georgia stock strain (Midwest Culture Service no. 500) maintained in Elliot's medium no. 2 (3).

The initial experiments of this study document

the ability of *L. pneumophila* to infect and multiply intracellularly in *T. pyriformis* in coculture. Cocultures were prepared by suspending both strains in sterile tap water and incubating this suspension at 35°C. The numbers of *L. pneumophila* present at 0, 3, and 7 days were determined by culturing samples of the suspensions on buffered charcoal-yeast extract agar (5). The number of *L. pneumophila* increased 3 to 4 logs within 1 week in coculture. *L. pneumophila* did not multiply in axenic culture in sterile tap water at 35°C. The bacteria required live protozoa to multiply; tests revealed that *L. pneumophila* could not multiply in a suspension of lysed *T. pyriformis* cells suspended in sterile tap water or in a cell-free extract of a spent *T. pyriformis*

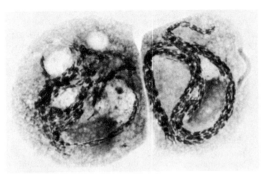

FIG. 1. Two heavily infected *T. pyriformis* protozoa, showing serpentine chains of *L. pneumophila* throughout their cytoplasm. Gimenez stain; magnification, ×1,650.

culture. Our belief that the *L. pneumophila* organisms were multiplying intracellularly was best supported by microscopic examination. All cocultures were examined daily by observing slide preparations stained with either the Gimenez technique or immunofluorescent antiserum to *L. pneumophila* serogroup 1 (2, 4). Both methods were able to show large numbers of *L. pneumophila* cells either in food vacuoles or in serpentine chains throughout the cytoplasm of the protozoan (Fig. 1).

One of the objectives of this study was to develop an enrichment procedure for *L. pneumophila* utilizing protozoa. The following experiments were therefore concerned with various procedures designed to increase the multiplication of *L. pneumophila* in coculture. Because there was a difference in the optimal temperatures of the two organisms, cocultures were tested at various temperatures. Cocultures incubated at 35°C produced the highest concentrations of *L. pneumophila*; however, *T. pyriformis* survives best at 25°C. It is possible that *L. pneumophila* multiplies more readily at a higher temperature which compromises the protozoa. In addition, temperature should also affect the interaction of *Legionella* spp. and similar protozoa in nature. Two other manipulations, (i) increasing the concentration of *T. pyriformis* and (ii) changing the temperature of incubation from 25 to 35°C after 2 h of incubation, resulted in higher final concentrations of *L. pneumophila*.

The reasoning behind both of these procedures was to allow more protozoa to ingest *L. pneumophila* organisms before being adversely affected by the increase in temperature.

The final experiment of this study incorporated the previous findings in an attempt to use *T. pyriformis* as an enrichment procedure for *L. pneumophila* in environmental water samples. Ten water samples from an outbreak of Legionnaires disease were inoculated with a suspension of *T. pyriformis*. These samples were incubated at 25°C for 2 h and then placed at 35°C for the remainder of the experiment. Five of the 10 samples containing naturally occurring *Legionella* species were found positive for the bacterium by using acid treatment and direct plating procedures before enrichment (1). After 7 days of incubation with *T. pyriformis* an additional 4 samples were found positive for *Legionella* species, resulting in 9 of 10 positive samples (P = 0.0625).

In conclusion, under the conditions of our study *L. pneumophila* bacteria can infect a free-living protozoan in an aquatic system, and our coculture system can serve as a laboratory ecological model. Furthermore, it is plausible that the coculture method can be used as a laboratory enrichment of *Legionella* spp. in environmental samples containing small numbers of *Legionella* spp. or in which other microflora are predominant.

LITERATURE CITED

1. Bopp, C. A., J. W. Sumner, G. K. Morris, and J. G. Wells. 1980. Isolation of *Legionella* spp. from environmental water samples by low-pH treatment and use of a selective medium. J. Clin. Microbiol. 13:714–719.
2. Cherry, W. B., and R. M. McKinney. 1979. Detection of the Legionnaires' disease bacteria in clinical specimens by direct immunofluorescence, p. 92–103. *In* G. L. Jones and G. A. Hebert (ed.), "Legionnaires' ": the disease, the bacterium and methodology. Center for Disease Control, Atlanta, Ga.
3. Elliot, A. M. (ed.). 1973. Biology of *Tetrahymena*. Dowden, Hutchinson and Ross, Stroudsburg, Pa.
4. Gimenez, D. F. 1964. Staining of rickettsia in yolk-sac cultures. Stain Technol. 39:135–140.
5. Pasculle, A. W., J. C. Feeley, R. J. Gibson, L. G. Cordes, R. L. Myerowitz, C. M. Patton, G. W. Gorman, L. L. Carmack, J. W. Ezzell, and J. N. Dowling. 1980. Pittsburgh pneumonia agent: direct isolation from human lung tissue. J. Infect. Dis. 141:727–732.
6. Rowbotham, T. J. 1980. Preliminary report on the pathogenicity of *Legionella pneumophila* for freshwater and soil amoebae. J. Clin. Pathol. 33:1179–1183.

Intracellular Growth of *Legionella pneumophila* Within Freshwater Amoebae

EDSEL P. HOLDEN, DAVID O. WOOD, HERBERT H. WINKLER, AND EDWIN D. LEINBACH

Department of Microbiology and Immunology, University of South Alabama College of Medicine, Mobile, Alabama 36688

Several lines of evidence have suggested that freshwater amoebae may play an important role in the environmental biology of *Legionella pneumophila*. Rowbotham (3) was the first to show that heavy inocula of a number of laboratory strains and clinical isolates of *L. pneumophila* inhibited the migration of a variety of species of the genera *Acanthamoeba* and *Naegleria* on seeded agar plates. Amoebae isolated from the *L. pneumophila*-amoeba interface regions of these plates appeared to contain large numbers of bacteria. Nagington and Smith (2) subsequently found that *Acanthamoeba polyphaga* could use *L. pneumophila* as a food source. Tyndall and Domingue (5) extended these studies and showed that when *L. pneumophila* was cocultured with *Acanthamoeba royreba* or *Naegleria lovaniensis* the number of viable bacteria first declined sharply, but subsequently increased over an extended period, although the bacteria could not survive in the amoeba culture medium alone. These authors also found no apparent effect of this interaction on the pathogenicity of either the amoeba or the *L. pneumophila* in mice and guinea pigs.

Neither the cell biology of this fascinating interaction nor the environmental consequences thereof have been explored. As a first step, we have studied the interaction between the well-characterized Neff strain of *Acanthamoeba castellanii* (ATCC 30010) and the virulent Philadelphia 1 strain of *L. pneumophila*. The latter was isolated from a guinea pig spleen homogenate obtained from the Centers for Disease Control, Atlanta, Ga., and passaged no more than three times on buffered charcoal-yeast extract agar.

In a repeat of Rowbotham's agar plate experiment (3), we found that the presence of a heavy inoculum of *L. pneumophila* immediately adjacent to a similar inoculum of *Escherichia coli* had no effect on the ability of the acanthamoebae to migrate through the *E. coli* inoculum line. The amoebae did not, however, penetrate the *L. pneumophila* line, but rather halted their migration and rounded up. This result suggests that it is unlikely that *L. pneumophila* produces a diffusible factor which is capable of altering the behavior of the acanthamoebae.

The possibility, suggested by the results of Tyndall and Domingue (5), that amoebae may produce a soluble factor(s) which enhances *L. pneumophila* growth was tested with a parabiotic chamber which consists of two glass compartments separated by a 0.4-μm Nuclepore membrane. When *L. pneumophila* was added to a compartment containing amoeba-conditioned medium but separated from the amoebae by the membrane, the viable bacterial counts declined to undetectable levels within 48 h. In contrast, when *L. pneumophila* was inoculated into the

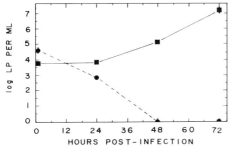

FIG. 1. Growth of virulent *L. pneumophila* in direct association with *A. castellanii*. The two components of a parabiotic chamber were separated by a 0.4-μm Nuclepore membrane filter. Amoebae were infected with legionellae for 1 h, washed, and then placed in fresh acanthamoeba medium in one compartment (■). The other compartment contained legionellae in amoeba-conditioned medium only (●). Samples were withdrawn at 24-h intervals and cultured on buffered charcoal-yeast extract medium after disruption of the amoebae by nitrogen cavitation.

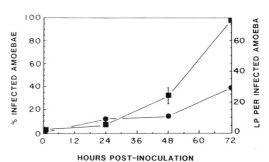

FIG. 2. Growth of virulent *L. pneumophila* and spread of infection in cultured *A. castellanii*. Amoebae were infected in suspension, washed, and suspended in fresh acanthamoeba medium. At 24-h intervals samples were withdrawn and stained by the Gimenez method. Each point represents the mean of replicate counts from groups of 100 amoebae. (■) Percent infected amoebae; (●) mean *L. pneumophila* (Lp) count per infected amoeba.

same compartment as the amoebae the viable bacterial counts associated with the amoebae increased by three orders of magnitude over a 72-h period (Fig. 1). These results make it extremely unlikely that *A. castellanii* produces a diffusible factor which promotes *L. pneumophila* growth.

The alternative explanation, that *L. pneumophila* is capable of growing within acanthamoebae, was approached by using a differential fixation technique in conjunction with fluorescent-antibody staining. To permeabilize the amoebae to the fluorescent antibody, they were fixed with cold acetone before staining (4). When penetration of the antibody was not desired, the cells were fixed with 2.5% glutaraldehyde. When acanthamoebae cocultured with virulent *L. pneumophila* were examined by these procedures, the number of fluorescent bacteria in cold acetone-fixed cells increased continuously over a 72-h period, as did the percentage of infected amoebae. Amoebae fixed with glutaraldehyde did not show this increase. Amoebae fixed with methanol and stained by the Gimenez method (1) showed an increase in *L. pneumophila* which paralleled the increase seen in cold acetone-fixed, fluorescent-antibody-stained cells (Fig. 2). Disruption of the infected amoebae by nitrogen cavitation, followed by culture of the resulting suspension on buffered charcoal-yeast extract agar, demonstrated that most of the intracellular bacteria were viable. Thus it appears that virulent *L. pneumophila* is capable not only of surviving but of actively growing

within acanthamoebae.

Whether interactions of the type described in this study actually occur in *L. pneumophila*-containing environments is still unknown. We have confirmed Rowbotham's observation that *L. pneumophila*-infected amoebae are very fragile, at least in the later stages of the infection (3). This fragility may complicate the isolation of infected amoebae from environmental samples.

The apparent cytopathic effect of virulent *L. pneumophila* on acanthamoebae as seen in the agar plate experiments also remains to be explained. Similarly, the effect of the encystment stage of the amoeba life cycle on the growth and survival of *L. pneumophila*, the ecological implications of this effect, and the possible relationships between these interactions and the pathogenesis of legionellosis are open avenues for investigation.

LITERATURE CITED

1. Gimenez, D. F. 1964. Staining rickettsiae in yolk sac cultures. Stain Technol. 39:135–140.
2. Nagington, J., and D. J. Smith. 1980. Pontiac fever and amoebae. Lancet ii:1241.
3. Rowbotham, T. J. 1980. Preliminary report on the pathogenicity of *Legionella pneumophila* for freshwater and soil amoebae. J. Clin. Pathol. 33:1179–1183.
4. Shramek, G., and L. Falk, Jr. 1973. Diagnosis of virus-infected cultures with fluorescein-labeled antisera, p. 532–535. *In* P. F. Kruse, Jr., and M. K. Patterson, Jr. (ed.), Tissue culture methods and applications. Academic Press, Inc., New York.
5. Tyndall, R. L., and E. L. Domingue. 1982. Cocultivation of *Legionella pneumophila* and free-living amoebae. Appl. Environ. Microbiol. 44:954–959.

Intracellular Replication of *Legionella pneumophila* in *Acanthamoeba palestinensis*

CHANDAR ANAND,† ANDREW SKINNER, ANDRE MALIC, AND JOHN KURTZ

The Virus Laboratory and Department of Electron Microscopy, John Radcliffe Hospital, Oxford, England OX3 9DU

Although the members of genus *Legionella* are essentially intracellular organisms during their growth in human lung, yolk sac, macrophages, monocytes, and various cell cultures, they are capable of in vitro growth provided specialized culture media, optimum pH (6.6 to 6.9), and optimum temperature (35°C, with a range between 25 and 42°C) are furnished. In contrast, legionellae are widespread in the environment and have been isolated from a variety of aquatic habitats in which the pH and the temperature of the water ranged, respectively,

from pH 5.5 to 8.3 and from 6 to 63°C. The waters also contain other bacteria that either are inhibitory to or parasitize legionellae. An association between *Legionella* spp. and photosynthesizing blue-green algae (cyanobacteria) has been reported, but legionellae have also been isolated from waters without algae and from waters not exposed to light.

In view of these facts, many investigators believe that *Legionella* spp. may not be free living in nature. It has been suggested that free-living amoebae of the genera *Acanthamoeba* and *Naegleria*, which are ubiquitous in water, may be natural hosts for *Legionella* spp. To determine this and to delineate the intracellular repli-

† Present address: Provincial Laboratory of Public Health, Calgary, Alberta, Canada T2P 2M7.

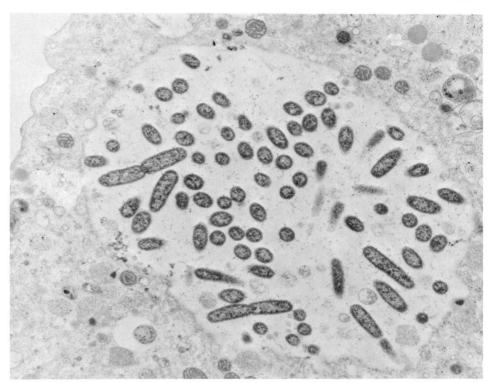

FIG. 1. Electron micrograph of a large *A. palestinensis* phagosome full of dividing and nondividing legionellae. Culture (16 h) of *A. palestinensis* and *L. pneumophila* at 35°C (×12,480).

cation of legionellae in free-living amoebae, we studied the interaction at two temperatures between *Legionella pneumophila* serogroup 1 and *Acanthamoeba (Moyerella) palestinensis*, which we had isolated simultaneously from a local cooling tower.

A. palestinensis was isolated on Stoianovitch malt yeast extract agar (2) layered with heat-killed *Escherichia coli*, cloned, and maintained in axenic culture in 4% Neff medium (1). *L. pneumophila* was maintained on buffered charcoal-yeast extract agar containing alpha-ketoglutarate. A suspension of *L. pneumophila*, prepared in distilled water from a 48-h growth on this medium, was added to give a final concentration of approximately 10^7 CFU/ml to two 12-ml volumes of each of the following media contained in 250-ml tissue culture flasks (Nunc): (i) Neff medium; (ii) Neff medium conditioned by growth of *A. palestinensis* for 3 days; (iii) *A. palestinensis* in log-phase growth at a concentration of 2×10^5 amoebae per ml in Neff medium; (iv) Neff medium containing sonicated *A. palestinensis*. The mixtures were incubated with gentle agitation, one of each pair being set at 35°C and one at 20°C, for up to 2 weeks. Viable *Legionella* counts were performed at 0, 8, and 24 h and daily thereafter.

For electron microscopy, the amoeba-legionella mixture was fixed in glutaraldehyde, post-fixed in osmium tetroxide, and embedded in Spurr epoxy resin. Sections were stained with uranyl acetate and lead citrate.

Sections 1 μm thick, cut from resin blocks used for electron microscopy, were placed on multispot slides, dried overnight, deresinated with sodium ethoxide, and conditioned to water by being rinsed through a series of solutions of decreasing alcohol concentration in water. Immunofluorescent-antibody staining was carried out using a monoclonal antibody to *L. pneumophila* serogroup 1.

At 35°C in media which did not contain live amoebae (media i, ii, and iv), the numbers of legionellae were rapidly reduced until bacteria were no longer isolated (days 1, 3, and 4, respectively). In contrast, legionellae increased in numbers in medium iii, which contained viable *A. palestinensis*, from an initial count of 10^7 CFU/ml to 10^{10} CFU/ml on day 5. The subsequent decrease to 10^6 CFU/ml on day 12 corresponded to a decrease in the trophic forms and an increase in the cystic forms of the amoebae.

At 20°C, starting with a count of 10^6 to 10^7 CFU/ml, legionellae also dropped rapidly in number and could not be isolated from media i,

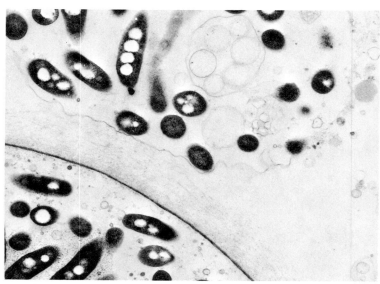

FIG. 2. Electron micrograph of an early cyst with developing wall (lower left) and an extruded phagosome (upper right). Both contain legionellae. Culture (96 h) of *A. palestinensis* and *L. pneumophila* at 35°C (×16,240).

ii, iii, or iv after 1, 2, 4, or 6 days, respectively. On day 6, however, *L. pneumophila* was isolated in low numbers and persisted consistently in medium iii, which contained *A. palestinensis*.

At 35°C, bacteria with an ultrastructure consistent with that of *Legionella* sp. could be seen by electron microscopy singly or in small numbers in food vacuoles after 4 h, and at 16 h many amoebae contained large phagosomes full of legionellae, some of which could be seen dividing (Fig. 1). By 24 h some amoebae were rounded up, and bacteria were seen free in their cytoplasm. Phagosomes full of legionellae could also be seen free outside the amoebae (Fig. 2). At 96 h occasional rounded-up amoebae containing legionellae showed commencement of cyst wall formation with a deposition of exocyst (Fig. 2). At 20°C, only single or small numbers of legionellae were seen in phagosomes at various times. Concentric whorls representing products of digestion of bacteria were seen in lamellosomes in some amoebae. By immunofluorescent-antibody staining, intracellular organisms observed by electron microscopy fluoresced with the monoclonal antibody to *L. pneumophila* serogroup 1.

At higher temperature (35°C), cocultivation of *L. pneumophila* and *A. palestinensis* resulted in intracellular multiplication of the bacteria as demonstrated by growth kinetics, electron microscopy, and fluorescent-antibody studies. The decrease in the numbers of legionellae, after a peak at day 5, was probably due to the reduction in the number of trophozoites available to be infected in the closed experimental system used,

as well as the limited survival time of the free bacteria in the medium. At room temperature (20°C), limited multiplication of legionellae occurred because the acanthamoebae phagocytosed and digested the legionellae incompletely.

The fact that the *Legionella* and *Acanthamoeba* species that we examined were isolated from the same water sample and interacted on coculture suggests that legionellae infect amoebae in nature and multiply in them at a rate influenced by temperature. If so, sites colonized by amoebae could then act as a focus for multiplication of *Legionella* spp.

Assuming that fully formed cysts of *A. palestinensis* containing legionellae can develop from the early cysts observed, any legionellae within them may be able to survive more adverse conditions than they could as free bacteria. Hypothetically, *Acanthamoeba* sp. cysts resistant to levels of chlorination used to treat domestic water supplies could protect legionellae harbored within them from the chlorine and could release them into domestic piped water supplies.

We thank Ian Watkins, Oxford University, for providing the monoclonal antibody, and D. Warhurst, London School of Tropical Medicine and Hygiene, for identification of *A. palestinensis*.

LITERATURE CITED

1. **Curson, R. T. M., T. J. Brown, and E. A. Keys.** 1980. Effects of disinfectants on pathogenic free-living amoebae in axenic conditions. Appl. Environ. Microbiol. **40:**62–66.
2. **Page, F. C.** 1976. An illustrated key to freshwater and soil amoebae. Scientific publication no. 34. Fresh Water Biological Association, Ambleside, Cumbria, England.

Control of Endemic Nosocomial Legionellosis by Hyperchlorination of Potable Water

IAN M. BAIRD, WILLIAM POTTS, JUDITH SMILEY, NANCY CLICK, SHERRI SCHLEICH, CHARLES CONNOLE, AND KIM DAVISON

Riverside Methodist Hospital, Columbus, Ohio 43214

Nosocomial legionellosis was recognized in Riverside Methodist Hospital, Columbus, Ohio, in 1977. Since that time, ongoing surveillance has demonstrated that this hospital is a focus of endemic disease. Between April 1977 and March 1982, we recognized 116 cases of legionellosis, 91 (78%) of which were nosocomial disease. In March 1982, based on the observation of Shands et al. (K. N. Shands, J. L. Ho, G. W. Gorman, R. D. Meyer, P. H. Edelstein, S. M. Finegold, and D. W. Fraser, Clin. Res. **29:**260A, 1981), we instituted a program of chlorination of our potable water in an attempt to control the nosocomial nature of this disease.

The potable water supply was chlorinated at 4 ppm (4 mg of free chlorine per liter) by a chlorination and flow-metering system. Water samples were collected repeatedly from 91 sites throughout the hospital. Specifically, 45 faucets, 25 shower heads, 8 cooling towers when operational, 8 hot water tanks, and 5 miscellaneous sites were cultured on at least 14 different occasions. These water specimens were acid treated by the method of Bopp et al. (1) and cultured for *Legionella pneumophila* by recommended methods (3). The diagnosis of legionellosis was confirmed by the presence of a typical clinical illness, radiographic findings of a pneumonia, and recommended serological methods (2), using indirect fluorescent-antibody reagents and direct fluorescent-antibody reagents against *L. pneumophila* serogroups 1 through 4. The period of chlorination of the potable water extended from late March 1982 until 31 March 1983. We compared the incidence of nosocomial legionellosis before and after the chlorination program.

The chlorination equipment satisfactorily maintained free chlorine levels of 4 ppm from 16 April 1982 until the completion of the study. Before the institution of chlorination, 10 of 70 faucets and shower heads were culture positive for *L. pneumophila*. Of the 45 faucets, 9 (20%) were positive, and 1 of 25 shower heads (4%) was culture positive. During the 12-month period of chlorination, the 45 faucets and 25 shower heads were cultured 16 times. Of the 720 faucet evaluations, 9 were positive for a positivity rate of 1.25% ($P < 0.01$; Pearson's chi-square analysis). Of the 400 shower head evaluations, we found 5 positives, again a positivity rate of 1.25% ($P < 0.05$; Pearson's chi-square analysis). *L. pneumophila* was not eradicated from hot water tanks during the chlorination period. The organism was present in cooling tower water in small numbers during the chlorination period.

During the 6-year surveillance period, the number of diagnostic tests for legionellosis increased (Table 1). In the 5-year period before chlorination, we recognized 91 patients with nosocomial legionellosis and 25 patients with community-acquired legionellosis. In the 12 months of chlorination, we recognized seven patients with nosocomial legionellosis and five with community-acquired legionellosis. The decrease in nosocomial legionellosis is statistically significant ($P < 0.02$; Pearson's chi-square analysis). There was no difference in the incidence of community-acquired disease.

The chlorination program, although initially expensive ($25,000 for equipment and installation), is inexpensive to maintain. We anticipate that the overall 10-year cost will be approximately $4,000 per year.

We conclude that chlorination of the potable water supply at 4 ppm of free chlorine is an effective and inexpensive method of decreasing the incidence of nosocomial legionellosis at our institution.

TABLE 1. Diagnosis of legionellosis during study period

Time period[a]	No. of DFA[b] tests/no. positive	No. of IFA[c] tests/no. positive	Total patients	Nosocomial	Community acquired
1977–78	NK[d]/7	17/9	16	16	0
1978–79	49/4	95/10	13	9	4
1979–80	75/1	151/17	17	12	5
1980–81	47/15	170/25	26	19	7
1981–82	329/44	170/17	44	35	9
1982–83	512/8	79/6	12	7	5

[a] Each indicated time period is from April to March.
[b] DFA, Direct fluorescent antibody.
[c] IFA, Indirect fluorescent antibody.
[d] NK, Not known.

LITERATURE CITED

1. **Bopp, C. A., J. W. Sumner, G. K. Morris, and J. G. Wells.** 1981. Isolation of *Legionella* spp. from environmental water supplies by low-pH treatment and use of selective medium. J. Clin. Microbiol. **13:**714–719.
2. **Meyer, R.** 1983. Legionella infections: a review of five years of research. Rev. Infect. Dis. **5:**258–278.
3. **Pasculle, A. W., J. C. Feeley, R. J. Gibson, L. G. Cordes, R. L. Myerowitz, C. M. Patton, G. W. Gorman, C. L. Carmack, J. W. Ezzell, and J. N. Dowling.** 1980. Pittsburgh pneumonia agent: direct isolation from human lung tissue. J. Infect. Dis. **141:**727–732.

Continuous Hyperchlorination of a Potable Water System for Control of Nosocomial *Legionella pneumophila* Infections

R. MICHAEL MASSANARI, CHARLES HELMS, RODNEY ZEITLER, STEPHEN STREED, MARY GILCHRIST, NANCY HALL, WILLIAM HAUSLER, JR., WILLIAM JOHNSON, LAVERNE WINTERMEYER, JOAN S. MUHS, AND WALTER J. HIERHOLZER, JR.

Department of Internal Medicine and Program of Epidemiology, University of Iowa Hospitals and Clinics, University Hygienic Laboratory, and Department of Microbiology, University of Iowa College of Medicine, Iowa City, Iowa 52242; Children's Hospital Medical Center, Cincinnati, Ohio 45229; Iowa State Department of Health, Des Moines, Iowa 50308; and Veterans Administration Hospital, Iowa City, Iowa 52240

Epidemics and clusters of Legionnaires disease (LD) pneumonia occurring in association with *Legionella pneumophila* colonization of potable water systems have been reported in the United States and Europe (1, 2, 4, 6). Information regarding effective, long-term control of *L. pneumophila* in this reservoir is limited. Elevation of temperature in the hot water system temporarily reduces the number of organisms, but colonization and nosocomial *L. pneumophila* infections recur unless the intervention is repeated at regular intervals (3; M. Best, V. L. Yu, J. Stout, A. Goetz, and R. Muder, Program Abstr. Intersci. Conf. Antimicrob. Agents Chemother. 22nd, Miami Beach, Fla., abstr. no. 88, 1982). We recently reported an outbreak of LD in association with *L. pneumophila* colonization of potable water in a large university hospital (5). Flow-adjusted chlorine injectors were installed in both hot and cold water systems to maintain chlorine levels at >5 ppm (mg/liter). This report describes the successful eradication and control of *L. pneumophila* in a potable water system for 16 months after installation.

The University of Iowa Hospitals and Clinics is a 1,076-bed tertiary-care facility located in Iowa City. The hospital comprises three divisions (A, B, and C) which are served by five separate water systems. Division C, the site of the epidemic, was opened in February 1981. Potable water in Division C is supplied by the University Water Treatment Plant through a single intake exclusive for that division. A portion of incoming cold water is diverted into two steam-heated tanks connected in parallel and heated to 41°C. Hot and cold water are distributed throughout the division in copper conduits. Hot water circulates continuously; cold water circulates on demand. There is no cross-circulation of water between Division C and other water systems in the hospital.

Concurrent surveillance of nosocomial infections is conducted for 11 months of each year on selected inpatient units of the University Hospital. Between February and December of 1981, 24 patients hospitalized in Division C were identified who fulfilled criteria for the diagnosis of nosocomial LD. These criteria included radiographic evidence of pneumonia associated with

documented *L. pneumophila* infection occurring 2 or more days after admission or within 14 days of discharge from University Hospital. Documentation of *L. pneumophila* infection required demonstration of a fourfold or greater rise in antibody titer ($\geq$128) to *L. pneumophila* by direct immunofluorescent-antibody staining, detection of *L. pneumophila* in respiratory secretions or tissues by positive direct fluorescent-antibody staining, culture isolation, or any combination of these.

Environmental surveillance was initiated in December 1981, at which time specimens were obtained from hot and cold water outlets including shower heads and sinks and inoculated on buffered charcoal-yeast extract medium as previously described (5). In December 1981, *L. pneumophila* serogroup 1 was isolated from 21% (10/47) of outlets sampled at weekly intervals from several different sites in Division C (Table 1). The organism was isolated almost exclusively from the hot water system. Simultaneous determinations of free chlorine revealed levels of 2.6 ± 0.2 ppm in cold water with only negligible levels in hot water (0.2 ± 0.2 ppm).

On 16 December 1981, chlorine levels in water supplied by the water treatment plant to Division C were increased to 5 ppm. Chlorine levels in hot water were not appreciably altered, however. On 18 December 1981, both the hot and cold water systems in Division C were shock chlorinated to 15 ppm for 12 h. The temperature in the hot water system was subsequently elevated from 41 to 64°C for 41 days. The system was flushed after each intervention. Despite these efforts, 3 of 35 outlets were still positive in January 1982.

In January 1982, continuous flow-adjusted chlorinators were installed in both the hot and cold water systems. Chlorine levels have been maintained at 8.0 ppm (range, 2.5 to 15.0 ppm) in hot water and 7.3 ppm (range, 2.0 to 15.0 ppm) in cold water during the 16 months since installation. After intervention, environmental cultures have been collected every 2 to 4 weeks from randomly selected hot and cold water outlets in Division C. Of 355 samples collected during this 16-month period, only 2 grew *L. pneumophila*. No *L. pneumophila* has been isolated from 221

TABLE 1. Frequency of isolation of *L. pneumophila* from peripheral water outlets in Division C

Date	No. of outlets sampled[a]	No. of outlets positive for *L. pneumophila*	Free chlorine residual in water (ppm)[b]		No. of cases of nosocomial LD
			Hot	Cold	
1981					
December	47	10	0.2 ± 0.2	2.6 ± 0.2	3
1982					
January[c]	35	3	9.9 ± 4.9	ND[d]	0
February	33	0	10.1 ± 4.3	ND	0
March	30	0	8.7 ± 4.6	8.3 ± 1.5	0
April	18	0	7.4 ± 2.9	6.7 ± 3.8	0
May	35	2	7.0 ± 2.9	8.1 ± 1.9	0
June	26	0	11.8 ± 1.5	4.6 ± 4.1	0
July	28	0	7.9 ± 0.6	5.4 ± 5.0	0
August	27	0	7.5 ± 1.2	5.7 ± 1.3	0
September	28	0	6.0 ± 1.0	5.0 ± 1.7	0
October	28	0	6.7 ± 0.6	4.3 ± 0.8	0
November	28	0	8.5 ± 1.6	4.5 ± 0.7	0
December	14	0	6.3 ± 0.7	7.0 ± 1.3	0
1983					
January	14	0	6.1 ± 0.4	6.1 ± 0.2	0
February	14	0	7.5 ± 0.6	4.5 ± 1.4	0
March	14	0	3.2 ± 0.4	4.7 ± 1.0	0
April	14	0	ND	ND	0
May	14	0	8.0 ± 0.6	6.6 ± 1.5	0

[a] Total number of sink and shower taps sampled.
[b] Mean ± standard error of the mean.
[c] Chlorinators were installed in hot and cold water systems by 14 January 1982.
[d] ND, Not done.

samples examined during the 12 months since May 1982 (Table 1).

Those specimens from which *L. pneumophila* was isolated in May 1982 were obtained from contiguous rooms and were collected after the rooms had been vacant for at least 32 days. It was presumed that residual chlorine had dissipated from the distal portion of the system because the water was not flushed regularly. To assure the safety of incoming patients to Division C, water outlets in rooms vacant for more than 48 h are flushed immediately before admission to the room.

No new cases of nosocomial LD have been detected in Division C since hyperchlorination of the water system (Table 1). Two cases of community-acquired LD have been admitted to the division during this time. Sporadic cases of nosocomial LD continue to occur in Divisions A and B, which are not served by the hyperchlorinated water system. This rate has remained constant over several years, however.

Engineers responsible for maintenance in Division C report no increased demand for maintenance of the water system during the 16 months since installing the chlorinators. Furthermore, at present levels of free chlorine in the water system (5 to 10 ppm), water use has been acceptable and chlorine levels have been unnoticed by staff and patients.

Guidelines and recommendations regarding long-term management of *L. pneumophila* colonization in potable water systems are limited. Hyperchlorination, ozone, and elevated hot water temperatures have all been tried as intervention techniques (3; Best et al., 22nd ICAAC, abstr. no. 88; P. H. Edelstein and C. L. Howell, Abstr. Annu. Meet. Am. Soc. Microbiol. 1982, Q90, p. 225). Although hyperchlorination has been used successfully in cold water systems (3), only elevation of water temperature to 76°C has been reported to reduce colonization by *Legionella* species in hot water systems (3; Best et al., 22nd ICAAC, abstr. no. 88). This intervention has not been successful in eradicating *L. pneumophila* from the system, however. Furthermore, water temperatures of 64 to 76°C impose a risk of injury for patients and staff. Despite reports of mechanical problems and user unacceptability, we have found hyperchlorination to be a highly effective method for both eliminating and controlling *L. pneumophila* colonization in a potable water system. Because of concern for the long-term effects of elevated chlorine levels on the distribution system, we are developing improved methods for controlling chlorine levels in hot and cold water systems which will then allow determination of the absolute minimum levels of chlorine necessary to maintain the system free of *L. pneumophila*.

In conclusion, continuous hyperchlorination of a recently constructed potable water system has proven an effective method for eliminating and preventing recurrent colonization by *L. pneumophila*. After installation, there have been no new cases of LD, even among severely immunocompromised patients. Hyperchlorination of the water system has required no increased maintenance of the system, and water has been acceptable to both staff and patients.

LITERATURE CITED

1. **Brown, A., V. L. Yu, E. M. Elder, M. H. Magnussen, and F. Kroboth.** 1980. Nosocomial outbreak of Legionnaires' disease at the Pittsburgh Veterans Administration Medical Center. Trans. Am. Assoc. Physicians **93:**52–59.
2. **Cordes, L. G., A. M. Wiesenthal, G. W. Gorman, J. P. Phair, H. M. Sommers, A. Brown, V. L. Yu, M. H. Mag-**nussen, R. D. Meyer, J. S. Wolf, K. N. Shands, and D. W. Fraser. 1981. Isolation of *Legionella pneumophila* from hospital shower heads. Ann. Intern. Med. **94:**195–197.
3. **Fisher-Hoch, S. P., C. L. R. Bartlett, J. O'H. Tobin, M. B. Gillett, A. M. Nelson, J. E. Pritchard, M. G. Smith, R. A. Swann, J. M. Talbot, and J. A. Thomas.** 1981. Investigation and control of an outbreak of Legionnaires' disease in a district general hospital. Lancet **i:**932–936.
4. **Haley, C. E., M. L. Cohen, J. Halter, and R. D. Meyer.** 1979. Nosocomial Legionnaires' disease: a continuing common-source epidemic at Wadsworth Medical Center. Ann. Intern. Med. **90:**583–586.
5. **Helms, C. M., R. M. Massanari, R. Zeitler, S. Streed, M. J. R. Gilchrist, N. Hall, W. J. Hausler, Jr., J. Sywassink, W. Johnson, L. Wintermeyer, and W. J. Hierholzer, Jr.** 1983. Legionnaires' disease associated with a hospital water system: a cluster of 24 nosocomial cases. Ann. Intern. Med. **99:**172–178.
6. **Tobin, J. O'H., J. Beare, M. S. Dunnill, S. Fisher-Hoch, M. French, R. G. Mitchell, P. J. Morris, and M. F. Muers.** 1980. Legionnaires' disease in a transplant unit: isolation of the causative agent from shower baths. Lancet **ii:**118–121.

Disinfection of Hospital Hot Water Systems Containing *Legionella pneumophila*

LINDEN E. WITHERELL, LILLIAN A. ORCIARI, ROBERT W. DUNCAN, KENNETH M. STONE,
AND JAMES M. LAWSON

Vermont State Department of Health and Medical Center Hospital of Vermont, Burlington, Vermont 05401

In Vermont, *Legionella pneumophila* has been isolated from hospital hot water storage tanks, shower heads, and water taps. These isolates have been obtained during periods of legionellosis outbreaks and also during periods without known outbreaks. The presence of *L. pneumophila* in hospital hot water systems under any circumstances is of concern because the pathogen may infect large numbers of susceptible individuals (K. N. Shands, J. L. Ho, R. D. Meyer, and D. W. Fraser, Program Abstr. Intersci. Conf. Antimicrob. Agents Chemother. 20th, New Orleans, La., abstr. no. 501, 1980). Due to the potential public health hazards associated with *L. pneumophila* in hospital hot water systems, it was decided to eliminate or reduce the levels of the organisms to below detectable limits.

The temperature normally maintained in hospital hot water systems (37 to 46°C) corresponds to levels which have been found in environmental studies to favor the presence of *L. pneumophila*. Because of this and the fact that these temperatures are detrimental to the maintenance of chlorine residuals at levels normally maintained for disinfection in community water distribution systems, it was decided that supplemental disinfection was required and should be done at the hospital rather than to the community's water system.

We considered elevating the temperature in the hot water system to 77°C for 72 h every 2 months (M. Best, V. L. Yu, J. Stout, A. Goetz, and R. Muder, Program Abstr. Intersci. Conf. Antimicrob. Agents Chemother. 22nd, Miami Beach, Fla., abstr. no. 88, 1982) but did not because of the possible safety hazards associated with this method. After *L. pneumophila* was recovered from the hospital's hot water system, we decided in favor of on-site supplemental chlorination. However, it was recalled that a previous attempt at on-site chlorination with constant feed units had resulted in corrosion damage to the hospital's hot water distribution system because high levels of chlorine occurred in the system during periods of low demand, such as at night. To avoid this, chlorine was added to the cold water makeup lines that supplied the hot water systems, in proportion to the demand on the systems. This was done by installing meter-paced, electrically powered solution feed pumps. Sodium hypochlorite solution was used as the disinfectant. An initial free available chlorine residual of 3.0 mg/liter was maintained in the hot water system for 10 days. Thereafter, a free available chlorine residual of 1.5 mg/liter was used.

Chlorine levels in the hot water system were measured throughout the hospital by maintenance or nursing personnel during each work shift. Chlorine residuals were measured colorimetrically with a HACH model CN-66 (*N,N*-

diethyl-*p*-phenylenediamine DPD method [1]) free and total chlorine test kit. Water samples for *L. pneumophila* analyses were collected during and after initiation of disinfection procedures from the hot water system and examined for *L. pneumophila* by direct culture methods developed at the Centers for Disease Control (2).

The corrosivity of the hot water was evaluated by calculating the Langelier index of the water (1a). This required that the pH, alkalinity, hardness, and temperature be measured. The total dissolved copper, lead, iron, and zinc in the hot water were also determined. This was done after institution of disinfection on water samples collected from the hospital's hot water systems and on water samples collected from the hot water system of a nearby building that served as a control and did not receive booster chlorination.

L. pneumophila was not detected culturally in any of the samples collected after institution of disinfection. A small but statistically significant difference in the aggressivity of the hot water in the hospital and the nearby control building was found. The hot water in the hospital was considered mildly aggressive (Langelier index = -0.3 ± 0.08) as compared with the hot water in the control building, which was considered mildly

nonaggressive (Langelier index = $+0.2 \pm 0.12$). The hospital's hot water probably was more aggressive because of the combination of lower pH and lower temperature than in the control building. However, the only difference in the levels of trace metals was a slight elevation of copper and iron in the hot water in the control building. Consequently, no corrosion control activities were required.

Our study showed that on-site supplemental chlorination of a hospital hot water system can effectively reduce levels of *L. pneumophila* in water to below the limits of detection. This method is inexpensive and simple, and if careful attention is given to water quality parameters in the design of the on-site supplemental chlorination system, water corrosiveness can be controlled.

LITERATURE CITED

1. **American Public Health Association.** 1980. Standard methods for the examination of water and wastewater, 15th ed., p. 292–293. American Public Health Association, Inc., Washington, D.C.
1a. **Crawford, H. B., and D. N. Fischel.** 1971. Water quality and treatment, 3rd ed. McGraw-Hill Book Co., New York.
2. **Gorman, G. W., and J. C. Feeley.** 1982. Procedures for the recovery of *Legionella* from water. Centers for Disease Control, Atlanta, Ga.

Laboratory and Field Applications of UV Light Disinfection on Six Species of *Legionella* and Other Bacteria in Water

RICHARD W. GILPIN

The Medical College of Pennsylvania and Hospital, Philadelphia, Pennsylvania 19129

Continuous UV light disinfection of circulating water systems was evaluated for effectiveness against *Legionella* spp. and *Pseudomonas aeruginosa*. These bacteria have been directly associated with acquisition of overt clinical diseases, legionellosis and folliculitis, respectively, from appliances such as heat exchangers and whirlpools (2).

The levels of free chlorine needed to kill *Legionella pneumophila*, *P. aeruginosa*, and *Flavobacterium aquatile* were established by incubating these bacteria in deionized water containing calcium hypochlorite. Free chlorine levels, determined by the *N,N*-diethyl-*p*-phenylenediamine method, ranged from 0 to 12 mg/liter. All three genera were killed in 10 min with a free chlorine level of 2.5 mg/liter. Killing within 1 min required much greater levels of free chlorine: 12, 7, and 10 mg/liter for *L. pneumophila*, *P. aeruginosa*, and *F. aquatile*, respectively. However, greater amounts of chlorine were needed to accomplish a similar rate of killing in whirlpools because of several variables that reduced chlorine effectiveness (1).

I tested the UV sensitivity of six species of *Legionella* and *P. aeruginosa*. Bacteria were suspended in 3 mM potassium phosphate buffer (pH 7.0) and exposed to 34 μW of UV radiation per cm² for various times. The *Legionella* spp. were killed at different rates (Fig. 1). Times required for 50% kill at a UV intensity of 1 μW/cm² ranged from 5 min for *L. longbeachae* serogroup 2 to 30 min for *L. gormanii* with *L. pneumophila* at an intermediate time of 17 min. *P. aeruginosa* was killed at a rate similar to *L. micdadei* (8 min). Exposure times for 99% kill ranged from 33 min for *L. longbeachae* to 63 min for *L. gormanii*. These levels of UV radiation are in the moderate dose range as compared to those reported for other gram-negative bacteria (3).

The effectiveness of UV inactivation of bacteria in circulating water systems has not been reported. *L. pneumophila* or *P. aeruginosa* were circulated at a flow rate of 85 liters/h through a 3-liter closed system connected to a stainless-steel jacketed apparatus containing a biocidal UV lamp with a measured UV output of 7 ergs/mm²

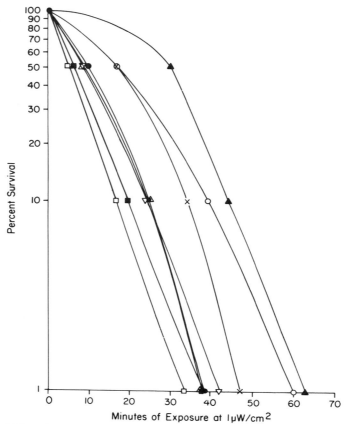

FIG. 1. UV sensitivity of *Legionella* spp. and *P. aeruginosa*. Samples were irradiated as described in the text. Each point represents the geometric mean from three independent experiments. Time (minutes) of exposure to a UV intensity of 1 μW/cm² to produce indicated percent survival of 10⁷ CFU of bacteria per ml. Symbols: ●, *L. bozemanii*; ○, *L. dumoffii*; ▲, *L. gormanii*; △, *L. micdadei*; ×, *L. pneumophila* serogroup 1; ■, *L. longbeachae* serogroup 1; □, *L. longbeachae* serogroup 2; ▽, *P. aeruginosa*.

per s per 100 cm at 254 nm. After 10 min of equilibration the UV lamp was turned on, and samples were taken at the inlet to the UV apparatus. *L. pneumophila* and *P. aeruginosa* were killed within 15 s. Clearly, the test organisms were killed on their first pass through the system.

A 340-liter stainless-steel hydrotherapy whirlpool was connected to a similar UV system. Before each patient used it, the whirlpool was drained, scrubbed down with Povadyne (Chaston Medical and Surgical Products, Inc., E. Farmingdale, N.Y.), rinsed, and filled with fresh tap water at 40°C. The whirlpool water was not supplemented with chemical biocides because of possible irritation to patients with open wounds. Results of plate count determinations on water collected before and after patient use showed that patients using the whirlpool for 20 min contributed approximately 4.5×10^3 CFU of bacteria per liter of the water (Table 1). During

25 of 27 experiments when water was collected from the whirlpool being used by patients and was continuously being treated with UV, no bacteria could be recovered from duplicate 250-ml water samples. On only two occasions, 5 and 10 CFU/liter were isolated.

A more rigorous test was performed using an undersized UV treatment system to treat evapo-

TABLE 1. Disinfection of whirlpool water with UV light

Whirlpool operating	UV light	Total CFU/liter of water	
		Mean	Range
Without patients	Off	1.5×10^1	5.0×10^0–1.2×10^2
With patients	Off	6.8×10^4	7.6×10^3–9.0×10^5
With patients	On	0.56×10^0	0–1.0×10^1

rative condenser return water. Approximately 13% of the return water was directed through the UV system on each pass. Water samples were collected daily at the inlet to the UV system for 10 weeks with the UV lamp off to establish the base-line number of bacteria, and for 6 weeks with the lamp continuously on. The counts of bacteria in condenser water collected during continuous UV treatment were significantly lower ($P = 0.001$, Student's t-test) than those in untreated water.

The results of this study indicate that continuous UV disinfection of circulating water systems may provide an effective alternative or supplement to chemical biocide treatment of water in whirlpools and hot tubs. Further field investigations are currently in progress.

LITERATURE CITED

1. **Centers for Disease Control.** 1979. Rash associated with use of whirlpools—Maine. Morbid. Mortal. Weekly Rep. 28:182–184.
2. **Centers for Disease Control.** 1982. Water-related disease outbreaks. Annual Summary 1981, issued September 1982, p. 11–13. Centers for Disease Control, Atlanta, Ga.
3. **Zelle, M. R., and A. Hollaender.** 1955. Effects of radiation on bacteria, p. 365–430. *In* A. Hollaender (ed.), Radiation biology, vol. 2. McGraw-Hill Co., New York.

Disinfection of *Legionella pneumophila*-Contaminated Whirlpool Spas

LINDEN E. WITHERELL, LILLIAN A. ORCIARI, KENNETH C. SPITALNY, RAYMOND A. PELLETIER, KENNETH M. STONE, AND RICHARD L. VOGT

Vermont State Department of Health, Burlington, Vermont 05401

An outbreak of 34 cases of Pontiac fever and an outbreak of 7 cases of Legionnaires disease occurred recently at an inn (Inn A) in W. Dover, Vt. Environmental samples were collected, and *Legionella pneumophila* was isolated from water collected from the recreational whirlpool spa at this inn. After this initial finding, additional samples were collected from six other whirlpool spas at nearby inns. *L. pneumophila* was recovered from two of the units by using the methods developed at the Centers for Disease Control (3).

The term "recreational whirlpool spa" is used to denote a shallow pool containing warm water circulated by hydrojets and air induction systems, which is used for physiological and psychological relaxation. These pools are not meant for swimming or diving. They are not drained, cleaned, and refilled after each use. Recreational whirlpool spas should not be confused with home fill-and-drain tub spas or hospital hydrotherapy units.

Since the two outbreaks of legionellosis at Inn A were epidemiologically and bacteriologically associated with the inn's contaminated whirlpool spa, this facility was closed to the public. It and the other two whirlpool spas contaminated with *Legionella* spp. were disinfected.

Disinfection was performed by the method described in a publication of the Centers for Disease Control (2). Continuous chlorination was used as the disinfection method. One of the whirlpool spas used an automatic solution feed pump. The other two whirlpool spas used erosion-type chlorinators. A residual of 1.0 to 3.0 mg of free available chlorine per liter and a pH of 7.2 to 7.8 were maintained.

Field measurements of temperature, pH, chlorine residual, and turbidity were taken at the time of sample collection. Temperature was obtained with a calibrated thermometer. Chlorine residuals were analyzed by use of a HACH model CN-66 DPD Free and Total Chlorine Test Kit (0 to 3 mg/liter). pH measurements were made with a HACH model 17-H Phenol Red pH Test Kit (pH 6.5 to 8.5). Turbidity was measured by use of a Bausch & Lomb "Spectronic Mini 20" spectrophotometer with the nephelometric unit 33-09-11 attached. Water samples were collected from each whirlpool spa for standard plate count, coliform, and *L. pneumophila* analyses by aseptic technique in sterile, 118-ml, glass screw-cap bottles.

The Vermont State Department of Health Laboratory used direct culture methods developed by Gorman and Feeley of the Centers for Disease Control for the analysis of *Legionella* samples (3). The standard plate count and coliform samples were analyzed by standard methods (1).

The treatment method employed resulted in the reduction of *L. pneumophila* to below detectable limits within the first week of treatment. Increased temperatures and aeration had minimal impact on the maintenance of chlorine residuals in unoccupied whirlpool spas. However, sharp reductions in free available chlorine residuals (1 to 2.5 mg/liter) were noted within 15 min after whirlpool spas were occupied.

After the institution of continuous disinfection, no cases of legionellosis were noted among users of the two whirlpool spas that remained in operation.

LITERATURE CITED

1. **American Public Health Association.** 1980. Standard methods for the examination of water and wastewater, 15th ed. American Public Health Association, Washington, D.C.
2. **Centers for Disease Control.** 1981. Suggested health and safety guidelines for public spas and hot tubs. Centers for Disease Control, Atlanta, Ga.
3. **Gorman, G. W., and J. C. Feeley.** 1982. Procedures for the recovery of *Legionella* from water. Centers for Disease Control, Atlanta, Ga.

Field Trial of Biocides in Control of *Legionella pneumophila* in Cooling Water Systems

JOHN B. KURTZ, CHRISTOPHER BARTLETT, HILARY TILLETT, AND URSULA NEWTON

Department of Virology, John Radcliffe Hospital, Oxford OX3 9DU, Communicable Disease Surveillance Center, Colindale, London NW9, and Technical Division, Houseman (Burnham) Ltd., Slough SLA 7LS, England

Cooling towers provide a variety of habitats suitable for the growth of a wealth of organisms. Many species of bacteria, algae, and protozoa coexist, some in symbiosis, some perhaps as parasites, as Rowbotham (5) has suggested is the relationship between *Legionella pneumophila* and some freshwater amoebae. *L. pneumophila* is often present in this situation. For example, recently in a small survey of cooling towers in England we isolated the organism from 17 of 26 (65%) cooling water systems (unpublished data).

Several outbreaks of pneumonia due to *L. pneumophila* have implicated cooling towers (1, 2) as the source of infection. It is therefore important to look at factors favoring the presence of the organism and at methods to eliminate or, more realistically, to reduce its numbers in these sites. In a survey in the summer of 1981 of 14 cooling systems in London (3), there appeared to be an association between the presence of the organism and the exclusion of light and higher water temperatures, but there was no correlation with the total bacterial counts of the water. Amoebae of various species were present in every sample. This study also suggested that a biodegradable (approximate half-life in lake water, 24 h), chlorinated phenolic thio-ether of low toxicity might effectively reduce *L. pneumophila* numbers temporarily (3).

A trial of two biocides was therefore carried out in the summer of 1982 in 16 London cooling systems. After initial sampling, the systems were subdivided into three groups, matched as closely as possible with regard to design of system, capacity, and the presence of *L. pneumophila*. One group (five systems) was treated

TABLE 1. Isolation of *L. pneumophila* from cooling towers after biocide treatment[a]

System	Treatment	Before treatment	July 14	July 16	July 19	July 21	July 23	July 26	July 28	July 30	Aug 2	Aug 4	Aug 6	Aug 9	Aug 23	Sept. 20
1	LP5	+	−	−	−	0	0	−	−	−	−	−	−	−	−	−
2	LP5	+	−	−	−	0	−	−	−	−	−	−	−	−	−	−
3	LP8	+	+	+	+	+	0	+	−	+	+	+	+	+	−	+
4	LP8	+	+	+	+	+	+	+	+	+	+	+	+	+	−	−
5	Control	+	+	+	+	+	+	+	+	0	+	+	+	+	+	−
6	Control	+	−	−	−	−	+	−	−	0	−	+	+	−	+	+
7	LP5	+						+							−	−
8	LP5	+						+							−·	−
9	LP5	+						+							−	−
10	LP8	−						−							−	−
11	LP8	+						+							−	+
12	LP8	+						+							−	−
13	LP8	−						0							−	−
14	Control	−						−							−	−
15	Control	+						−							−	−
16	Control	−						−							+	−

[a] Treatment was commenced 14 July for systems 1 to 6 and 28 July for systems 7 to 16. +, *L. pneumophila* isolated; −, not isolated; 0, no specimen received.

TABLE 2. Action of biocides on *L. pneumophila* and *A. palestinensis*

Biocide	Concn (ppm)	Log_{10} CFU^a of *L. pneumophila* serogroup 1 at time (h):			Log_{10} CFU^a of *L. pneumophila* serogroup 6 at time (h):			Log_{10} viable *Acanthamoeba* trophozoites^a at time (h):		
		0	1	3	0	1	3	0	1	3
LP5	10	11	8	6	11	9	6	5	3	2
	100	10	5	<2	11	5	<2	5	<1	<1
LP8	10	10	10	9	11	10	10	4	4	4
	100	11	10	10	11	10	9	4	4	4
Control	0	10	9	9	11	11	10	5	5	5

^{a} Per milliliter of reaction mixture.

with the chlorinated phenolic thio-ether Hatacide LP5 (Houseman [Burnham] Ltd.; U.S. Patent application 468250/83) to achieve a level 4 h after dosing of 200 ppm (mg/liter). The second group (six systems) was similarly treated with a polymeric quaternary ammonium chloride (Hatacide LP8, Houseman [Burnham] Ltd.) to achieve levels of 50 ppm. Five systems served as controls. Treatment was commenced in July and repeated at weekly intervals until the end of September. Samples of water for analysis were taken, just before the next dosing, at monthly intervals from each system. In addition, to see whether there was only a very transient effect, six systems, two from each group, were sampled three times a week, on days 2, 4, and 7 after dosing, for 4 weeks.

Isolation of *L. pneumophila* was done by guinea pig inoculation as previously described (3) and by direct plating of 0.1 ml of the concentrated deposit both with and without prior acid treatment (0.01 M HCl, pH 2.1, for 45 min at 37°C) on ACES [*N*-(2-acetamido)-2-aminoethanesulfonic acid]-buffered charcoal-yeast extract agar with added glycine (3 g/liter), vancomycin (1 μg/liter), and polymyxin (50 U/liter). Isolates were confirmed by gas-liquid chromatography and serogrouped by immunofluorescent microscopy. Chemical analysis and total bacterial counts of the samples were carried out using standard methods.

Laboratory tests of the biocides against *L. pneumophila* serogroups 1 and 6 were done as previously described (3), and their action against a strain of *Acanthamoeba palestinensis* was studied. The method, briefly, was as follows: *A. palestinensis* was grown as an axenic culture in Neff medium (6) at 35°C for 48 h. The resulting amoebae (mainly trophozoites) were concentrated by centrifugation, and 0.2 ml of the concentrate was added to 4.8 ml of the biocide being tested. At 0, 1, and 3 h, 1-ml samples were withdrawn, added to 19 ml of distilled water, washed by centrifugation, and suspended in 1 ml of distilled water. Serial 10-fold dilutions were

made, and 0.1 ml of each was inoculated onto Stoianovitch malt extract-yeast extract agar (4), previously seeded with a lawn of dead *Escherichia coli*. Counts of viable amoebic trophozoites were done at 24 and 48 h.

At the initial sampling, *L. pneumophila* was isolated from 12 of the 16 systems, including all those subsequently treated with Hatacide LP5 (Table 1). *L. pneumophila* was not isolated from any of the 31 samples taken during treatment with Hatacide LP5, but it was present in 24 of 35 samples from Hatacide LP8-treated systems and from 19 of 32 samples from the control group. All the isolates were serogroup 1 except those from system 3, which consistently yielded serogroup 6 strains.

The laboratory tests of the biocides showed that 100 ppm Hatacide LP5 reduced the counts of both *L. pneumophila* serogroups 1 and 6 by >9 log_{10} within 3 h, but that a similar concentration of Hatacide LP8 did not materially affect the counts (Table 2). The effect of Hatacide LP5 against trophozoites of *A. palestinensis* was also impressive, the viable count being reduced from 10^5/ml to below detectable levels within 1 h.

It is interesting that the efficacy of the biocides in the laboratory was reflected in the field trial results. In this trial, in which Hatacide LP5 appeared to effectively control *L. pneumophila* in contaminated cooling water systems, none of the 16 sites had been implicated in an outbreak of Legionnaires disease. In another situation in which a cooling system had been implicated as the source of an outbreak of Legionnaires disease, Hatacide LP5 was also demonstrated to be effective in controlling *L. pneumophila*.

If recirculating cooling water systems require a biocide treatment program it would be prudent to include in it one that is effective against *L. pneumophila* and its possible host.

LITERATURE CITED

1. **Band, J. D., M. LaVenture, J. P. Davis, G. F. Mallison, P. Skaliy, P. S. Hayes, W. L. Schell, H. Weiss, D. J. Greenberg, and D. W. Fraser.** 1981. Epidemic Legionnaires'

disease: airborne transmission down a chimney. J. Am. Med. Assoc. 245:2404–2407.

2. Dondero, T. J., R. C. Rendtorff, G. F. Mallison, R. M. Weeks, J. S. Levy, E. W. Wong, and W. Schaffner. 1980. An outbreak of Legionnaires' disease associated with a contaminated air-conditioning cooling tower. N. Engl. J. Med. 302:365–370.

3. Kurtz, J. B., C. L. R. Bartlett, U. A. Newton, R. A. White, and N. L. Jones. 1982. Legionella pneumophila in cooling water systems. J. Hyg. 88:369–381.

4. Page, F. C. 1976. An illustrated key to freshwater and soil amoebae. Scientific Publication No. 34. Freshwater Biological Association, Ambleside, Cumbria, England.

5. Rowbotham, T. J. 1980. Preliminary report on the pathogenicity of Legionella pneumophila for freshwater and soil amoebae. J. Clin. Pathol. 33:1179–1183

6. Stevens, A. R., and W. D. O'Dell. 1973. The influence of growth medium on axenic cultivation of virulent and avirulent Acanthamoeba. Proc. Soc. Exp. Biol. Med. 143:474–478.

Susceptibility of *Legionella pneumophila* to Cooling Tower Microbiocides and Hospital Disinfectants

H. KOBAYASHI AND M. TSUZUKI

Surgical Center, University of Tokyo Hospital, 7-3-1 Hongo, Bunkyo-ku, Tokyo, Japan 113

Since the first observed case of Legionnaires disease in Japan in November 1980 (1), its etiological agents have been isolated from the environment in almost all areas of Japan. The objective of this study was to determine the susceptibility of *Legionella pneumophila* to cooling tower microbiocides and hospital disinfectants.

The strains of *L. pneumophila* tested in this study were six ATCC type strains, two of serogroup 1 and one each of serogroups 2 through 5 (ATCC 33152, 33153, 33154, 33155, 33156, and 33216).

The microbiocides employed in this study were as follows: three cooling tower microbiocides, poly(hexamethylene biguanidine) hydrochloride (recommended concentration for use, 0.1 to 0.3%), sodium hypochloride, and dichlorophene (recommended concentration, 0.1% [wt/vol]); six hospital disinfectants, glutaraldehyde, povidone-iodine, chlorhexidine, benzalkonium chloride, 77% (vol/vol) ethanol, and 0.2% chlorhexidine in 80% (vol/vol) ethanol.

The test strains were cultured on buffered charcoal-yeast extract agar at 23°C for 4 days and suspended in phosphate-buffered saline at a pH of 7.2, as described by Skaliy et al. (2). The inoculum was adjusted to 10^5 to 10^6 CFU/ml. In the test of cooling tower microbiocides, 0.1 ml of the suspension was introduced into 9.9 ml of microbiocide and mixed for 1 or 24 h at 23°C. In the test of hospital disinfectants, 1 ml of the suspension was introduced into 3 ml of disinfectant and mixed for 30 s or 1 min at 23°C. The suspension was counted on buffered charcoal-yeast extract agar, and an additional sample was introduced into phosphate-buffered saline for the control.

Samples of cooling tower microbiocide mixture (0.05 ml) or hospital disinfectant mixture (0.02 ml) were inoculated into five 10-ml volumes of buffered charcoal-yeast extract broth and cultured at 35°C for 4 days or 1 week. Each culture was then plated onto buffered charcoal-yeast extract agar and blood agar.

Poly(hexamethylene biguanidine) hydrochloride at 2,000 ppm (0.2%) was effective after 1 h of treatment (Table 1). Exposure at 1,000 ppm (0.1%) for 24 h permitted survival of serogroup 3, though the same treatment for 1 h was inhibitory. Similar results were observed for serogroup 4.

The effective concentration of sodium hypochlorite for serogroups 1, 2, 3, and 4 was 0.47 ppm for 1 h of exposure time.

Dichlorophene, which is one of the *bis*-phenols, was effective for all five serogroups at 50 ppm for 1 h.

In studies of environmental, instrumental, and hand-washing disinfectants in hospitals and laboratories, effective concentrations against all six strains from five serogroups with exposure for 1 min were 0.75% for glutaraldehyde, 53 ppm for povidone-iodine, 0.38% for chlorhexidine, and 0.38% for benzalkonium chloride. In all procedures, controls permitted survival of all five serogroups.

Both 77% (vol/vol) ethanol and 0.2% chlorhexidine in 80% (vol/vol) ethanol with emollients were effective against the five serogroups after exposure for 30 s.

The results of poly(hexamethylene biguanidine) hydrochloride against serogroups 3 and 4, in our opinion, suggest that this microbiocide may be effective for most of the strains of *L. pneumophila*. It is possible that some strains were resistant to this concentration because the microbiocide was inactivated by killed organisms, allowing more resistant cells to grow when subcultured in the broth diluted 1:200 (0.05 ml/10 ml). Although the recommended commercial concentration of this microbiocide, between

TABLE 1. Susceptibility of *L. pneumophila* to poly(hexamethylene biguanidine) hydrochloride ($n = 5$)

Sero-group	Inoculum size (per ml)	Time (h)	Growth[a] at concn (ppm):				
			0	250	500	1,000	2,000
1	4×10^5	1	+	+	−	−	−
		24	+	+	−	−	−
2	2×10^5	1	+	−	−	−	−
		24	+	−	−	−	−
3	4×10^4	1	+	−	−	−	−
		24	+	+	+	+	−
4	5×10^4	1	+	−	−	−	−
		24	+	+	+	−	−
5	1×10^3	1	+	+	−	−	−
		24	+	−	−	−	−

[a] +, Growth; −, no growth.

0.1 and 0.3%, appears to be effective for *L. pneumophila*, we believe that this is too narrow a safety range for legionellae.

Our results for sodium hypochlorite are almost the same as the effective concentration for serogroup 1, 0.65 ppm, that was reported by Wang et al. (3). The great advantage of hypochlorite is that it is one of the more economical microbiocides. Unfortunately, it is corrosive when its concentration is kept high enough that it is not inactivated by organic material in cooling towers.

Since the commercially recommended concentration for dichlorophene (1,000 ppm) is 20 times larger than the concentration inhibitory for legionellae, it appears that dichlorophene should prove to be one of the most useful cooling tower microbiocides for *L. pneumophila*.

We believe that among environmental, instrumental, and hand-washing disinfectants for hospitals and laboratories, glutaraldehyde, chlorhexidine, and benzalkonium chloride can be effective at lower concentrations; we are currently conducting a study to determine this. Ethanol is one of the best hand disinfectants for preventing cross-contamination in hospitals, but it unfortunately damages skin when it is used frequently and continuously. Since damaged skin invites infection, emollients should be present in hand disinfectants to prevent skin microabscess and possible cross-contamination. Chlorhexidine (0.2%) in ethanol with emollients is usually more available than ethanol for hospital infection control, and it is effective against *Legionella* spp. For these reasons, it should be used for hand cleaning.

Although our study showed that dichlorophene should be a good disinfectant for cooling towers and that chlorine can be used in treatment of potable water, we believe that the safest and most complete treatment for potable water and hot shower systems is filtration or UV irradiation or both, at both ends of the system. We suggest that cross-contamination of *L. pneumophila* in hospitals and laboratories can be prevented by the use of glutaraldehyde, chlorhexidine, povidone-iodine, or ethanol.

LITERATURE CITED

1. **Saito, A., T. Shimoda, M. Nagasawa, H. Tanaka, N. Ito, Y. Shigeno, K. Yamaguchi, M. Hirota, M. Nakatomi, and K. Hara.** 1981. The first case of Legionnaires' disease in Japan. Kansenshogaku Zasshi **55:**124–128.
2. **Skaliy, P., and H. V. McEachern.** 1979. Survival of the Legionnaires' disease bacterium in water. Ann. Intern. Med. **90:**662–663.
3. **Wang, W. L. L., M. J. Blaser, J. Cravens, and M. A. Johnson.** 1979. Growth, survival, and resistance of the Legionnaires' disease bacterium. Ann. Intern. Med. **90:**614–618.

Bacteriological, Chemical, and Physical Characteristics of Samples from Two Hot Water Systems Containing *Legionella pneumophila* Compared with Drinking Water from Municipal Water Works

A. C. HOEKSTRA, D. van der KOOIJ, A. VISSER, and W. A. M. HIJNEN

Municipal Dune Water Works of the Hague, 2501 CS 's-Gravenhage, and The Netherlands Water Works Testing and Research Institute, KIWA Ltd., 2280 AB Rijswijk, The Netherlands

Water samples obtained from the hot water systems (HWS) of two hospitals (hospital I, a 10-year-old building, and hospital II, a 3-year-old building) were investigated for the presence of *Legionella* species by enumeration on buffered charcoal-yeast extract agar. A number of additional microbiological, physical, and chemical characteristics were also determined for the samples. The HWS of the hospitals were supplied with drinking water originating from the Municipal Dune Water Works of The Hague, The Netherlands. This water has been prepared from dune-infiltrated river (Meuse) water by the addition of powdered activated carbon followed

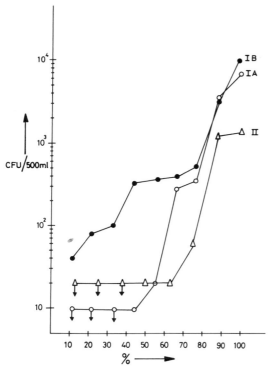

FIG. 1. Colony counts of *L. pneumophila* arranged in order of increasing magnitude in water samples obtained from the HWS of two hospitals. Samples of series IA and IB were obtained from hospital I at different dates, and samples of series II were obtained from hospital II. Symbols with arrow indicate counts below the detection limit.

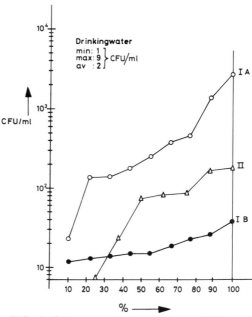

FIG. 2. Colony counts (incubation at 37°C for 2 days) of aerobic heterotrophic bacteria in samples from the HWS of two hospitals (see legend to Fig. 1) compared with counts observed in FPDW during 1 year.

by rapid and slow sand filtration. No postdisinfectant is used.

The numbers of legionellae in the HWS samples varied from <10 to 10^4 CFU/500 ml in hospital I (series IA and IB) and to 1.4×10^3 CFU/500 ml in hospital II (Fig. 1). The great majority of the observed legionellae were identified as *Legionella pneumophila* serogroup 6, with only a few isolates belonging to serogroup 1. Legionellae were not observed in samples of drinking water.

Numbers of aerobic heterotrophic bacteria (colony counts on Lab Lemco [Oxoid Ltd.] agar plates incubated at 37°C for 48 h) ranged from <10 to about 3,000 CFU/ml in the HWS samples (Fig. 2). These numbers were significantly above the counts of these organisms in freshly prepared drinking water (FPDW). Fungi, actinomycetes, and aerobic sporeforming bacteria were detected in small numbers (usually less than 100 CFU/100 ml) in the HWS samples. These numbers did not exceed the counts of these organisms in FPDW. Citrate-utilizing pseudomonads

were not observed in 100-ml volumes of the HWS samples, but these organisms (not *Pseudomonas aeruginosa*) were always observed in 100-ml samples of FPDW (average count, 49 CFU/100 ml; maximum count, 320 CFU/100 ml) (1). The bacteriological composition of the hot water differed from that of drinking water by the presence of *L. pneumophila*, the increased colony counts (at 37°C) of aerobic heterotrophic bacteria, and the absence of citrate-utilizing bacteria (Table 1).

A number of the physical and chemical characteristics of the HWS samples of hospital I

TABLE 1. Microorganisms in hot water and in drinking water

Organism	Hot water	Drinking water
Legionellae[a]	+	−
Aerobic heterotropic bacteria[b]	++	+
Citrate-utilizing pseudomonads[c]	−	+
Actinomycetes[c]	+	+
Aerobic sporeforming bacteria[c]	+	+
Fungi[c]	+	+

[a] Samples of 500 ml; one sample of 100 liters of drinking water was tested.
[b] Samples of 1 ml.
[c] Samples of 100 ml.

TABLE 2. Physical and chemical data on samples of a water system containing legionellae and on samples of FPDW

Parameter	HWS[a]	FPDW[b]
Temp (°C)	36 to 59	9 to 13
pH	7.85 to 8.00	7.60 to 7.80
Conductivity (mS/m)	61 to 63	53 to 60
Turbidity (FTU)	0.22 to 5.2	0.03
UV extinction (m^{-1})	3.1 to 4.0	4.3 to 8.1
Total organic carbon (mg of C/liter)	2.2 to 2.9	1.5 to 5.6
NO_2 (mg of N/liter)	0.01	0.01
NH_4 (mg of N/liter)	0.01	0.01
Fe (mg/liter)	0.01 to 0.04	0.01 to 0.07
Ca (mg/liter)	69 to 81	73 to 85
Mg (mg/liter)	8.0 to 15	7.1 to 9.4
Cu (µg/liter)	310 to 790	1 to 12
Pb (µg/liter)	1 to 6	3 to 7
Zn (µg/liter)	5 to 175	1 to 18
Cd (µg/liter)	0.1 to 3.2	0.5 to 1.3

[a] Fourteen samples from hospital I (series IB).

[b] Minimum and maximum values observed during 1 year (weekly observations).

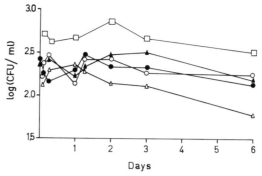

FIG. 4. Behavior of a strain of *L. pneumophila* serogroup 6 in pasteurized water containing different amounts of copper. Symbols: ○, FPDW with less than 1 µg of Cu per liter; ●, FPDW with 1,000 µg of Cu (added as $CuSO_4$) per liter; △, FPDW with 2,500 µg of Cu per liter; □, water from HWS containing 330 µg of Cu per liter; ▲, water from HWS containing 790 µg of Cu per liter (no additions). Samples (300 to 600 ml) were incubated at 25°C.

differed from those of FPDW (Table 2). Compared with FPDW, water of the HWS had a much higher temperature, a slightly higher pH, a slightly better conductivity, a clearly higher turbidity, slightly higher concentrations of iron, and clearly higher concentrations of copper (300 to 800 µg of Cu per liter of HWS and 1 to 12 µg of Cu per liter of FPDW). The UV extinction and calcium concentration of the HWS samples were slightly below those of FPDW.

For the assessment of the concentration of easily assimilable organic carbon, growth experiments were conducted with strains of *Pseu-*

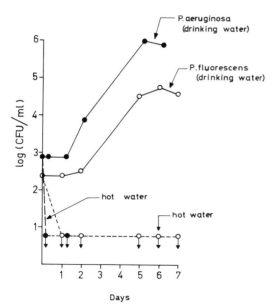

FIG. 3. Behavior of a strain of *P. fluorescens* and a strain of *P. aeruginosa* in pasteurized samples of water from the HWS of hospital I and in FPDW. The water samples (300 to 600 ml) were incubated at 25°C. Symbols with arrows indicate counts below the detection limit (7 CFU/ml).

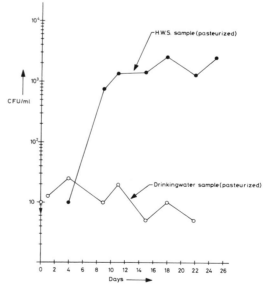

FIG. 5. Behavior of a strain *L. pneumophila* serogroup 6 in a pasteurized sample of water from the HWS of hospital I and in pasteurized FPDW. The samples (600 ml) were incubated at 37°C.

domonas *fluorescens* and *P. aeruginosa* in the various samples (2). The HWS samples were toxic for these organisms, but growth of *P. fluorescens* and *P. aeruginosa* was observed in FPDW (Fig. 3). Addition of copper to the (nontoxic) samples of FPDW revealed that copper concentrations as observed in the HWS samples were toxic for these pseudomonads. The absence of citrate-utilizing pseudomonads in HWS samples may therefore be due to the presence of copper. *L. pneumophila* obviously is not influenced by the levels of copper observed in HWS samples. Furthermore, survival of the organism in water samples supplemented with copper at a concentration of 2.5 mg/liter strongly suggests that the organism is copper tolerant (Fig. 4).

L. pneumophila serogroup 6 multiplied to about 2×10^3 CFU/ml in an HWS sample after pasteurization and incubation of the samples at 37°C. In a pasteurized FPDW sample, no increase of *L. pneumophila* inoculated to about 10 to 20 CFU/ml was observed (Fig. 5). This sug-gests that a specific component(s) required for growth may be introduced in the water in the HWS. Further investigations are being conducted to identify these compounds to define measures for controlling the growth of legionellae in HWS.

We are much indebted to D. G. Groothuis and P. L. Meenhorst for techniques and materials for enumeration and identification of *L. pneumophila*.

The contribution of KIWA to the described research was part of the research program of KIWA Ltd., assigned and financed by the Netherlands Waterworks Association (VEWIN).

LITERATURE CITED

1. **van der Kooij, D.** 1979. The occurrence of pseudomonads in surface water and in tap water as determined on citrate media. Antonie van Leeuwenhoek J. Microbiol. Serol. **43**:187–197.
2. **van der Kooij, D., A. Visser, and W. A. M. Hijnen.** 1982. Determining the concentration of easily assimilable organic carbon in drinking water. J. Am. Water Works Assoc. **74**:540–545.

Bactericidal Effect of Commonly Used Disinfectants on *Legionella pneumophila*

ENDRE MERO

Microbiology Laboratory of Medizinisches Landesuntersuchungsamt, Stuttgart, West Germany

Improvement in culture techniques has resulted in more routine isolation of *Legionella pneumophila* and related species and has presented, therefore, an increased opportunity for studying the different characteristics of this organism. Little work has been published, however, on the susceptibility of *L. pneumophila* to commonly used antiseptics and disinfectants (1, 4, 5). This may be due to the following problems: (i) *L. pneumophila* is able to grow only in special media, whether liquid or solid; (ii) the growth of this bacterium is relatively slow and it is influenced by many factors such as temperature, nutrients, inhibitors, etc.; (iii) the neutralizers (e.g., Tween 80, soy lecithin, sodium thiosulfate) used against the residual disinfectant in the liquid culture media are also inhibitory to the growth of *L. pneumophila*; (iv) studies with *L. pneumophila*, especially in liquid medium, are difficult because of the higher risk of contamination with faster growing and less fastidious organisms; and (v) to verify the reproducibility of the results, it is necessary to make many parallel examinations.

At present the most effective liquid medium for cultivation of *L. pneumophila* is yeast extract broth supplemented with α-ketoglutarate (YEB-α) (2, 3). Because of the inhibitory effect of neutralizers on *L. pneumophila* it was not possible to use any of them in our experiments. That is the reason I modified the YEB-α medium as follows. Finely ground beef heart (500 g), free from fat, was heated to boiling in 1,000 ml of alkalinized distilled water (6.3 ml of 4 N NaOH per 1,000 ml of water). After the mixture was cooled and the remaining fat was removed, the liquid was strained through muslin. The cooked meat was dried at a temperature below 50°C overnight, tubed about 1 cm deep, and sterilized by autoclaving. Finally, the distributed dried meat was covered with 10 ml of filtration-sterilized YEB-α (pH 6.9 ± 0.05). This medium was called cooked meat-yeast extract broth supplemented with α-ketoglutarate (CMYEB-α).

CMYEB-α has some advantages. This medium is able, without neutralizers, to inactivate the residual disinfectants in the medium and to produce good conditions for growing reversibly damaged *L. pneumophila* (Table 1). Also, the cooked meat in the liquid medium has probably the same effect as activated charcoal and can absorb harmful metabolic products of *L. pneumophila* and so ensure the constant growth of the organism in this fluid medium. Growth

TABLE 1. Comparison of disinfectant-inactivating effect of YEB-α and CMYEB-α[a]

Disinfectant concn (%)	Inactivation after the indicated exposure time (min) with:							
	YEB-α				CMYEB-α			
	5	15	30	60	5	15	30	60
0.5	−	−	−	−	−	−	−	−
0.25	−	−	−	−	−	−	−	−
0.10	−	−	−	−	+	+	−	−
0.05	−	−	−	−	+	+	+	+
0.025	−	−	−	−	+	+	+	+
0.0125	−	−	−	−	+	+	+	+
0.00625	+	+	+	+	+	+	+	+
0.003125	+	+	+	+	+	+	+	+

[a] The disinfectant was Isothiazolon (Ebotec Aqua; Bacillolfabrik Dr. Bode and Co., Hamburg, West Germany).

curves obtained with CMYEB-α and YEB-α were very similar up to 6 days, but by 8 days the number of viable organisms remained the same in the former but had dropped by about 4 logs in the latter. It was possible, therefore, to examine the bactericidal effect on *L. pneumophila* of some disinfectants commonly used in the hospital, microbiology laboratory, and air-conditioning system.

The results of the qualitative suspension test with agents used for instruments and surface and hand disinfection, and the proposed concentrations and exposure times, are shown in Table 2.

The results obtained for two disinfectants used in evaporating condensers (Table 3) indicate that the polymere quaternary ammonium compound plus Isothiazolin-on-derivate (Phobrol; Schülke and Mayr GmbH, Norderstedt, West Germany) is effective only in much higher concentration than the one presently proposed. The other disinfectant, Isothiazolon (Ebotec Aqua; Bacillolfabrik Dr. Bode and Co., Hamburg, West Germany) has a bactericidal effect against *L. pneumophila* in the presently proposed concentration.

TABLE 2. Bactericidal effect of different disinfectants on *L. pneumophila* in the qualitative suspension test

Disinfectant[a]	Concn (%)	*L. pneumophila* viability after indicated exposure time (min)[b]				Proposed concn/exposure time
		5 (0.5)	15 (1)	30 (2)	60 (5)	
Gigasept[c] (instrument)	3.0	−	−	−	−	
	2.0	−	−	−	−	
	1.0	+	−	−	−	3%/60 min
	0.5	+	+	+	+	
	0.25	+	+	+	+	
Buraton 10 F[d] (surface)	1.0	−	−	−	−	
	0.5	−	−	−	−	
	0.25	+	−	−	−	0.5%/60 min
	0.10	+	+	+	+	
	0.05	+	+	+	+	
Mikrozid liquid[e] (surface)	100	−	−	−	−	
	50	−	−	−	−	
	25	−	−	−	−	100%/30 min
	20	+	+	+	−	
	10	+	+	+	+	
Desderman[f] (hand)	100	−	−	−	−	100% (3 ml)/0.5 min[g];
	50	−	−	−	−	100% (2× 5 ml)/5 min[h]
	25	+	+	+	−	
	20	+	+	+	+	
	10	+	+	+	+	

[a] All from Schülke and Mayr GmbH, Norderstedt, West Germany.
[b] Times given in parentheses are those for Desderman only.
[c] Contains 6.8 g of succinic dialdehyde, 4.5 g of dimethoxytetrahydrofurane, and 4.5 g of formaldehyde in a 100-ml solution.
[d] Contains 14.0 g of glyoxale, 9.0 g of formaldehyde, 2.0 g of glutaraldehyde, and 0.5 g of 2-ethylhexanal in a 100-ml solution.
[e] Contains 25 g of 94% ethanol, 35 g of propanol, 40 mg of glutaraldehyde, and 10 mg of 2-ethylhexanal in a 100-g mixture.
[f] Contains 78.2 g of 95.3% ethanol and 0.1 g of 2,3,4,5-tetrabromo-6-methylphenol in a 100-g mixture.
[g] For hygienic hand disinfection.
[h] For surgical hand disinfection.

TABLE 3. Bactericidal effect on *L. pneumophila* of two disinfectants used for treating water in air-conditioning systems[a]

Disinfectant[a]	Concn (%)	*L. pneumophila* viability after indicated exposure time (min)				Proposed concn/exposure time
		5	15	30	60	
Phobrol[b]	7.5	−	−	−	−	0.2%/automatically
	5.0	+	+	+	+	regulated dosing
	3.0	+	+	+	+	
	1.0	+	+	+	+	
	0.2	+	+	+	+	
Ebotec Aqua[c]	0.5	−	−	−	−	0.1 to 0.15%/in water of
	0.25	−	−	−	−	air-conditioning system
	0.10	+	+	−	−	
	0.05	+	+	+	+	
	0.025	+	+	+	+	

[a] From Schülke and Mayr GmbH; contains polymere quatenary ammonium compound plus Isothiazolin-on-derivate.

[b] See footnote to Table 1.

The determination of the concentration of the above-mentioned disinfectants effective on *L. pneumophila* is in progress, using the quantitative suspension test.

England et al. could isolate *L. pneumophila* from cooling towers treated with disinfectants deemed effective in earlier laboratory tests (1). Our results may indicate a reason for this finding.

Because the evaporating condensers in air-conditioning systems are widely used in hospitals, hotels, and other facilities, and because they can be a source of *Legionella* infection, it is important to do more extended studies in this field.

LITERATURE CITED

1. England, A. C. III, D. W. Fraser, G. F. Mallison, D. C. Mackel, P. Skalyi, and G. W. Gorman. 1982. Failure of *Legionella pneumophila* sensitivities to predict culture result from disinfectant-treated air-conditioning cooling towers. Appl. Environ. Microbiol. 43:240–244.
2. Ristroph, J. D., K. W. Hedlund, and R. G. Allen. 1980. Liquid medium for growth of *Legionella pneumophila*. J. Clin. Microbiol. 11:19–21.
3. Ristroph, J. D., K. W. Hedlund, and S. Gowda. 1981. Chemically defined medium for *Legionella pneumophila* growth. J. Clin. Microbiol. 13:115–119.
4. Skalyi, P., T. A. Thompson, G. W. Gorman, G. K. Morris, H. V. McEachern, and D. C. Mackel. 1980. Laboratory studies of disinfectants against *Legionella pneumophila*. Appl. Environ. Microbiol. 40:697–700.
5. Wang, W. L. L., M. J. Blaser, J. Cravens, and M. A. Johnson. 1979. Growth, survival, and resistance of Legionnaires' Disease bacterium. Ann. Intern. Med. 90:614–618.

ROUND TABLE DISCUSSION

Microbial Analysis of Environmental Specimens

Moderator: JAMES C. FEELEY

Participants: J. M. Barbaree, P. J. L. Dennis, C. B. Fliermans, R. McKinney, J. Tobin, R. L. Tyndall, R. W. Gilpin, and R. B. Yee

This round table on environmental sampling and testing was organized to address the following questions. Is current methodology adequate for detecting, isolating, and quantitating legionellae in environmental specimens? What new methods and techniques need to be developed? A summary of the major discussions that occurred is presented.

Water specimens. Most of the panelists believed that legionellae could be effectively isolated from most potable water and cooling-tower water specimens by currently available culture methods. The consensus was that direct plating procedures have greater sensitivity for recovering legionellae than the inoculation of guinea pigs. The majority of participants also thought that most microbiologists who worked in a well-equipped laboratory could isolate legionellae from environmental specimens. However, there was some concern about the effectiveness that the existing selective media have for isolating legionellae from extremely "high-count" water. One panelist pointed out that potable and cooling-tower water are not really "true" environmental specimens. He said that although the current methods were quite good for medically oriented specimens, they were inadequate for true environmental specimens, such as water from rivers and lakes. Several panelists emphasized that no single medium should be used exclusively and that methodology must be matched with the specimen. They pointed out that one medium or pretreatment method might work very well for one water specimen but not for another. None of the panelists could foresee major advances in the formulation of selective media in the near future. One panelist viewed the present state of the art as being at a high plateau and believed that no further major improvements could be made without advances in understanding of the physiology of the various *Legionella* species.

There was lengthy discussion of the sensitivity and specificity of the direct fluorescent-antibody staining technique. One panelist estimated its sensitivity for detection of legionellae at $\geq 10^5$/ml. An expert in the audience pointed out that the direct fluorescent-antibody conjugates were developed primarily for demonstrating legionellae in clinical specimens with excellent specificity. He said that only conjugates for *Legionella pneumophila* serogroups 1 to 4 had been extensively checked for specificity against a wide variety of bacteria and that these conjugates cross-reacted with some strains of soil and water bacteria belonging to the genera *Pseudomonas*, *Alcaligenes*, and *Flavobacterium*. Therefore, it should not be surprising that cross-reacting organisms would be detected in environmental specimens. He specifically emphasized caution in the interpretation of positive direct fluorescent-antibody reactions in the absence of supporting cultural isolation data.

Air specimens. This discussion of air specimens consisted mainly of negative reports. Several reasons were offered for the lack of success in this area. The main explanation was that air sampling has proven to be very difficult for the detection of any organism. Air sampling procedures are very harsh on microorganisms, even spore formers, and they cause extreme dehydration and severe impact injury to most organisms. The second reason suggested was that the numbers of legionellae in air are usually very low as compared with other organisms. Despite these problems, some people in the audience reported successful recoveries of legionellae from air. Some of the samplers that these investigators used with success were the multistage plate type, the Litton High Volume sampler, and the all-glass liquid impinger. The matter of air sampling efficacy is still unresolved, and most participants believed that a great deal more developmental work needs to be done in this area.

Marker systems. A number of different typing systems, such as monoclonal antibody, outer membrane protein, and plasmids, were reported at the Symposium. One of the panelists said that the combining of marker systems in the characterization of a *Legionella* isolate should enable very precise strain identification. He thought that this could be very helpful epidemiologically.

The same panelist also voiced concern about the possible confusion that would result from all of the different reagents, especially monoclonal antibodies, that were being produced. He suggested that a reference center be established to try to ensure some order and enable comparison of one marker system with another.

An attendee in the audience asked for a show of hands to indicate whether there was sufficient interest for exchange of reagents. About 50 people raised their hands, and a meeting was scheduled after the round table.

Most of the participants agreed that typing of organisms was extremely important and that more work was needed to expand them. It was also believed that simple tests were needed to determine the virulence of *Legionella* strains.

Quantitation. The need for good methods for quantitation of legionellae was voiced. Currently only estimates of the numbers of legionellae can be made. Discussion revealed some of the reasons why accurate quantitation is not being accomplished. One reason was that legionellae are being injured, especially in air during collection and processing. Pretreatments of specimens with acid and heat and exposure to selective agents in the media are extremely detrimental to recovery of the legionellae. Ideally the use of all selective and injurious agents should be avoided, but this may not be practical. Many thought that the direct fluorescent-antibody technique combined with a vital stain could accomplish this, provided that the problems of cross-reactivity were eliminated. Others thought that the use of a genetic probe or the autoimmune radiography technique were possible solutions.

Guidelines and standardizations. At numerous times throughout the round table, investigators expressed the need for guidelines on how an environmental water specimen should be processed. Many of the participants felt that they could not interpret or compare results of colleagues because the sensitivities of methods being used were not reported. This point was emphatically made by the statement "not all BCYE [buffered charcoal-yeast extract] agar is the same." With this remark, the round table was concluded.

ROUND TABLE DISCUSSION

Transmission and Control

Moderators: W. JAKUBOWSKI, C. V. BROOME, E. E. GELDREICH, AND A. P. DUFOUR

Participants: C. L. R. Bartlett, P. Berger, P. Edelstein, C. B. Fliermans, D. W. Fraser, G. W. Gorman, L. E. Witherell, and R. B. Yee

The purpose of this round table was to address (i) transmission of legionellosis, (ii) control methods, and (iii) the need and responsibility for implementing control of *Legionella* spp. in the environment.

Transmission (C. V. Broome, moderator). The initial discussion raised the issue of whether current efforts to control legionellosis should emphasize attempts to eradicate the organism whenever it is identified in the environment or whether control measures should be directed toward documenting methods of transmission and designing control measures specifically to interrupt transmission. The opinion was offered that eradication was not feasible in view of the ubiquity of the organism in the environment and that directed control would be more appropriate. It was stated that differentiating among reservoirs, amplifiers, and disseminators of the organism might help in focusing control efforts. For example, it may be more feasible to minimize the presence of amplifiers and disseminators, rather than to eradicate legionellae from a site. However, further research needs to be done in defining the methods of amplification and dissemination.

Since a directed approach to control measures is contingent on identification of cases and investigation of the method of transmission, surveillance for both individual cases and potentially linked clusters of cases should be emphasized. Even a hospital sensitized to the occurrence of legionellosis and with a laboratory capable of making the diagnosis could still miss cases if the appropriate specimens were not obtained. Since legionellosis does not have any pathognomonic clinical features, widespread use of specific laboratory diagnostic testing should be encouraged. After diagnosis, cases should be reported to appropriate public health authorities so that epidemiological studies can be undertaken as appropriate. In the United Kingdom, the cases diagnosed through the Public Health Service Laboratories are reported to the epidemiology group, facilitating the surveillance and investigation of legionellosis.

The discussion then focused upon the role of potable water in transmission of legionellosis. Circumstantial evidence and evidence from intervention studies have implicated hot water systems; an association with cold water as a possible source of legionellosis has not been shown.

The engineering details of two buildings at the University of Ohio, both of which had legionellae in the water supply but only one of which had a problem with nosocomial legionellosis, were discussed. No obvious differences could be identified in the plumbing systems, suggesting that the different attack rates may have been due to a difference in virulence between the two strains. This difference in strain virulence was also postulated by other investigators who found that only strains without plasmids were isolated from patients with nosocomial legionellosis in the Pittsburgh Veterans Administration hospital, although both plasmid-containing and plasmidless strains were present in the water supply. These strains had been further studied, and both plasmid-containing and plasmidless strains retain their characteristics through multiple generations, both on solid media and on animal passage. It was agreed that the question of differences in strain virulence was an important one which required further clarification through microbiological, animal model, and epidemiological studies to assess environmental *Legionella* isolates and their potential as human pathogens.

The possible role of the concentration of organisms in determining risk from different sources was raised. It was felt that current methodology for quantitation had not been sufficiently developed to permit assessment of this variable at this time. Further research in methods of quantitation, taking into account viability and possible strain differences in virulence, should be done. An additional parameter which could be studied in epidemiologically defined situations is the particle size of aerosols generated, preferably combined with an assessment of the presence of viable legionellae in such particles.

Although several studies have suggested a role for potable water in transmission of legionellosis in institutions, based on a decrease in cases after interventions to the hot water system, less information is available on the risk of disease from domestic water systems. Information was presented suggesting that domestic hot water systems may be culture positive for legionellae. Preliminary serological evidence suggested that use of hot water in room humidifiers might be linked to seroreactivity to legionellae. Of the two culture-positive residential water heaters in one study, one had been set at 150°F (ca. 66°C); however, this had a side heater unit, so that the temperature was unlikely to have been uniform throughout the tank. There was general agreement that further information is needed on the potential risk of disease from residential hot water systems. Recently, the state of Washington has recommended reducing the temperature settings on home water heaters to 120°F (ca. 49°C), both as an energy conservation measure and to reduce the number of deaths and severe injuries from scalding. An average of three deaths due to tap water scalding and a large number of severe injuries occur annually in Washington. With this demonstrable risk occurring with higher temperature settings, it is crucial to quantify the risk, if any, of legionellosis occurring due to a reduction in temperature settings.

Finally, panelists and audience were asked for summary conclusions regarding appropriate control measures for water systems given the current knowledge of risk of transmission, virulence, and quantitation. All concurred that recommendations to deal with the presence of legionellae in residential water systems were not appropriate at this stage. Conversely, any hospital with a nosocomial cluster of cases should conduct an epidemiological investigation and institute appropriate control measures. It was pointed out that this could range from use of sterile water in respiratory therapy devices as had been done in Chicago, Ill., to hyperchlorination and temperature elevation in hot water systems as had been used in Pittsburgh, Pa., and Los Angeles, Calif. However, it was noted that it was not possible to standardize recommendations for management because there was not a set of recommendations that was universally appropriate. For example, cases of *L. micdadei* pneumonia continued to occur at a hospital after heating water to 130°F with use of 150°F temperatures bi-monthly, even though there was a decrease in the number of organisms cultured from the water.

Discussion focused on the issue of whether it was appropriate to institute regulations for all hospitals, considering that hospitalized patients tend to represent a group which may have increased susceptibility to acquisition of legion-

ellosis. It was stated that hospitals should maintain good engineering "housekeeping practices," especially when infrequently used tanks or distribution systems are brought into service. Both the British Communicable Disease Surveillance Center and the Centers for Disease Control are systematically collecting information on the engineering and design features of buildings which have had nosocomial legionellosis. They hope to determine whether any features of system design, construction, or maintenance are common to facilities with nosocomial legionellosis, and are absent in facilities which have not had the disease identified.

Although other potential means of transmission for legionellosis, such as cooling towers, recreational whirlpools, or industrial coolant systems, were not discussed because of lack of time, further research is needed to assess the relative importance of these means of transmission as causes of sporadic and epidemic legionellosis.

Control (E. E. Geldreich, moderator). The state of knowledge on methodology for studies of immunology, epidemiology, and ecology of *Legionella* spp. has been significantly advanced in the past few years. We now have the tools to measure occurrence in a variety of environmental areas. However, has engineering technology for controlling this opportunistic pathogen kept pace with these discoveries to minimize the risk to public exposure? Where are the voids in information that need critical attention in environmental control?

Since water supply may be an important vehicle in the movement of legionellae from aquatic sources to sites of colonization and human exposure, it is important to consider whether the multiple-barrier control concept used for environmental control of other waterborne pathogens is also effective for legionellae. These barriers to microbial passage involve suppression of pathogens in source waters, application of appropriate water supply treatment processes, and protected transmission of potable water through the distribution system and building plumbing without providing colonization sites in the pipe networks or in attached devices.

Most participants in this round table consider legionellae to be ubiquitous. One participant from the audience reported that in Puerto Rico legionellae have been isolated in guinea pigs injected with a wide variety of waters, including water collected from leaves on trees in a tropical rain forest at a 4,000-m elevation. It was pointed out that surface waters may be more prone to have legionellae than groundwaters. In a study now in progress in the United Kingdom on domestic water supplies that rely solely on upland waters, river supplies, or bore hole supplies, investigators found *Legionella* in all three sources. Data from another study in progress on *Legionella* occurrences in public water supplies in Vermont will also be analyzed for any correlation to source water type (surface or groundwater). *Legionella* strains have been found in thermal springs in Yellowstone National Park and in waters from cooling towers that were supplied with groundwater, while other cooling towers in the area, using surface water, did not contain detectable levels of legionellae. No detectable *Legionella* organisms were found in the effluent of stripping columns used to remove organic compounds in groundwaters. The question of whether *Legionella* is a contaminant of groundwaters or a native bacterium appears unresolved at the moment. There was a concern that well drilling practices, sampling protocols, and fractured rock strata may introduce these organisms into high-quality groundwater.

There does not appear to be a suitable surrogate indicator for legionellae identified at this time that might be used in water quality monitoring. No positive correlations were reported for fecal coliforms, standard plate count, or 12 different physical and chemical constituents in several investigations. In one study on cooling-tower effluents, the only physical quality that had a positive correlation with presence of legionellae was ambient air temperature.

Several peaks in *Legionella* occurrence were reported in source waters to power plant cooling towers during two summer periods of a 3-year study. No correlations with either ambient water temperature elevations or algal blooms could be established. While cooling-tower effluents may be a major point discharge, it was suggested by one panelist that this input of legionellae to receiving waters would have little impact on the existing aquatic population of this organism. Such a position would nullify efforts directed towards diminishing *Legionella* occurrences through environmental regulations that might be proposed to obtain better ambient water qualities for downstream users (e.g., for water supply, recreation, irrigation). Efforts to control legionellae through source water quality protection and water treatment may be impractical because the focal point of amplification appears to be the distribution network and the water supply systems of buildings.

Water treatment operations do not produce or deliver sterile water. They do produce a public water supply that meets community, state, and national standards for water of defined sanitary quality for drinking purposes. However, there may be some undefined minimal risk that may be of special concern to the medical community when normally insignificant microorganisms in potable water are amplified in therapeutic equip-ment or procedures, e.g., whirlpools, dialysis units, etc. The role of water treatment processes in the development of an ecological niche for *Legionella* spp. needs to be investigated. Treatment practices involving lime softening created a high pH in one university hospital system in Iowa that made effective disinfection difficult and may have been a factor in a nosocomial outbreak in that facility.

There is no base of information currently available on treatment processes that would be effective in *Legionella* control. Chlorination, which has been used in control of outbreaks, may result in increased trihalomethane formation. Processes to reduce precursors for trihalomethane formation when chlorination is applied are being evaluated. Granular activated-carbon filtration, changing the point of disinfectant application, use of an alternative disinfectant, and volatile organic stripping are examples of methods to minimize trihalomethane production in treated drinking water. No information is available on what part these processes play in the amplification of legionellae. However, further reduction in organic compounds in drinking water should reduce the colonization potential of many organisms that persist in distribution systems and building water supply lines through decreasing the availability of nutrients and promoting more effective disinfectant residual action. The role of amoebae and ciliates in harboring legionellae where there may be subsequent protected passage through water supplies with limited treatment is unknown. However, there is some evidence that coliforms engulfed by amphipods are protected from disinfection. The engulfed coliforms are subsequently released when the macroinvertebrates are crushed by water flow hydraulics in the pipe network.

One environmental engineer suggested that elimination of legionellae through water treatment technology would be very difficult and not the most cost-effective place to create the barrier. Several participants agreed that hospitals may have special water quality needs beyond the availability of a safe, sanitary drinking supply from the water plant and that satisfying these needs should be the responsibility of the user, i.e., the hospital.

Public water supply distribution systems have the same problems noted for hospitals and high-rise buildings but on a larger scale, as they transmit greater volumes of water over many kilometers and often operate with pipe sections over 100 years old. These problems include colonization sites in dead ends; sediments and tubercles in pipes, which reduce flow; rapid disappearance of available chlorine to organic compounds, which may also protect surviving organisms; cross-connection contamination;

negative line pressures; corrosion sites; poor flushing programs; and passage of organisms during periods of inadequate treatment or releases of entrapped biological films as a result of sudden hydraulic surges.

Investigations in the United Kingdom on water distribution and building plumbing systems have revealed that legionellae and other organisms may colonize some materials (e.g., polyurethane, plasticized polyvinyl chloride, polysulfide sealant, etc.) used in the construction of "O" rings or as joint sealants. Often, growth in packing and rubber gaskets is due to variations in raw ingredients and the need to use plasticizers in the formulations. Biocidal agents added to these materials may prevent the occurrence of microbial growth, but these agents are often leached out into waters held in static lines. Hot water passage accelerates the leaching of the biocides, and they may be destroyed through chemical attack by chlorine residuals. A microbiological protocol to test the suitability of these materials has been developed by the Water Council in the United Kingdom for use by industries manufacturing these products or by individual laboratories.

The use of shock or periodic hyperchlorination in the hospital water supply to control legionellae was discussed in the context of the potential for additional trihalomethane production in the water that might increase the exposure to this carcinogen. It was suggested that a disinfection protocol should not produce significant additional trihalomethane because the reaction of chlorine with residual organic compounds should be essentially complete, if a free residual is present, by the time the water arrives at the hospital from the treatment plant. Any additional trihalomethane production should, therefore, involve only the organic compounds produced in biological films and debris present in the hospital plumbing system. The potential health hazard to patients and personnel from ingesting hyperchlorinated water for brief periods of time appears to be minimal. However, the effects of chlorine inhalation or contact at levels used to decontaminate a plumbing system are less well-defined and should be considered, especially where water supplies to respiratory and dermatology wards may be involved.

Where aggressive waters have caused building plumbing corrosion problems, the use of sodium silicate was suggested as a control measure to create a film on the walls of the pipe so that water does not come in contact with the metal pipe surfaces.

Need and Responsibility for Control (A. P. Dufour, moderator). An association of legionellosis and the presence of legionellae in water has been established. Explosive outbreaks of legionellosis have been traced to water aerosols contaminated with legionellae generated from heat-exchange cooling towers. Sustained outbreaks in hospitals have been associated with water in building plumbing systems contaminated with legionellae. Waters contaminated with legionellae in recreational whirlpool spas, hospital nebulizers, and respiratory therapy equipment have been identified as sources of infection. Most of the systems, apparatus, or equipment associated with these outbreaks can be described as amplifiers or disseminators because they facilitate growth of legionellae from low concentrations or distribute the organism to susceptible hosts. Water storage tanks and hot water plumbing systems are examples of amplifiers, while shower heads and nebulizers are examples of disseminators. Most of these amplifiers and disseminators receive water from potable water distribution systems, and these delivery systems may be the ultimate source for the legionellae. However, to date legionellae have not been isolated from any potable water distribution system.

It was pointed out that *Legionella pneumophila* or related species have been isolated from similar amplifiers and disseminators in hospitals where no cases of legionella pneumonia have occurred. One report, in fact, indicated that two buildings in close proximity to each other, which received water from the same source and housed similar patient populations, had appreciably different attack rates of legionellosis, despite the fact that legionellae were isolated as frequently from one building as from the other.

The discussion then focused on the need for control of environmental *Legionella* spp. in hospital settings. One frequently expressed viewpoint was that the finding of legionellae in the absence of overt disease was not sufficient justification to initiate decontamination procedures of unknown effectiveness, toxicity, and cost. In contrast, others believed that the finding of legionellae in hospital plumbing systems was a sufficient basis to begin decontamination procedures. Proponents of this view emphasized that there could be a considerable delay between recognition of disease and initiation of intervention measures. There was some agreement that buildings used to house organ transplant patients or other high-risk patients and buildings that were being brought into service after a period of nonuse might warrant special precautions.

Controlling *Legionella* spp. in cooling towers is necessary when cooling towers are linked to outbreaks of disease, but the need for routine treatment against legionellae is less clear, as the effectiveness and toxicity of continuous decontamination of cooling towers are unknown. It was pointed out that some biocides, while they efficiently retarded slime formation, actually favored the growth of legionellae.

Two outbreaks of nonpneumonic legionellosis have been associated with use of recreational whirlpool spas. Further epidemiological information is required to establish the significance of this means of transmission. Two nosocomial outbreaks of pneumonic legionellosis have been ascribed to the presence of legionellae in hospital nebulizers and respiratory therapy equipment. These disseminators can be easily controlled by using sterile water in place of tap water in reservoirs of respiratory therapy equipment.

In summary, control of legionellae in water should be immediately implemented in situations where new cases of illness are occurring which are epidemiologically associated with water systems. Cooling towers should be decontaminated whenever they have been associated with disease. Surveillance for cases should be actively maintained in high-risk populations, such as renal transplant patients. Active surveillance should also be instituted in buildings that have been brought back into use after long periods of nonuse. No action should be taken to eradicate legionellae in the absence of associated disease, even if legionellae were isolated from the water system, as the need for eradication, as well as the effectiveness and toxicity of available treatments, has not been demonstrated. Clearly, studies should be undertaken to define these parameters.

It was obvious from the general direction of the discussion that no single overall protocol for the control of legionellae in water environments would be appropriate. Instead, separate strategies were proposed depending on the source, amplifier, or disseminator of the organism. This practice will likely continue until the health significance of all routes of exposure can be determined.

The tone of the discussions would indicate that there is no sound basis for regulating the presence of legionellae in water environments at this time. The need for more epidemiological and laboratory information was stressed with regard to making decisions about the control and regulation of this organism. Until we have more information on the virulence of different strains of *Legionella*, including ecology and epidemiological significance, it will be difficult to determine whether this organism should be regulated.

SUMMARY

Ecology, Transmission, and Control†

RAMON J. SEIDLER

Department of Microbiology, Oregon State University, Corvallis, Oregon 97331-3804

This report is a summary of the approximately 29 presentations given in the areas of ecology, transmission, and control of *Legionella* spp. The various presentations are divided here into state-of-the-art lectures; organismic interactions; sampling, recovery, and detection systems; water delivery systems, including power plant and cooling-tower environments and potable water; and round table discussions dealing with regulatory considerations.

STATE-OF-THE-ART LECTURES

In the two state-of-the-art lectures on ecology and environmental control, J. C. Feeley and C. B. Fliermans summarized advances made since the First International Symposium on Legionnaires Disease on the detection, distribution, and enumeration of *Legionella* spp.

Feeley indicated that the greatest advance in microbial methodology is the use of selective media whereby viable legionellae can be detected directly from clinical or environmental specimens. Buffered charcoal-yeast extract (BCYE) medium with environmental specimens should be used with some caution, however, since we do not know the plating efficiencies with the various *Legionella* species already reported, let alone with the new species, which may number up to 23 according to D. J. Brenner.

Current methodologies for detecting legionellae from environmental specimens will probably not be as effective in detecting legionellae from clinical specimens. A combination of methods should now be used for isolating legionellae from environmental specimens. These methods include injection of concentrated water specimens in guinea pigs, direct fluorescent-antibody (DFA) examination, and plating onto BCYE containing α-ketoglutarate and ACES buffer. Caution is advised because of the potential for cross-reactivity of DFA reagents with certain non-*Legionella* strains, such as *Pseudomonas* species. A single enrichment, plating, and detection method may not apply to all samples. For example, special quantitation problems are to be expected from so-called "high-count" (high total microbial count) water specimens. With the

present technology the recommendation is to use acid treatment enrichment (pH 2.2, 15 min) followed by neutralization and plating onto BCYE containing antibiotics. With water containing low counts of legionellae, the cells should be concentrated by membrane filtration and sonicated off the filter before plating. How quantitative a recovery system this is still remains to be investigated.

One area where information is lacking is the existence of suitable methods for detection and enumeration of legionellae from air samples. Since inhalation is the route of infection, methods which detect airborne legionellae would be an important contribution to our understanding of the agent's physiological ecology and to possible quantitation of human health hazards in the work environment.

Fliermans summarized the benefits obtained from the available culture techniques for studying the occurrence of *Legionella* spp. in the environment. Direct cultivation techniques have revealed the ubiquity of *Legionella* spp. in diverse natural aquatic habitats. Environmental samples have also shown the existence of numerous new species of *Legionella*. Isolations of legionellae from aquatic habitats have been made at in situ temperatures as high as 63°C from hot springs and also from pristine subalpine lakes at much cooler temperatures. In other studies legionellae have not been found in estuarine or saline environments but only in freshwater.

Fliermans indicated that the use of the 2-iodophenyl-3-nitrophenyl-5-phenyltetrazolium chloride (INT) procedure will have an important role in physiological ecology. The electron transport system of respiring microorganisms reduces INT to insoluble crystals which precipitate out intracellularly. The crystals can be observed microscopically. The INT procedure can be coupled with the serogroup-specific DFA staining technique to monitor in situ respiring strains of *Legionella* in the environment. This technique has revealed that temperatures of about 65°C are optimum for INT precipitation, but the optimum temperatures of growth of the tested strains of *Legionella* appear closer to 45°C. Additional studies are needed to assess the

† Technical paper no. 6998, Oregon Agricultural Experiment Station, Corvallis, OR 97331-3804.

correlation among cell viability, physiological cell activities, and respiratory INT-precipitating activity.

ORGANISMIC INTERACTIONS

Several studies dealt with the intracellular growth of various *Legionella* species in protozoans. *Tetrahymena pyriformis* is a widely distributed, freshwater, ciliated protozoan found in domestic wastewaters and other freshwater ecosystems, where it feeds on bacteria. Fields et al. discovered that *Legionella pneumophila* serogroup 1 will actually multiply within the ciliate in tap water suspensions. *L. pneumophila* failed to grow in *T. pyriformis* spent culture medium or in lysed cell suspensions of the protozoan.

Cell suspensions of *L. pneumophila* and *T. pyriformis* in sterile tap water revealed that the bacterium would increase some 500-fold in 7 days when incubated at 35°C. *L. pneumophila* grew within the protozoan vacuoles in serpentine or snake-like chains. The concentrations of the protozoan and the bacterium as well as the incubation temperature appear to influence the amount of growth of *L. pneumophila*.

Investigations indicated that *T. pyriformis* could provide selective enrichment for *L. pneumophila* from natural water samples containing low *Legionella* counts. In one series of experiments, the protozoan was incubated in 10 water samples which were analyzed for legionellae at time zero and after 7 days of incubation. The incidence of detectable viable legionellae increased from 50% of the samples at zero time to 90% of the samples after 7 days of incubation with *T. pyriformis*.

T. J. Rowbotham showed that *L. pneumophila* would grow and exhibit a very characteristic morphological sequence of events in amoebae such as *Acanthamoeba polyphaga*. These bacteriovorus amoebae are ubiquitous in the same aquatic environments as legionellae. Rowbotham likened the amoebae to human macrophages in lungs; thus, the physiological and morphological events in amoebae may provide a model for studying the intracellular growth of legionellae.

The various *Legionella* species exhibit a host range for *A. polyphaga*. For example, virulent *L. pneumophila* serogroup 5 and agar-attenuated *L. pneumophila* serogroup 1 do not multiply in this amoeba, but virulent strains of *L. pneumophila* serogroups 1 and 6 as well as strains of *Legionella jordanis*, *Legionella bozemanii*, and *Legionella gormanii* do infect and multiply in the strain of *Acanthamoeba* studied.

Legionellae are apparently attracted to the amoeba, are phagocytized, and enter a vacuole. During the course of intracellular *Legionella* growth, amoeba ribosomes migrate to the circumference of the vacuole containing legionellae. At a later stage of infection, host cell mitochondria are seen in close proximity to the infected vacuole. After about 22 h at 35°C the amoeba is filled with legionellae. Vesicles containing large numbers of legionellae are sometimes released as intact particles from the infected host cell. The acanthamoebae normally form heat-resistant cysts. It is not known whether the dormant cysts contain legionellae.

Enrichment for legionellae has been achieved after incubation of the amoeba in clinical specimens such as sputum. This enrichment is followed by the acid enrichment procedure and by plating onto selective medium. The amoeba enrichment has been useful in isolating *L. pneumophila* from specimens which are negative by direct culture.

In related studies, Holden et al. investigated the interactions of a virulent *L. pneumophila* strain with the amoeba *Acanthamoeba castellanii*. On agar plates, direct cell contact with virulent *L. pneumophila* stopped the migration of the amoeba. Investigators suggested that direct cell contact rather than a diffusible metabolite was required to stop the migration of the amoeba. After 1 h of incubation, *Legionella* cells were removed from suspensions by the amoeba. After 72 h of incubation, viable *Legionella* counts increased 1,000-fold. Similar observations were made with an avirulent *Legionella* strain.

Hume and Hann studied the growth and interactions of legionellae with several species of green algae. Growth of legionellae with algae was compared with that of suspensions of *Pseudomonas aeruginosa* and *Escherichia coli*. All bacterial species investigated grew similarly when incubated in a basal salts medium at 22°C for 5 days in the presence of *Gloeocystis*, *Chlorella*, and *Scenedesmus* spp. *Legionella*, *P. aeruginosa*, and *E. coli* grew in close proximity, forming microcolonies, among cells of *Chlorella* and *Oocystis*. The intimate relationship with these algae did not occur in cocultures with *Scenedesmus* cells.

Energy-dispersive X-ray analysis and atomic adsorption spectrophotometry revealed that the various microbial components were differentially sequestering metal ions. The bacterial components sequestered S, Cl, K, Cu, and P, while the algae sequestered Mg and Cu. The mechanism by which algae promote the multiplication of legionellae is not yet understood.

SAMPLING, RECOVERY, AND DETECTION SYSTEMS

Several groups reported ongoing efforts to improve *Legionella* detection and enumeration techniques. For example, Calderon and Dufour evaluated a variety of media for the direct enumeration of legionellae from environmental water samples. They studied BCYE formulations and media with a modified base, including one in which peptone replaced yeast extract. Medium performance was evaluated for accuracy, specificity, and selectivity. Samples tested included seeded water samples and natural waters from cooling towers, hot water tanks, and potable waters.

Accuracy in recovery was determined by seeding natural water with known densities of laboratory-cultivated legionellae. Accuracy of recovery varied widely when the samples were pretreated by acid enrichment and plated onto various media. Some 1% of the dosed legionellae were recovered with one standard medium formulation. It was indicated that standard media were developed for clinical use, and their application to environmental water samples for the isolation of *Legionella* spp. may not be satisfactory. The authors indicated that an accurate and reliable method for quantifying legionellae from environmental waters is still not available.

Dennis et al. investigated in parallel several of the existing conventional media with BCYE as the base formula for detecting legionellae from contaminated environmental water specimens. They examined the effect of acid pretreatment and 50°C heat in conjunction with plating onto BCYE, BCYE with cefamandole, polymyxin, and cycloheximide, and BCYE containing glycine, vancomycin, polymyxin, and cycloheximide. All samples were concentrated by membrane filtration. In their hands the best selection medium was BCYE with glycine, vancomycin, polymyxin, and cycloheximide, and they found that pretreatment with acid buffer or heating was useful in reducing many background microorganisms. No significant difference in isolation rates was noted with acid pretreatment versus heating, although the latter was more convenient to perform.

J. Gilbart presented a study involving the detection of *Legionella* species in environmental samples based on the unique patterns exhibited by derivatized branched-chain fatty acids. The procedure relies on the patterns exhibited by the cell fatty acids when detected by a gas chromatograph equipped with an electron capture detector. Over 500 pure cultures of *Legionella* from a variety of sources (clinical and environmental) revealed close cluster patterns for each *Legionella* species, indicating a 95% probability for discrete species identification. Other bacterial genera tested did not exhibit patterns which could be confused with those of *Legionella*.

A survey was conducted by Gilbart for legionellae in environmental samples, comparing the gas chromatography detection technique with that of direct cultivation. It was found that legionellae could be detected in environmental samples concentrated by filtration with a sensitivity limit of 10^3 to 10^4 cells per ml of concentrated sample. Some 30 to 40 environmental samples from institutional water systems have been analyzed for legionellae. Analysis of culture-positive specimens by gas chromatography revealed characteristic branched-chain fatty acids while culture-negative specimens contained no such fatty acids.

Tuovinen et al. investigated several ways of increasing the efficiency of large-volume sampling and enumeration methodologies. They found that the current conventional membrane filtration and direct plating procedures from large-volume samples were not suitable due to inadequate flow rates, poor recovery or lack of detection of legionellae, or both. A modified filtration unit was tested with a high-volume pump and several brands of membrane filters and was found to be more effective in detecting legionellae than a conventional membrane filtration unit. A hollow fiber filtration unit used in conjunction with a high-performance pump was also satisfactory in handling large (20-liter) samples. These high-performance filtration units were faster than a continuous centrifugation process.

Brief discussions occurred in several sessions concerning the use of new technologies such as monoclonal antibodies and gene probes as sensitive and specific tools for detecting *Legionella* spp. Although studies are still in progress, some investigators believed that a pool of monoclonal antibodies would increase the sensitivity for detecting legionellae and minimize the risk of nonspecific cross-reactions with other genera. There is also the possibility of producing a single monoclonal antibody that is genus specific for *Legionella*.

D. E. Kohne presented a summary statement dealing with a gene probe he had constructed for detecting *Legionella* spp. For proprietary reasons, data were not presented. He did, however, report that the probe was able to detect all 23 currently known *Legionella* species and that it did not cross-react with any other bacterial genera tested. Detection with the probe takes about 8 h. Application to both clinical diagnosis and detection of *Legionella* spp. in environmental samples was envisioned.

POTABLE AND INDUSTRIAL WATER SYSTEMS

Fisher-Hoch et al. reported on the occurrence of legionellae in the plumbing system of Kingston Hospital in the United Kingdom. It was demonstrated that despite chlorination of water supplies and raising of hot water temperatures, Legionnaires disease outbreaks among patients still occurred. Legionellae approaching densities of 10^9/liter were isolated from the sludge of a stagnant hot water cylinder. In the water heaters studied, temperature stratification was remarkable. Temperatures of 60°C were recorded in water near the central position of the tank, while temperatures near the sludge at the bottom were only about 10°C. The temperature of the water heaters was raised to boiling to eliminate viable legionellae. Taps and showers around the institution remained sporadically positive for legionellae despite the treatments given to the hot water cylinders. Sequential testing of 25-ml aliquots of tap water revealed that the highest *Legionella* counts were in the first three or four samples. Subsequent samples were generally negative, indicating that the faucet itself may have been shedding the cells. Dismantling of faucets and shower heads revealed that black rubber washers were colonized with legionellae. The investigators replaced the washers with ones made of material approved by the National Water Council and subsequently have been unable to isolate legionellae from the taps.

Baird et al., in a study involving the control of endemic nosocomial legionellosis, found that hyperchlorination of the potable water supply was sufficient to reduce cases to a level seen with community-acquired legionellosis. During 1977 to 1982, approximately 93 cases of nosocomial legionellosis were reported (mean, 1.55 patients per month). In March 1982 the institution was provided with an on-site automatic chlorination unit which was adjusted to maintain 4 ppm of chlorine. Monitoring revealed fluctuations down to 1.5 to 2 ppm of chlorine residuals at certain times of the day at the taps. Before the in-house chlorination procedures, some 14% of the faucets sampled were positive. This was reduced to 1.25% after chlorination; however, some 7.1% of the 1,276 sites monitored were still positive for legionellae. Nevertheless, there was a significant drop in cases of nosocomial legionellosis to a level comparable to that recorded for community-acquired cases.

In similar studies, Massanari et al. reported their experiences with hyperchlorination to control a nosocomial outbreak of 22 cases of legionellosis in a midwestern university hospital. They initially shock chlorinated both the hot and cold water at over 15 ppm. This was accompanied by water flushing and elevation of the hot water temperature to 147°F (ca. 64°C) for 41 days, with subsequent lowering to 105°F (ca. 41°C). In addition, a continuous-flow proportional chlorination unit was installed that served both the hot and cold water supplies. Free chlorine residuals were maintained at mean levels of 8 and 7.3 ppm in the hot and cold water, respectively.

Initially, 10 of 47 sites sampled were found culture positive for legionellae. This was reduced to 3 of 33 sites after the initial chlorine shock, flushing, and raising the temperature of the hot water to 147°F. After the installation of the automatic chlorinator, culture-positive sites were reduced to only 2 of 229. The two positive sites were taken from two adjacent rooms which were vacant about 1 month before sampling. Since no new cases of legionellosis have been identified to date in areas of the hospital that received hyperchlorinated water, these investigators concluded that hyperchlorination is an effective method for treating hospital water that has been epidemiologically and bacteriologically associated with nosocomial cases of legionellosis.

Witherell et al. summarized their experiences in Vermont with the control of legionellae in hospital water supplies. They reported that the low hot water temperatures typically maintained in hospitals (38 to 46°C) may favor or promote the colonization of legionellae. They pointed out that it is very difficult to maintain effective levels of chlorine in both hot and cold water after it enters hospitals. They reported that raising water temperatures in hospitals increases the chance of scalding of patients, and supplemental on-site chlorination may result in excessive corrosion of pipes and other equipment. However, they emphasized that properly monitored on-site supplemental chlorination could indeed be used to effectively reduce legionellae to below detectable limits and not be corrosive, provided that the supplemental chlorination units are properly located and monitored regularly.

Stagnated water and faucet obstructions in hospital plumbing systems were demonstrated by Ciesielski et al. to be the sites of *L. pneumophila* serogroup 1 colonization in a Colorado hospital. In the initial phases of the study, five of five hot water storage tanks and six of eight shower heads sampled grew *L. pneumophila*. Faucets with aerators, vacuum breakers, and backflow preventors were also found to be sources of *Legionella*. *L. pneumophila* was isolated from 22 of 30 aerated faucets, but not from 26 unobstructed faucets. Swab cultures from disassembled faucets gave mean counts of 278 CFU of legionellae per swab for the aerator, 171 CFU of legionellae per swab for the pipe, and 83 CFU of legionellae per ml of water. Nine aerators were sterilized and reassembled onto the fixtures and periodically examined for viable legionellae. Within 2 months three of nine aerators and their screens were sampled, cultured, and found positive for legionellae. These researchers concluded that elimination of stagnation of hot water in pipes would lower the contamination of potable water with legionellae. They believed that aerator screens should be removed because they are sites of *Legionella* colonization and obstruct the flow of tap water.

In a series of related reports coauthored by investigators from Oak Ridge National Laboratories, the University of Tennessee, and the Savannah River Laboratory, scientists studied thermally altered electric power plants for densities, types, and infectivity of *Legionella* species. Water samples were collected seasonally at comparable locations in each of nine power plants and from source waters at each site. Samples were concentrated 500-fold by centrifugation, and *Legionella* cell densities were determined by DFA. The INT stain that detects a functional electron transport system was used to estimate cell viability. Infectivity of water samples was determined by injecting, in duplicate, 2 ml of concentrated samples into guinea pigs and observing them over a 10-day period. Animals showing signs of illness were sacrificed, and swabs of peritoneal fluid and organs were taken and cultured for legionellae.

Of 149 samples examined, *L. pneumophila* serogroup 1 was the most prevalent, being found in 20 plant-altered and 5 source water samples at six of the nine sites. In addition, *L. pneumophila* serogroups 2, 3, 4, and 6 as well as *L. gormanii, L. bozemanii,* and *L. oakridgensis* were isolated. A previously undescribed and unnamed *Legionella* species was also isolated from a thermally altered power plant at one test site.

The highest cell densities of *Legionella* (mean, 8×10^6/liter) were found during winter and spring in source waters near closed-cycle power plant water operations. *Legionella* densities decreased with passage of the water through the power plant. Infectivity of water samples for guinea pigs was higher in electric power plants with closed-cycle water cooling systems (80% of samples infective) than in those with a once-through cycle (20% of samples infective). About 20% of the source water samples were infective to guinea pigs. Numbers of *Legionella* in spring, summer, and fall in closed-cycle plant water systems were higher than in the source waters. Contrary to what the investigators had anticipated, infectivity could not be related to *Legionella* density or viability. It was concluded that the practice of using *Legionella* densities greater than 10^8/liter in cooling towers as the sole criterion for instituting control measures may not be appropriate for all systems.

Kurtz et al. conducted a field trial in England on the incidence and control of *L. pneumophila* in cooling water systems. An initial survey found that 12 of 16 systems investigated contained *L. pneumophila* in the cooling water. The density of *Legionella* in the 16 systems was monitored following the addition of biocides. Five systems were dosed weekly with a chlorinated phenolic thioether to a final concentration of 200 ppm, six were dosed with a polymeric quaternary ammonium chloride to 50 ppm, and five were unamended. Although *L. pneumophila* was present initially in all systems dosed with the thioether, no legionellae were recovered by direct plating or by guinea pig inoculation after treatment. On the other hand, legionellae were present in 24 of 37 samples taken during treatment with the quaternary ammonium compound and still persisted in 2 of 6 systems at the end of the trial. Investigators concluded that in this study the chlorinated phenolic thioether appeared to effectively control *L. pneumophila* in cooling waters.

In response to two outbreaks of Legionnaires disease in Vermont, Witherell et al. conducted an inventory and inspection program of wet-type heat rejection units. A total of 185 units were studied, including 125 cooling towers, 50 evaporative condensers, and 10 heat rejection units in other categories (such as evaporative pad coolers). The investigators examined 151 attributes for each unit. There were 185 samples analyzed, and 18 (9.7%) were *L. pneumophila* positive by the culture method of Gorman and Feeley. A statistical analysis was conducted, examining *L. pneumophila* positivity from the units, and no correlation was noted between microbiocidal treatment and the presence or absence of legionellae.

Grow et al. reported on the environmental factors influencing levels of legionellae in biocide-treated cooling towers in Alabama. Eleven operating cooling towers of over 200-ton capacity were examined for numbers of DFA-INT-positive and culture-positive *L. pneumophila*. The highest counts observed ranged from 1,000 to 4,500/ml. Four cooling towers were studied in depth over a 1-year period. All four towers studied were in the same geographic area, all were receiving the same biocide treatment (thiocarbamates), and all contained legionellae. It was found that filtered water from cooling towers did not contain residual biocidal capacity for legionellae. Although these studies are still in progress, it appears that relative humidity is the only current environmental parameter which correlates most significantly with levels of viable *Legionella* in water of cooling towers that are in continuous operation.

R. W. Gilpin reported on the laboratory and field applications of UV disinfection in the control of *Legionella* species in recirculating water systems. Six *Legionella* species and *Pseudomonas aeruginosa* were tested for their sensitivity to UV irradiation. Studies indicated that *Legionella* species required approximately 5 to 30 min for about a 50% kill and 17 to 44 min for a 90% kill after exposure to 1 μW of radiation per cm². Disinfection was also efficient in a 340-liter whirlpool bath. The author concluded that UV disinfection of water is an appropriate supplement or alternative to chemical treatment for the control of legionellae and other bacteria.

Witherell et al. reported on the efficacy of chlorine disinfection of whirlpool spas in the elimination of legionellae. The study was conducted after outbreaks of Pontiac fever and Legionnaires disease at a Vermont inn. *L. pneumophila* was epidemiologically and environmentally associated with whirlpool spa use. Three of five whirlpool spas examined were positive for *Legionella*. Disinfection by continuous chlorination was undertaken at the culture-positive sites. A residual of 1 to 3 mg of free available chlorine per liter and a pH of 7.2 to 7.8 was maintained. All chlorine treatments resulted in a reduction of *L. pneumophila* to below detectable limits. Although increased temperatures and aeration had little impact on disinfection concentration, sharp reductions in chlorine residuals were noted shortly after spa use. After the establishment of continuous disinfection to maintain free chlorine residuals, *L. pneumophila* was not recovered and no cases of legionellosis were noted among users of the two whirlpool spas that remained in use.

Desplaces et al. investigated the occurrence of legionellae in 140 water samples taken from the Paris area. Overall, 42 samples were collected from six air-conditioning systems, and 98 samples were taken from plumbing fixtures in 18 buildings (62 from cold tap water, 17 from hot tap water, 11 from hot water tanks, and 8 from mains supplying buildings). Cultivation of legionellae was by procedures recommended by the Centers for Disease Control and included the use of BCYE with and without antibiotics, α-ketoglutarate, and glycine. Overall, 32% (45 of 140) of the samples contained at least one culturable *Legionella* strain. The positive samples were found in 4 of 6 air-conditioning systems and in 5 of 18 buildings, including 11 of 11 hot water tanks. Legionellae were not isolated from any of the 8 mains or from 19 building distribution water entry points. Nevertheless, 9 of 62 cold tap water samples were culture positive, and one sample contained 10^5 CFU of legionellae per liter. Sixty percent of the positive samples contained two or three different *Legionella* strains. The most common strain isolated was *L. pneumophila* serogroup 1 (48% of total isolates). Three *Legionella*-like organisms were also isolated, and these are being characterized further.

Spino et al. investigated the occurrence of legionellae and other bacteria in chemically contaminated groundwater before and after passage of the water over a series of redwood slat aerators. Various culture procedures were used in the attempt to isolate legionellae, including direct plating of concentrated samples as well as acid or heat pretreatment. This study was undertaken to evaluate the influence of the aeration-chemical decontamination process on the microbiological quality of the water. Levels of heterotrophic bacteria were higher in the water subjected to this process, indicating possible colonization of the redwood. Levels of heterotrophic bacteria were on the order of 10^3 to 10^4 CFU/ml, and coliforms of the genus *Klebsiella* were common. All cultural attempts to recover legionellae from both treated and untreated water and from the biomass from the redwood were negative. These investigators concluded that the treatment of groundwater by aeration and chemical decontamination does not pose a substantial risk for the introduction of legionellae into the water. The authors also indicated that their negative results for legionellae substantiate that legionellae, if present, are in extremely low numbers in water collected from habitats where temperatures are <15°C.

ROUND TABLE DISCUSSIONS

In the round table there was a great amount of discussion about the role of potable water as a vehicle for transmission of legionellosis. It was noted that water had been clearly incriminated both epidemiologically and bacteriologically in several outbreaks. In these outbreaks, an apparatus or piece of equipment was identified as an amplifier or disseminator of legionellae. The equipment and apparatus either are environments for the regrowth of legionellae (amplifiers such as hot water tanks) or function to generate aerosols and disseminate the organism to susceptible hosts (disseminators such as cooling towers and showers). A large unknown in the epidemiology of *Legionella* transmission by the amplifiers is the lack of proof as to whether the agent is transmitted via public water supplies and, if so, whether current engineering treatment practices and subsequent water distribution play a role in the dissemination of legionellae to the amplifiers. Although legionellae have been isolated from potable water lines within institutions, they have not been isolated from any potable water collected from mainlines before the lines enter these institutions or from water collected immediately after treatment at the municipal water treatment plant.

There is a lack of information about which kinds of treatment practices are effective for *Legionella* control. Some water treatment practices which are effective against coliforms and other bacteria could aid in the development of legionellae by releasing nutrients from these organisms which subsequently accumulate in slow-flow sections or in hot water tanks. Other treatment practices, such as lime softening, create elevated water pH and may make chlorine disinfection ineffective. Poor distribution system maintenance allows sediments to accumulate in slow-flow sections and in reservoirs which subsequently become niches for microbial growth, especially in areas where pipes have corroded and tubercles have formed. Whether *Legionella* spp. can colonize these areas or whether the release of biofilms from these areas becomes a source of nutrients for legionellae in the amplifiers is unknown.

A question was raised as to whether there is a need to formalize control of legionellae in these water environments and whether the control should be regulated. The general consensus was that at present there is insufficient knowledge to support the issuance of a general all-encompassing regulation. However there was a strong consensus that control procedures for legionellae should be practiced under special conditions such as start-ups of plumbing systems in new buildings or restarts of plumbing systems that have not been used in existing buildings; institutional wards which contain organ transplant patients; and institutions where new cases of legionellosis are occurring which are epidemiologically associated with water systems.

Some participants believed that the detection of viable legionellae in the potable water system of an institution, regardless of the presence or absence of human illness, justifies the initiation of a decontamination procedure. Many participants expressed concern regarding the potential legal liability of hospitals in which legionellae were known to be present in the plumbing system but no corrective control measures were instituted. A case was made for control of legionellae in hospital nebulizers and respiratory therapy equipment and went unchallenged. It was pointed out that control of legionellae in these disseminators could be achieved by using sterile water in place of tap water.

A brief discussion was held on the feasibility and desirability of controlling legionellae in cooling towers. The consensus of opinion was that regulatory control measures should not be instituted because of the lack of epidemiological data supporting the need for such regulations. In addition, some participants believed that biocides effective for other microbes and also thought to be effective for legionellae would not be able to continuously control legionellae. Furthermore, it was thought that increased numbers of legionellae might result from the elimination of other microbes from that water environment.

The overall tone of the discussion indicated that at this time there is no scientific basis for regulating the presence of legionellae in most water environments. There were also repeated statements on the necessity for gathering of more epidemiological information before decisions could be made concerning the control and possible regulation of legionellae. It was also the consensus that no single protocol would emerge for the control of legionellae in water environments.

AUTHOR INDEX

SUBJECT INDEX

Acanthamoeba castellanii
  L. pneumophila growth within, 329
Acanthamoeba palestinensis
  biocides, 340
  L. pneumophila, intracellular replication, 330
Acanthamoeba polyphaga
  host for legionellae, 325
Aerosols
  survival of *L. pneumophila* in humidity, 301
Africa
  southern, legionellosis in, 247
Age
  and immunity to *L. pneumophila*, 186
Alveolar macrophages
  inflammatory response to *Legionella* infection, 115
Amino acid metabolism
  ampicillin and erythromycin, effect on *L. pneumophila* growth and morphology, 103
  and oxygen sensitivity of legionellae, 61
  for identification of legionellae, 74
  state of the art lecture, 61
Antibiotics
  activity against legionellae in vitro, 99
  activity against *L. pneumophila* during sepsis, 99
  beta-lactam, activity against legionellae, 100
  influence on intracellular *L. pneumophila*, 159
Antibodies
  Legionella, prevalence in young men in Taiwan, 254
Antigens
  guinea pig humoral response, 168
  guinea pig susceptibility, 145
  immunostimulatory, 145
  Legionella spp., reactions among serogroups, 153
  serogroup-specific, 153
Antigens, *L. pneumophila*
  flagellar, 42
  immune responses to after Legionnaires disease, 190
  immunostimulation by, 187
  relatedness of human and monkey antisera to, 46
  stability, 44
Aquatic bacteria
  L. pneumophila growth relationships, 323

Beta-lactamases
  of legionellae, inhibition and hydrolysis, 100
Biocides
  L. pneumophila in cooling water systems, 340
Bull's-eye lesion
  of *L. pneumophila*, 127

Carbohydrates
  metabolism by legionellae, state of the art lecture, 61
  profiling, for classification of legionellae, 71
Catalase
  for classification of legionellae, 61, 82
Cell envelopes
  L. pneumophila serogroup 1 strains, immunochemical analysis, 268
Cellular immunity
  guinea pig, development after exposure to *L. pneumophila*, 178

human, in patients with Legionnaires disease and other pneumonias, 180
Charcoal
  and oxygen toxicity in legionellae, 61
  mechanisms of action, 61
Chlorine
  hyperchlorination in hospitals, 333, 334
Chlorine dioxide
  inactivation of *L. pneumophila*, 68
Chlorophyta
  and *L. pneumophila*, growth relationships, 323
Classification of *Legionella* spp.
  by amino acid metabolism, 74
  by carbohydrate profiles, 71
  by catalase and peroxidase, 82
  by DNA relatedness, 55
  by phenotypic properties, 55
  division into genera, 55, 71
  state of the art lecture, 55
Continuous culture
  for growth of *L. pneumophila*, 68
Cooling towers
  factors influencing *L. pneumophila* growth in, 316
  Vermont, *L. pneumophila* in, 315
Cysteine
  and oxygen toxicity in legionellae, 61
  metabolism, 61
  requirement for growth of legionellae, 61
Cytolytic endotoxin
  activity in legionellae, 93

Denmark
  seroreactors to legionellae, frequency in patients with pneumonia, 258
Direct fluorescent-antibody testing
  commercial reagents, 27
  review of CDC data, 21
  state of the art lecture, 3
Disinfectants
  commonly used, effect on *L. pneumophila*, 346
Disinfectants, hospital
  L. pneumophila susceptibility, 342
Disinfection
  hospital hot water systems containing *L. pneumophila*, 336

Electron-capture gas chromatography
  Legionella spp. in water, 296
Endemic nosocomial legionellosis
  control by hyperchlorination of water, 333
Environmental *Legionella* specimens, microbiological analysis
  isolation from air, 281
  isolation from water, 281
  media, 281
  quantitation, 281
Enzymes
  beta-lactamase, of legionellae, 100
  catalase and peroxidase, and classification of legionellae, 61, 82
  hemolysin, *L. pneumophila*, characterization, 97
  in metabolic pathways of legionellae, state of the art lecture, 61